Springer Praxis Books

Space Exploration

A History of Human Space Exploration

Other Springer-Praxis books by Ben Evans in this Series

Escaping the Bounds of Earth: The Fifties and Sixties
2009
ISBN: 978-0-387-79093-0

Foothold in the Heavens: The Seventies
2010
ISBN: 978-1-4419-6341-3

Partnership in Space: The Mid to Late Nineties
2014
ISBN: 978-1-4614-3277-7

Tragedy and Triumph in Orbit: The Eighties and Early Nineties
2012
ISBN: 978-1-4614-3429-0

At Home in Space: The Late Seventies into the Eighties
2012
ISBN: 978-1-4419-8809-6

Partnership in Space: The Mid to Late Nineties
2014
ISBN: 978-1-4614-3277-7

Ben Evans

The Twenty-first Century in Space

 Springer

Published in association with
Praxis Publishing
Chichester, UK

Ben Evans
Space Writer
Atherstone, United Kingdom

SPRINGER-PRAXIS BOOKS IN SPACE EXPLORATION

Springer Praxis Books
ISBN 978-1-4939-1306-0   ISBN 978-1-4939-1307-7 (eBook)
DOI 10.1007/978-1-4939-1307-7

Library of Congress Control Number: 2014956555

© Springer Science+Business Media, LLC 2015
No part of this work may be reproduced, stored in a retrieval system, or transmitted in any form or by any means, electronic, mechanical, photocopying, microfilming, recording or otherwise, without written permission from the Publisher, with the exception of any material supplied specifically for the purpose of being entered and executed on a computer system, for exclusive use by the purchaser of the work.

Printed on acid-free paper

Springer is part of Springer Science+Business Media (www.springer.com)

Contents

Author's preface... ix
Acknowledgements xi

1. **Shuttle at the crossroads** 1
    A "new" Space Shuttle 1
    An ambitious mission 19
    Glimmers of the future 38
    Mission for Japan... 54
    "Weird science" 67
    Record rendezvous 84
    A precursor to Space Station 97
    Record-setters... 110

2. **A new partnership** 125
    Voyage of destiny... 125
    An American 'living' in space................................... 149
    "Atlantis in free drift" 181
    An 'international' space station................................ 195
    A 'permanent' American presence?............................... 215

3. **Preparing for Space Station**................................... 255
    Visiting a national icon.. 255
    Broken ankle, broken plans, broken record...................... 275
    Technology for Space Station 288
    Last chance spacewalks... 301
    Science for Space Station 318
    End of the old era 332
    Powerful X-ray eyes 349
    Revisiting a national icon 362

4. Steps to the future 379
   Fire! 379
   A troubled summer 392
   Arrival of 'The Wolfman' 411
   The final voyages 425
   A glimpse of the future 438

5. New millennium 453
   Brave new world 453
   The end of Mir 464
   A new dawn 481
   Epilogue: The new millennium 500

Bibliography 507
Index 517

To Michelle
For everything

Author's preface

When I set out to write a five-volume series to commemorate the first 50 years of our adventure in space, it seemed a big project, though relatively straightforward and something that I have always wanted to do. An obsessive space enthusiast for as long as I can remember, I received my first space book at the age of five, as was given a toy Space Shuttle as a birthday present soon afterwards and by the time I reached my seventh birthday I had watched in astonishment as Enterprise – mounted atop a Boeing 747 carrier aircraft – hurtled above my primary school in Birmingham, England, during a sports day. It caused me to drop the egg from my spoon, unfortunately, but the sight hooked me for life. The Moon landings excited me, and still do, beyond compare, and I began writing articles at the age of 15 for the British Interplanetary Society's *Spaceflight* magazine and, later, for *Countdown* and *Astronomy Now*. As I grew older, it became a goal of mine to someday write a 'meaty' history of the human exploration of space, which I continued to believe ranks highly as one of the greatest adventures ever undertaken by our species.

However, each space history book that I read seemed to 'lack' something. Some were overloaded with facts and figures, whilst others were devoted wholly to a popular audience, and still more simply ignored the detail and human interest factor. I cannot promise the reader that my series fulfils any of these gaps, but what I can say with certainty is that I have spent an enjoyable six years exploring the story of our adventure, told through news sources, books, the memoirs of those involved, magazines, press kits and oral histories and have learned a huge amount. The reader may love or loathe my work – they may find it hard to put down or may simply find it a useful additional castor for their sofa – but I have derived great joy from every aspect of researching and writing it.

It has been impossible for me to track entire decades within the pages of each volume. The first, *Escaping the Bonds of Earth*, had to take into account some of the achievements of the 1950s, as a prerequisite to focusing on 'its' decade, the 1960s. In a similar vein, the second volume, *Foothold in the Heavens*, needed the focus to fall in considerable depth upon some of the most remarkable events of the Space Age, including the Apollo 11 lunar landing, at the expense of covering a whole decade. The third instalment, *At Home in Space*, tackled the 1970s and 1980s, an era of the first space stations and the origins of the Space Shuttle, whilst the fourth volume, *Tragedy and Triumph in Orbit*, covered the devastating loss of Challenger and the success of the Soviet Union's Salyut 7 and Mir space stations.

By this time, it became clear that I was punching above my weight. My determination to cover each mission with the level of detail that it deserved, including biographies of each crew member, turned *Tragedy and Triumph in Orbit* inside something much longer; so long, in fact, that it barely covered the 1980s and was already approaching 600 pages in extent. As a result, with the gracious approval of Clive Horwood at Praxis, the series expanded to encompass a sixth volume, to cover as much of the first half-century of our exploration of space as possible. *Partnership in Space*, the fifth volume, covered the early 1990s, leaving this final text, *The Twenty-First Century in Space*, to explore the final decade of the last century and provide an overview of the explorations of the future.

I have learned much about the human space programmes of both Russia and the United States and, equally importantly, I have learned a great deal about the political events which shaped their progress. Starting with Yuri Gagarin's pioneering voyage in April 1961, the journey has carried me through a handful of dramatic decades, punctuated by conflict and reconciliation, meddling and political manoeuvring, and has seen the first men walk on the Moon, the first men occupy an orbiting space station, the first men pilot a reusable spacecraft beyond the atmosphere, the first men from nations other than Russia and the United States and, of course, the first representatives of *womankind* to carry their dreams and aspirations into the heavens. The human side element has always been profoundly important to me and the remarkable tales of astronauts and cosmonauts, engineers and managers, journalists and spectators, add a richness to the story. In fact, the tale is not simply about who spent the most time in space, who made the most spacewalks or who flew the most missions. Rather, it is a collection of stories: the stories of how a few hundred remarkable people, all of whom accomplished an uncommon goal, forever altered our perspective of the world in which we live and fixed our eyes, our minds and our imagination on the cosmos around us.

Ben Evans
Atherstone, Warwickshire, July 2014

Acknowledgements

This book would not have been possible without the support of a number of individuals, to whom I am enormously indebted. I must firstly thank my wife, Michelle, for her constant love, support and encouragement throughout the time it has taken to plan, research and write this manuscript. As always, she has been uncomplaining during the weekends and holidays when I sat up late, typing on the laptop, or poring through piles of books, old newspaper cuttings, magazines, interview transcripts, press kits or websites. It is to her, with all my love, that I would like to dedicate my final Springer-Praxis book. My thanks also go to Clive Horwood of Praxis for his enthusiastic support over the past 12 years and to David M. Harland for reviewing each of my manuscripts and offering a wealth of guidance and advice. I deeply appreciate not only their support, but also their patience, in what has been an overdue project and one which has proven far harder to write than I had imagined. Additional thanks go to Ed Hengeveld, who has been hugely gracious with his time in identifying suitable images for this book, including many 'unfamiliar' ones, which will surely bolster the text. Others to whom I owe a debt of gratitude are my parents-in-law Sandie Dearn and Malcolm and Helen Chawner. To those friends who have encouraged my fascination with all things 'space' over the years, many thanks: to Dave Evetts and Mike Bryce, to Andy Rowlands, to Rob and Jill Wood and to the late Andy Salmon, who sadly passed away in June 2013. Our two golden retrievers, the ever-hungry Rosie and the attention-seeking Milly, have provided a ready source of light relief and a regular opportunity for me to leave the laptop and either play with them, take them for walkies or give them a biscuit.

1

Shuttle at the crossroads

A "NEW" SPACE SHUTTLE

Native to the lands of the Americas, the Northern Flicker (*Colaptes auratus*) is a medium-sized migratory woodpecker, with a particular penchant for constant, noisy drumming on trees or metallic objects as a means of communicating or declaring territory. Grey-capped, beige-faced, and with yellow-gold splashes on their tails, underwings and primaries, they are one of many hundreds of species that populate the expansive wildlife refuge of NASA's John F. Kennedy Space Center (KSC) on Merritt Island in Florida. Since the dawn of the Space Age, more than five decades ago, this region of the Atlantic coast has resounded to the roar of rocket engines, carrying humans and our mechanised emissaries into the heavens. Over Memorial Day Weekend in late May 1995, it resounded also to the jack-hammering of the Northern Flicker, which mistook the Space Shuttle's giant, brown-hued External Tank (ET) . . . for a tree.

By that point in time, the Shuttle had reached the halfway mark in what would prove to be its 30-year career as the world's first reusable piloted space vehicle. It had completed 68 missions since April 1981, transporting dozens of men and women from numerous nations into low-Earth orbit to launch, retrieve and repair satellites, operate scientific experiments and prepare for the establishment of a permanent International Space Station (ISS) at the cusp of the new millennium. And in May 1995, two missions – Space Shuttle Discovery on STS-70 and her sister Atlantis on STS-71 – stood ready to launch within weeks of each other. The first was tasked with deploying an important communications and data-relay satellite for NASA, as well as testing a new Space Shuttle Main Engine (SSME), whilst the second would inaugurate a series of historic rendezvous and docking missions between the United States and Russia's Mir space station. On 2 May, NASA announced that delays to the launch of the critical, US-financed Spektr scientific research module for Mir would push STS-71 into late June and that STS-70 would fly first, hopefully on the 8th.

"Both of these flights are very important in NASA's spaceflight effort," said former astronaut Brewster Shaw, then serving as the agency's head of Space Shuttle Operations. "STS-70 represents the first flight of the new Block I Space Shuttle Main Engine with the new Phase II+ powerhead, single-coil heat exchanger and new high-pressure oxidiser turbopump. The STS-71 mission represents a significant leap forward in our co-operative effort with the Russians and also the development of the ISS. By flying the missions in this order, we are able to make the best use of the workforce, Shuttle processing resources and the ability to meet our future manifest assignments."

By this point in its operational service, the Shuttle had changed markedly in appearance and goals from its original incarnation. Developed shortly after the Apollo Moon landings, the notion of transporting astronauts into orbit aboard a vehicle the size and shape of a conventional airliner was quite at odds with the United States' previous philosophy of launching ballistic capsules atop expendable boosters. It was intended to make space access more "routine", with a fleet of reusable orbiters capable of weekly or fortnightly launches, and one of its key supporters was the US Air Force, which had a dominant role in its design and development. In fact, the military envisaged the Shuttle as a critical means of launching large classified satellites into space.

The orbiter itself was similar in dimensions to the DC-9 airliner, roughly 36 m long with its wings spanning 24 m from tip to tip. Its habitable area consisted of a two-tiered cockpit – a 'flight deck' for operations and a 'middeck' for experimental work, eating and sleeping – which backed onto the 20 m long payload bay and an aft compartment to house the three main engines and two Orbital Manoeuvring System (OMS) pods and support a vertical stabiliser tail fin. Forty-four tiny Reaction Control System (RCS) thrusters in the Shuttle's nose and tail would provide attitude control and additional manoeuvrability whilst in space. The graphite-epoxy payload bay doors were, at the time, the largest aerospace structures yet built from composite material and *had* to be opened within a couple of hours of reaching orbit, to enable the radiators lining their interior faces to shed excess heat from electrical systems into space. The five-piece doors were hinged at either side of the mid-fuselage, mechanically latched at the forward and aft bulkheads and thermally sealed at the centreline. Ordinarily, they were driven 'open' and 'closed' by electromechanical power, but if they were unable to be opened, it was declared that the vehicle must return to Earth at the earliest opportunity. Conversely, if the doors did not *close* properly at mission's end, two crew members were trained to operate the mechanism manually on a spacewalk. Each orbiter was notionally designed to make a hundred missions before major refurbishment would become necessary, with five vehicles – Columbia, Challenger, Discovery, Atlantis and Endeavour – ultimately built. However, even the fleet leader, Discovery, had made only 39 voyages when it was retired in March 2011.

The unusual appearance of the Shuttle 'stack' on the launch pad has been described by veteran astronaut Story Musgrave as "like bolting a butterfly onto a bullet". The bullet was the ET, which resembled an enormous aluminium zeppelin, standing on end, some 46.6 m tall. It comprised two tanks for liquid oxygen and liquid hydrogen, separated by an 'inter-tank' for instrumentation and umbilicals.

The oxygen tank at the top housed up to 542,640 litres of oxidiser and the hydrogen tank held some 1.4 million litres of fuel, both of which were fed through a pair of 43 cm lines into disconnect valves in the belly of the Shuttle's aft compartment and thence into the combustion chambers of the three main engines.

At the start of each Shuttle mission, the main engines burned for about eight minutes and were shut down a few seconds before the ET was jettisoned, right on the edge of space. Each engine measured 4.2 m in length, weighed 3,400 kg and was 'throttleable' at one-percent incremental steps from 65 percent to 104 percent rated performance. The throttle was controlled by the Shuttle's General Purpose Computers (GPCs) and throttling back reduced stress on the vehicle during periods of maximum aerodynamic turbulence and also served to limit the G-loads in the final phase of ascent. Despite the immense power generated by each engine and the colossal amount of propellant needed to run them, they provided only about 20 percent of the muscle to reach space. The remainder came from the two 45.4 m tall Solid Rocket Boosters (SRBs), the first solid-fuelled rockets ever used in conjunction with a manned spacecraft. Loaded with a powdery aluminium fuel, mixed with an oxidiser of ammonium perchlorate, the boosters, built by Morton Thiokol (now ATK Thiokol) of Utah, were mounted like a pair of large Roman candles on either side of the ET.

Preparing for each Shuttle flight required several years, but the actual bringing together of the components began with setting up the boosters on a Mobile Launch Platform (MLP) in the gigantic Vehicle Assembly Building (VAB) at the swampy KSC launch site. Each booster comprised six blocks, called 'segments', positioned by overhead cranes with pinpoint accuracy, one atop the next, and joined by a ring of bolts. To prevent a leakage of searing gases whilst operating, a series of rubberised O-rings sealed the joints between the segments. After propelling the Shuttle and ET to an altitude of about 45.7 km, about two minutes after liftoff, pyrotechnics separated the boosters, auxiliary rockets at their nose and tail pushed them away and parachutes were deployed to lower them to a gentle splashdown in the Atlantic Ocean. They were then recovered, stripped down and refurbished for reuse. When the assembly of the SRBs was complete, the ET was moved into position between them and connected by a series of spindly, but strong, attachment struts. After checks to verify its mechanical and electrical compatibility, the Shuttle itself was moved from the nearby Orbiter Processing Facility (OPF), tilted by crane onto its tail and mated to the tank.

The transfer of the 1.8 million kg stack from the VAB to either Pad 39A or 39B – a distance of 5.6 km – took six hours, with the aptly named 'crawler' inching the $2.2 billion national asset along a track made from specially imported Mississippi river gravel. Once the platform was 'hard down' on the pad surface, the crawler withdrew and a servicing structure rotated into place around the vehicle. Further checks were conducted, payloads installed and the crew participated in a Terminal Countdown Demonstration Test. This was essentially a full dress rehearsal of the final part of the countdown, after which there was a simulated main engine shutdown and emergency evacuation exercise.

As for the *return* of the orbiter from space, in October 1974 NASA decided that

1-01: The STS-70 stack is rolled out to the launch pad.

the first missions would land at Edwards Air Force Base in California; according to Jim Fletcher, the NASA Administrator, this offered "the added safety margins and good weather conditions" needed on those early flights. These margins were particularly evident in the nature of the two primary Edwards landing strips: Runway 17, which crosses a salt flat and measures 13 km long, and the all-concrete Runway 22, whose 4.5 km length was extended by a 500 m asphalt overrun at each end. However, when the vehicle became fully operational, it was expected to routinely use a specially built 4.5 km runway, the Shuttle Landing Facility (SLF) at KSC. Indeed, when Shuttle construction work got underway in Florida in 1974, this runway was one of the first structures to be built.

On the launch pad, the ET was arguably the Shuttle's largest and most visible structure and it was perhaps for this reason that the Northern Flicker woodpeckers were attracted to the STS-70 stack in late May 1995. Discovery had been rolled out of the VAB and transported to Pad 39B on the 11th and was tracking an opening launch attempt on 8 June, with a crew of five expected to deploy the seventh Tracking and Data Relay Satellite (TDRS-G) atop an Inertial Upper Stage (IUS) booster. Their mission was expected to run for between five and eight days, a standard flight duration in the first half of the Shuttle era. Then, on 31 May, NASA engineers were called to assess 71 "excavations" and numerous other beak and claw marks in the ET's foam insulation material. The holes had been produced by the nesting woodpeckers and measured between 1.2 and 10.1 cm in diameter. Two days later, after unsuccessful attempts to repair the damage at the pad, it was decided that the damage was sufficiently severe to warrant the return of the STS-70 stack to the VAB for repairs. In readiness for the rollback, the IUS batteries were disconnected and those aboard TDRS-G – an important communications asset to support future Shuttle missions and US scientific satellites, including the Hubble Space Telescope (HST) – were discharged. By 6 June, work began to remove the TDRS-G/IUS combo from Discovery's payload bay, for temporary accommodation in Pad 39B's Payload Changeout Room.

By 15 June, the STS-70 stack had been returned to the VAB, repaired and restored to the launch pad. However, with STS-71 and Atlantis now further along in their processing campaign, it was decided to fly her mission first on 22 June and postpone STS-70 until no earlier than 13 July. This also preserved the option to launch a third mission, STS-69, aboard Space Shuttle Endeavour, on 30 July. The change came as a disappointment to the five-member STS-70 crew, under the command of veteran astronaut Tom Henricks. Since their assignment to the mission in August 1994, Henricks' team had prided itself as the first (almost) "all-Ohio" crew, for with the exception of New York-born pilot Kevin Kregel, they all originated from the Buckeye State. With the damage wreaked by the Northern Flickers, they also adopted a new crew mascot: Woody Woodpecker, who took principal place in an unofficial, tongue-in-cheek second take on their official, Ohio-themed crew patch. In the meantime, to protect Discovery and the ET from further attacks, model owls and predator-eye balloons were set up in the vicinity of the launch pad.

Terence Thomas Henricks, in command of STS-70, was making his third Shuttle flight, having previously occupied the pilot's seat on two missions. He was born in

Bryan, Ohio, on 5 July 1952 and earned his degree in civil engineering from the Air Force Academy. After graduation in 1974, Henricks underwent flight instruction at Craig Air Force Base in Alabama and trained on the F-4 Phantom. He flew in fighter squadrons in England and Iceland and was selected to attend test pilot school at Edwards Air Force Base in 1983, then remained in California as an F-16C Falcon test pilot and headed the 57th Fighter Weapons Wing Operating Location until his selection by NASA as an astronaut candidate in June 1985. In addition to his flying accomplishments, Henricks was a master parachutist, with 749 jumps, and earned a master's credential in administration from Golden Gate University.

Fellow Ohioans on the STS-70 crew included Mission Specialists Donald Alan Thomas and Nancy Jane Currie, both making their second Shuttle flights, and "rookie" spacefarer Mary Ellen Weber. Thomas came from Cleveland, where he was born on 6 May 1955, and attended high school in his home state, before entering Case Western Reserve University to study physics. He earned his degree in 1977 and went on to gain a master's and a doctorate in materials science from Cornell University, in 1980 and 1982, with an emphasis upon the evaluation of crystalline defects and sample purity in the superconducting properties of niobium. Clutching his PhD in hand, Thomas joined AT&T Bell Laboratories in Princeton, New Jersey, as a senior member of the technical staff, helping to develop advanced materials and processes for high-density interconnections of semiconductors. In 1987, he moved to Lockheed to review materials used in Shuttle payloads and, a year later, he joined NASA's Johnson Space Center (JSC) in Houston, Texas, for the first time, as a materials engineer. Thomas' particular emphasis was upon the projected lifetime of advanced composites being considered for use aboard Space Station Freedom. Following an unsuccessful attempt to join the 12th astronaut class in June 1987, Thomas was selected as a member of the 13th intake in January 1990. He first flew on the STS-65 mission in July 1994.

As for Nancy Currie, her career path toward the astronaut corps could not have been more different from that of Thomas. Born Nancy Jane Sherlock on 29 December 1958 in Wilmington, Delaware, she grew up and spent many of her formative years in Ohio, and later admitted that a career in the space programme was unthinkable in her childhood. "Women *weren't* military pilots," she told the NASA oral historian. "Women *weren't* astronauts. It really wasn't a concrete goal of mine until much later in my life." Having said this, she considered it a natural progression to enter the military and pursue an education which ultimately opened professional routes which might previously have been closed to her. Only at the end of her time in high school did the military begin to accept female aviators and only whilst she was at Ohio State University, studying the biological sciences, did NASA begin to accept female astronauts. Upon receipt of her degree in 1980, she entered the Army with the dream of becoming an aviator and perhaps eventually a military doctor. However, women were gradually being introduced into combat positions and after rotary-wing pilot training she served as an instructor pilot at the Army Aviation Center. Other positions followed – section leader, platoon leader and brigade flight-standardisation officer – and in 1985 she earned a master's degree in safety engineering from the University of Southern California.

Eventually, she became a Master Army Aviator and by the end of her military career in 2005 she had logged 3,900 hours in the air aboard various rotary-wing and fixed-wing aircraft. Years later, she remembered her first day in a helicopter and the instant thought: *This is for me. This is what I want to do.* Her decision to pursue a master's in safety engineering is an interesting side note. "When I was in flight school," she explained, "we actually had an accident that killed my instructor pilots and two of the guys that I flew with every day. It was just kind of a strange coincidence that I wasn't in the aircraft. It was at that time that I decided to devote a portion of my career and my academic life to safety engineering." Later, during her NASA life, she went on to earn a doctorate in industrial engineering, with emphasis upon human factors. "Catastrophic things *can* happen," she reflected, sadly, "due to human error in the cockpit or human error combined with a malfunction in the aircraft."

Her arrival at JSC came in September 1987 as a flight simulation engineer on the Shuttle Training Aircraft. She had earlier applied unsuccessfully for admission into that year's astronaut candidate class. That same year, her daughter Stephanie was born. Three years later she was selected as an astronaut and for the first half of the 1990s until she met her future husband, Sherlock was a single parent. "It was somewhat difficult for a mom to have a lifestyle like that," she said of her early NASA career, "gone an awful lot on travel and studying a lot at home." That study was particularly intense in the months leading up to STS-70, for Sherlock served as the flight engineer, responsible for assisting the pilots to monitor Discovery's systems during ascent and re-entry. By the end of her career, she had served as flight engineer on four Shuttle missions, more times than any other female astronaut.

With Henricks, Thomas and Currie providing the prior flight experience, STS-70's two first-time astronauts were Mary Ellen Weber and Kevin Kregel. Both experienced quite different backgrounds – the former a scientist, the latter a test pilot – whose careers had converged on the space programme and culminated in a journey into orbit in the summer of 1995. Like Don Thomas, Weber came from Cleveland, where she was born on 24 August 1962. "In school, I liked problem-solving," she told a NASA interviewer, years later. "I liked trying to figure out how things worked, why things worked, trying to understand the details, the physics behind things." After high school, she entered Purdue University to study chemical engineering and graduated in 1984, then proceeded to the University of California at Berkeley to earn her doctorate in physical chemistry in 1988. Her PhD work specialised in the physics of gas-phase chemical reactions involving silicon. Weber joined Texas Instruments to research new processes for manufacturing computer chips and was assigned to the SEMATECH consortium of semiconductor companies, and later to Applied Materials, to create a revolutionary reactor for next-generation chips. She was selected into NASA's 14th group of astronaut candidates in March 1992.

An avid skydiver, Weber won silver and bronze medals at the US National Skydiving Championships and joined the world-record-setting largest completed freefall formation in 2002, which boasted no fewer than 300 participants. "Skydiving was the first aspect of aviation that I got involved with," she recalled, "and it was exhilarating. It was exciting. It was very challenging. It's a sport that you can do for many years and there's always room for improvement and always another challenge around the corner. That's what I liked about both fields, science and aviation. I first wanted to jump out of [aircraft] in college and it wasn't until I was around aviation and I really got exposed to it and realised how exciting flying was. After graduate school, I got my licence doing that. Like Nancy Currie, Weber had little inclination to consider the astronaut business as a career option when growing up. "That was not something women did," she told the NASA interviewer. "It wasn't until I was in graduate school in chemistry and I was very much into science and research. I'd also gotten involved in a lot of aspects of aviation and space just seemed like the perfect adventure and the perfect mix of all the things that I loved."

Aviation proved a similar calling for Kevin Richard Kregel, born in Amityville, New York, on 16 September 1956. For him, it was Long Island-based Grumman Aerospace – the company that built the Apollo lunar modules – which inspired him to follow space exploration and someday become an astronaut. "It's something I've always wanted to do since I was a little child," Kregel told a NASA interviewer. "I aspired to go the same route as the original astronauts and that is why I became an Air Force test pilot, got an engineering degree and applied to the programme." After high school, he graduated from the Air Force Academy in 1978 with a degree in astronautical engineering, then entered active military service and won his pilot's wings in August 1979 at Williams Air Force Base in Arizona. Assigned to RAF Lakenheath in England during 1980-1983, Kregel flew the F-111 Aardvark ground-attack and tactical strike aircraft and, upon his return to the United States, flew A-6E Intruder all-weather strike aircraft as an exchange officer with the US Navy at Naval Air Station Whidbey Island in Seattle, Washington. He also completed 66 carrier landings during a cruise of the Western Pacific aboard the USS *Kitty Hawk*.

Kregel attended the Navy's test pilot school at Patuxent River, Maryland, as an Air Force exchange candidate, and after graduation worked at Eglin Air Force Base in Florida on weapons and electronic systems testing on the F-111, the F-15 Eagle tactical fighter and the initial certification of the F-15E Strike Eagle multi-role fighter. In 1988 he earned a master's degree in public administration from Troy State University and resigned from the Air Force to join NASA in April 1990 as an aerospace engineer and instructor pilot. Based at Ellington Field in Houston, Texas, he provided flight instruction on the Shuttle Training Aircraft (STA) – a modified Grumman Gulfstream business jet, with controls modified to mirror many of the orbiter's handling characteristics – and supported the initial testing of avionics upgrades to the T-38 Talon training aircraft. Two years later, in March 1992, Kregel was selected alongside Weber as a member of NASA's 14th astronaut group. As the sole member of the STS-70 crew not to have been either born or raised in Ohio, in the months before launch he was proclaimed an "Honorary Ohioan" by then-Ohio Governor George Voinovich.

As Discovery's pilot, one of Kregel's primary duties during ascent was monitoring the performance of the three liquid-fuelled main engines from his seat on the right side of the flight deck. On STS-70, however, one of those Rocketdyne-built engines was somewhat different from its two companions, for it was the first flight of a new

1-02: The STS-70 crew departs the Operations & Checkout Building on 13 July 1995. Commander Tom Henricks leads Mary Ellen Weber, Kevin Kregel, Nancy Currie and Don Thomas.

Block I design, incorporating various enhancements to increase its reliability and safety. Designated SSME No. 2036, it was situated in the No. 1 position at the 'top' of the pyramid of engines on the aft fuselage, and included Pratt & Whitney's new 23,700 rpm Alternate High Pressure Oxidiser Turbopump (AHPOT), produced by a casting process that eliminated all but six of the 300 welds of the earlier design. It also featured a two-duct powerhead to improve fluid flow through the engine to decrease pressures and loads and a single-coil heat exchanger to remove all but seven welds. The new powerhead replaced three smaller fuel ducts in the earlier design with two enlarged ducts to improve engine performance.

Overall, the Block I design was expected to reduce wear and maintenance on the engine components. Whereas the earlier turbopumps needed to be removed after each flight, the new Block I AHPOT eliminated the need for detailed inspections until it had flown ten times. On STS-70, the No. 2 and 3 engines were of the earlier design, but NASA intended from STS-73 – then scheduled for launch in September 1995 – that the orbiters would begin flying with a full suite of three Block I powerplants. Extensively tested in an "aggressive" 20,000-second programme at the John Stennis Space Center in Mississippi, whose demands equalled the stresses imposed by 40 Shuttle launches, the AHPOT was formally declared ready for flight by NASA on 15 March 1995. "Completing flight certification of the Alternate High Pressure Oxidiser Turbopump is a major milestone in the Space Shuttle Main Engine program," said Otto Goetz, then the SSME deputy project manager for development at NASA's Marshall Space Flight Center (MSFC) in Huntsville, Alabama. "The Alternate Turbopump is now ready for its first flight and for nine flights thereafter." The Block I AHPOT also included a new ball bearing material of silicon nitride, which proved superior to the steel bearings used previously: it was 30 percent harder and its ultra-smooth finish allowed for less friction wear during operations.

These engine developments formed part of a $1 billion effort to overhaul the Shuttle's main propulsion system, with expectations that by September 1997 a further upgraded Block II powerplant would bring Pratt & Whitney's new High-Pressure Fuel Turbopump (HPFT) for the liquid hydrogen hardware into service. According to Tim Furniss, writing in *Flight International* in May 1995, the original intention was to enhance the Shuttle's safety and reliability, but the objective expanded with the need to support high-inclination, high-energy missions to Russia's Mir space station and to the future ISS, both of which operated in orbits of 51.6 degrees. This differed markedly from the 28.5 degrees originally planned for the US-led Space Station Freedom. In fact, the Block I and II programme had been halted by NASA in December 1991 when it ran $260 million over budget. Furthermore, the Advanced Solid Rocket Motor (ASRM) programme to upgrade the SRBs, enhance their safety characteristics and assist with transporting larger payloads into orbit was cancelled in October 1993. The Block I and II programme was resumed in May 1994 and the Block I tests were completed in just nine months, but there were doubts about its performance and the 150 kg weight increase of the new engine hardware over its predecessor added to worries about the Shuttle's critical payload-to-orbit margins. However, Dennis Jenkins has noted that the incorporation of the new engine "offered a significant decrease in the risk of losing a vehicle during ascent", adding that the computed loss-of-vehicle probability in the earlier engines was one in 262 flights, compared to a reduction of one in 335 flights for Block I.

As these efforts progressed to enhance the safety of the Shuttle, the management of the four-strong fleet of orbiters was also expected to change as it reached its 15th year of operations. In November 1995, NASA announced that it would pursue a non-competitive agreement with United Space Alliance (USA) – a joint venture between Rockwell International and Lockheed Martin Corporation, which together held a 69-percent stake in the dollar value of all Shuttle-related prime contracts – to "eventually" assume responsibility for Shuttle operations. "This clearly is the appropriate path to take," NASA Administrator Dan Goldin said at the time. "It will allow us to ensure the safe operations of the Space Shuttle, meet the flight manifest and maintain our commitment to launch the first space station elements in late 1997." Commenting on the previous experience of Rockwell and Lockheed Martin, he added that "the task of combining the existing separate contracts under the consolidated Shuttle contract will be greatly simplified" and noted that the non-competitive nature of the Space Flight Operations Contract (SFOC) was justified because "there was no other company that could possibly meet our safety, manifest and schedule requirements".

Over the course of the following year, a Statement of Work was produced and USA prepared definitive technical and cost proposals, with a key emphasis on the reduction of contract requirements, facilities and workforce. Since NASA was not expected to be heavily involved in the management of day-to-day Shuttle operations, it was rationalised that fewer of its civil servants would be required to manage the programme. By April 1996, two Novation Agreements were signed, in an effort to ease the transition into the SFOC. Under the provisions of these agreements, USA was designated the Prime Contractor for Shuttle processing (previously undertaken by Lockheed Space Operations Company at KSC) and for Shuttle operations (hitherto executed by Rockwell Space Operations Company at the Johnson Space Center in Houston, Texas). "This agreement basically constitutes a name change for contracts that will essentially remain intact for the near term," remarked Space Shuttle Program Manager Tommy Holloway. "However, it is significant in that it is the first phase of a transition to a single contract and it constitutes the first official contract between USA and NASA. This allows us to get an early start on implementing a single Space Flight Operations Contract approach and it supports our requirement that the transition to that contract is efficient and, above all, safe."

On 30 September, NASA formally announced a six-year, $7 billion SFOC base contract with the potential for a pair of two-year extensions, producing the option for a total estimated value of $12 billion over the following decade. "The contract assigns greater responsibility to the contractor," NASA explained, "reducing the government's role in overseeing day-to-day, routine Shuttle operations." However, the agency retained ultimate responsibility for safety and continued to direct high-level management of the programme and its missions, with the final decision on all problem resolutions kept with NASA in all cases. A variety of mechanisms, including structured surveillance and audits, reviews of unusual, 'out-of-family' technical problems and safety performance grading which were linked to USA's award fees, remained in place to ensure continued safety. "The safety grade will be one of the factors used each six months to determine the amount of fee that are awarded," it was stressed. "Any safety grade less than 'very good' also would eliminate the cost-reduction incentive feature for the past six months, a provision designed to guard against the possibility of an over-zealous contractor cost-reduction effort impacting the safety of Shuttle operations." In précis, SFOC replaced 12 previous individual contracts, including the largest pair, previously performed by Lockheed Martin in Florida and Rockwell in Houston, in favour of just one. It was stressed at the time that a potential second phase of the contract could incorporate a further 16 contracts, covering the supply of SSMEs, ETs and SRBs.

In mid-December 1996, responding to a request from the Office of Science and Technology Policy, the Aerospace Safety Advisory Panel concluded a six-month review and determined that the implementation of SFOC had not increased risks to the Shuttle programme. However, the panel made 22 recommendations, including the need for NASA to ensure the availability of a skilled and experienced workforce in sufficient numbers to meet ongoing safety demands. "I'm very pleased that the panel has given the Space Shuttle Program a clean bill of health," said Administrator Goldin after the publication of the review, "but the panel also pointed out some areas for continuing emphasis and the need to improve. NASA concurs with these findings and I have instructed the institutional management and the Space Shuttle Program to implement them as soon as possible."

On the other hand, the SFOC transition was met with dismay in other quarters. Former astronaut Bryan O'Connor, then serving as Director of the Space Shuttle Program at NASA Headquarters in Washington, DC, resigned his post in protest February 1996. In his oral history interview, O'Connor – a veteran of two Shuttle missions – expressed astonishment at the "high risk" of the agency's backing-off from many of its responsibilities of oversight and insight. However, he acquiesced that he felt it was practical, as long as it was not implemented too rapidly, and he felt that such critical features as main engines and SRBs should not be tendered out to a commercial operator. More importantly, though, O'Connor's primary concern was that communications channels between projects and programme managers were being hampered. "When you get the Center Director between a project manager and his programme manager, sometimes you lose something in the communication," O'Connor reflected. "We were trying to streamline that communication." Lack of adequate communication between management levels had been one of the criticisms levelled at NASA by the Rogers Commission, in the aftermath of the Challenger disaster. "Communications failures cause safety problems," said O'Connor. "Reorganisation in a way like this will cause communications problems. That's a risk to safety that we don't need to do." In retrospect, he admitted that his fears about the safety of the SFOC and the communications issues in the near term were largely unfounded, but the fact remained that NASA intended to cut its annual $14 billion budget by a third by 2001 and reduce its contractor and civil-servant workforce by 27,000, an eighth of whom would be Shuttle-related personnel. "The $3.2 billion Shuttle budget operations budget," wrote Tim Furniss in *Flight International* in February 1996, "has to be shaved to $2.5 billion."

Propelled by the first of a new generation of main engine components, and with a new Shuttle operations contract just around the corner, STS-70 was tasked with deploying TDRS-G as part of a network of communications platforms in geostationary orbit, 35,000 km above Earth. The 2,200 kg satellite was built at TRW's Redondo Beach facility in California. Since the beginning of the manned space programme, astronauts had relied heavily upon ground stations and tracking ships for communications and it was only possible to maintain contact for around 20 percent of each 90-minute orbit. Moreover, during re-entry, super-heated plasma around the vehicle caused a period of radio 'blackout'. With the arrival of TDRS, it would be possible to communicate with astronauts during 85-98 percent of each orbit and throughout re-entry.

As the Shuttle gained momentum in the mid-1970s, it was envisaged that a pair of these powerful satellites – one stationed over the equator, just off the north-eastern corner of Brazil, known as 'TDRS-East', and a second over the central Pacific Ocean, near the Phoenix Islands, known as 'TDRS-West', would fill this urgent communications and tracking need. TDRS-A was launched in April 1983, but was almost lost when its IUS booster failed to insert it into its proper orbit. Only by using the satellite's own hydrazine thrusters were controllers able to gradually manoeuvre it into its final 'East' location, although the result was that its operational lifetime was shortened. Ongoing problems with the IUS meant that it was almost three years before the 'West' satellite, TDRS-B, could be launched ... and that was the primary payload aboard the ill-fated Challenger on 28 January 1986. Two more TDRS satellites (C and D) were launched in September 1988 and March 1989. Unfortunately, TDRS-C also succumbed to anomalies which affected its Ku-band relay capability and TDRS-E, launched in August 1991, took its place. A further satellite, TDRS-F, was launched in January 1993. With TDRS-G, NASA's requirement for having two fully operational satellites and two fully operational 'ready reserve' satellites would be complete and TDRS-A and TDRS-C were relegated to the status of on-orbit 'spares'. Yet TDRS was no miracle worker. It could not process or adjust communications traffic in either direction. Rather, it operated as a 'bent pipe' repeater, relaying signals and data between its Earth-circling users and the highly automated ground terminal. Signals processing, therefore, was done on the ground and the satellite's sophistication was devoted to its very high throughput. Located in the New Mexico desert, White Sands provided a clear line of sight with both satellites and its limited amount of annual rainfall meant that weather conditions would not interfere with their Ku-band uplink or downlink channels.

By the early summer of 1995, as NASA prepared to launch TDRS-G into orbit, plans were advanced to develop a more sophisticated second generation of satellites, which it was anticipated would enter operational service around the turn of the millennium. As long ago as January 1990, NASA had released a Request for Proposals for an Advanced TDRS (ATDRS) system. It was described by William Guion of the TDRS Project Office at the Goddard Space Flight Center (GSFC) in Greenbelt, Maryland, as differing from the first system by showcasing "a new frequency band with a high-data-rate capability of 650 million bits per second and an

1-03: The Tracking and Data Relay Satellite (TDRS)-G payload is successfully deployed from Discovery's payload bay on 13 July 1995.

enhanced multiple-access system which will increase the data rate provided each of several simultaneous users from the current 50,000 bits per second to three million bits per second". At that time, the four-satellite ATDRS constellation was expected to see its maiden launch in July 1997 and would be fully operational by 2001. Over the course of the next few years, the programme matured and in February 1995 NASA awarded a $481.6 million contract to Hughes Aircraft Company to design, build and perform on-orbit checkout of three new TDRS platforms, with Ka-band capability, in addition to Ku-band and S-band. The first launch was rescheduled for 1999.

In the aftermath of the woodpecker incident, STS-70 was postponed until no earlier than 13 July 1995, being leapfrogged by the first Shuttle-Mir docking mission, STS-71. Early that morning, the five astronauts boarded Discovery and proceeded through an exceptionally smooth countdown. The clock was briefly halted for less than a minute at T-31 seconds, just prior to Autosequence Start – the point at which the Shuttle's computers assumed primary control of all of the vehicle's critical functions – due to fluctuations on the ET's automatic gain control range safety system receiver. The hold was called by Booster Range Safety Engineer Tod Gracom, seated in the Launch Control Center, and when the issue had been cleared, the countdown was resumed and Discovery roared into orbit at 9:41:55 am EDT.

After configuring their vehicle from a rocket into a spacecraft, and opening the payload bay doors, the astronauts set to work preparing TDRS-G and its solid-fuelled IUS booster for deployment, which was targeted to occur six hours and 13 minutes into the mission. Housed inside Discovery's payload bay, TDRS-G was huge, even though still in its 'launch' configuration with its communications hardware hidden from view, its umbrella-like antennas closed and its two electricity-generating solar arrays each folded into three parts. The satellite comprised three main segments: an equipment module, a communications payload and a battery of relay antennas. At this stage, however, it bore little resemblance to the enormous 'windmill' into which it would transform in geostationary orbit. Mounted on the base of TDRS-G, the IUS booster was held securely in Discovery's payload bay by the doughnut-shaped 'tilt table' known as the Airborne Support Equipment (ASE). As well as providing the crew with an ability to hoist the entire 14 m long stack from a horizontal position to the deployment angle of 58 degrees, it incorporated electronics, batteries and cabling to enable the astronauts to issue commands during the lengthy checkout of both the satellite and booster. The ASE included a low-response spreader beam and torsion bar mechanism to limit the dynamic loads on the spacecraft to less than a third of what might otherwise have been imposed. It was secured into the payload bay by means of six standard, non-deployable attachment fittings, which mated to the ASE's forward and aft frames, and two payload-retention latch actuators.

Leading the complex deployment effort on STS-70 were Mission Specialists Don Thomas and Mary Ellen Weber. They firstly raised the IUS tilt table to an angle of 29 degrees above the payload bay, where they conducted preliminary systems tests, placed the payload onto IUS internal power and disconnected orbiter umbilical cables. Next, they increased it to its deployment angle of 58 degrees. A spring-loaded ejection system was employed to release TDRS-G/IUS, precisely on time, at 3:55 pm, and the stack drifted into the inky blackness of space at a rate of 0.1 m/sec. Fifteen minutes after deployment, Tom Henricks and Kevin Kregel fired Discovery's manoeuvring thrusters to create a safe distance from the payload, ahead of the ignition of the IUS first stage engine.

Originally intended as a temporary substitute for a reusable 'space tug' when it was designed in the 1970s, the IUS was dubbed the 'Interim' Upper Stage and later became 'Inertial' in recognition of its sophisticated internal guidance system. Losing its 'interim' status also reflected a growing awareness, when the space tug was cancelled in late 1977, that the IUS' services would be needed throughout the 1980s. In fact, not until the early years of the present century did it fly for the last time as a 'standalone' booster. Prime contractor for the IUS was Boeing, which began developing the two-stage vehicle in August 1976 and supported its first launch aboard a Titan 34D rocket six years later. Measuring 5 m long and a little under 3 m in diameter and weighing some 14,740 kg, the cylindrical booster – made from Kevlar-wound aluminium – was capable of delivering 2,270 kg payloads to geostationary altitudes. Its first stage carried 9,700 kg of solid propellant and a large motor, capable of firing for up to 145 seconds; this made it the longest burning solid-fuelled engine ever used in space applications. Meanwhile, the second stage carried 2,720 kg of propellant. Both the first and second stage nozzles, commanded by redundant electromechanical actuators, could steer the former by up to four degrees and the latter up to seven degrees. Although solid rockets were known to generate a harsh impulse, the separation mechanism between the first and second stages employed a low-shock ordnance device in order to avoid damaging its payload. Moreover, solid propellant was chosen over a liquid-fuelled booster because of its simplicity, safety, high reliability and low cost. Hydrazine-fed reaction control thrusters provided the IUS with additional stability during the 'coasting' phase between the first and second stage firings, as well as ensuring accurate roll control and assisting with the satellite's insertion into geostationary orbit. Situated between the two stages was an equipment section with avionics systems to provide guidance, navigation, control, telemetry and data management services to TDRS-G.

During its period in free flight, the IUS' on-board computer issued signals to begin mission events by manoeuvring to the required thermal attitude and removing the pyrotechnic inhibits for the first stage motor. Ignition took place at 4:55 pm and the motor burned for 146 seconds as planned. About 5.5 hours later, the second stage ignited for 108 seconds to continue the process of boosting TDRS-G into its 35,000 km geostationary orbit, initially positioning the satellite at 179.88 degrees West longitude. With the IUS providing stabilisation, TDRS-G deployed its solar arrays and space-to-ground communications system, then the spent IUS was jettisoned to execute a collision-avoidance manoeuvre and deactivate itself. Meanwhile, the satellite underwent a month of checkout, before being manoeuvred to its operational location at 171 degrees West, high above the central Pacific Ocean, close to the Gilbert Islands.

Following the deployment, the Mission Control Center in Houston, Texas, was transferred from the old control room – used for more than three decades, since the Gemini era – to the new $500 million White Flight Control Room (FCR). Although it was only down the hall from its predecessor, the White FCR (pronounced 'Flicker') boasted advanced workstations and high-speed computing terminals, supported by 200 km of fibre optic cables to produce what was claimed to be the largest and fastest Local Area Network in the world at that time. The White FCR assumed primary responsibility for all on-orbit functions on STS-70, with the exception of Discovery's re-entry and landing, which were conducted from the old control room. Developed by Loral Space Information Systems at $75 million *under* budget, it was expected that the new room would enable NASA to reduce its Shuttle operating costs by about $20 million per annum, with a full takeover of all Shuttle operations anticipated by 1997. At this event, it was intended that the old control room would be preserved as a national monument.

With the deployment of their primary payload complete, on the evening of 13 July 1995, Tom Henricks and his crew settled down to a planned eight days of scientific and technological experiments, housed in the payload bay and aboard Discovery's middeck. Perhaps in order to kick off the mission in an appropriate fashion, Mission Control awakened the five astronauts early on 14 July with the theme tune for 'Woody Woodpecker', in a light-hearted nod at their previous misfortune.

One of the most visible experiments was the Hand-held, Earth-oriented, Real-time, Co-operative, User-friendly, Location-targeting and Environmental System, which its Department of Defense sponsors had crafted into the impressive acronym of 'HERCULES'. Flown on two previous Shuttle missions before STS-70, it represented the third generation of a space-based geolocation system and consisted of a XYBION multispectral video camera and a battery of equipment provided by the Naval Research Laboratory. Its objective was to calculate and tag each frame of high-resolution video with latitude and longitude co-ordinates, with an accuracy of about 5 km. Unfortunately, the astronauts' attempts to acquire imagery with HERCULES were frustrated by difficulties with its inertial measurement unit, since it required the location of two guide stars with the camera in different orientations. They sent Mission Control several images of Florida and the Bahamas and investigators later remarked that the video provided "excellent insights" into potential improvements for the system. During 'daytime' observations, they used any Nikon-compatible lens, but during nighttime passes they attached an image intensifier which had been developed by the US Army's Night Vision Laboratory. Elsewhere, the Window Experiment (WINDEX) sought to gain an understanding of the chemistry and dynamics of the environment around Discovery, partly in order to prevent damage to sensitive systems, such as the solar arrays planned for the ISS. Early on 14 July, Henricks fired the Shuttle's manoeuvring thrusters as Nancy Currie observed their effect in terms of the shimmering glow produced around the extremities of the OMS pods and vertical stabiliser tail fin. Later in the mission, observations were also made of waste water dumps and releases from the flash evaporator system.

In the middeck, Mary Ellen Weber tended to the Bioreactor Development System (BDS), which grew orange-hued colon cancer tissue samples to test the performance of the device. It was hoped that the peculiar microgravity environment of low-Earth orbit would enable researchers to grow cells into three-dimensional tissue pieces, which was not achievable in terrestrial gravity. From her perspective, Weber described the samples as developing well and looking far better than similar samples cultured on Earth. Continuing the bioscience bias of the STS-70 experiments, the Microencapsulation in Space (MIS) investigation involved the 'microencapsulation' of the drug ampicillin within a biodegradable polymer. It was hoped that in medical applications on Earth, such polymers would degrade in the body, thus releasing the antibiotic at a controlled rate. Previous work aboard the Shuttle in 1992 yielded samples which were more perfectly shaped, more homogeneous and purer than

Earth-made microencapsulated drugs. Additionally, the crew oversaw the growth of crystals of human alpha interferon protein, used widely to combat human viral hepatitis B and C, whilst the hormonal system and muscle formation of the tobacco hornworm and the cell division of the daylily were monitored in a pair of Biological Research in Canisters (BRIC) investigations. The National Institutes of Health sponsored a range of experiments to examine the effects of space travel upon the behaviour, muscle, nerve and bone development, and circadian timing of rats. In addition, NASA and the Walter Reed Army Institute collaborated on a study of the embryogenesis of the Medaka fish egg.

In readiness for their return to Earth, the crew donned their liquid-cooled underwear prior to climbing into their bright orange Launch and Entry Suits. During one of the breaks in the checklist for the process of configuring Discovery for her irreversible de-orbit burn, Kregel turned to Henricks with a quizzical look on his face. "Has anybody mentioned that your arms get longer in orbit?" Henricks knew that the spine had a tendency to extend slightly in conditions of microgravity, and, indeed, the sleeves of Kregel's long johns did seem to be too short. "We decided we'd just tell the docs when we get back," Henricks rationalised in a Smithsonian interview, years later. "Maybe your arms *do* get longer." All at once, petite Mary Ellen Weber floated up to the flight deck. She was concerned that she could not find her liquid-cooled underwear. Kregel offered to help find it ... only to discover that he had put hers on by mistake. "During the private medical conference with the ground," Henricks joked, "I had to tell the doctor that we'd discovered another side effect of space flight: a tendency to cross-dress!"

Landing conditions at KSC on 21 July appeared initially favourable, with forecasters carefully watching for the formation of cloud layers and the presence of ground fog. Flying the Shuttle Training Aircraft, astronaut Steve Oswald noted that he could not see the SLF runway from his vantage point and Flight Director Rich Jackson took the decision to wave off the first landing attempt. Next day, the 22nd, there were three opportunities to bring Discovery home to Florida, with the first at 6:26 am EDT called off in the hope that already acceptable weather would continue to improve in time for the second opportunity at 8:02 am. At length, Henricks and Kregel were cleared to fire the OMS engines for the irreversible de-orbit burn at 7:00 am, and after hypersonically knifing through the atmosphere they guided Discovery to a smooth touchdown on Runway 33 at the Cape at 8:02. The STS-70 mission had lasted just a couple of hours shy of nine full days and marked the end of Discovery's 22nd flight.

It was already scheduled that she would be transported to prime contractor Rockwell International's facility in Palmdale, California, for several months to receive extensive upgrades and major modifications. As long ago as March 1994, NASA's Space Shuttle Director Tom Utsman had noted that the work in Palmdale would entail the installation of the Orbiter Docking System (ODS), compatible with Russia's Mir space station. Discovery's internal airlock module would be removed from the middeck and a new external airlock installed in the forward section of the payload bay, preparatory for Mir and ISS operations. After standard post-mission servicing, Discovery was mated atop a Boeing 747 Shuttle Carrier Aircraft (SCA) and departed KSC on 27 September 1995, bound for California. Her stay on the

West Coast lasted nine months and she was back in Florida by the end of June 1996, having also benefitted from repairs to her thermal protection system, upgraded payload bay floodlights and structural corrosion inspections.

AN AMBITIOUS MISSION

With the completion of STS-70, the next mission, STS-69, was originally scheduled for launch on 30 July, to begin a complex, 11-day flight to deploy and retrieve two satellites and perform a spacewalk to evaluate techniques and tools for ISS assembly and maintenance. However, in the days after Discovery's landing, disassembly of the SRBs revealed a tiny air pocket, known as a "gas path", in the nozzle internal joint No. 3, at the base of the right-hand booster. It extended from the motor chamber, through the Room Temperature Vulcanising material and up to the primary O-ring. The latter formed part of a critical sealing mechanism to prevent 2,760 degrees Celsius gas leaking through the booster casing and, although the primary O-ring was not breached, it did exhibit a "slight heat effect", three singe marks and small concentrations of soot. Also observed were heat-affected insulation and eroded adhesive. The incident raised unpleasant memories of the O-ring failure which doomed Challenger in January 1986 and, significantly, a similar gas-path issue, with four instances of singing, had also been identified in one of Atlantis' SRBs, following the STS-71 launch on 27 June 1995. "Gas paths or small air pockets are the result of nozzle fabrication involving backfilling of the joint with insulation material," explained NASA's anomaly report. "Similar paths had been expected or observed following previous flights, but missions STS-71 and STS-70 marked the first time slight heat effect was noted on the primary O-ring."

As investigators set to work on the problem, the STS-69 launch met with delay. Originally targeted for launch on 30 July, Endeavour and her five-man crew first fell foul to the ravages of Hurricane Erin, which was threatening the coast of Central Florida. Having been initially rolled out to Pad 39A on 5 July, the Shuttle stack was rolled back to the VAB on 1 August as a protective measure and returned to the pad on the 8th. However, by this stage, as the scale of the SRB anomaly became more apparent, the launch was postponed until no earlier than the end of August. Although the primary O-ring was not breached and the secondary O-ring was untouched in the STS-71 and STS-70 ascents, it became clear that similar instances of exhaust burning through to reach the seals had occurred on 11 previous Shuttle flights. Quoted by *Flight International* a few days later, NASA noted that there were "no design issues" with the SRBs and that the STS-71 and STS-70 crews were in "no added danger", the investigation team was studying a procedure which could allow for ultrasound inspections and minor adjustments to the injection of an insulating putty. It was hoped that this "may reduce" the possibility of gas paths reaching the primary O-ring on future missions. "Working inside the nozzle of the boosters," explained *Flight International* on 30 August, "technicians are replacing the insulating material protecting the nozzle O-ring joints with a new material. Seven sets of [SRBs] will undergo this work and the new material will be introduced on future

1-04: Members of the STS-69 crew at their stations in the Shuttle flight deck simulator. In the pilot's seat at left is Ken Cockrell, with flight engineer Jim Newman at right. Payload Commander Jim Voss, in short-sleeved shirt, is visible at centre.

production." Although it was considered unlikely that STS-69 would fly until at least mid-September 1995, engineers and managers ultimately concluded at the Flight Readiness Review that the issue did not pose a significant safety-of-flight concern and the mission was rescheduled for launch on 31 August.

However, this attempt was scrubbed by the Mission Management Team at 3:30 am EDT on the 31st, just prior to the loading of liquid oxygen and hydrogen aboard the ET, when a temperature 'spike' was recorded in Endeavour's No. 2 fuel cell. Since mission rules dictated that all three of the orbiter's electricity-generating cells had to be fully operational for a launch to go ahead, it was elected to replace the unit and the countdown resumed on the afternoon of 4 September, aiming for a liftoff on the 7th. The STS-69 crew departed their quarters and arrived at Pad 39A at 8:10 am on that morning, with Newman taking a moment on the gantry to snap a photograph of his four crewmates. They took their seats aboard the Shuttle, but after the middeck side hatch had been closed and sealed, it failed the standard leak check. The hatch was opened, its seals checked and it was closed and retested without incident. Later in the countdown, a pressure issue cropped up with a water spray boiler, but was cleared, since it did not violate any Launch Commit Criteria and imposed no constraints. In the final minutes, Dave Walker stretched out his gloved hand to his three comrades on the flight deck – Cockrell, Voss and Newman – to express how much he had enjoyed working and training with them. Endeavour thundered safely into orbit at 11:09 am EDT.

For the first time in the Shuttle programme, two satellites would be deployed *and* retrieved, using the Canadian-built Remote Manipulator System (RMS) mechanical arm. The first was the Wake Shield Facility (WSF), flying its second mission, and the other was a payload whose cumbersome title (the 'Shuttle-Pointed Autonomous Research Tool for Astronomy') had been crafted by NASA's finest acronym-makers into a name which sat well on the tongue ('SPARTAN'). Yet when veteran NASA astronaut Jim Voss was assigned as payload commander of STS-69 in August 1993, the cargo was quite different, consisting instead of the fourth Spacehab laboratory and the Shuttle Pallet Satellite (SPAS)-III. "Spacehab is a complement of commercial experiments flown in a pressurised module in the Shuttle's cargo bay as a supplement to the middeck of the orbiter," a NASA press release announced, "and SPAS-III is a group of instruments which will measure the atmosphere around the orbiter and the background clutter in the Earth's atmosphere, calling for a complex flight plan." One of the key instruments aboard SPAS-III was the Infrared Background Signature Survey (IBSS), developed by the Strategic Defense Initiative (SDI). Distinguished by its large, barrel-shaped cryostat, IBSS had previously flown on the STS-39 mission in April 1991 and sought to obtain scientific data in support of missile defence applications. Flying freely from the Shuttle aboard SPAS-III at various ranges, IBSS would make spectral, spatial and temporal radiometric observations of the orbiter's exhaust plumes and replications of booster firings. Interactions of these plumes with the ionosphere would be analysed, as would the region around the engine nozzles.

In September 1991, *Flight International* reported that Fairchild Space had been awarded a $14 million systems integration contract for SPAS-III by the SDI's Sensor Technology Directorate. However, SPAS-III and IBSS met with delay and eventually with cancellation, and by the time the remainder of the STS-69 crew was named in July 1994 it had vanished from the payload complement entirely. So, too, had the Spacehab module, which was remanifested onto a later Shuttle mission. In their place was WSF-2 and a SPARTAN satellite equipped with a battery of experiments for NASA's Office of Aeronautics and Space Technology, and known as 'OAST-Flyer'. As the months wore on, OAST-Flyer was shifted to a subsequent flight and replaced with the SPARTAN-201 payload of two instruments for solar science research. As payload commander, it was Voss' responsibility to develop and co-ordinate the crew's training plan, liaise with the mission's payload sponsors and principal investigators, attend relevant meetings and oversee pertinent hardware and software changes, ahead of the announcement of the other crew members. In essence, Voss was responsible for the conduct and accomplishment of STS-69's scientific objectives, and for this role he was amply qualified. In fact, a childhood love of science fiction eventually guided James Shelton Voss into space. Born in Cordova, Alabama, on 3 March 1949, he was a keen wrestler and footballer and studied aerospace engineering at Auburn University. After graduation in 1972, he entered the Army, but was permitted to delay his entrance into active duty until he had earned a master's degree in aerospace engineering sciences from the University of Colorado. Voss' military career began well: he was the Distinguished Graduate of the Army's Infantry Basic Course and received the Honor Graduate and Leadership

Award upon completion of the Ranger Course. (The latter has been variously described as "the toughest combat course in the world" and "the most physically and mentally demanding leadership school the Army has to offer".) Voss' accomplishments, even at this early stage in his Army career, give an impression of his character.

Yet an astronaut career eluded him, for virtually all of the astronauts then serving with NASA were test pilots. "And though I was in the military," he told a NASA interviewer, "I was an engineer and I didn't have vision that was good enough to be in the programme then." Everything changed in January 1978, when the first selection of Shuttle astronauts was made and included a range of military and civilian personnel, engineers and scientists, pilots and physicians. Voss tendered his first unsuccessful application to NASA in that very year and it would take no fewer than *six* attempts before he was finally accepted into the astronaut corps in June 1987. He flew aboard two Shuttle missions in November 1991 and December 1992, prior to his STS-69 assignment.

Having said this, Voss was not wholly consumed by NASA. His Army work was something he pursued with a passion – "The things that I did in my military career ... were things I *wanted* to do," he said – and he spent time as a platoon leader, intelligence staff officer and company commander in West Germany, attended the Infantry Officer Advanced Course in 1979 (making the Commandant's List) and taught mechanics as a professor at the Military Academy at West Point, receiving the William P. Clements Jr Award for Excellence in Education. Graduation from Naval Test Pilot School in 1983 brought with it the Outstanding Student Award, after which he entered the Armed Forces Staff College and served as a flight test engineer with the Army's Aviation Engineering Flight Activity. He began working for NASA in November 1984 as a vehicle integration test engineer, seconded from the Army, and supported four Shuttle missions, including the final mission of Challenger. In the wake of that disaster, Voss participated in the Rogers investigation. Selection by NASA in the summer of 1987 was unquestionably one of the high points of his life, but, fundamentally, he was driven by a love of aviation ... and space flight was simply a logical extension of aviation. Years later, he would pay tribute to a number of individuals for guiding his steps: his grandparents, Jim and Millie Wright, who raised him, together with Major Jack Damewood, his Reserve Officer Training Corp (ROTC) instructor at Auburn University, whose model of a good soldier and officer steeled Voss to follow his example. Others included his wresting and football coaches, Bobby Barrett Ray Campbell and Swede Umbach.

Although Voss' duty on STS-69 encompassed the payloads, the responsibility for mission safety and success fell on the shoulders of the mission commander, David Mathiesson Walker. Nicknamed 'Red Flash', in honour of his sandy hair, Walker came from Columbus, Georgia, where he was born on 20 May 1944. After attending high school in Florida, he entered the Naval Academy, received his degree in 1966, and immediately underwent flight instruction at the Naval Aviation Training Command in Florida, Mississippi and Texas. Walker was designated as a naval aviator in December 1967 and served two tours in Vietnam, flying the F-4 Phantom from the USS *Enterprise* and USS *America*, during which he gained the

Distinguished Flying Cross. Following his return to the United States in 1970, he entered test pilot school at Edwards Air Force Base and later served as an experimental and engineering test pilot at the Naval Air Test Center in Maryland, where he worked on the evaluation of the new F-14 Tomcat fighter. Walker also performed tests of a leading-edge slat modification of the Phantom jet and, in 1975, after replacement pilot training on the Tomcat, he served as a fighter pilot on two overseas deployments to the Mediterranean aboard the *America*. As an aviator, Walker was top-notch, but in the words of Mike Mullane, who flew with him as a member of the T-38 chase team in support of the first Shuttle launch, STS-1, "he was *too* cocky, the type of pilot who thinks he's bulletproof even when he's sober". Crewmate Bob Cabana later paid tribute to Walker as "kind of a throwback to the 1960s". By the time of STS-69, Walker had flown three Shuttle missions, two in command, and had been the progenitor of the 'Dog Crew' heritage.

In preparing for his STS-53 mission – whose crew also included Jim Voss – Walker's training team nicknamed them the 'Dogs of War' and the astronauts quickly received their own 'dog names' and 'dog tags'. The training team, in turn, received dog names and tags. In recognition of his hair, Walker became 'Red Dog' and Voss assumed the moniker of 'Dog Face'; for STS-69, their next flight together, they continued the tradition and their fellow crewmates acquired their own alter-egos: Ken Cockrell became 'Cujo', Jim Newman became 'Pluto' and Mike Gernhardt, the only rookie astronaut, became 'Underdog'. On launch morning, 7 September 1995, Walker and his men presided not over plates and mugs and cutlery at the breakfast table … but over a quintet of dog bowls, each emblazoned with their respective dog names. As they departed the Operations & Checkout Building at KSC they were greeted by a chorus of tongue-in-cheek barks and woofs from the assembled journalists, family members and friends. Even in orbit, the astronauts could not escape: on the morning of 10 September 1995, three days into their mission, they were awakened to the sound of 'Bingo Was His Name', sung by Cockrell's five-year-old daughter, Madeline …

Seated alongside Walker in the pilot's seat aboard Endeavour's flight deck, Kenneth Dale Cockrell, a native of Austin, Texas, was making his second space mission on STS-69. Born on 9 April 1950, his eyes were focused upon the skies from a very young age – "I remember at age five, seeing an airplane fly over our house," he told a NASA interviewer, "and this airplane, for some reason, struck a chord in me" – and implanted the seed for the dream of aviation. Much of Cockrell's subsequent schooling was self-directed toward the goal of joining the military as a pilot: he earned a degree in mechanical engineering from the University of Texas at Austin in 1972 and was commissioned into the Navy later that year. Initial flight instruction at Naval Air Station Pensacola in Florida was followed by an assignment to the carrier USS *Midway*, during which time he flew the A-7 Corsair in the Western Pacific and Indian Oceans. He also gained a master's degree in aeronautical systems from the University of West Florida in 1974. Cockrell was selected for test pilot school in 1978 and after graduation he was based at the Naval Air Test Center for the next four years, working on a variety of aircraft. Subsequent assignments included officer positions on the staff of the commander of the USS *Ranger* and the USS *Kitty Hawk* Battle Group. He then returned to operational duty in an F/A-18 Hornet squadron and completed two cruises on the USS *Constellation* in 1985-1987.

Dovetailed into this naval career, Cockrell had a keen eye on NASA's astronaut programme and the arrival of the Shuttle – a unique *winged* spacecraft – really inspired him. Years later, he joked that "I think I'm tied for the record for the number of times applying and to get the job". He interviewed alongside future astronauts-to-be Steve Oswald and Franklin Chang-Díaz. "It's kind of interesting," said Cockrell, who flew into space with both Oswald and Chang-Díaz, "that Franklin and I both got off the same airplane to attend our first interview together. We arrived in Houston and were standing in the terminal building for the car from the hotel to pick us up and struck up a conversation, realised what we were both here to do, and *that* was in 1979!" Although Chang-Díaz made the cut the following year, and Oswald in 1985, Cockrell was unsuccessful. He failed to secure an interview in 1984 and was interviewed in 1985 and 1987, both without success, but eventually made the cut in 1990. Perseverance definitely paid off and in his subsequent roles at NASA, including chief of the astronaut corps in 1997-1998, Cockrell advised potential applicants to "pick a discipline ... that interests you a lot" and regarded the astronaut business as "a secondary profession" for individuals who already excelled in their primary professions.

Cockrell resigned from active naval duty in 1987 and joined the Aircraft Operations Division at JSC as an aerospace engineer and research pilot, performing air-sampling and other high-altitude research. In later life, he would credit Steve Oswald with having helped him secure the Aircraft Operations post and guiding him through the transition from post-Navy into civilian and NASA life. Although selected as a pilot candidate, Cockrell served as a mission specialist on his first flight. After flying in the pilot's seat on STS-69, he would go on to command no fewer than three Shuttle missions between 1996 and 2002.

Behind and between Walker and Cockrell, in the flight engineer's seat, was civilian physicist James Hansen Newman. Born in the Trust Territory of the Pacific Islands (today part of the Federated States of Micronesia) on 16 October 1956, he completed high school in San Diego and earned his degree in physics from Dartmouth College in New Hampshire in 1978. These were quickly followed by master's and doctoral degrees from Rice University in 1982 and 1984. Newman completed a year's post-doctoral work in atomic and molecular physics and was appointed as an adjunct professor in Rice's Department of Space Physics and Astronomy. The space programme beckoned from an early age for Newman. "My uncle recommended that I go into the military and become a pilot and learn to fly and get into the astronaut office that way," he told a NASA interviewer, years later, "but I decided that my real love was for science and technology and that the right thing for me to do was to enter the space programme as a civilian scientist and to use those skills that I had developed in the laboratory to bring into NASA as part of the team that makes up a successful space flight." In 1985 he arrived at JSC to work on flight crew and flight control team training for Shuttle propulsion, guidance and control issues and was a simulator supervisor when selected as an astronaut candidate in January 1990. Newman flew his first mission on STS-51 in September 1993, during which he performed a lengthy spacewalk.

In fact, by the time STS-69 launched, Newman was the only member of the crew to boast an Extravehicular Activity (EVA, or spacewalk) to his credit. That was expected to change, for NASA had implemented a series of EVA Development Flight Test (EDFTs) to evaluate tools and techniques for the construction of the ISS. Already, in September 1994 and February 1995, Shuttle spacewalkers had evaluated a variety of new systems, and on STS-69 Jim Voss and Mike Gernhardt would perform a 6.5-hour spacewalk to press the envelope still further. Planned for late in the 11-day mission, the excursion would support four Detailed Test Objectives (DTOs): to evaluate space suit design modifications to guard against extremely cold temperatures, to perform specific assembly and maintenance tasks, to evaluate an Electronic Cuff Checklist and to support ground crews in making improvements to EVA operations. Specifically, it was intended that Voss and Gernhardt would perform an hour apiece at a 'task board' on the starboard side of Endeavour's payload bay, working with hand rails, mechanical fasteners, electrical and fluid connectors and mock-up Orbital Replacement Units (ORUs). The men would also install a pair of 'thermal cubes' in the payload bay – one at the end of the RMS mechanical arm, the other at the task board work site – to gather temperature data. Additional measurements would be taken from sensors inside their space suits

1-05: Jim Voss is pictured at the end of Endeavour's Canadian-built Remote Manipulator System (RMS) mechanical arm during his EVA with Mike Gernhardt.

throughout the EVA. To differentiate between the two men, Voss was designated 'EV1' and wore red stripes on the legs of his suit, whilst Gernhardt ('EV2') wore a pure-white suit.

As the only member of the STS-69 crew not to have flown before, the mission would kick off a dazzling career for Michael Landon Gernhardt. He was born in Mansfield, Ohio, on 4 May 1956, the son of a keen fisherman. "My first love was the ocean," he told a NASA interviewer. "I got fascinated with fishing and scuba diving at a very young age." His father often took him fishing at Florida's Marco Island and Gernhardt became intrigued by the mystery of the water, "of really not knowing what you were going to catch" and "this desire to learn how to dive and to get underneath the sea and learn all that I could about it". The tangible link between diving and space exploration came in the early 1970s, when America's Skylab workshop orbited high above Earth and the Tektite underwater habitat resided deep beneath the sea. "I put the two things together in my mind at that time," he recalled. "Living and working under the ocean is very similar to living and working in space." Indeed, this would guide him through a career as a professional deep-sea diver to a spacewalker and ultimately to command of the NASA Extreme Environment Mission Operations (NEEMO) undersea research platform, off the Florida Keys. After high school, Gernhardt attended Vanderbilt University, from where he earned a degree in physics in 1978. His early career saw him working as a professional deep-sea diver and project engineer on a variety of subsea oilfield construction and repair activities, logging more than 700 dives around the world. "The first job I had after I got out of school was down in Peru," Gernhardt said. "We were rebuilding a series of platforms that were ... completely rusting apart and falling over. I was the engineer and diver on the job and we were basically rebuilding them, one brace at a time, underwater."

In the meantime, his mother knew of his ultimate desire to become an astronaut and encouraged him to accrue as many degree qualifications as possible. Whilst working on his master's degree in biophysics at the University of Pennsylvania, Gernhardt developed a new theoretical decompression model, based on tissue gas bubble dynamics, and participated in the design and field implementation of new decompression tables. After gaining his master's degree in 1983, he managed and later served as vice president of special projects for Oceaneering International, leading the development of a telerobotic system for subsea platform cleaning and inspection, as well as a new system of diver and robot tools. In 1988, he persuaded Oceaneering's board of directors to allow him to start a company called Oceaneering Space Systems to transfer subsea technology applications into supporting NASA's plans to develop Space Station Freedom. In the meantime, Gernhardt earned a doctorate in bioengineering from the University of Pennsylvania in 1991 and was working on the development of human- and robot-compatible tools for Freedom when he was selected by NASA as an astronaut candidate in March 1992.

With a background which carried him from fisherman's son to deep-sea diver to astronaut, it is unsurprising that Gernhardt's career should culminate in him spacewalking hundreds of kilometres above Earth. In a very real sense, he considered diving and spacewalking to carry many parallels. "We had umbilical hoses, astronauts have tethers," he reflected. "You have to keep your umbilical hose and your tether clear and the thought processes are almost identical. You try to design the task to make it as easy as you can, because it's a difficult environment to work in. Physically, they're very much different. In the case of diving, we're exposed to the ambient pressure conditions, typically wearing a wet suit or a hot-water suit, so you've got no delta pressure across the suit, so it's much easier to swim and work. Then again, underwater, you're dealing with currents, very low visibility when you're working down on the bottom of the ocean, so that makes it more challenging." On the other hand, working in a space suit (whose pressure Gernhardt compared to that of an inflated football) offered the difficulty of having to fight against the rigidity of the pressure. "The term *spacewalk* is probably a misnomer, because we're actually walking with our fingers," he said, "like you do in the *Yellow Pages*, but so that the forces are different. You don't have weight like you do on Earth, but you have mass ... and that inertial mass has to be well-controlled. If you start moving fast, you'll go tumbling out of control in a second. You have to be very light with your touch on your fingertips and you have to constantly sense what your body's doing. If you feel like you're pitching up with your feet, you need to reach out and touch and null that rate." Years later, Gernhardt would pay tribute to the EDFT group of spacewalks – of which five were executed between September 1994 and December 1997 – in that they helped to establish the basic techniques and building blocks of hardware to be used to assemble the ISS in orbit.

Near the end of STS-69, on 16 September 1995, Voss and Gernhardt prepared for the first spacewalk of their respective careers. The Shuttle's cabin pressure was firstly lowered from its normal 101.3 kPal to around 70.3 kPal to prepare the men for the 29.6 kPal pure oxygen of their space suits, as well as clearing nitrogen from their bloodstreams to avoid an attack of the bends. Before clambering into the two-piece suits, electrical harnesses were attached to their 'hard' upper torsos to provide biomedical and communications links through their backpacks. The next step was the connection of a soft black-and-white communications hat – famously nicknamed the 'Snoopy cap' since Apollo days – to the top of the torso. Physically, the suits were nothing less than $2.5 million miniature spacecraft in their own right, consisting of 'upper' (above-waist) and 'lower' (below-waist) segments, together with helmet, gloves and backpack, known as the Portable Life Support System (PLSS). The suits had been developed under a series of contracts with Hamilton Standard of Connecticut. Voss and Gernhardt pulled themselves into the lower torso, which featured joints at its hips, knees and ankles and a metal body seal closure for connecting to a ring on the upper torso. It also included a large bearing at its waist for greater mobility and allowed the astronauts to twist whilst their feet were held firmly in restraints.

After donning the trousers of the suit, their next step was to plug the airlock's service and cooling umbilical into a display and control panel on the front of the upper torso. This would provide coolant water, oxygen and electrical power from the Shuttle until shortly before they were scheduled to go outside, thereby conserving the limited consumables available in their backpacks. The two men finally entered the airlock, where the upper torsos were 'hanging' on opposing walls. With arms outstretched, and with Ken Cockrell (designated as STS-69's intravehicular crewman) nearby to assist, they slipped into the upper torsos through a half-diving, half-squirming motion and their waist rings were brought together, connecting the coolant water tubing and ventilation ducting of the long underwear and the biomedical sensors to their backpacks. Cockrell helped them to lock the body seal closure rings at their waists. The hard upper torso was essentially a fibreglass shell under several fabric layers of a thermal and micrometeoroid garment. On its back was the life-support system, and on its chest was the display and control unit by which the spacewalker would manage oxygen, coolant and other consumables.

Next step: the gloves. Snapped into place on the wrist rings of the upper torso, these had silicone rubber fingertips to provide a measure of tactile sensitivity when handling tools in Endeavour's cavernous payload bay. Finally, the enormous polycarbonate bubble helmets were lifted over the astronauts' heads and clicked into place on the neck rings of their upper torsos. Over the top of each helmet was an assembly containing manually adjustable visors to shield their eyes from solar glare, together with two EVA lamps to illuminate work areas out of range of the Sun or the Shuttle's own payload bay floodlights. Mobility in the neck rings was unnecessary, because the helmets were easily big enough to allow the astronauts to move their heads around. Unlike previous Apollo space suits, the modularised Shuttle ensemble, with its waist closure ring, eliminated the need for pressure-sealing zips and therefore had a much longer shelf life. Additionally, the use of newer, stronger and more durable fabrics enabled space suit engineers to design joints with better mobility, resulting in lower weight and a reduction in overall cost. Jim Voss and Mike Gernhardt, by now floating motionless in Endeavour's tiny airlock, were, in effect, small spacecraft in their own right. However, they were not yet 'self-contained', as their oxygen, electricity and coolant water were still being provided by the Shuttle's systems; not until shortly before the two men ventured outside would they transfer to their suits' life-support consumables.

At length, a go-ahead came from Mission Control and Voss commenced the final depressurisation of the airlock and pushed open the outer hatch into the payload bay. Sunlight flooded into both of their faces and, beyond, the enormous expanse of the 20 m bay. The EVA officially started at 4:20 am EDT and the spacewalkers' first task was to install the thermal cubes at the end of the RMS and at the task board on the starboard wall of the bay. They then removed a debris shield from the work site, manipulated a duplicate of a computer control box for ISS hardware and tested new helmet lights and suit heaters. Taking turns at the end of the RMS, they were manoeuvred by Jim Newman away from the radiated warmth of the payload bay, exposed to temperatures as low as minus 84 degrees Celsius. Earlier in 1995, spacewalkers Mike Foale and Bernard Harris had ended their EVA early when suit modifications to counter cold temperatures proved ineffective. However, on STS-69, Voss and Gernhardt benefitted from battery-powered fingertip heaters in their gloves and a system that enabled them to totally shut down their suits' liquid-cooling system and rely upon body heat for warmth. The men continually provided subjective ratings on their comfort levels to flight controllers and experienced no difficulties during their 'cold soak' evaluations. They returned inside Endeavour after an EVA lasting six hours and 46 minutes. *Flight International* noted a few weeks later that in spreading the spacewalking load across several missions, NASA was in effect training a cadre of astronauts for the estimated 650 hours of EVAs required to build and maintain the ISS, whose construction phase was then expected to run from 1997-2002. Indeed, of the five Shuttle missions during which EDFT excursions were conducted in the 1994-1997 timeframe, half of the spacewalkers involved went on to perform EVAs in support of ISS construction or maintenance tasks.

Although the spacewalk was a key highlight of STS-69, the two primary payloads had already been successfully deployed and recovered by the time that Voss and Gernhardt ventured outside. In fact, the mission marked the very first occasion on which two satellites had been deployed *and* retrieved on the same Shuttle flight. The SPARTAN-201 payload was a 1,000 kg cube-shaped spacecraft, designed to accommodate instrumentation for astrophysics or solar physics research. Built by GSFC at the relatively inexpensive price of $3.5 million, it measured $3.2 \times 1.07 \times 1.22$ m and was designed to be reused and flown at intervals of approximately six to nine months. Data from its instruments would be stored on tape recorders and pointing and stabilisation achieved through a three-axis attitude control mechanism. Deployed and retrieved by means of the Shuttle's RMS arm, SPARTAN first flew aboard the Shuttle in June 1985 with a battery of high-energy astrophysics detectors and was headed for orbit a second time aboard Challenger's ill-fated final flight in January 1986 to perform ultraviolet observations of Halley's Comet.

A replacement for the lost SPARTAN was constructed in the wake of the Challenger disaster and the carrier flew several times in the 1990s. The 'SPARTAN-201' experiment package focused upon studies of the Sun and comprised two instruments: an Ultraviolet Coronal Spectrometer (UVCS), built by the Smithsonian Astrophysical Observatory of Harvard University, to determine the velocities of coronal plasma within the 'solar wind' and a White Light Coronagraph (WLC), assembled by the High Altitude Observatory of the National Center for Atmospheric Research, to measure visible light in order to determine the temperatures and densities of coronal electrons. Once deployed from STS-56 in April 1993, SPARTAN-201 undertook a pre-programmed mission; it lacked command and control capability and was effectively left to undertake its tasks, retrieved by the Shuttle after two days and returned to Earth for the recovery of data and refurbishment in readiness for its next flight. SPARTAN-201 flew again on STS-64 in September 1994 and would thus be making its third mission on STS-69. In total, with two further missions ahead of it, the satellite flew on five occasions between April 1993 and November 1998 and its spacecraft 'bus' is today displayed in the National Air & Space Museum in Washington, DC.

Preparations to deploy SPARTAN-201 got underway almost immediately after Endeavour and her crew settled into orbit on the morning of 7 September 1995. Following a couple of minor glitches with a circuit breaker and a sluggish carbon dioxide scrubber, both the solar physics satellite and the WSF were powered up and underwent initial health checks. Astronauts Jim Newman and Mike Gernhardt activated the RMS, which would be employed to deploy and retrieve both satellites. The $100 million mechanical arm was Canada's contribution to the Shuttle

1-06: The SPARTAN-201 payload, pictured in free flight during STS-69.

programme – a contribution that dated back to 1974, when Spar Space Robotics Corporation was contracted by the country's National Research Council to build a device for deploying and retrieving satellites from orbit and, ultimately, assembling the components of a space station. In May 1979, the first contracts were signed with NASA for the production of three RMS units and their supporting hardware and software. The challenges involved in building an arm of such complexity and dexterity were enormous: it needed to operate both autonomously and under manual control and meet strict weight and safety requirements. Moreover, nothing quite like it had ever been built or used in space before, which made Spar's task yet more difficult. Although a functional floor rig was built to test its joints, the first real demonstration did not come until it was actually uncradled on the STS-2 mission in November 1981. Measuring 15.2 m long in order to be able to reach the far end of the payload bay, it consisted – just like a human arm – of shoulder, elbow and wrist joints, linked by two graphite epoxy booms. Other components were constructed from titanium and stainless steel. To protect it from thermal extremes in space, the arm was covered in white insulation and fitted with heaters to maintain its temperature within required limits. Without a payload attached, it could move at up to 60 cm/min, but this was reduced to a tenth of that speed when fully loaded.

Ingeniously, the means by which the arm 'picked up' and 'put down' objects was achieved by the so-called 'end effector' – essentially a hand that employed a kind of three-wire snare to capture a prong-like grapple fixture attached to a deployable or retrievable payload. The Hubble Space Telescope (HST), at that time scheduled for launch in the mid-1980s, had an in-built grapple fixture that would enable it to be deployed, retrieved and serviced by future Shuttle crews. During 'operational' missions, astronauts would use two television cameras on the arm's wrist and elbow to guide the end effector over a target's grapple fixture, before commanding the three metal ties of the snare to close around it at precisely the right instant. When this was done, it would impart a force of 500 kg onto the grapple fixture, establishing a grip sufficient for the RMS to move the target. Although the arm was controlled by the Shuttle's GPCs, its motions were directed by an astronaut using a joystick on the aft flight deck. As each instruction was issued, the computers examined it and determined which joints needed moving, their direction and their speed and angle. Meanwhile, the computers also looked at each joint at 80-millisecond intervals and, in the event of a failure, automatically applied a series of brakes and notified the crew.

STS-69's first night in space was a disturbed one for Dave Walker, who was awakened twice by a pair of alarms that indicated a problem in the data path between Endeavour's GPCs and the Ku-band communications system. Walker reset the system and Newman later rebooted it. Next morning, 8 September, the deployment of SPARTAN-201 got underway. Already, a trio of RCS thruster firings had positioned the orbiter in the proper orientation for the deployment and Gernhardt released the satellite at 11:42 am EDT. Within minutes of departing the mechanical arm, SPARTAN-201 executed a planned 45-degree pirouette, to verify that its internal attitude control system was functioning correctly. Walker and Cockrell then performed two separation burns to manoeuvre Endeavour away from the payload, creating a mean distance of about 64 km and leaving SPARTAN-201 alone for two days of dedicated solar science operations.

In the meantime, the five astronauts settled down to their other experiments, one of which was the International Extreme Ultraviolet Hitchhiker (IEH)-1. This was designed to measure and monitor long-term variations in the magnitude of absolute extreme ultraviolet flux from the Sun and to examine extreme ultraviolet emissions from Jupiter's plasma 'torus' – which originates from its volcanic moon, Io. The experiment consisted of two instruments: the University of Southern California's Solar Extreme Ultraviolet Hitchhiker (SEH) and the joint US/Italian Ultraviolet Spectrograph Telescope for Astronomical Research (UVSTAR). Previously flown aboard a sounding rocket flight in September 1990, SEH was designed to accurately measure solar flux at extreme ultraviolet wavelengths, whilst the twin telescopes and imaging spectrographs of UVSTAR marked a collaboration between US and Italian scientists to examine not only the Jupiter/Io system at far and extreme ultraviolet wavelengths, but also to acquire spectrally resolved images of stars. One particular focus was to complement and extend observations planned by the Jupiter-bound Galileo spacecraft, which arrived at the giant planet in December 1995. However, UVSTAR ran into difficulties within hours, due to pressure issues and problems commanding an elevation gimbal which enabled the instrument to swivel backwards and forwards. The SEH payload, on the other hand, performed as expected.

Other niggling troubles centred on the Electrolysis Performance Improvement Concept Study (EPICS) experiment. This sought to validate the electrochemical process of the Static Feed Electrolyser (SFE), manufactured by Life Systems, Inc., and originally baselined for Space Station Freedom as a means of electrolysing water in space to generate oxygen and hydrogen and thus reduce the annual resupply weight of the orbital complex. Since tests of the system on Earth were hampered by the complications of gravity and buoyancy, it was necessary to evaluate it in the space environment. Unfortunately, all three of the self-contained electrolysis units aboard EPICS automatically shut themselves down early in the STS-69 mission, and although some hope existed to restore power to two of them, mission managers ultimately opted to completely power down the experiment.

More success was forthcoming from the third flight of the Shuttle Glow Experiment (GLO-3), developed by the Air Force's Phillips Laboratory and the University of Arizona. Its task was to observe the mysterious shroud of luminosity (known as the 'glow phenomenon') observed by astronauts on earlier missions. One theory argued that it was due to atmospheric gases on the 'ram' side of the Shuttle colliding and interacting with gaseous thruster products and contaminant molecules. "The glow intensity is weak, decreases with altitude and requires some special conditions for good detection," explained NASA in its STS-69 press kit. "Both the Sun and Moon must be below the horizon, for example, so the spatial extent of the glow will be mapped precisely [to] 0.1 degrees." GLO-3 comprised an imager and a spectrograph to unambiguously identify the source of the glow spectrum and map its spatial extent.

With the retrieval of SPARTAN-201 planned for early on 10 September, Walker and Cockrell performed a thruster firing on the 9th to refine Endeavour's rate of closure and prepare for the rendezvous. By the morning of the 10th, they had manoeuvred the Shuttle to a distance of just 100 m from the payload. Gernhardt was scheduled to capture SPARTAN-201 at 10:24 am EDT, but this opportunity was missed due to the payload entering an unexpected attitude during proximity operations. "Concern about whether the SPARTAN had operated correctly was raised when the spacecraft was to be retrieved," noted NASA in a 29 September news release. "At that time, the crew reported that SPARTAN was rotating slowly and its batteries seemed to have been drained." Preliminary indications suggested that the payload placed itself into a 'safe mode' and shut down its power systems, which prevented it from achieving its intended rendezvous attitude. Consequently, Walker and Cockrell manually flew a 180-degree manoeuvre 'around' their quarry to line up the RMS grapple fixture with the mechanical arm's end effector and Gernhardt successfully captured it at 11:02 am. Approximately 19 minutes later, he had lowered the satellite onto its berth in the payload bay.

At the close of its third mission, SPARTAN-201 was hailed as a huge success by NASA. In late September, after STS-69 had landed, Project Scientist Dick Fisher of GSFC reported that its data tapes indicated that the payload operated as planned throughout its flight. The attitude issue during the final rendezvous and proximity operations could not be properly investigated until the payload was back on Earth. Nor could a full assessment of its scientific data be made until it was back in the hands of principal investigators. One of the key mission objectives had been to observe the Sun's north pole, since the mission coincided with the passage of the Ulysses spacecraft over this region. As Ulysses performed *in-situ* measurements of the physical properties – including temperature, density, ionic composition and magnetic and velocity fields – of electrons, protons and ions in the solar wind, SPARTAN-201 complemented it by completing observations from low-Earth orbit. Specifically, researchers were interested in making collaborative observations of the source of the solar wind and developing a clearer understanding of physical circumstances of the Sun's outer corona. Summing up, the investigators concluded that the third mission of SPARTAN-201 had been an enormous success. "The White Light Coronagraph instrument obtained spectacularly good data over 95 percent of the planned observing sequence," explained Fisher, adding that preliminary findings from the Ultraviolet Coronal Spectrometer pointed to excellent data.

With the SPARTAN-201 operations thus behind them, the STS-69 crew focused attention on the Wake Shield Facility, whose RMS-assisted deployment and retrieval was to be undertaken by Jim Newman. The payload was developed by the University of Houston, in conjunction with Dr Alex Ignatiev, head of the Space Vacuum Epitaxy Center (SVEC). Today known as the Texas Center for Superconductivity at the University of Houston (TcSUH), this institution has for more than two decades explored thin-film deposition, processing and characterisation of semiconducting, high-temperature superconducting and ferroelectric oxide material systems. As long ago as the 1970s, NASA engineers published papers arguing that a satellite sailing through space would leave an 'ultra-vacuum' in its wake, but the absence of practical applications at the time meant that the idea was left unexplored until Ignatiev and his team revived it. In 1986, they joined forces with nine other companies to form a Center for the Commercial Development of Space – one of several industry-academia partnerships sponsored by NASA's Office of Advanced Concepts and Technology – with the plan to build the free-flying Wake Shield Facility. It was conceived as a 3.6 m stainless steel disk, to be deployed using the RMS arm for the purposes of generating an 'ultra-vacuum' within which to grow thin films for future advanced electronics applications. The purely functional shape of the WSF made it appear like a factory cast-off: of dull, silver-grey colour, it was a clumsy arrangement of boxes, rods, tubing and angular shapes. It was designed, built and managed by SVEC, along with its industrial partner, Space Industries, Inc., of League City, Texas. In March 1989, SVEC and Space Industries, Inc., formally partnered with NASA and carried the payload from the drawing board to the launch pad in less than 60 months with a budget of around $12 million. From the outset, it was presented as a major 'first', since the generation and characterisation of an ultra-vacuum in space had never been attempted. With up to four flights of the WSF planned, it first rose into space aboard STS-60 in February 1994, but as discussed in the previous volume of this History series, technical difficulties prevented its deployment. Instead, it was 'hung' at the end of the RMS arm and validated its basic concept, by successfully growing five thin semiconducting films and gathering associated data. In the aftermath of STS-60, an advisory committee analysed the flight anomalies and a NASA independent review board evaluated the satellite's systems and unanimously agreed that WSF-2 was ready to fly.

The primary objectives of the satellite's second mission were to spend about 50 hours in free flight, characterising the ultra-vacuum environment and employing the techniques of molecular and chemical beam epitaxy to grow a thin film of gallium arsenide (GaAs) on a prepared substrate. Upon this substrate, atomic or molecular beams of arsenic and gallium formed thin films in layers to create a 'wafer' with an ultra-high-purity top region. The WSF was designed to grow epitaxial films on seven different substrate wafers and it was expected that the free-flyer would produce at least one 'thick' GaAs film of up to nine micrometres deep. The use of this material in digital cellphones, high-speed transistors, high-definition television and fibre-optic communications and optoelectronics was already known to present "a very promising economic advantage". Gallium arsenide was thought to yield electronic devices eight times faster than silicon, whilst at the same time requiring a mere tenth of the power demand ... but could not be easily produced on Earth of sufficiently high purity and quality. "If improved GaAs material were available," noted NASA's press kit for STS-60, the WSF-1 mission, "it could significantly impact the global semiconductor market", whose worldwide consumption in 1990 alone reached $56.8 billion and whose projection for 1994-1995 had risen to $109 billion. "Within this giant market," continued the press kit, "GaAs currently holds only a 0.5 percent niche. It is predicted that the niche for GaAs should grow to 2 percent (or about $2.2 billion) by 1995."

Although a moderate, 'natural' vacuum was known to exist in low-Earth orbit, it was hoped that the forward-facing (or 'ram') side of the WSF would 'push' even the few atoms present out of the way, leaving a unique ultra-vacuum in its wake and producing conditions between a thousand and ten thousand times better than those attainable in the best ground-based vacuum laboratories. Despite ostensibly representing the relatively 'dirty' side of WSF, the ram could be used to support other experiments, including technology payloads, with a total 'real estate' of around 6 m^2. To make the most of this facility, four payload-attach points were placed on the ram side, capable of holding up to 90 kg of hardware. A fully equipped spacecraft in its own right, the 1,800 kg free-flyer had cold-gas propulsion thrusters to effect a satisfactory separation distance from the Shuttle and silver-zinc batteries to provide 45 kilowatt-hours of power for the thin-film growth cells, heaters, process controllers and other instruments. Mounted on the 'wake' side of the spacecraft, in addition to the epitaxy process control equipment, were pressure gauges, mass spectrometers, potential analysers and a video camera, together with the attitude-control system and batteries. Meanwhile, attached to the ram face of the disk were the avionics hardware and associated support equipment.

In addition to the GaAs epitaxial growth hardware, a number of other co-operative experiments were aboard WSF-2. The University of Toronto's Institute for Aerospace Studies provided a space exposure package, Baylor University Space Science Laboratory supplied a cosmic dust and orbital debris monitor to better characterise the particle environment around the satellite and NASA's Jet Propulsion Laboratory (JPL) of Pasadena, California, conducted a passive hyper-velocity debris capture and exposure experiment that had collectors mounted aboard WSF-2 and its cross-bay carrier. A new Earth reference attitude determination system was provided by Honeywell Satellite Systems. Gregory Jarvis High School in Mohawk, New York, supplied an experiment to study variations in Earth's magnetic field. The Air Force's Phillips Laboratory supplied the Charging Hazards and Wake Studies (CHAWS) experiment to examine interactions and inherent hazards between the space environment and space systems. Of specific interest to the Phillips investigators were measurements of ambient, low-energy positively charged particles on the wake and ram sides of WSF-2 and studies of the magnitude and directionality of current collected by negatively charged objects. Several CHAWS objectives were achieved whilst the free-flyer was in the grasp of the RMS arm. Also aboard were a series of highly sensitive accelerometers, the Shuttle Plume Impingement Experiment, a pair of 'SmartCans' housing materials science investigations and a pair of student-supplied plant-growth studies.

Since WSF-2 was mounted 'horizontally' atop a bridge-like carrier, it occupied roughly a quarter of the payload bay on STS-69 and plans called for it to grappled by the RMS, under Jim Newman's control, on 10 September. He would leave it in that position, still latched to its carrier platform, overnight, ahead of the scheduled unberthing and deployment on the 11th. In readiness for WSF-2 operations, Walker and Cockrell had executed a pair of OMS engine firings to raise the Shuttle's orbit by 24 km. Already, Gernhardt and Newman had begun to slightly shift their sleep schedules, to ensure that one or both of them would be awake during all critical mission events. Newman successfully unberthed the payload and manoeuvred it into its so-called 'ram-cleaning' position for three hours, which was essential for the atomic oxygen of low-Earth orbit to 'scour' its ram side and atomically clean it in readiness for two days of semiconductor processing. In stark contrast to the problems experienced on STS-60, the payload's Attitude Determination and Control System (ADACS) checked out without incident. The actual deployment was postponed by almost two hours, as flight controllers wrestled with communications dropouts between WSF-2 and the data relay and telemetry equipment aboard its carrier platform in the payload bay. Eventually, Newman released the satellite over the starboard side of the bay at 7:25 am EDT, as Endeavour flew high above West Africa, and WSF-2 executed a perfect firing of its own cold-gas thruster to manoeuvre itself away from the orbiter and avoid the risk of becoming contaminated by waste water dumps, fuel cell purges and thruster firings. This was the first occasion that a payload had manoeuvred away from the Shuttle; previously, the orbiter had taken the 'active' role in such operations.

Nine hours later, at 4:30 pm, as it trailed Endeavour by about 10 km, WSF-2 successfully began its first thin-film semiconductor processing run. Described by investigators as a 'dirty' run, it lasted about three hours and served to remove any residual contaminants from the containers housing the GaAs sample growth materials. Earlier in the afternoon, Walker and Cockrell had fired the Shuttle's RCS thrusters towards the satellite to slow their relative separation rate, with the intention that the two spacecraft would be about 45 km apart by the time of rendezvous and retrieval on 13 September. The RCS firing also allowed investigators to better understand the effects of thrusters on a free-flying payload. By the morning of its second day of operations, WSF-2 had successfully grown three thin films. However, just before the fourth film run was about to commence, the satellite pitched forward slightly, after sensing a temperature increase, and placed itself into safe mode as a precautionary measure. Operations resumed about 12 hours later, after the ADACS had been given time to cool down, and it was decided to incorporate further cooling periods into the WSF-2 timeline and extend the satellite's time in free flight by 24 hours. This produced a revised retrieval time of late on the morning of 14 September.

Another delay cropped up when controllers were rendered unable to trigger the flow of arsenic from aluminium source cells on WSF-2 onto a substrate platform. The science instruments were again shut down in the hope that allowing them to cool would permit additional attempts to grow further thin-film samples. It later became clear that the shutter on the source cell had failed to close on command, but investigators doubted that this would negatively affect the quality of the sample. In total, four of the seven planned thin films were satisfactorily grown, with plans for a fifth eventually called off when the payload controllers noticed a low reading in one of the satellite's four batteries. Meanwhile, on 13 September, Walker and Cockrell manoeuvred Endeavour to slightly narrow the distance between the Shuttle and the payload to about 42 km, preparatory to the following day's rendezvous and retrieval. Extending WSF-2 operations any further was not an option, for Jim Voss and Mike Gernhardt were scheduled to perform their EVA on 16 September, and they were already at work configuring their space suits and equipment. Voss also received the welcome news that he had been selected by the Army for promotion from lieutenant-colonel to full colonel.

By the morning of the 14th, Endeavour had closed to within 130 m of WSF-2, whereupon Walker and Cockrell performed a series of 14 RCS firings to assess the impact of thruster plumes on the satellite's structure. The ADACS performed well in response to these stresses and its sensors measured the force and pressure of the plumes. Newman grappled WSF-2 with the RMS arm at 9:59 am EDT and berthed it back onto its carrier platform in the bay at 11:18 am. Endeavour's orbit was subsequently adjusted, ahead of Voss and Gernhardt's EVA, and STS-69's final research experiments consumed the final days of the mission. A few final tasks with WSF-2 included a second unberthing by Newman on 15 September, after which it was 'hung' over the starboard payload bay wall for about five hours in support of the CHAWS experiment.

In addition to SPARTAN-201 and WSF-2, a multitude of experiments, encompassing a wide range of disciplines, occupied Endeavour's payload bay. One cross-bay Hitchhiker 'bridge' carried the IEH-1 payload, whilst a smaller 'Hitchhiker Jr.' carried a Complex Autonomous Payload to study the growth of organic non-linear optical crystals and thin films. Elsewhere, a Getaway Special Bridge Assembly (GBA) housed a group of scientific and space technology investigations. The Capillary Pump Loop (CAPL)-2 experiment provided a full-scale unit to evaluate the heat-transport requirements of liquid ammonia in readiness for a thermal control system proposed for NASA's Earth Observing System (EOS). Also affixed to the GBA was the Thermal Energy Storage (TES)-2 payload, which sought to explore the long-duration behaviour of lithium fluoride-calcium fluoride eutectic (a thermal energy storage salt) after repeated cycles of melting and freezing in the microgravity environment. TES-2 occupied a single 180 kg, dustbin-sized Getaway Special (GAS) canister. Since their first Shuttle flight in March 1982, GAS canisters had become frequent items aboard the reusable orbiters, with as many as 13 flown at a time. They were part of a drive to encourage universities, government agencies, foreign nationals and even private individuals to develop their own scientific experiments for carriage into orbit. In addition to the GAS canister housing TES-2, four more canisters carried experiments from providers as diverse as the European Space Agency (ESA) and NASA's Langley and Lewis Research Centers to students from McDowell High School in Erie, Pennsylvania. On the middeck, other experiments focused on the loss of bone density, the gravity-sensing mechanisms within mammalian cells, the commercial processing of the building blocks for future pharmaceuticals and biomedicines and a series of plant growth studies.

Even as their mission entered its final days, the 'Dog Crew' of STS-69 could not escape the daily reminders of their canine camaraderie, with 'He's A Tramp', from the cartoon movie 'Lady and the Tramp', providing their wake-up music one morning. With weather conditions considered near-perfect at KSC's SLF runway, and only scattered clouds, light north-easterly winds and a slight chance of ground fog, Dave Walker was given the green light for the irreversible de-orbit burn at 6:35 am EDT. A little over an hour later, precisely on time, Endeavour alighted on Runway 33 at 7:38 am, concluding a mission of slightly less than 11 full days in orbit.

GLIMMERS OF THE FUTURE

Whilst in space, Walker had taken the opportunity for a unique ship-to-ship call with fellow astronaut Ken Bowersox, who was to command the next Shuttle flight, STS-73, then planned for launch on 28 September 1995. His mission would fly aboard NASA's oldest orbiter, Columbia, and would come after several months of modifications and enhancements at prime contractor Rockwell International's Palmdale plant in California. By this stage, NASA had already decided that future improvements would all be performed at Rockwell's facility. Previously, some work had been done at KSC to save the time and money needed to ship the orbiters across the United States from coast to coast, but it was realised that employing 300 Rockwell staff for the modifications would free up KSC's 7,000-strong workforce to focus on readying the vehicles for flight. Originally, after flying the STS-65 mission in July 1994, Columbia was to have begun a modification period at KSC early the following year, which would have enabled her to fly the STS-67 mission with the ASTRO-2 astronomy payload in December 1994 and be ready in time for STS-73 in September 1995. After the procedural changes and the decision (made in March 1994) to use Palmdale staff exclusively for the Shuttle modifications, it became increasingly clear that the time needed to remove the ASTRO-2 hardware, fly Columbia out to California, conduct six months of work and prepare her for a September 1995 mission would not be practical. ASTRO-2 was therefore shifted onto Columbia's sister ship, Endeavour, and was flown successfully in March 1995. This provided the workforce with a broader window to accomplish the work on Columbia and in October 1994 she was transported atop a Boeing 747 SCA to Palmdale. The journey required five days and two refuelling stops in Huntsville, Alabama, and Ellington Field, Texas.

She underwent 66 modifications to enhance her performance, meet new mission requirements and reduce turnaround times. Among them, Columbia received new wiring to allow future crews to monitor downlinked data on laptop computers, new filters for her hydrogen flow control valves to reduce the likelihood of contamination and efforts were made to better monitor structural corrosion. Moderate problems with corrosion had already been noticed, particularly on the leading edges of her wings, and improvements were added to rectify the situation. It was speculated that the amount of time she had spent on the launch pad prior to the installation of a weather-protection system had subjected her airframe to significant amounts of salty Atlantic spray. Whilst in California, she underwent 460 X-ray and 19 visual inspections and showed herself to be in excellent condition after 17 space missions and a cumulative 146 days spent orbiting Earth between 1981 and 1994.

Aside from the modifications and inspections, two other key differences would characterise Columbia and her crew on the STS-73 mission. One of these was a full set of the new Block I main engines, one of which had been test-flown alongside two old-style engines aboard STS-70 in July 1995. The second difference became apparent when the seven-strong crew – mission commander Ken Bowersox, pilot Kent Rominger, payload commander Kathy Thornton, mission specialists Catherine 'Cady' Coleman and Mike Lopez-Alegria and payload specialists Fred Leslie and Al

1-07: Columbia patiently waits on the pad for STS-73.

Sacco – departed the KSC astronaut quarters for the launch pad: they all wore new suits. Since the resumption of flight operations in the wake of the Challenger disaster, all Shuttle crews had worn pumpkin-orange David Clark Company partial-pressure suits. By 1990, work was underway to introduce a full-pressure garment, known as the Advanced Crew Escape Suit (ACES) and its first flight units were delivered to NASA in May 1994. They offered a simple, lightweight and low-bulk suit that astronauts could don and doff easily, quickly and without help from others. Following her return to Florida from California in April 1995, Columbia was ensconced in one of KSC's three OPF bays to begin processing for STS-73, which was planned as a 16-day science mission to carry the second United States Microgravity Laboratory (USML-2), a reflight of a Spacelab research flight in mid-1992 which had concentrated on fluid physics, materials science, biotechnology, combustion science and commercial space processing. With the long mission duration and with her seven-member crew working in two 12-hour shifts to operate the USML-2 experiments around-the-clock, STS-73 would also offer an excellent baseline for work aboard the ISS.

However, the efforts involved in getting Columbia off the ground would turn into a gargantuan task and reinforced the orbiter's reputation for being an immovable bear when it came to launching her into orbit. On no fewer than six occasions would Bowersox and his crew prepare themselves for liftoff, only to be given the disappointing news of a delay as scrub after scrub plagued them. The STS-73 stack was rolled out to Pad 39A in late August and, despite initial fears of a week-long delay due to the SRB gas-path issue, the analysis of STS-69's boosters were clear and NASA pressed ahead with an opening launch attempt for Columbia on 28 September. Following a smooth countdown, the clock was halted early that morning, when a sensor detected a hydrogen leak in one of the main engine fuel valves as technicians began loading the ET with cryogenic propellants. The valve was replaced and launch was rescheduled for 5 October, but even this second attempt to get Columbia off the ground came to nothing, due to fears of high winds, thunderstorms and lightning associated with Hurricane Opal. Weather forecasters expected Opal to make landfall near Pensacola on the evening of the 4th, which might jeopardise the fuelling of the ET and STS-73's launch window the following day. Another try on 6 October was also foiled when it transpired that fluid had been mistakenly drained from part of the Shuttle's nose-wheel steering hydraulics.

In the event, there was barely a 30 percent probability of acceptable weather, due to the prospect of gusting winds, rain and clouds associated with Opal. Yet another attempt on 7 October was scrubbed by Launch Director Jim Harrington at T-20 seconds, following an indication of a fault with one of two Master Events Controllers (MEC), which were responsible for sending commands to fire pyrotechnics to break the SRBs' hold-down bolts, as well as separating the spent boosters and ET from the orbiter at the appropriate time during ascent. Since the MEC was situated deep inside Columbia's aft compartment, it was necessary to drain the ET of propellants before replacing it, and this led to an unavoidable postponement until 14 October. This date slipped by 24 hours for additional inspections of the main engines' oxidiser ducts, following the discovery of a crack in a similar duct which was then undergoing tests at NASA's Stennis Space Center in Mississippi. Ultrasound checks were performed on welds on each of the engine's high-pressure oxidiser turbopump discharge ducts to confirm that they were of the proper thickness. Also during this delay, one of Columbia's five GPCs failed and had been replaced and tested. The 15 October launch date had, by now, also come under threat from the deteriorating weather and it was scrubbed at T-5 minutes, due to low clouds and rain. Another attempt was impossible before the 19th, since the Eastern Test Range's tracking assets were already supporting an Atlas rocket from Cape Canaveral Air Force Station. The Atlas was originally slated to fly on 17 October, but was postponed a day due to bad weather, and this in turn pushed STS-73 back to the 20th. Even as Bowersox and his crew steeled themselves for a record-tying seventh attempt to get into space, the weather prospects seemed grim.

None of the delays seemed to have dampened the enthusiasm of the crew, five of whom had not flown into space before. Assignments for STS-73 got underway in March 1994, when veteran astronaut Kathryn Cordell Thornton was announced as the payload commander and first-timer Catherine Grace Coleman was assigned as a mission specialist. Despite sharing a place on one of the longest Shuttle flights in history, the two women's careers had precious few other features in common. Thornton, by then a veteran of three previous Shuttle missions, was born on 17 August 1952 in Montgomery, Alabama, and after high school entered Auburn University to study physics. She received her degree in 1974 and a PhD from the University of Virginia, five years later. After completing her research on a NATO-awarded post-doctoral fellowship at the Max Planck Institute for Nuclear Physics in Heidelberg, West Germany, she was employed by the Army as a physicist at the Foreign Science and Technology Center in Charlottesville, Virginia, in 1980, and entered NASA's astronaut corps in May 1984. Coleman, on the other hand, came from Charleston, South Carolina, where she was born on 14 December 1960, although she completed high school in Virginia. It was not until she was an undergraduate in chemistry at Massachusetts Institute of Technology that she decided that an astronaut career was for her. "I'm not one of those kids that wanted to do this ever since they were six," Coleman told a NASA interviewer. "It took me quite a bit longer. I was in college when I went to a talk by Dr Sally Ride, who was the first American woman astronaut. I listened to what she had to say and realised that this was the job that I wanted." Having said this, the notion of *exploration* had been with her since childhood. Her father was a Navy man who designed undersea habitats and Coleman was fascinated not only by the science of the laboratory, but also by the real-world lure and adventure of the military. Having gained her bachelor's degree in 1983, Coleman was commissioned as a second lieutenant in the Air Force through the Reserve Officers Training Corps (ROTC) programme and "worked out a deal" to commence PhD research in polymer science and engineering at the University of Massachusetts. She worked for four years as a research chemist at the Materials Directorate of the Air Force's Wright Laboratory, synthesising model compounds for optical applications, such as advanced computers and data storage, and received her doctorate in 1991. In her pre-NASA career, she also worked as a surface analysis consultant for the Long Duration Exposure Facility and volunteered as a test subject for the centrifuge programme at the Armstrong Aeromedical Laboratory. She was selected as an astronaut candidate in March 1992.

In a similar vein to the early assignment of Jim Voss to STS-69, the announcement of Thornton and Coleman some 18 months ahead of launch, and several months before NASA named the remainder of the STS-73 crew, had much to do with the need to co-ordinate training plans, to liaise with the USML-2 Payload Operations Control Center (POCC) at MSFC and to oversee hardware and software changes. "The training time that the responsible mission specialist should put into that needs to be longer than the pilot and commander probably need to put into basic orbital operations," said former astronaut Kathy Sullivan in a NASA oral history, "so you're going to want to slot in mission specialists 18-24 months in advance, so they can build the relationships that are necessary with the scientific team and the payload operations team."

STS-73 centred on a bus-sized Spacelab module, housed in Columbia's payload bay. The origin of 'Spacelab' can be traced back to the first few months after Neil Armstrong and Buzz Aldrin landed on the Moon. In late 1969, NASA outlined a number of key directions for the US space programme after Apollo. President John Kennedy's challenge to land a man on the lunar surface before the end of the decade had been met, but it was not until this time that serious consideration was given to what would come next. The establishment of some kind of permanent, or at least 'frequent', human presence in space was of major importance – hence the Shuttle – and in December 1972, the European Space Research Organisation (ESRO) – which, in 1974, merged with the European Launcher Development Organisation (ELDO) to establish today's European Space Agency (ESA) – agreed at a ministerial conference in Brussels to develop a modular, multi-purpose laboratory to fly in the payload bay of the winged orbiter. Originally described as a 'sortie module', it was later renamed 'Spacelab'. To NASA, the word 'sortie' merely reflected the low-cost, short-duration nature of the research facility ... but in Europe, and particularly France, the similarity of the word with the verb 'to leave' – *sortir* – generated some distaste. Indeed, the French word *sortie* is equivalent to the English word *exit*. In the United States, even the name 'Spacelab' was initially accepted with some hesitation, for the agency had only recently concluded its 'Skylab' series of missions and was worried that the two might become confused. "However," wrote Douglas Lord in *Spacelab: An International Success Story*, "despite NASA's objections, once the Europeans had committed to the programme, they unilaterally decided to use the name 'Spacelab', and 'Spacelab' it became." The deal called for Europe to develop the facility in exchange for flying its own astronauts on specific missions. Spacelab, it seemed, would permit NASA to neatly sidestep the biggest obstacle in its financial battles with Congress: how to have both the Shuttle *and* a temporary space station in a decade of decreasing budgets. Unfortunately, the reality proved an unhappy prelude to the project's eventual success.

Numerous payload specialists from foreign governments, industry and scientific institutions flew aboard the reusable orbiters and aboard Spacelab missions between 1983 and 2003. In June 1994 a quartet of distinguished researchers in the fields of materials science and fluid physics were selected to train for two positions on the

USML-2 crew. 'Prime' candidates were meteorologist and fluid dynamicist Dr Fred Weldon Leslie of MSFC and chemical engineer Dr Albert Sacco Jr of Worcester Polytechnic Institute in Worcester, Massachusetts. Leslie was born in Ancon, Panama, on 19 December 1951, and attended high school in Texas, later graduating from the University of Texas with a degree in engineering science in 1974. He went on to earn a master's degree in 1977 and a doctorate in 1979, specialising in meteorology with a minor in fluid mechanics, both from the University of Oklahoma. A keen parachutist, with more than 5,500 jumps, and a multiple world-record-holder in large freefalling and skydiving formations, Leslie served as a postdoctoral research associate at Purdue University after gaining his PhD. He worked on fluid vortex dynamics and participated as a visiting scientist at MSFC in 1980, before joining NASA as a research scientist in Marshall's Space Science Laboratory. He participated in the development of several early Spacelab experiments and was co-investigator for the Geophysical Fluid Flow Cell (GFFC) payload, which flew aboard Spacelab-3 in April 1985 and later aboard USML-2. Leslie subsequently served as chief of MSFC's Fluid Dynamics Branch and was the mission scientist for the joint US/Japanese Spacelab-J mission. At the time of his selection to the USML-2 crew, he had been newly appointed as deputy chief of MSFC's Earth System Science Division. His fellow payload specialist on USML-2, Al Sacco, came from Boston, Massachusetts, where he was born on 3 May 1949. He earned his degree in chemical engineering from Northeastern University in his hometown in 1973 and received a PhD in the same discipline from MIT in 1977. Upon receiving his doctorate, Sacco joined the faculty of Worcester Polytechnic Institute's Department of Chemical Engineering and gained the headship in 1989. Aside from his career in academia, Sacco was an avid scuba diver and certified instructor and, with his father and brother, ran a Boston restaurant business for many years. Backing up Leslie and Sacco, and ready to step into their shoes if needed, were materials engineer Dr David Matthiesen and physicist Dr Glynn Holt.

Spacelab's pressurised module comprised two components: a 'core' segment, which housed data-processing equipment, a workbench and a set of air-conditioned research racks lining its walls, and an 'experiment' segment, providing additional room for scientific operations. Although the core could be flown on its own, this configuration was ultimately never used and all module flights employed both segments joined together, with a pressurised volume of 75 m^3, to form a 'long module'. When one considers the dimensions of the short module, it is clear why NASA opted to use the longer version for a total of 16 missions between November 1983 and May 1998: the long module was over 7 m long and virtually doubled the amount of 'rack' space in which to carry experiments. Racks were refrigerator-sized facilities which could be 'rolled' into the module's cylindrical shell and the facility also offered a central aisle of floor space, onto which experiments could be affixed, and provided a pair of ceiling openings for viewing windows or scientific airlocks. The racks contained air ducts to cool experiments and power-switching panels. On the first dedicated Spacelab mission, STS-9, the module contained 12 racks, of which the pair nearest the entrance to the module were devoted to control subsystems. The ceiling of the core section provided a 0.3 m opening for a high-optical-quality Scientific Window Adaptor Assembly

(SWAA), through which Earth observation cameras could be directed, and that of the experiment section provided a Scientific Airlock (SAL), into which samples requiring exposure to space could be inserted and retrieved. As with other payload bay hardware, the exterior of Spacelab was covered with a layer of passive thermal protection material to protect it from the extremes of sunlight or frigid orbital darkness. Closed at each end by a pair of truncated cones, the module was held in place by three longeron fittings on the payload bay walls and one in its floor at the midpoint of the bay to avoid violating the Shuttle's centre-of-gravity constraints during ascent and re-entry. It was linked to the crew cabin by a 5.8 m tunnel, built by McDonnell Douglas. Since the Spacelab hatch was 1.5 m 'higher' than that of the middeck airlock, a 'joggle' section was included in the tunnel to provide this vertical offset.

Supporting USML-2's expansive research programme, the Extended Duration Orbiter (EDO) capability offered provisions for a lengthy mission, beyond the standard flight of seven to ten days. Extended Shuttle flights of up to 28 days had been in the back of NASA's mind since the pre-Challenger era. They were considered a useful means of evaluating astronaut group dynamics in a closed environment, as well as providing the scope to perform scientific research for longer than previously achievable on the Shuttle. By the spring of 1990 it was decided that Columbia would be decommissioned for about six months in the summer of the following year, with an expectation that the initial EDO missions would fly for 13-16 days, extendable to 28 days by the mid-1990s. A Memorandum of Understanding was signed between NASA and the Shuttle's prime contractor, Rockwell International, to produce a pallet of additional cryogenic reactant tanks for accommodation in the payload bay. "NASA will reimburse Rockwell in three yearly instalments after the pallet has been delivered in December 1991," the space agency announced in an April 1990 news release. "NASA will offer the use of the EDO mission kits as an optional service to all Shuttle customers." Early in January 1991, NASA modified its contract with Rockwell by an additional $93.5 million to accommodate the EDO work. "Under the terms of the modification," it was noted, "Rockwell is required to modify the Columbia orbiter to extend the mission duration of flights from ten days to 16 days, plus a two-day contingency." The improvements included new environmental control and life-support systems, better waste collection facilities and additional gaseous nitrogen and crew stowage provisions.

In addition to the 1,630 kg EDO pallet, Columbia received a new Regenerative Carbon Dioxide Removal System (RCRS). The EDO pallet measured 4.6 m wide and was designed to occupy an upright position at the rear end of the payload bay. It carried four liquid oxygen tanks, two liquid hydrogen tanks and two helium tanks, with the option to add more for yet longer flights. When fully loaded, the pallet could store 1,420 kg of liquid oxygen and 167 kg of liquid hydrogen, which pushed its total weight to almost 3,180 kg. Columbia's sister ship, Endeavour, had also been equipped with EDO provisions during her assembly, together with the option to mount *two* pallets in her payload bay to support 28-day missions, although this capability was later removed for weight-saving reasons, ahead of flights to Mir and the ISS. Endeavour did, however, fly a single EDO mission, lasting more than 16 days, in March 1995.

Meanwhile, the RCRS provided a method for removing the crew's exhaled carbon dioxide from the cabin in a more efficient manner than had been previously achievable with lithium hydroxide canisters. It worked by means of adsorption, instead of absorption. "It uses amine, coated on very small beads, almost like a powder," said Frank Samonski, then-chief of NASA's Environmental Control and Life Support Systems Branch, "but it pours like sand into a multi-layered bed, like a heat exchanger. When you pass gas containing carbon dioxide and moisture over it, the moisture activates this coating and the carbon dioxide molecules stick to the coating. It's not a chemical bond and can be broken by exposure to vacuum. One bed is 'online', collecting CO_2, and the other is 'desorbing' to space vacuum. Through a series of valves, you switch those beds and dump the CO_2 overboard in a cyclic manner. It's not good for long-term missions like the space station, but it's just right for the Shuttle." As well as the pallet, two extra nitrogen tanks were added to Columbia during those six months to support the crew cabin and several more stowage lockers were installed in the orbiter's middeck. Between June 1992 and July 1994, Columbia flew four EDO missions, the last of which peaked at almost 15 days in duration, with the planned 16-day STS-73 anticipated to be her longest orbital voyage to date.

In November 1994, veteran astronaut Kenneth Dwane Bowersox was named to command STS-73, having served as pilot on two previous Shuttle missions, including USML-1. Bowersox came from Portsmouth, Virginia, where he was born on 14 November 1956. For several years, his family lived in Oxnard, California, since his father was stationed at Port Hueneme naval base. Growing up in the early days of America's manned space programme, Bowersox was quickly hooked. Childhood friend John Jarvis later recalled that the boys would beg their parents to be allowed to skip school in order to watch the Gemini missions on television. For Bowersox, it stretched back even before Gemini. "When I was about seven years old," he told a NASA interviewer, "I was driving around in a car with my father, listening to the radio, and on the radio was a broadcast, describing John Glenn orbiting the Earth. *That* sounded like a pretty neat thing to do. From that point, I've just wanted to be an astronaut." Achieving such an exalted goal was far easier said than done, of course, but Bowersox had a helping hand in the form of a school reading assignment sheet, which listed the education and experience requirements. At the time, these revolved around military aviation, test pilot training and advanced engineering credentials. "And so I said, well, that's what I'm going to do, and I took that path," Bowersox said. In 1978, he graduated from the Naval Academy with a degree in aerospace engineering and entered active naval service. A master's degree in mechanical engineering from Columbia University followed in 1979. Designated a naval aviator two years later, he flew the A-7E Corsair off the USS *Enterprise* and logged over 300 carrier-arrested landings. He completed test pilot school in 1985, flew the A-7E and F/A-18 aircraft at the Naval Weapons Center at China Lake, California, and was selected by NASA as a pilot astronaut candidate in June 1987.

Joining Bowersox aboard Columbia's flight deck for STS-73 were fellow naval aviators Kent Vernon Rominger in the pilot's seat and Spanish-born Michael Eladio Lopez-Alegria as the flight engineer. Rominger, who would later rise through the ranks to become chief of the astronaut office, was born on 7 August 1956 in Del

Norte, Colorado. In his case, flying was everything. "I'm here because of my love for flying," Rominger once told a NASA interviewer. "From the time I was about five years old and my father took me flying, I loved it." After gaining a degree in civil engineering – "I always really enjoyed math and science and engineering was the application of that" – from Colorado State University in 1978, this love of aviation drove him into the Navy through the Aviation Reserve Officer Candidate programme. Rominger qualified as an aviator in September 1980 and trained initially in the F-14 Tomcat, flying for the next four years aboard the USS *Ranger* and USS *Kitty Hawk*. During this period, he attended the Navy's Fighter Weapons School (famously nicknamed 'Top Gun') and later earned a master's degree in aeronautical engineering from the Naval Postgraduate School in 1987. "Really, I set my goals on becoming an astronaut once I was a fighter pilot," he recalled. "From that point on, that was really my motivation to become a test pilot, because that's one of the prerequisites for becoming an astronaut pilot." Rominger later carried out carrier suitability sea trials of the upgraded F-14B and logged its first aircraft carrier arrestment and catapult launch. Designated a Distinguished Graduate from Naval Test Pilot School and the Naval Air Test Center's Test Pilot of the Year for 1988, he later deployed to the Arabian Gulf aboard the USS *Nimitz* in support of Operation Desert Storm. Rominger was selected by NASA as an astronaut candidate on his second application in March 1992.

Seated behind and between Bowersox and Rominger on the flight deck was a man who would go on to become one of the world's most experienced spacewalkers. In fact, at the time of writing, Mike Lopez-Alegria – nicknamed 'L.A.' in the astronaut corps – ranks second on the list of all-time EVA experience, with no fewer than ten spacewalks and more than 67 hours to his credit. Born in Madrid, Spain, on 30 May 1958, his mother worked for NASA and frequently brought home information for her young son. With Italian ancestry on his mother's side and Spanish on his father's side, his parents imbued their son with equal values of pragmatism and the thoughtfulness of the dreamer. Lopez-Alegria grew up in Mission Viejo, California, and earned a degree in systems engineering from the Naval Academy in 1980. Lopez-Alegria became a military aviator the following year and worked as a T-34 Mentor flight instructor and later as a pilot and commander of the EP-3E Orion signals reconnaissance aircraft. Assigned to the Naval Postgraduate School in 1986, he studied for a two-year master's degree in aeronautical engineering, after which he was detailed to Naval Test Pilot School in Patuxent River, Maryland. "I was not in a tactical jet squadron," he told a NASA interviewer, years later. "I was in a heavy propeller reconnaissance squadron. There are no people out of my particular aviation community that have either gone to test pilot school or come to NASA, so that was a kind of a change in focus." In fact, it was whilst reading about becoming a test pilot in *Naval Aviation News* that he realised a possible career path into NASA. "I wasn't smart enough to invent a cure for cancer," he once said. "I couldn't play the piano like a concert pianist, I couldn't do a lot of things ... really well, but I could do a handful of things, reasonably well!" After graduation, Lopez-Alegria served as an engineering test pilot at the Naval Air Test Center, ahead of selection by NASA as an astronaut candidate in March 1992.

1-08: Columbia roars into orbit on 20 October 1995.

Emerging from KSC's Operations & Checkout Building on the morning of 20 October 1995, the seven STS-73 astronauts wore back-to-front baseball caps. As Al Sacco later explained, they wanted to demonstrate that science was *not* for geeks. As the crew were being strapped into their seats, the closeout team at Pad 39B were startled by a fire alarm, although fortuitously there were no indications of a real blaze having broken out. A short, three-minute delay was enforced by a computer glitch when the range command destruct system momentarily experienced a communications dropout, but after six previous foiled attempts it seemed that nothing would thwart Columbia from spearing into orbit. "Patience and perseverance are a couple of real good virtues to have in this business," said the head of the Mission Management Team, former astronaut Loren Shriver, after the Shuttle's typically rousing liftoff at 8:53 am EST, "along with a couple of real crack weathermen and a superhuman launch team!"

Overall, the STS-73 ascent was nominal, with the exception of a minor oil temperature problem with one of the water-spray boilers, which required Bowersox and Rominger to shut down an Auxiliary Power Unit (APU) sooner than planned. In anticipation of the kind of research that the crew would perform during their 16 days in orbit, at 10:28 am – just 90 minutes after launch – the astronauts opened Columbia's payload bay doors. They positioned the port-side door at a 62-degrees-open position, rather than fully open. The Shuttle would operate in a 'gravity gradient' attitude, with her tail to Earth and left wing facing into the direction of travel, to provide a stable microgravity environment for the sensitive USML-2 experiments. Consequently, the partially open port-side door helped to minimise the chance of micrometeorites or other orbital debris hitting the delicate radiator panels. On the afternoon of 25 October, Bowersox opened the port-side door for about an hour to provide sufficient clearance for a waste water dump from Spacelab's condensate tank. Such dumps, which occurred through a nozzle on top of the module's forward end cone, had to be performed every few days to get rid of water from the dehumidifiers. After completing the operation, Bowersox returned the door to its 62-degrees-open position.

It was already common practice on research missions of this type for the crew to work in two 12-hour shifts to maintain operations around-the-clock. The Red team comprised Bowersox, Rominger, Thornton and Sacco, whilst the Blue team consisted of Lopez-Alegria, Coleman and Leslie. As with USML-1, which carried two world-class scientists, STS-73 proved no exception: Al Sacco had developed several of the mission's zeolite crystal growth investigations, whilst Fred Leslie helped to design the GFFC experiment, which was intended to mimic the dynamics of planetary and stellar atmospheres. "We're primarily researchers and scientists," Leslie said of the role of the payload specialist, "and the disadvantage is that often you only fly once." The advantages, however, were that he received a flight assignment more quickly than 'career' NASA astronauts. "I went into the programme in 1994 and flew in '95, so it was fast!" He also added that, since he was officially on government business, he received extra pay, but after transport and accommodation deductions, this amounted to just a couple of dollars per day. "You know the government," Leslie grinned. "You can't go anywhere without travel orders!"

During USML-2, Leslie had ample opportunity to work with GFFC, which consisted of two 'hemispheres' – a baseball-sized one, made from stainless steel, mounted inside a larger, transparent one of sapphire – both of which were affixed to a turntable. A thin layer of silicone oil filled the gap between the two hemispheres. During operations, the temperatures of both hemispheres, together with the rotation of the turntable, were minutely adjusted by the experiment's computer, which also introduced thermally driven motions into the oil. This enabled physicists to simulate and model fluid flows in the atmospheres of rotating stars and planets. A similar experiment flew on Spacelab-3 in April 1985 and revealed several types of convection difficult to study on Earth, as well as enabling researchers to observe their structures, instabilities and turbulence. During USML-2, the experiment was employed to model conditions in Earth's mantle – a region of predominantly magnesium-rich silicate rock – as well as plasma flows on the surface of the Sun. After activation, late on 21 October, Principal Investigator John Hart and his team at the University of Colorado at Boulder controlled the experiment remotely from the ground. Time-lapse movies were collected from a series of still images, produced every 45 seconds, as fluids swirled between GFFC's two rotating hemispheres. Despite minor problems relaying temperature parameters, Leslie was happy with its performance, calling it "a planet in a test tube". Later in the mission, efforts were made to simulate activity in Jupiter's atmosphere, which was of particular interest as it radiates significantly more heat than it receives from the Sun, contracting and liberating gravitational potential energy as heat. During each experiment run, voltage, rotation-rate and temperature parameters were adjusted to create unstable and turbulent flows to better explore ocean and atmospheric dynamics.

Elsewhere in the 10,310 kg USML-2 module were several other experiments modified or improved in the wake of their first flight. The European Space Agency's Glovebox had been outfitted with a larger working area and better lighting and supported seven separate investigations. Experiments on USML-1 in June 1992 had yielded protein crystals of much higher quality than had been previously achieved and by 21 October Al Sacco had already gotten his first zeolites up and running. Zeolites are used on Earth for the purification of fluids in life-support systems, as well as in the petroleum refinery process and in waste-management and biological fields. They act as 'molecular sieves' to separate specific molecules from solutions and may someday enable gasoline, oil and other petroleum products to be refined less expensively. In addition to operating experiments in the Spacelab module, Sacco tended to his own Zeolite Crystal Growth (ZCG) furnace in Columbia's middeck, which processed 38 sample containers to create large, near-perfect crystals. Results from USML-1 had shown that zeolite crystals whose nucleation and growth were carefully controlled from the outset produced greater levels of purity than Earth-grown ones.

Other experiments stored in middeck lockers included the Diffusion-Controlled Crystallisation Apparatus for Microgravity (DCAM), which pioneered a method of autonomously running long-duration protein crystal growth experiments aboard the ISS. More than 1,500 protein samples were processed during the USML-2 mission, despite a requirement to lower the Shuttle's cabin temperature after one of the thermoelectric coolers slightly overheated. Another middeck experiment was Astroculture, which was flying its final 'test' mission to grow and provide nutrients for potato plants. It grew ten small tubers to evaluate the extent to which microgravity affected starch accumulation. Starch is an important energy-storage compound, but evidence from previous missions had suggested that its accumulation was more restricted in space. This did not appear to be the case, at least in the first few days of the mission, but by 1 November video footage showed that their leaves were beginning to wilt. However, Astroculture successfully demonstrated its ability to provide proper nutrients, light, water and humidity levels and after USML-2 its hardware was made available commercially for sale or lease.

"This is a pathfinder for the kind of investigations we'll have on the space station," Kathy Thornton told journalists during a space-to-ground news conference on 2 November. "There are very complicated experiments on-board, but they're working beautifully." As with most Spacelab missions having a microgravity research emphasis, the module was outfitted with devices to measure the influence of Shuttle accelerations on very sensitive experiments. As Thornton powered up USML-2 after arriving in orbit, Sacco activated two accelerometers, one of whose sensor heads had been deliberately placed next to the Surface Tension Driven Convection Experiment (STDCE). This allowed scientists to view in great detail the behaviour of fluid flows in microgravity, where temperature differences existed along their interfaces. Such 'thermocapillary' flows occur in many industrial processes on Earth and when they manifest themselves during melting or resolidification they can create defects in crystals, metals, alloys and ceramics. Gravity driven convection overshadows the flows in ground-based tests, making them difficult to measure. High above Earth, on the other hand, it became possible to explore them in greater depth. Fluid physicists expected this knowledge to provide clearer insights into bubble and drop migration, which in turn could aid the development of better fuel-management and life-support apparatus. STDCE used silicone oil as the test fluid and a laser diode and several cameras monitored its transition from a 'steady', two-dimensional flow into an 'oscillatory', three-dimensional one. Smaller, ground-based tests had highlighted periodic variations in the fluid's motion and temperature.

By comparing conditions for the onset of this oscillation in microgravity and on Earth, it was hoped to identify its cause. New optics provided precise images of oil surface shapes and flow patterns. Early in the mission, Fred Leslie reported the onset of oscillations during the very first STDCE run, as ground controllers watched three different views simultaneously downlinked by Spacelab's new television system. By 23 October, the astronauts drew down the volume of silicone oil to create a concave surface. As a laser gradually heated the surface, investigators identified the transitional point at which oscillations began to occur in each run, prompting Principal Investigator Simon Ostrach to speculate that the onset was affected by heat sources, temperature distributions on the surface and the dimensions of the container. In subsequent STDCE runs, Cady Coleman lowered and raised oil levels to create deeply concave, then convex, surfaces. The 'size' of the laser beam on the oil surface was also adjusted to determine its effect on the direction and nature of fluid flows, as well as lifting its temperature to introduce oscillations. This led to several

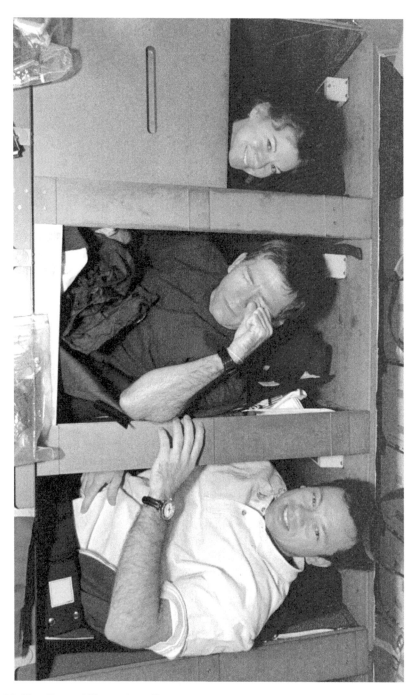

1-09: Blue Team shift members (from top) Cady Coleman, Fred Leslie and Mike Lopez-Alegria in their sleep stations aboard Columbia's middeck.

instances of new phenomena. During one test, erratic flows with no apparent organisation or pattern appeared as Thornton increased oil temperature beyond the point at which flows began to oscillate. Later, Al Sacco handed control to the ground, which ran the experiment via telescience.

Sacco also demonstrated 'surface tension' in space by squeezing orange juice from a tube. It promptly formed a ball, thereby highlighting how surface tension makes fluids assume spherical shapes when they no longer have to contend with terrestrial gravity. The six-channel, simultaneous video footage from STDCE was only possible thanks to the new High-Packed Digital Television Technical Demonstration (HI-PAC). Prior to USML-2, video from Spacelab could only be downlinked on a single channel, but as part of testing the new system the STS-73 crew was able to acquire live images of their colleagues sitting at their consoles in Mission Control.

Other fluid physics investigations were carried out in the Drop Physics Module (DPM), another facility carried over from the USML-1 mission. This sought to explore future ways of conducting 'containerless' processing and encapsulation of living cells for pharmaceutical research. One investigation was the Drop Dynamics Experiment, provided by Spacelab-3 astronaut Taylor Wang, which gathered high-quality data in support of such applications. During USML-2, the experiment investigated the breaking-apart of distorted drops under a fluid of varying viscosity, as well as attempting to encapsulate and create 'compound' drops. It was hoped that such tests might lead to methods to insert living cells for the treatment of hormonal disorders into polymer shells to protect them from immunological attack and to provide timed releases, perhaps in case of diabetes. A second DPM experiment examined the influence of 'surfactants' – substances which alter a fluid's properties by aiding or inhibiting the way it mixes with other substances – on the behaviour of single and multiple drops. Early in the mission, investigators watched video of the first deployment as Thornton released a 2 cm glob of water and used precisely controlled sound waves from a pair of loudspeakers to manipulate its movements. Later, Fred Leslie succeeded in encapsulating an air bubble inside a floating water droplet; whilst Al Sacco manipulated drops treated with an organic surfactant known as bovine serum albumin, deploying two drops and bringing them perfectly together until they coalesced into a single blob; and Cady Coleman stretched a 'bridge' of liquid between the DPM's two injector tips and centred a drop of water inside a glob of silicone oil, whilst the interfaces of both were moving in opposite directions. After more than 18 months in the Spacelab simulator, practicing the encapsulation of drops, Coleman reported that "it's sure nice to see it for real".

Elsewhere, the Crystal Growth Furnace (CGF) proved capable of achieving very high operating temperatures, in excess of 1,000 degrees Celsius and processing large semiconducting crystals, together with metals and alloys. During USML-2, it was employed to grow samples of cadmium-zinc-telluride, mercury-cadmium-telluride, gallium arsenide and mercury-zinc-telluride, having achieved great success in producing surprisingly defect-free specimens on USML-1 three years earlier. For the second mission, the primary CGF sample chamber was outfitted with a spring-loaded piston that moved to reduce the volume of the cylinder as the material contracted whilst cooling. This helped to eliminate air voids in the crystal and ensured that it maintained an 'even' contact with the container walls along its entire surface. Cadmium-zinc-telluride is routinely used as a base on which infrared-sensing mercury-cadmium-telluride crystals can be grown. USML-1 had produced cadmium-zinc-telluride crystals 1,000 times purer than those grown on Earth. The materials flown on the second mission carried potential terrestrial applications in high-speed, low-power digital circuits, optoelectronic-integrated circuits and solid-state lasers. All samples were processed satisfactorily, together with a fifth specimen – of germanium, with a trace of gallium – to demonstrate the influence of changes in the Shuttle's attitude. In total, eight crystals were grown, one of which saw the CGF operating at its highest temperature (1,250 degrees Celsius) and another which took just 90 minutes to complete. In total, in the middeck and the Spacelab module, more than 1,500 protein samples were processed during USML-2. These included feline calcivirus, akin to a virus responsible for causing digestive problems in humans, which proved to be so well-formed that it elicited applause from team members on the ground. Sacco also set up initial conditions for the growth of a collagen-binding domain protein, which is important in the study of arthritis and joint disease, and Coleman activated a protein known as duck delta crystallin, which is similar to the protein responsible for causing a rare, but deadly disease in humans.

Combustion science was another focus of STS-73's Glovebox research and, in recognition of his efforts, Fred Leslie received a collective 'high-five' for his work on 2 November, according to team members of the Fibre-Support Droplet Combustion experiment. "Fred gets an A+ in combustion theory!" they exulted. He had successfully placed droplets of fuel onto a thin fibre, using needles in the experiment module, then igniting them with a hot wire. Despite early difficulties with clogged drop accumulators, he performed several 'burns', varying the quantities of methanol and methanol-water fuel on each run.

Columbia was performing perfectly on her 18th mission. Her crew took time out on 25 October to tape the ceremonial first pitch for Game Five of the baseball World Series. Uniquely, it marked the first time that the thrower – Ken Bowersox, in this case – was not actually in the ballpark for the pitch. The video from orbit was replayed on the enormous Jacobs Field Jumbotron screen in Cleveland, Ohio, ahead of the game. Before throwing the slow-spinning pitch, Bowersox wished both the Atlanta Braves and Cleveland Indians good luck. The STS-73 crew later signed their on-board baseballs and gave them to Major League Baseball to be enshrined in the Baseball Hall of Fame in Cooperstown, New York.

As the mission's second week in space wound down, the astronauts were asked to place the Shuttle into several different attitudes to increase temperatures across her belly. Flight controllers were concerned that her tyre pressures were lower than acceptable for landing. Thermal conditioning of the belly brought them up to within the required minimum landing pressure limit. Early on 5 November 1995, the final experiments aboard the Spacelab module were shut down, the payload was deactivated and the hatch closed. Right on cue, Bowersox and Rominger fired Columbia's OMS engines at 5:46 am EST to begin the hour-long return to Earth. With double sonic booms echoing across the marshy Florida landscape, the vehicle touched down at 6:45:22 am, completing a mission of 15 days, 21 hours, 52 minutes and 22 seconds, marking Columbia's longest flight to date and the second-longest in the programme. Her sister ship, Endeavour, had achieved a mission of 16 days and 15 hours earlier in 1995, but that record would not stand for long. By the end of 1996, two missions by Columbia would have established empirical duration records which would endure until the end of the Shuttle era.

MISSION FOR JAPAN

In December 1994, NASA announced the crew for Endeavour's STS-72 mission, then scheduled for launch in November of the following year. The six-man team was notable in that it included Koichi Wakata, the first Japanese astronaut ever selected for mission specialist training. As part of its preparations for the Space Station Freedom programme – later to become the ISS – European, Japanese and Canadian candidates were brought into NASA's existing astronaut corps to train for positions on Shuttle and eventually station crews. Wakata's assignment to STS-72 came about partially because the nine-day flight was tasked with retrieving a Japanese satellite, the Space Flyer Unit (SFU), from low-Earth orbit. This 3,580 kg, octagonal-shaped satellite had been launched by the National Space Development Agency of Japan (NASDA), atop an H-II rocket from the Tanegashima Space Centre on 18 March 1995, carrying a wide range of automated experiments in materials science, space biology, engineering and astronomy. In addition to key involvement from NASDA – which, in October 2003, became today's Japan Aerospace Exploration Agency (JAXA) – the SFU project also featured co-operation from the Institute of Space and Astronautical Science (ISAS) and the Ministry of International Trade and Industry.

Its objectives included providing scientists and engineers with a group of long-duration research facilities on a reusable spacecraft, which, it was expected, would fly at least two missions. Aboard the SFU when it rose into orbit in March 1995 were the Infrared Telescope in Space (IRTS) to explore the history of the Universe and the structure of the Milky Way, the Space Plasma Diagnostics Package (SPDP) and several other experiments to evaluate the assembly of large structures and high-voltage solar arrays in space, to demonstrate electric propulsion and fuel-management technologies, to observe the behaviour of liquids in the microgravity environment and to monitor the hatching and evolution of eggs in space. After retrieval by the STS-72 crew, it was intended that the SFU would be serviced and readied for a second mission. However, none of NASA's Shuttle manifests from 1992-1997 featured a second retrieval flight and after STS-72 the SFU never flew again. Today it is housed in Tokyo's National Museum of Nature and Science.

With a Japanese crewman aboard Endeavour for the mission, it is hardly surprising that Wakata was assigned as primary operator of the Shuttle's RMS arm to recover the SFU. His career as an astronaut would carry him into the construction of the space station and he would go on to become the first Japanese to fly a long-duration mission in 2009 and the first Japanese to command the ISS in 2014. He would also serve as chief of Japan's astronaut corps from 2010 until 2012. Wakata

1-10: Japan's Space Flyer Unit (SFU) is recaptured by Endeavour's mechanical arm, after ten months in orbit.

was born on 1 August 1963 in the city of Ômiya, within Saitama Prefecture of Japan's main island, Honshu. "I started to have a strong longing for going to space when I saw the Apollo lunar landing," he told a NASA interviewer, "although at the time it seemed to me that only the people in the US and also the former Soviet Union could work in space. It seemed to me that to go into space was something ... beyond reach for me as a Japanese, since we did not have a human space programme at the time." Nevertheless, Wakata grew up with a love of aviation, building and flying model aircraft and as an undergraduate participating in competitions to construct and test gliders. In addition to studying mathematics and science, he learned English, "because many of the technical books in aerospace engineering were written in English back in those days". Wakata gained a degree in aeronautical engineering from Kyushu University in 1987, followed by a master's credential in 1989, then entered Japan Airlines as a structural engineer.

Initially assigned to the Base Maintenance Department in Narita, near Tokyo, he later worked in the Airframe Group at the Systems Engineering Office of Japan Airlines' engineering division. He was closely involved in the research of structural integrity of transport aircraft, including fatigue fractures, corrosion prevention programmes and the environmental effects on the polished aluminium fuselage of various commercial aircraft. An astronaut career drew closer for Wakata, due to Japan's involvement in the space station programme and its intention to supply a large research laboratory for the orbital complex. "In 1991, NASDA announced that they [would] select a couple of astronauts to participate in the NASA astronaut training," he explained, "and also to eventually fly in the assembly flights for the Japanese module on the International Space Station. Fortunately, I was selected." His selection by NASDA came in April 1992 and he began training alongside NASA's 14th group of astronaut candidates at JSC the following August.

Commanding STS-72 was veteran astronaut Brian Duffy, making his third Shuttle mission. Born in Rockland, just south of Boston, Massachusetts, on 20 June 1953, Duffy grew up watching the contrails of jet aircraft from nearby South Weymouth Naval Air Station and aspired for a career in aviation. After completing high school in Massachusetts, he entered the Air Force Academy and earned a degree in mathematics in 1975. "To be a 19-year-old ... in the back seat of an F-4 [Phantom]," he said, "going 600 miles an hour made me realise *that* was something I really wanted to do." Undergraduate pilot training followed at Columbus Air Force Base, Mississippi, and Duffy spent the next several years flying the F-15 Eagle fighter out of Langley Air Force Base, Virginia. He was deployed overseas to Japan, earned a master's credential in systems management from the University of Southern California in 1981, completed test pilot school as a Distinguished Graduate in 1982 and directed F-15 testing at Eglin Air Force Base in Florida. By his own admission, Duffy loved it. Then, one Friday afternoon in the late spring of 1985, at the end of a weekly pilots' meeting the squadron commander concluded with exciting news: NASA had issued a call for Shuttle astronaut candidates. As Duffy looked over the rigorous academic and professional requirements, he was astonished to realise that he satisfied them all. "And I thought, I *couldn't* let that eight-year-old kid from so many years ago down to find out that I was qualified to apply and not do it," he recalled. "Would I be able to live with myself?" Selected in June 1985, he brought his family to Houston later in the summer and by the end of the following January the Duffys had built a new home in El Lago and had just retrieved their boxes from storage. At around the same time, Challenger exploded, stalling the Shuttle for almost three years. As circumstances transpired, his third mission – and his first command – coincided with the tenth anniversary of that accident.

Alongside Duffy, the only other veteran crewman on STS-72 was Leroy Chiao, who came from Milwaukee, Wisconsin. He was the child of Chinese parents who fled the Communist takeover in the aftermath of the Second World War by first settling in Taiwan and then emigrating to the United States in the 1950s. (In his later astronaut career, Chiao selected 'Shandong' for his radio callsign, to honour the coastal province in eastern China where his mother and father were born.) Chiao was born on 28 August 1960 and became interested in the space programme as a child, when he saw Neil Armstrong and Buzz Aldrin walking on the Moon. "I remember being in my parents' home a hot summer day," he recalled in a NASA interview, "and then, later that evening, actually take the first steps on the Moon. To me, that was *really* a big event!" However, unlike most children of his generation, who sought after the astronaut goal for a while, then gave it up, Chiao kept it at the back of his mind through high school and his university education. "Mechanical things and electrical things and chemical things" intrigued him as a youngster, together with a fascination for mathematics and the sciences, which led Chiao to select chemical engineering for his degree choice at the University of California at Berkeley. Receiving his undergraduate award in 1983, he proceeded to the University of California at Santa Barbara and earned his master's and doctoral degrees in 1985 and 1987. Chiao initially worked at Hexcel Corporation – "a composite materials supplier" – and at Lawrence Livermore National Laboratory, working on the fabrication of filament-wound and thick-section aerospace composites. In September 1989, after submitting an application to join NASA's astronaut corps, he was invited to Houston for a week of interviews and physical and physiological tests. Four months later, he was selected. It was not easy. "The thing I tell people who ask me how do you become an astronaut," he explained later, "[is] there really is no one way to do it . . . If you're a person who puts all your eggs in one basket and says, well, I'm going to study something I'm not really interested in and hope to become an astronaut, you're in for a lot of disappointment. My advice . . . always is to study something that you're interested in and work in a field that interests you. *Then* apply." Chiao first flew aboard STS-65 in July 1994 and was making his second Shuttle mission on STS-72.

Commander Brian Duffy brought a great deal of expertise to the SFU retrieval effort, for he had previously served as pilot aboard STS-57 in June 1993, which recovered the EURECA satellite from orbit after ten months and brought it back to Earth. Accomplishing the rendezvous demanded a very precise liftoff from KSC's Pad 39B, timed to occur within a 49.5-minute "window" that opened at 4:18 am EST on 11 January 1996. The launch would thus position Endeavour onto a course that would continually close the distance to the SFU, setting the astronauts up to conduct the retrieval about 48 hours into their mission. The STS-72 countdown operations began on 8 January, tracking an on-time liftoff on the 11th. Early on launch morning, the six-man crew suited-up and boarded Endeavour for a relatively smooth final couple of hours, punctuated only by minor problems with facilities or ground servicing equipment. Shortly before the countdown entered a planned, built-in hold at T-20 minutes, it became clear that the Shuttle's main engines were running a little cool and the flow of gaseous nitrogen was adjusted. Later, a communications configuration glitch was detected between Mission Control in Houston and ground stations in Florida, prompting a 23-minute delay into the first half of the launch window. None of these problems prevented Endeavour from rocketing into the darkened sky at 4:41 am EST, turning night into day across the marshy Florida landscape, and reaching a preliminary orbit just nine minutes later.

At the point of orbital insertion, Endeavour trailed the SFU by 14,740 km, closing this distance by up to about 1,390 km during each circuit of Earth, and the crew spent their first day in space checking out their rendezvous tools and equipment. With Wakata designated as the primary RMS operator, he was responsible for powering up and testing the Canadian-built mechanical arm, whilst Duffy carried out tests of controls on the aft flight deck. Other crew members also activated numerous of STS-72's secondary payloads, including protein crystal growth studies on the middeck and two GAS experiments in the payload bay to explore the influence of vibrations on a pair of cantilevered flexible beams and to grow crystals of the horse heart myoglobin and bovine pancreatic ribonuclease S proteins. The Shuttle Laser Altimeter (SLA) in the payload bay was making its first flight and was designed to measure the distance between Earth's surface and the orbiting Shuttle, transmitting a series of short laser pulses and precisely measuring the reflected echoes from land, ocean, vegetation and cloud-tops.

By the evening of 11 January, halfway through the first day of the mission, Endeavour had continued to reduce the distance between herself and the SFU. The Japanese flight controllers verified that their satellite was in good shape, despite having lost functionality in two of its reaction-control thrusters, which were not considered a threat to the retrieval. Early on the 12th, Duffy and his pilot, first-time astronaut Brent Jett, pulsed the Shuttle's RCS thrusters to slightly adjust their orbit and avoid an unwanted close encounter with the US Air Force's MISTY military satellite. Had they not taken the action, it was estimated that Endeavour would have passed within an uncomfortably close 1.2 km of the dead satellite. As circumstances transpired, the two spacecraft passed within 8 km of each other. By that evening, following another orbit-raising manoeuvre, the Shuttle was just 480 km behind the SFU, closing at a slower rate of about 96 km per orbit.

The final rendezvous in the early hours of 13 January began about three hours ahead of the scheduled 4:26 am EST retrieval, with the Terminal Initiation (TI) burn of Endeavour's RCS thrusters being made about 14.8 km from her quarry. The Shuttle approached the SFU along the so-called 'R-Bar' corridor, tracing an imaginary line directly 'upwards' from the centre of Earth to reach the SFU. At this stage, Wakata powered up the RMS and extended it high above the payload bay, ready to capture the satellite. As they drew to within about 0.8 km of each other, about 90 minutes ahead of the scheduled retrieval, the Shuttle's rendezvous radar locked onto the SFU. With a constant stream of range and rate-of-closure data at his disposal, Duffy assumed manual control of the orbiter. Drawing nearer, at 180 m, he switched the Shuttle's RCS to 'Low-Z' mode, which employed offset thrusters in the nose and tail for braking, rather than risking contamination by using the thrusters facing directly towards the satellite. During this period, range and rate-of-closure data was provided by Leroy Chiao, courtesy of a hand-held laser ranging device. By the time Endeavour reached about 45 m from the SFU, he entered a 45-minute station-keeping phase as the Japanese flight controllers began to command the satellite's twin solar arrays to close and latch into place. It was during these activities that STS-72's first gremlin reared its head.

The satellite had earlier been placed onto internal battery power, ahead of the commands for it to close the arrays. Although they retracted correctly, they did not properly latch into place. As a result, it was decided to discard both array canisters. The astronauts and flight controllers had trained for just such an eventuality in their contingency simulations. As Endeavour and the SFU orbited high above Africa, the canisters were jettisoned (the first at 4:35 am, the second at 4:47 am) from the satellite to ensure that they did not complicate the retrieval. Finally, the go-ahead was issued to approach the SFU for retrieval and Duffy executed a small yaw manoeuvre to align the RMS grapple fixture with the satellite's capture mechanism. With the two spacecraft now above the Gulf of Mexico, near the western tip of Cuba, Wakata grappled the SFU at 5:57 am, about 91 minutes later than planned. He then lowered it gently into the payload bay, locking it into place at 6:39 am by closing four retention latches. The internal batteries were bypassed and a remotely operated electrical cable was connected to the side of the satellite. To honour their success, the crew were awakened for their next day in space by the traditional Japanese song, 'Sea in Springtime'.

Success had been accomplished by a crew which balanced only two experienced astronauts with four first-timers. "The crew experience was night and day between the two," Leroy Chiao admitted, but stressed that "We were a very cohesive crew." Seated in the pilot's position and assisting Duffy with the SFU rendezvous was Brent Ward Jett Jr. Born in Pontiac, Michigan, on 5 October 1958, Jett received his early schooling in Florida and was drawn by the allure of aviation. "As a child, I always wanted to fly," he told a NASA interviewer, "and at some point as an early teenager, that sort of focused on becoming a Navy pilot and then my goal was to become a Navy carrier pilot. *That* eventually evolved, later in life, to a goal of wanting to be a test pilot. After I achieved that goal, that's the first time I actually even thought about becoming an astronaut. I did not dream of it as a child. When we saw the Apollo astronauts walk on the Moon, it just seemed like an unreachable goal." Jett initially spent a year at the University of Florida, where one of his professors – a Navy lawyer – encouraged him to enter the Naval Academy. Jett went on to earn his degree in aerospace engineering from the Naval Academy in 1981, graduating first in his 976-strong class and qualifying as a naval aviator in March 1983. After initial training in the F-14 Tomcat at Naval Air Station Oceana, Virginia, he was assigned to Fighter Squadron (VF)-74, flying the F-14 Tomcat during deployments to the Mediterranean Sea and Indian Ocean aboard the USS *Saratoga*. During this period,

1-11: Pictured during the EVA, this view shows the gold-coated OAST-Flyer satellite in Endeavour's payload bay.

he was designated an airwing-qualified Landing Signal Officer and attended the Navy's Fighter Weapons School ('Top Gun'). Selected for a co-operative education programme run by the Naval Postgraduate School and Naval Test Pilot School, Jett gained a master's degree in aeronautical engineering in 1989 and was designated the Distinguished Graduate as a test pilot in June 1990.

It was whilst checking out of his squadron and preparing to drive to Monterey,

California, for his postgraduate studies that one of Jett's superiors handed him a set of Navy Instructions. "I think you're going to need this someday," he said. Jett looked at the papers and noticed that they were the Navy Instructions on applying to NASA for the astronaut selection process. "I hadn't really thought of it up to that point," Jett admitted, years later, "because I was really focused on going to test pilot school, but I stuck it in my briefcase and I kept it. A couple of years later, I pulled it out and looked at it and got the updated version and ended up applying." After completing test pilot school, he served as a project test pilot in the Carrier Stability Department of the Naval Air Test Center's Strike Aircraft Test Directorate, returning to the operational Navy in September 1991 as an F-14B pilot, based on the *Saratoga*. Six months later – on his first application to NASA – Jett was selected as one of only four Shuttle pilot candidates in the space agency's 14th astronaut intake.

In addition to retrieving a Japanese satellite, the nine-day STS-72 mission featured a multitude of scientific and technological payloads, including a deployable SPARTAN free-flying spacecraft of identical design to that flown aboard STS-69 in September 1995. Unlike its predecessor, however, this SPARTAN was equipped not with solar physics instruments, but with a package of experiments sponsored by NASA's Office of Aeronautics and Space Technology (OAST). Designated 'OAST-Flyer', the deployment of the 1,200 kg satellite was conducted early on 14 January. Following a thorough systems check, latches holding it in the payload bay were released and Koichi Wakata grappled OAST-Flyer with the RMS arm and lifted it to its pre-deployment position. He released the satellite at 6:32 am EST for two days of autonomous activities.

Aboard OAST-Flyer were four experiments, three sponsored by NASA and a fourth by the Amateur Radio Association at the University of Maryland. The Return Flux Experiment (REFLEX) tested the accuracy of computer models predicting the effect of contaminants on sensitive surfaces, such as lenses, sensors and scientific instruments, in the space environment. In addition to measuring the erosion of surface coatings, the experiment investigated molecular 'backscattering' as the spacecraft issued tiny particles of dirt, which went on to collide with other particles and bounce back onto the spacecraft. The Solar Exposure to Laser Ordnance Device (SELODE) sought to evaluate five new techniques of firing pyrotechnics using fibre optic laser pulses, rather than electricity. The University of Maryland's SPARTAN Packet Radio Experiment (SPRE) supported amateur radio applications, with GPS data provided through the fourth OAST-Flyer payload, the Global Positioning System Attitude Determination and Control Experiment (GADACS). This latter investigation determined the attitude of the satellite, its location and velocity, and provided accurate timing to execute SPARTAN thruster firings. It marked the first occasion on which a spacecraft had been controlled using GPS technology and a similar receiver was also located aboard Endeavour herself.

After deployment, OAST-Flyer executed a 45-degree pirouette to confirm that its attitude-control system was functioning properly. Pilot Brent Jett then manoeuvred the Shuttle to a distance of about 72 km, with periodic thruster firings over the following two days to maintain a distance of about 167 km between the two spacecraft. At the conclusion of its operations, OAST-Flyer manoeuvred itself into the correct orientation and awaited the return of Endeavour, early on 16 January. About 15 km 'behind' the satellite, and two hours ahead of capture, Duffy pulsed the RCS thrusters for a TI burn which positioned the Shuttle onto an R-Bar rendezvous, approaching OAST-Flyer from 'below'. As with the earlier SFU operation, the rendezvous radar and hand-held laser ranging device offered critical support during the proximity operations. At a distance of 0.8 km, and just 45 minutes before the scheduled retrieval time, Duffy assumed manual control, firing Endeavour's thrusters directly towards the satellite for braking, until he reached 66 m, at which point he transitioned to Low-Z mode to avoid contaminating OAST-Flyer's experiments. Closing to within 10 m, Duffy maintained his position, enabling Wakata to grapple the satellite, precisely on time, at 4:47 am EST. Shortly thereafter, it was returned to its berth in the payload bay for return to Earth.

Betwixt the deployment and retrieval of OAST-Flyer, the first of STS-72's two EVAs was conducted in the early hours of 15 January. In a manner not dissimilar to the spacewalk performed by Jim Voss and Mike Gernhardt a few months earlier, it formed part of NASA's EDFT initiative to evaluate tools and techniques for the construction and maintenance of the ISS. When the crew was assigned in December 1994, it was intended that Leroy Chiao and fellow mission specialist Dan Barry would perform both spacewalks, but as their training developed, it was decided to incorporate mission specialist Winston Scott into the plan, thereby increasing their all-round EVA expertise. As a result, Chiao (designated 'EV1', with red stripes on the legs of his suit) and Barry ('EV2', wearing a pure-white suit) would perform the first spacewalk, with Scott choreographing their every move from inside Endeavour's cabin. Two days later, Chiao and Scott (designated 'EV3', with broken red stripes on the legs of his suit) would venture outside on the second spacewalk, with Barry choreographing their actions from the flight deck.

At the time of STS-72, Chiao was 35 years old, somewhat younger than the 42-year-old Barry and the 45-year-old Scott. None of them had previously performed an EVA, but it made sense for Chiao – the only one of them with prior Shuttle flight experience – to serve as the chief spacewalker. It set him up for some interesting conversations with Barry. "Dan was an interesting guy," Chiao told the NASA oral historian. "He was flying his first mission and he felt like he should have had my job. He thought he should be the lead EVA guy ... and he let me know it, which set up an interesting relationship! There was a little bit of friction between Dan and me for that reason, but we worked through it."

For Barry and Scott, both making their first Shuttle flights, the opportunity to travel into orbit was sweetened with the promise of one spacewalk apiece. Winston Elliott Scott had risen from the segregation of the 1960s to become a Navy helicopter pilot and only the seventh African-American ever to be selected into NASA's astronaut corps. Born in Miami, Florida, on 6 August 1950, his early education was segregated and he initially attended George Washington Carver Senior High School for black students. Following the desegregation of all Dade County schools in 1966-1967, Scott entered Coral Gables High School and graduated the following year. He developed a keen interest in music and later played trumpet with various bands along the Cape Canaveral Space Coast and in a Houston-based big band. He gained a degree in music from Florida State University in 1972 and entered the Navy, where he completed training in fixed-wing and rotary-wing aircraft and became a naval aviator two years later. He served a tour of duty at Naval Air Station North Island in California, flying the SH-2F LAMPS helicopter. Scott was subsequently chosen for Naval Postgraduate School and in 1980 earned a master's degree in aeronautical engineering with avionics, then undertook jet training in the TA-4J Skyhawk and served as an F-14 fighter pilot at Naval Air Station Oceana. Designated an Aerospace Engineering Duty Officer in 1986, Scott was a production test pilot at the Naval Aviation Depot at Naval Air Station Jacksonville in Florida, flying the F/A-18 Hornet and A-7 Corsair aircraft. He later headed the Product Support (Engineering) Department and the Tactical Aircraft Systems Department at the Naval Air Development Center in Warminster, Pennsylvania. Additionally, Scott worked as an associate instructor of electrical engineering at Florida A&M University and Florida Community College and gained a 2nd degree black belt in Shotokan karate.

Scott's crewmate on STS-72, Daniel Thomas Barry, was born in Norwalk, Connecticut, on 30 December 1953, but attended schools in Louisiana and became interested in electrical engineering through his brother-in-law, Wayne Keote. "I had no idea what electrical engineering was," he admitted, "but he showed me what it was and took a real interest in teaching me about engineering and really inspired me to go on in that area." Barry earned his degree in electrical engineering from Cornell University in 1975. As a child, he had been fascinated by flying and, by his own admission, by "always finding the highest thing I could locate to jump off". Becoming an astronaut was a dream from as early as he could remember. "Every kid in first-grade school wanted to grow up to be an astronaut," he later told a NASA interviewer, "but as time went on, different people wanted different goals, but this was one I had with me my whole life." After receiving his undergraduate credential, Barry entered Princeton to receive a master of engineering and a master of arts in electrical engineering and computer science in 1977 and a PhD in the same area in 1980. Barry was a National Science Foundation postdoctoral fellow in physics at Princeton, before entering the Leonard M. Miller School of Medicine at the University of Miami and graduating in 1982. "There was an opportunity to go medical school in a programme that allowed people with PhDs to obtain an MD in two years," he explained. "I thought I would go right back to bioengineering, but got interested in rehabilitation medicine." He served his internship and a physical medicine and rehabilitation residency at the University of Michigan in 1985 and worked as an assistant professor in the Department of Physical Medicine and Rehabilitation and in the Bioengineering Program. For Barry, it appealed to him as a field of medicine which was oriented toward enabling patients to achieve personal goals, in spite of their physical or mental disabilities. For three summers in 1985-1987, he also worked at the Marine Biological Laboratory in Woods Hole, Massachusetts, focusing on skeletal muscle physiology and biological signal processing, whose potential applications included acoustic signals generated by contracting skeletal muscles, electrical signals from muscles and sounds from the heart. He also worked in prosthetic design. Barry was selected as an astronaut candidate by NASA, alongside Wakata, Jett and Scott, in 1992.

1-12: Backdropped by OAST-Flyer and the SFU, and the glorious panorama of Earth, Leroy Chiao works in Endeavour's payload bay during one of his EVAs.

So it was that at 12:35 am EST on 15 January 1996, Leroy Chiao and Dan Barry transferred their space suits' life-support utilities to internal battery power, pushed open the outer airlock hatch and floated into Endeavour's floodlit payload bay for what they expected to be around six hours of operations. Although he had one prior Shuttle flight to his credit, Chiao's breath was truly taken away by the sensation of actually *being* in space, with nothing but a visor between him and the cosmos. "Getting to do spacewalks was a surreal experience," he told the NASA oral historian, years later, "seeing the Earth with peripheral vision involved, really feeling like the Earth was a *ball*. Of course, looking through the window, you *know* it's a ball, but there's something about looking through the window and then being *out there* with an unrestricted view that really made it feel like it was three-dimensional for the first time." Flight Surgeon Phil Stepaniak later told Chiao that his heart rate spiked as soon as the airlock hatch opened, then quickly returned to normal as he began work.

With Jett and Wakata operating the RMS, and Scott choreographing their every move from the aft flight deck, the two men quickly set to work on their respective tasks. They assembled and tested the Lockheed Martin-built Portable Work Platform (PWP). Designed as a prototype for a mobile EVA worksite to be located at the end of the Space Station Remote Manipulator System (SSRMS), the PWP included a temporary equipment restraint aid to hold ORUs, a portable foot restraint stanchion with tool boards, sliding locks and rigid tether sockets and an Articulating Portable Foot Restraint (APFR). After assembling the PWP, Chiao and Barry installed it onto the end of Endeavour's RMS arm, allowing Jett to grapple various pieces of hardware intended to hold large modular components. In a sense, this work closely mirrored the manner in which equipment and avionics boxes would be moved about during ISS assembly.

Then the spacewalkers set to work on their next task, which was to unfurl the 113 kg Rigid Umbilical (RU), which simulated an umbilical connection of a type that might someday link ISS modules to the electricity-generating solar arrays and supporting truss structure. Stored on the port-side wall of the payload bay, the RU included a cable tray, which housed a range of avionic and fluid lines, and measured 2.4 m long in its folded configuration and 5.3 m long when fully deployed. Chiao and Barry mated one end of the umbilical to a connector (known as the Port Rigid Umbilical Mount) on the port side of the payload bay, and attached the other end to a connector on the starboard wall. The spacewalkers then installed a series of simulated electrical and fluid connections and later partially unfolded the RU to simulate various positions planned for its attachment on the ISS. Most tasks, Chiao reported, could be completed without difficulty, although he felt that the Electronic Cuff Checklist (ECC) – a wrist-mounted portable computer to supplement written checklists for spacewalkers – was a good idea that was poorly executed, since solar glare made its display difficult to read and it also interfered with the space suit controls.

Towards the end of the EVA, during a brief quiet spell, Barry managed to catch a glimpse of stars. "We had a test at the very end of the spacewalk to test the lights mounted on our helmets," he told a Smithsonian interviewer, years later. "To get the best test, we did something never done before during a spacewalk: we turned off *all* of the orbiter's lights, which are so bright that they prevent spacewalkers from seeing the stars." Barry found that his helmet lights functioned perfectly well, and he could continue to work without problems, but when he switched them off as well he was plunged into utter darkness. "It was so black," he explained, "I couldn't see my *hand*. I waited for my eyes to adjust and was rewarded by *millions* of stars, faint at first, but eventually turning into brilliant points."

Their work completed, the spacewalkers returned to the airlock and EVA-1 officially ended at 6:44 am EST, after slightly more than six hours. "We'll come out together [again], Leroy," called Barry as the EVA drew to a close. "There's going to be plenty of work to do for station." Indeed, both men would perform spacewalks to assemble and maintain the ISS, later in their careers. With the following day, 16 January, devoted to the retrieval of OAST-Flyer, the second spacewalk, by Chiao and Scott, got underway at 12:40 am EST on the 17th. Problems with donning their suits delayed the start time by about an hour, but once outside the astronauts stepped smartly through their tasks. Going outside for the second time was an interesting experience for Chiao. On his first EVA, his heart rate had spiked when the airlock hatch opened, but on this second occasion ... it did not change at all. Talking to the flight surgeon after the mission, it seemed that his body and senses had almost gotten used to the experience. "It's interesting how adaptable humans really are," Chiao said, "and you get used to things."

Scott entered a foot restraint and held himself still for 35 minutes as Endeavour manoeuvred into an orientation which would chill the payload bay, thus testing thermal modifications to his suit. These included battery-powered heating elements, just above the suit's wrists, which provided warming of the astronauts' fingers, together with thermal socks and toe caps and new, adjustable thermal 'mittens'. Additionally, the modifications enabled the spacewalkers to shut off cooling water to their liquid cooling and ventilation garments, should they become too cold. During this 35-minute period, temperatures dropped to minus 122 degrees Celsius and Scott reported that, although he was keenly aware of the cold, he remained comfortable. "If I felt *that* comfortable standing still," he later told a post-mission debriefing panel, "I think I'd have been even warmer as a result of being busy."

Pressing on with their work, the astronauts tested cable trays and clamps for ISS use. Leroy Chiao noted that cold temperatures threatened to make the handling of pressurised umbilicals difficult and, certainly, both spacewalkers felt that the clamps caused hand fatigue and that larger connectors were more awkward to use than small ones. At times, they rocked backwards and forwards in their foot restraints to determine the levels of vibrational stress which might be imposed on space station structures. They also evaluated the Body Restraint Tether (BRT), previously tested on STS-69, which it was hoped would provide a method of stability for a spacewalker in places where a foot restraint might be unavailable. Essentially, the BRT provided a 'third hand' for a spacewalker, whilst working. Tests of a new Utility Box – a slidewire and a cable caddy – were also performed, with NASA engineers reporting to journalists that the data from the STS-72 EVAs would be incorporated into the spacewalk plans for ISS assembly missions. Their work finished, Chiao and Scott returned inside Endeavour at 7:34 am EST, after six hours and 54 minutes.

By this point, on 17 January, the bulk of the STS-72 mission tasks had been triumphantly completed: the OAST-Flyer deployment and retrieval operations had gone flawlessly, as had the two spacewalks, and a battery of experiments were running smoothly. In the payload bay, the Shuttle Solar Backscatter Ultraviolet (SSBUV) instrument flew for the final time, as part of ongoing efforts to calibrate the ozone-monitoring sensors aboard the National Oceanic and Atmospheric Administration's NOAA-9 and NOAA-11 satellites, together with NASA's Nimbus-

7. Calibration 'drift' of solar backscatter ultraviolet data from these three satellites had proven a cause for concern, in terms of the reliability of their data over time, and SSBUV was designed to be flown at least once per year. Before each of its missions, SSBUV was calibrated to a laboratory standard and recalibrated after its return to Earth. Housed inside a pair of GAS canisters, it flew annually from October 1989 until its final mission on STS-72. Its most alarming results described the marked depletion of stratospheric ozone since the mid-1980s – the much-publicized 'ozone hole' – due partly to solar effects and atmospheric dynamics, but chiefly to the release of man-made carbon fluorides. If the effects observed over a four-year period were extrapolated over half a century, it could lead to a 60 percent ozone depletion that would pose a serious threat for life on Earth. Affixed to a GBA in the payload bay, the Thermal Energy Storage experiment focused on the long-duration behaviour of thermal energy storage fluoride salts which underwent repeated cycles of melting and freezing in the microgravity environment. Such salts were under consideration for use in advanced space-based solar power systems and a clear understanding of their performance was acutely necessary. Within Endeavour's crew cabin, protein crystal growth research and National Institutes of Health (NIH) studies of the development of newborn rats were performed.

Only the most minor of problems troubled the Shuttle herself, with flight controllers keeping a close watch over a component in the flash evaporator system, part of Endeavour's coolant network. In response to the concern, as Chiao and Scott began EVA-2, Mission Control repressurised the cabin to help warm the orbiter and successfully dislodged ice from the flash evaporator. Of greater concern were colder than expected temperatures observed on a fuel line of the SFU satellite, now that it was in the payload bay, with fears centring on the possibility of a dangerous hydrazine leak if it were to freeze. However, its thermostats were considered to be operating correctly and maintaining hydrazine temperatures at acceptable levels, although procedures to warm the SFU's fuel lines were called up later in the flight and Duffy accordingly manoeuvred Endeavour into a warmer attitude. With all mission objectives accomplished, Duffy and Jett guided Endeavour smoothly to a touchdown on Runway 15 at KSC at 2:41:41 am EST on 20 January, completing a mission of almost nine full days in length. NASA's first Shuttle mission of 1996 had demonstrated many of the requirements which would be critical in the construction of the ISS ... and in very good time. By year's end, the first crews to occupy and build the multi-national orbiting outpost would have begun training.

"WEIRD SCIENCE"

In July 1992, Space Shuttle Atlantis headed into orbit with one of the most unusual experimental payloads on record – which STS-46 mission specialist Marsha Ivins described as "weird science" – as the Tethered Satellite System (TSS). Originally conceived by the late Professor Guiseppe Colombo of Padua University, it was intended to demonstrate the 'electrodynamics' of a conducting tether in an electrically charged region of Earth's atmosphere, known as the ionosphere. It was envisioned that Colombo's idea might ultimately lead to systems that would use tethers to generate electricity for spacecraft, employing our planet's magnetic field as a power source. Furthermore, by reversing the direction of the current in the tether, the force created by its interaction with the magnetic field could potentially put objects into motion, thus boosting a spacecraft's velocity without the need for precious station-keeping fuel and counteracting the effects of air drag. In the late 1980s and early 1990s, this was particularly appealing for the designers of today's ISS, as a means of compensating for the effects of atmospheric drag on the colossal outpost. Additionally, it was hoped that the concept could lead to the development of devices to trail scientific platforms far below orbital altitudes in difficult-to-study zones, such as the fragile ozone layer over the South Pole. Other applications for tethers included serving as extremely low-frequency antennas, capable of penetrating land and seawater, and perhaps generating artificial gravity or delivering payloads into higher orbits.

During the eight-day STS-46 mission, the TSS underwent its first demonstration in Earth orbit. Unfortunately, it achieved only partial success, when its 20.5 km tether twice became snagged on a 6 mm bolt in the deployment reel mechanism and refused to unroll more than about 258 m. Nevertheless, the mission proved the concept sufficiently for a reflight to be proposed. In March 1994, NASA Administrator Dan Goldin and Italian Space Agency (ASI) Special Administrator Professor Giampietro Puppi confirmed that the reflight would occur in February 1996. "NASA and ASI long have planned this reflight, but a formal commitment awaited US Congressional approval for NASA to spend Fiscal Year 1994 funds," it was reported. "NASA and ASI completed a study last year of the jointly developed TSS and confirmed their judgement of its usefulness as a unique Shuttle-based experiment carrier." A two-week-long flight was required to allow additional time for deployment and several days of ionospheric research.

Most of the original STS-46 crew was kept together for the reflight. Andrew Michael Allen, who had flown as pilot of the first mission, was assigned to command its second flight. Born on 4 August 1955 in Philadelphia, Pennsylvania, he graduated from Archbishop Wood Catholic High School and entered Villanova University to study mechanical engineering. Upon receipt of his degree in 1977, Allen entered the Marine Corps, trained as a pilot and spent several years flying F-4 Phantom fighters in Beaufort, South Carolina. Subsequently selected for fleet introduction of the F/A-18 Hornet, Allen was reassigned to Marine Corps Air Station El Toro in California from 1983-1986, serving as a squadron operations officer and completing both the Marine Weapons & Tactics Instructor Course and the Navy Fighter Weapons School ('Top Gun') at Miramar. In 1987, shortly before joining NASA, Allen was completing his studies at the Naval Test Pilot School at Patuxent River, Maryland. Following his selection into the astronaut corps the following June, he went on to fly two Shuttle missions as pilot, before receiving the assignment to command STS-75.

Allen was joined by payload commander Franklin Chang-Díaz and mission specialists Jeff Hoffman and Claude Nicollier, all three of whom had also flown aboard STS-46. Hoffman had served as payload commander for the first flight and would exchange roles with Chang-Díaz on the second mission. "It was unbelievably embarrassing when we discovered the cause of the tether jam," Hoffman told a NASA oral historian, "so there was a significant amount of redesign." Having said this, he believed there was a "strong scientific case" to refly TSS and felt that the inclusion of himself, Allen, Chang-Díaz and Nicollier, together with payload specialist Umberto Guidoni – who served in a backup role on STS-46 – was critical to mission success. "The astronaut office decided that the core of the payload crew ... should refly," he said, "which was great, because we were all good friends. There's always a certain stress when husbands get assigned to another flight, but since the families all knew one another, I think that made it a lot easier and more pleasant for everyone." Additionally, NASA carried over the STS-46 Lead Flight Director, Chuck Shaw, onto the second mission. The crew assignments came in three phases, with Chang-Díaz announced as payload commander in August 1994, Guidoni joining the crew in October and the other five members named in January 1995.

Franklin Chang-Díaz, the first Hispanic-American astronaut, has also jointly held a world record since 2002 as one of only two human beings to have flown as many as seven space missions. With its increased emphasis on the recruitment of minorities in the late 1970s, NASA recognised that whilst blacks and women were applying for the astronaut programme, Hispanic applicants were fewer in number. In September 1979, the agency specifically called for interested volunteers. "Many qualified Hispanics are hesitant to apply," admitted Jose Perez, deputy chief of NASA's Equal Opportunity Programs Office. "I would like to encourage those persons, and others, to call or write NASA for an application." One young man who answered this call, Franklin Ramón Chang-Díaz, originated from Costa Rica in Central America. He was born on 5 April 1950 in San José, the son of Ramón Chang-Morales and María Eugenia Díaz De Chang. Though his parents were both Costa Rican citizens, his father was of Chinese descent, whilst his mother was wholly Hispanic, and the family had also lived in Venezuela for a time. After studying at La Salle School he moved to the United States as a teenager in August 1968, with $50 in his pocket, hoping to become an astronaut. "I was captivated by Sputnik as a child," he told an interviewer. "I felt that, someday, humans would travel to distant planets and I decided that I wanted to be one of those travellers. I would be a space explorer."

It seemed an impossible dream for a youth who spoke no English. "My family never prevented me from doing that," he continued, "but they couldn't really help me. We were not a well-to-do family. Even though my parents put us in the best, most expensive schools and we got a first-rate education, my parents were not rich. Neither one of them finished college, so I was expected to make my own way as soon as I finished high school. I couldn't expect to receive a college education on my father's dime, but I was expected to *have* a college education." Encouraged by his parents, and his grandfather, to go to the United States, Chang-Díaz enrolled at the public high school in Hartford, Connecticut, and learned English through total immersion in the language. He failed his classes in the first two quarters, but his third and fourth quarters were outstanding. One of his teachers, Alan Winter, took notice of his efforts and began coaching the teenager and preparing him for university. Chang-Díaz succeeded in securing a scholarship for the University of Connecticut as an engineering student, but his lack of US citizenship presented an obstacle. "Well, *that* was a bucket of cold water," he remembered. "I went back to the high school and related this story to the teachers, who apparently wrote a *petition*. The Connecticut legislature met and decided to offer me one year of the scholarship and let me pay the lower, in-state tuition, because they had already offered the scholarship." Chang-Díaz was obliged to take loans and work in the university's physics department to support himself through the remaining three years. To him, the story of those years was all about America – "the ability to get ahead by hard work" – and he got his bachelor's degree in mechanical engineering in 1973.

By this stage in his life, he was gravitating towards energy and nuclear fusion, figuring that *this* power source would be critical for getting future astronauts to Mars. It offered a small insurance policy in the event that he did not succeed in his aspiration to become an astronaut. Chang-Díaz entered graduate school at MIT and in 1977 earned a PhD in applied plasma physics, with a research focus on the problems of controlled thermonuclear fusion. After the completion of his doctorate, he joined the technical staff of the famed Charles Stark Draper Laboratory, working on the design and integration of control systems for fusion reactor concepts. Two years later, NASA announced its intention to select Hispanic candidates for the astronaut programme and he tendered his application. When he received the call, in May 1980, that he had been selected, Chang-Díaz "went running out the door and across the street. I almost got run over by a *cab*! I was in a *totally* different world. My

1-13: On STS-75, astronauts Jeff Hoffman (left) and Franklin Chang-Díaz became the first and second astronauts to exceed 1,000 hours apiece aboard the Shuttle.

life changed completely from that day on." At first, being a scientist, and not a test pilot, posed another hurdle. It seemed to Chang-Díaz that his lack of flying credentials made him a less attractive candidate to draw a mission assignment. "That didn't seem right to me," he said, "and I kept working to remain both a scientist *and* an astronaut. In the end, I won out." By the end of his career with NASA in 2005, he had flown seven missions, creating a joint record with fellow astronaut Jerry Ross. Their record stands to this day.

On STS-75, Chang-Díaz and crewmate Jeffrey Alan Hoffman were both embarking on their fifth Shuttle flights. Almost anyone who met Hoffman when he joined NASA in January 1978 would have described him as the stereotypical professor: bearded, riding a collapsible bicycle and carrying a lunch pail in hand were three of the attributes noted by his contemporary Mike Mullane in *Riding Rockets*. "To the very end," Mullane wrote, "Jeff remained an unpolluted scientist." Yet his life had encompassed far more than academia; he was a skydiver, an accomplished mountaineer and a skilled engineer. Hoffman came from Brooklyn, New York, where he was born into a Jewish family of physicians on 2 November 1944. "My parents took me all over the place to museums and concerts," he recalled in his NASA oral history, "and among the other places was the Hayden Planetarium." This quickly hooked the young boy on astronomy. He received his schooling in Scarsdale and entered Amherst College in Massachusetts to study astronomy, graduating *summa cum laude* – with highest honours – in 1966. He went to Harvard for his doctorate, which he received in 1971. His research focused on high-energy astrophysics, specifically cosmic gamma rays and X-rays, and he participated in the design, construction, testing and flight operations of a balloon-borne, low-energy gamma ray telescope. Years later, he believed that this probably attracted him to NASA and vice versa. "I did a lot of work with my hands," he said, "building electronics, machining stuff. That probably stood me in good stead with NASA, because when they're selecting astronauts, they want people who know how to work in a lab, who can fix things and build things." He then moved to England for three years to undertake post-doctoral work at the University of Leicester, serving as project scientist for the medium-energy X-ray experiment on the European Exosat mission. Whilst in England, he met his wife, Barbara, and became a father. The Hoffmans returned to the United States in 1975 so that he could take up a position at MIT as the project scientist in charge of the hard X-ray and gamma ray experiment aboard the first High Energy Astronomy Observatory. "That was probably the most interesting scientific time that I've ever spent, because ... we discovered a new phenomenon called X-ray bursts," he recalled. "I guess we wrote about thirty-five to forty papers about it. These are thermonuclear explosions on the surface of a neutron star; pretty wild stuff."

By his own admission, Hoffman had been drawn to astronomy and astrophysics through his fascination for space exploration, although the opportunity to do such things himself seemed out of reach. "The early astronauts were all military test pilots," he pointed out. "I was never particularly interested in that career. In fact, I wasn't particularly interested in airplanes, because they didn't go high enough or fast enough. I always liked rocket ships." The chance finally came in October 1977, when he was invited to Houston for a week of interviews and testing; at first, his wife thought he was joking when he told her about the astronaut application. It came as a shock to Hoffman that it would spell the end of his research career in astrophysics. "NASA made it very clear that they were *not* looking for people to come and be research astronomers," he explained. "They were looking for astronauts who had to be generalists, because there were a lot of different things we were going to have to learn how to do." It was disappointing, in a sense, but it marked a change in Hoffman's career.

"The most unusual thing about my application," he continued, was that "I very well could have been the only person who was selected as an astronaut, who admitted in their application to having been convicted of a *crime*." It had happened during his tenure in Leicester, when he and some friends took a converted coastal steamer across the North Sea to explore the Norwegian fjords. Unfortunately, the original captain of the trip cancelled at the last minute and Hoffman – despite lacking the proper certification – stepped in. Upon their return, the coast guard arrested them and charged them a £10 fine, which Hoffman's friend disputed. The case was upheld because not only did Hoffman lack the required certification, to captain a British flagged vessel it was necessary to possess British citizenship. At the time, Hoffman did not have this and the party were convicted. "We actually had to go to Crown Court," he said, "with the wigs and the whole deal." They were fined £250 and when Hoffman came to fill in his NASA application form, he hesitated before deciding to be honest and admit to his offence. Surely, NASA would not delve *that* deeply into his past. They did. Nor could the selection board prevent themselves from making light of the situation. At his interview in Houston, the first greeting from astronaut Joe Kerwin, on the selection panel, as Hoffman walked into the room was: "Here comes the criminal!" This *criminal* look was surely made complete by a beard, which Hoffman quickly needed to remove. "As soon as I got to the altitude chamber in preparation for T-38 flying, it became pretty clear that you can't make a good face seal with a full beard, so off came the beard. My wife *shrieked* when I walked through the door!"

Making his third space voyage on STS-75, Swiss-born Claude Nicollier had become the first non-US citizen to fly aboard the Shuttle as a fully-fledged mission specialist when he participated in the first flight of TSS in 1992. He also served aboard STS-61, which serviced the Hubble Space Telescope, a year later. "My first dream as a child was to become a pilot," he told a NASA interviewer years later. "My second dream was to become an astronomer." In his professional life, he did both. Born in Vevey, Switzerland, on 2 September 1944, the son of a civil engineer, Nicollier graduated from high school at the Gymnase de Lausanne in 1962. He studied physics at the University of Lausanne and earned his degree in 1970. Several years' worth of postgraduate research at the university's Institute of Astronomy and the Geneva Observatory followed and Nicollier earned a master's credential in astrophysics in 1975. By this time, he had also been a Swiss Air Force pilot for almost a decade and he joined the Swiss Air Transport School in Zurich in 1974 to fly the DC-9 commercial airliner. A research fellowship in airborne infrared astronomy at ESA's Space Science Department at Noordwijk in The Netherlands began at the end of 1976 and in July of the following year Nicollier was selected – alongside West German physicist Ulf Merbold and Dutch physicist Wubbo Ockels – as one of the first European astronauts. Initially assigned to prepare as a payload specialist for the first Spacelab mission, in May 1980 Nicollier began training as a mission specialist candidate.

Seated alongside Andy Allen on Columbia's flight deck for STS-75 was Scott Jay Horowitz, nicknamed 'Doc', the first Shuttle pilot to hold a PhD. Born on 24 March 1957 in Philadelphia, Pennsylvania, he attended high school in California and earned a degree in engineering from California State University at Northridge in 1978 and a master's credential in aerospace engineering from Georgia Institute of Technology the following year. Horowitz remained at Georgia Tech to gain his PhD in aerospace engineering in 1982 and received the Outstanding Doctoral Research Award. He worked initially as an associate scientist for the Lockheed-Georgia Company in Marietta, Georgia, focusing on aerospace technologies to validate advanced scientific concepts. Horowitz entered the Air Force and completed initial flight instruction at Williams Air Force Base in Arizona in 1983. He then spent three years performing research and development for Williams' Human Resources Laboratory and flew the T-38 Talon, earning recognition as an Outstanding T-38 Instructor Pilot in 1985 and as a Master T-38 Instructor Pilot in 1986. Horowitz next served as an operational F-15 Eagle fighter pilot in Bitburg, Germany, then entered test pilot school at Edwards Air Force Base in California in 1990, graduating as the Class 90A Distinguished Graduate. During this period, he also served as an adjunct professor at Embry Riddle University and California State University at Fresno, running courses in aircraft design and propulsion, rocket propulsion and mechanical engineering. After qualifying as a test pilot, Horowitz flew for the 6512th Test Squadron at Edwards Air Force Base and was selected by NASA in March 1992.

With Claude Nicollier hailing from Switzerland, and two other crewmen – mission specialist Maurizio Cheli and payload specialist Umberto Guidoni – from Italy, STS-75 became the first Shuttle flight to feature three ESA crew members. Unusually, the seven-man crew was split into three separate shifts for the TSS deployment operation, reverting to a Spacelab-style dual-shift system after the conclusion of activities with the tethered satellite. The 'red' team consisted of Horowitz, Cheli and Guidoni, the 'blue' team of Chang-Díaz and Nicollier and a unique 'white' team of Allen and Hoffman; the latter, staggered shift was added to enable the astronauts to operate a suite of instruments on the TSS to gather real-world data about how conducting tethers might be used to generate electrical power in space. After the deployment operations, it was planned for Allen to rejoin the blue team and Hoffman the reds.

The satellite which formed the primary focus of these three shifts was a 1.6 m sphere, weighing 517 kg, with an outer shell of aluminium alloy and coated with an electrically conducting layer of white paint. It was, however, far more than just an oversized metallic football. Piercing its shell were windows for Sun, Earth and charged particle sensors, a connector for the umbilical tether and doors providing access to its on-board batteries. Extending from one side of the TSS was a long, fixed instrument boom, whilst a shorter antenna sprouted from its other side. To assist with the thermal control of the satellite, the interior of the spherical shell was painted black. If one were to break open the TSS, like an egg, the interior comprised two compartments: a payload module, housing its scientific instruments, and a service module for its subsystems. Additionally, in the centre of the spherical shell was a pressurised nitrogen tank, which provided propellant for the satellite's 12 cold gas manoeuvring thrusters. If the satellite could be termed an engineering marvel, the 2 mm conducting tether which connected it to a support mast in Columbia's payload bay was no less. Surrounding its Nomex core was electrically conducting copper wire, insulated with Teflon and coated with ultra-strong braided Kevlar-29. Jacketing the latter was an outer coat of braided Nomex to protect it from abrasion and the corrosive effects of atomic oxygen in Earth's rarefied upper atmosphere. During deployment, the tether was unreeled from a 2,027 kg mechanism, affixed to a Spacelab pallet and the Mission-Peculiar Equipment Support Structure (MPESS) in the payload bay.

Essentially, the mechanism took the form of a four-sided erectable tower, resembling a small broadcasting pylon, which unfolded slowly out of its storage canister using a series of rollers. As the canister rotated, fibreglass batons popped out of their stowed positions to form cross members – 'longerons' – which supported the tower's vertical segments. The tower was extended to a height of 11.8 m above the payload bay, so that when the satellite was released there existed no risk of it hitting any part of the surrounding structure. ("The complexity of the experiment is extreme," Andy Allen noted before launch.) That, however, was the easy part. Deploying the tower and even the TSS itself had already been done by the STS-46 crew; what Allen and his men planned to do on their 14-day mission in February 1996 was finish the task by getting the satellite and tether to their full length in order to demonstrate Guiseppe Colombo's concept. In fact, no fewer than 12 experiments were planned during the reflight – designated TSS-1R – of which six had been provided by NASA, five by ASI and one by the US Air Force. Several of these experiments were mounted on the MPESS. Two investigated the dynamics of the tether during its deployment phase, another provided theoretical support in the area of electrodynamics, a couple of others employed ground-based equipment to measure electromagnetic emissions from the satellite and seven others stimulated or monitored the entire assembly as it reeled out of the payload bay. Nearly 22 km of cable occupied the deployment mechanism, although only 20.5 km would actually be unravelled.

"Arrivederci, au revoir, auf wiedersehen and adios," Allen radioed cheerily from Columbia's flight deck on 22 February 1996, as he and his crewmates lowered their visors and prepared for launch. "We'll see you in a couple of weeks." Without further ado, and after a perfect countdown, they thundered aloft precisely on time at 3:18 pm EST. Unfortunately, the first portion of their ascent did not prove to be quite as perfect. Four seconds after liftoff, Allen and Horowitz spotted a potentially serious problem on their instrument panel. One of the Shuttle's three main engines, it seemed, was running at just 40 percent thrust; far lower than the 104 percent it should have been producing in the seconds after launch. The pilots checked with Mission Control, who verified that their telemetry indicated that all three engines

1-14: The Tethered Satellite inches away from its mast at the start of deployment operations on STS-75.

were performing normally at full power, and Columbia continued safely into orbit. Of course, if the (ultimately erroneous) instrument readout had been for real, it would have required Allen and Horowitz to perform an emergency landing back at KSC. "We had a couple of moments there that we got a little adrenaline rush," Allen said later. "I said [to Horowitz]: 'This looks like a bad simulation run'."

Safely in orbit, the crew divided into their respective shifts and blue team members Chang-Díaz and Nicollier set to work activating the TSS-1R support equipment in readiness for a planned deployment of the satellite early on 24 February. Before turning in at the end of his first shift, Hoffman also tested the reel motor and latching mechanism which secured the spherical satellite onto its docking ring atop the still-folded deployment tower. By the 23rd, with less than 24 hours to go before deployment, the mission had encountered its first spate of problems. A computer relay, responsible for sending commands to the satellite, experienced an electrical overload. Known as a Smart Flexible Multiplexer-Demultiplexer, or 'Smartflex', the relay had to be switched to a backup component. Although the backup performed satisfactorily, Mission Control opted to spend several hours evaluating it before giving the go-ahead to begin TSS-1R deployment activities. Next, a laptop computer on the Shuttle's aft flight deck encountered difficulties and was exchanged for a spare, which itself behaved sluggishly. Nevertheless, in the early hours of 24 February, the methodical procedure to activate the experiments associated with the satellite got underway. Firstly, the experiments were switched on individually, then together, in an effort to isolate the computer problems. By 3:00 am EST, all experiments were up and running and data was successfully transmitted through the Smartflex to the laptop and thence to Mission Control.

It was at around this point that managers decided to postpone the deployment until 25 February to gain additional confidence and testing time with the Smartflex. Although it had remained stable thus far, it was desirable to allow engineers to better understand its behaviour after several unexpected crashes and restarts. The delay also gave them ample opportunity to develop work-around procedures, in case the Smartflex encountered difficulties during the deployment procedure. According to Mission Scientist Nobie Stone, it was like learning to walk before running. One of these confidence-building tests involving 'mapping' Earth's charged particle environment, which varied dramatically as Columbia orbited the globe in periods of sunlight and darkness every 45 minutes. During this time, Nicollier activated the Tether Optical Phenomena (TOP) experiment, developed by Lockheed Martin's Palo Alto Research Laboratory in California. This employed a hand-held, low-light-level television camera to provide visual data in support of tether dynamics and optical effects created by the satellite. Through the overhead flight deck windows, Nicollier acquired stunning views of atmospheric airglow over the South Pole. Elsewhere, the Theory and Modelling in Support of Tether (TMST) investigation provided theoretical electrodynamic assistance to TSS-1R. Meanwhile, as preparations got underway, other experiments underwent tests.

One of these was an electron gun, called the Shuttle Electrodynamic Tether System (SETS), which generated a beam in support of the science experiments. It provided voltage and current readings from the tether throughout the deployment process.

Multiple test beams were fired from the electron gun on 24 February to acquire data on the Shuttle's ionospheric environs. A day later than originally planned, at 3:45 pm EST, the deployment procedure got underway and proceeded without a hitch under the watchful eyes of Hoffman and Guidoni. The satellite pushed itself away from its docking ring on top of the mast, using its cold gas thrusters, and was expected to reach a maximum distance of 20.5 km over a 5.5-hour period. "The satellite is rock-solid," Hoffman reported excitedly. It was expected that, after reaching its maximum extent, the satellite and tether would remain extended for 22 hours of studies. A slow 'creep' back towards Columbia, precisely choreographed from Mission Control, would have produced a docking back onto the mast at 1:43 pm EST on 26 February.

Orbital dynamics resulted in TSS-1R initially deploying 'upwards' at a rate of just 15 cm per minute and at an angle 40 degrees 'behind' the Shuttle's path. Its motion was carefully controlled by electric motors, which reeled out the tether, and by the satellite's thrusters. An hour into the deployment, it eclipsed the 258 m limit of its predecessor. Gradually, the deployment rate increased to 1.6 km/h, then slowed briefly in order that the 40-degree angle could be reduced to just five degrees. This placed the satellite almost directly 'above' the crew cabin. Throughout the process, Columbia's RCS thrusters were disabled by Allen and Horowitz to avoid causing unwanted oscillations in the tether. Beginning at a distance of 610 m, the satellite undertook a series of very slow rotations in support of the science investigations. The deployment of the tether increased to a peak speed of 8.1 km/h around 8:00 pm EST, by which time the satellite had reached a distance of 15 km.

Shortly afterwards, things started to go wrong.

The intention was that, as the tether neared its 20.5 km maximum length, the deployment rate would be gradually reduced. "It was within 1 km of its final length," Hoffman told the NASA oral historian, "at which point we were going to put on the brakes and just let it sit there and start all the experiments. I was recording this huge arc in the tether through the camera, when I started to see little ripples in the tether." To Hoffman's eyes, it reminded him of the tether jam on STS-46 and a horrible sense of déjà vu dawned on him. However, at 8:29:35 pm EST on 26 February, at a tantalisingly close-to-target distance of 19.6 km, it became clear that the tension was due to something else: the tether had not jammed ... but *snapped*. The shocked astronauts recorded video footage of the incident and the breakage appeared to have occurred near the top of the mast. "The tether has broken at the boom!" Hoffman radioed urgently. "It is going away from us." In fact, the tether and the satellite were accelerating away from the Shuttle at a rate of about 670 km during each 90-minute orbit. By the morning of the 27th, it was trailing Columbia by 4,830 km, flying some 50 km 'above' the Shuttle. After winding the remaining 10 m of tether back into the mechanism by 1:58 pm EST, the astronauts retracted the mast to its original configuration. The breakage close to the top of the mast, rather than further outwards, close to the satellite, proved fortuitous. "If it breaks at the bottom, it will fly away from you and you're not in any danger," said Hoffman, "but if it breaks at the top you've got 20 km of tether coming snapping back at you. We had practiced for that eventuality in the simulator. You've got to then cut the tether at the bottom and fly away from it."

Although nearly 24 hours of electrodynamic measurements had been lost, the $154 million reflight was far from a total failure. Already, when the satellite was less than 6 km from the mast, the Deployable Core Experiment (DCORE) recorded its first data. This experiment was mounted in the payload bay on the MPESS and its task was to control the flow of electric current in the tether using a pair of electron guns. Before the break, its first performance run had successfully generated a current of 480 milliamps from the electrical charge that had collected on the satellite's surface. This was about 200 times greater than the levels obtained during STS-46. Other experiments in the payload bay continued to function in support of the satellite and tether until as late as 6 March. "We did get a lot of good data during the deploy," Hoffman told journalists during a space-to-ground news conference. Currents measured during the deployment were at least three times higher than predicted in analytical models. Voltages as high as 3,500 volts were developed across the tether, achieving current levels of 480 milliamps. It was also possible for researchers to study the interaction of gas from the satellite's thrusters with the ionosphere. A first-ever direct observation of an ionised shockwave around the satellite – impossible to study or model in laboratories on Earth – was also accomplished. Moreover, as the satellite and its trailing 19 km of tether sped through the ionosphere, it was possible to continue investigations in spite of the fact that it was no longer physically linked to the Shuttle. It did not detract from the disappointment, however. "If you don't ever get your nose bloodied," said STS-75 Lead Flight Director Chuck Shaw, paraphrasing Theodore Roosevelt's famous comment, "you're not in the game." Capcom Dave Wolf told the astronauts that, whilst it was too early to speculate on the cause of the tether breakage, an investigative board was already getting established to explore the anomaly. The board was headed by Kenneth Szalai, the director of NASA's Dryden Flight Research Center at Edwards Air Force Base in California.

On 27 February, as the satellite and tether flew above the Electronic Signal Test Lab at JSC in Houston, Texas, ground controllers transmitted commands to successfully reactivate three of its on-board experiments: the Research on Orbital Plasma Electrodynamics (ROPE), the Magnetic Field Experiment for TSS Missions (TEMAG) and the Research on Electrodynamic Tether Effects (RETE). With a possible two additional days of data-gathering capability now re-established before the satellite's batteries were predicted to expire on 1 March, teams scrambled to assemble a last-minute research timetable to squeeze as much as possible from their dying payload. The ROPE experiment sought to examine the behaviour of charged particles in the ionosphere, as well as those surrounding the satellite and tether, under a variety of conditions. Meanwhile, RETE – provided by the Italian National Research Council – measured the electrical potential in the plasma 'sheath' around TSS-1R and identified waves excited by the tether. Elsewhere, the Second University of Rome's TEMAG experiment mapped fluctuations in magnetic fields around the satellite. It was hoped that, even though Columbia and TSS-1R were now physically separated by thousands of kilometres, firing electron beams from guns mounted in the payload bay would still disturb the ionosphere and be detectable by the satellite's instruments. On 28 February, scientists were able to observe a sunlight-induced electrical charge on the satellite's surface as it moved through the daytime and nighttime portions of its orbit. They also succeeded in reactivating and acquiring valuable data from two other satellite-mounted experiments. It was even possible, according to Jeff Hoffman, for ground-based observers to see TSS-1R from the southern United States.

Since the tether breakage, orbital dynamicists had predicted that Columbia would approach to within retrieval distance of TSS-1R on 29 February, and such a scenario was briefly considered, but ultimately discarded due to insufficient propellant margins aboard the Shuttle. Had it been approved, a retrieval would have consumed up to six days of crew time. In anticipation of its rendezvous, the satellite's batteries were placed in a low-power mode from the late afternoon of 28 February until the morning of the 29th to keep it alive for long enough. Right on cue, at 12:15 pm EST on 1 March, Andy Allen spotted TSS-1R and its tether, from a distance of just 75 km. "All we can really see are pinpoints of light, real close together," he radioed to Mission Control. By this time, however, the satellite's batteries were rapidly failing. Very weak signals had been detected through the Merritt Island and Bermuda tracking stations earlier that same day and no further data was received after 1 March. Still, it endured for far longer than expected, prompting one manager to liken it to the Energiser bunny, for its capacity to keep going.

Clearly, a huge amount of data was gathered and Umberto Guidoni noted that the mission had "demonstrated that tether dynamic applications work and we can generate electricity". Guidoni, who had served as a backup payload specialist on the STS-46 mission, was making his first space mission on STS-75. He was assigned to the crew in October 1994, a month after Franklin Chang-Díaz had been named as payload commander. Today a member of the European Parliament for Italy, Guidoni was born in Rome on 18 August 1954 and grew up with a fascination for the sciences. "When I was a teenager, my uncle used to give me science books," he told a NASA interviewer, "and he also gave me the first telescope." It was whilst using the telescope that the young Guidoni decided to pursue a scientific career. "I still remember the first time I pointed the telescope at the sky," he said, "and I saw Saturn with the rings. It was a beautiful image. That really made up my mind to become a scientist and that was the first step in order to become an astronaut." Guidoni earned a bachelor's degree in physics and a PhD in astrophysics from the University of Rome. He received a post-doctoral fellowship from the Italian Nuclear Energy National Committee for 1979-1980, allowing him to pursue work in the field of thermonuclear fusion. Guidoni served as a staff scientist in the solar energy division of the Italian National Council for Renewable Energy and in 1984 became a senior researcher at the Space Physics Institute of the National Research Council and was involved in the RETE payload as a co-investigator and experiment project scientist. Guidoni was selected, alongside fellow Italian physicists Franco Malerba and Cristiano Batalli-Cosmovici, for STS-46 payload specialist training in February 1989.

Despite the measure of success gained on STS-75, the mood aboard Columbia remained sombre. "Every time I turn around and look through the window and I see this empty bay," said Maurizio Cheli, "it's like a part of myself has left." As the first

Italian to serve as a fully-fledged Shuttle mission specialist, Cheli came from Modena, where he was born on 4 May 1959. He earned a degree in aeronautical sciences from the Italian Air Force Academy in 1982 and entered active military service, eventually rising to the rank of lieutenant-colonel. He underwent pilot instruction at Vance Air Force Base in Oklahoma and fighter training in the F-104G Starfighter at Holloman Air Force Base in New Mexico, then returned to Italy to join the 28th Squadron, 3rd Recce Wing. Cheli graduated from the Italian Air Force War College in 1987 and qualified as a test pilot at the Empire Test Pilots' School in the United Kingdom in 1988, receiving the McKenna Trophy as the best student on his course. Whilst assigned to the Italian Air Force's Flight Test Centre in Pratica di Mare, Rome, Cheli served as a Tornado and Boeing 707 inflight-refuelling tanker project pilot. Selected as an ESA astronaut in May 1992, he reported to JSC for Shuttle mission specialist training and later married fellow selectee Marianne Merchez of Belgium. Cheli gained a master's degree in aerospace engineering from the University of Houston and in January 1995 was assigned to STS-75 as the flight engineer, seated behind and between Allen and Horowitz on Columbia's flight deck.

As STS-75's commander, and a veteran of the previous TSS mission, Andy Allen derived an additional blow from the problems endured by the tethered satellite. "Scientists on the ground have lost a lot and we feel for them," he said. "We were looking forward to demonstrating that we could actually retrieve a satellite from 20 km and we've put an amazing amount of work into it." Jeff Hoffman added that the tether loss felt "like getting hit in the stomach". Of course, as noted in scientific papers later presented at an American Geophysical Union conference, the main scientific breakthrough was the discovery of tether currents three times higher than theoretically predicted. It was speculated that this might indicate some degree of ionisation around TSS-1R, even when its cold gas thrusters were switched off. In fact, when the thrusters were activated, the current climbed even higher, to 580 milliamps. Overall, its current-collection and power-generation capabilities proved to be several times higher than predicted.

With the completion of 'direct' TSS-1R operations, late on 26 February, the astronauts returned to their dual-shift – red and blue – system of activities and focused on their other mission tasks. The most important of these was the third United States Microgravity Payload (USMP-3), which reflew several materials experiments previously carried aboard USMP-1 in October 1992 and USMP-2 in March 1994. Its payloads included two semiconductor processing facilities – the Advanced Automated Directional Solidification Furnace (AADSF) and the Materials for the Study of Interesting Phenomena of Solidification on Earth and in Orbit (which produced a French-language acronym which spelt 'MEPHISTO') – together with the Zeno xenon critical-point experiment and the Isothermal Dendritic Growth Experiment (IDGE), plus the Space Acceleration Measurement System (SAMS) and Orbital Acceleration Research Experiment (OARE). Unlike TSS-1R, the USMP-3 payload performed without problems and was operated autonomously by ground controllers via 'telescience'. Franklin Chang-Díaz activated its experiments late on 23 February and Maurizio Cheli kept watch over it during two weeks of research.

The USMP-3 experiments were mounted on a pair of MPESS structures, which straddled the payload bay, like a bridge. The 'front' MPESS provided electrical power, data, communications and thermal control services to the payload, whilst the experiments themselves were attached to the 'rear' one. Managed by MSFC, the USMP facility was effectively autonomous and required little crew interaction, other than switching it on and then off. Moreover, it also offered exciting opportunities for principal investigators on the ground to control their experiments remotely via 'telescience'. Although USMP had flown twice before, it had never been aboard a dual-shift Shuttle mission, and this presented a number of obstacles. "Because they were growing crystals, it required an extremely quiet Shuttle," recalled Jeff Hoffman. "When the scientists discovered that we were going to be a two-shift flight – so somebody would always be awake – they were pretty upset. Just because of the tight scheduling, it couldn't be moved to another flight." The crew promised that they would be quiet at critical periods, in order to minimise disturbances through the Shuttle's structures and their potential impact on the sensitive USMP-3 experiments. "Very quickly, we learned what activities were causing disturbances and we would stop those," Hoffman continued. "They told us after the flight that it was as quiet as they had ever seen it. They could see the vernier jets firing; *they* made more noise than *we* did! In order to accomplish this, they had to declare that these eight-hour periods, when the other shift was asleep, were the so-called 'quiescent periods'. They weren't allowed to give us any other experiments to do ... so for the best part of a

1-15: Columbia touches down at the Kennedy Space Center on 9 March 1996.

week, for eight hours a day, we just had to float and look out the windows. I felt as if I were a space tourist. It was really quite extraordinary!"

During the mission, AADSF grew a large crystal of infrared-detecting lead-tin telluride. The furnace employed 'directional solidification' to process its samples, moving them through three separate temperature 'zones' – from a 'hot' region at 870 degrees Celsius to a 'cool' region at 340 degrees Celsius – to slowly cool and solidify them. The result was a 'flatter' solidification front and crystals which could be studied with greater clarity. Meanwhile, the French MEPHISTO investigators monitored the solidification and remelting of three tin-bismuth samples, which are used on Earth for semiconductors and metal alloys. Their ability to command the experiments from the ground proved enormously successful. "Time is money," explained IDGE Project Scientist Ed Winsa. "In our limited time in orbit, typical pre-programmed experiments would have missed out on a tremendous amount of science return if we had not been able to adjust growth cycles."

IDGE grew strange, tree-like crystalline structures, known as 'dendrites'. Dendrites are known to evolve as materials solidify under certain conditions and by improving scientists' understanding of them, it was hoped to improve the industrial production of a wide range of materials, including steels and super-alloys, used in applications ranging from making tin foil to cars and jet engines. Each of these materials is formed under conditions which yield dendrites. During USMP-3, dendrites were grown and photographed. As each growth 'cycle' ended, the tiny tree-like structures were re-melted and another dendrite was produced at a different temperature. Cameras recorded the entire process and the resultant data revealed that dendrites grew much faster and to larger sizes in the absence of gravity than their terrestrial counterparts. Significantly, on 4 March, USMP-3 made history when IDGE received commands issued directly from scientists on a US college campus. The investigator concerned was Marty Glicksman of Rensselaer Polytechnic Institute in Troy, New York, and he delivered commands which formed part of a third science phase for the dendrite-growth team. Overall, Glicksman and his colleagues collected twice as much data as they needed, which helped to greatly enhance post-flight statistical analyses. It was hoped that data from IDGE might aid Earth-based processes such as metal casting, welding and the production of aluminium, steel and copper.

Elsewhere, Zeno sought to precisely identify the 'critical point' – the peculiar temperature-and-pressure point at which its characteristics rapidly drift between those of a liquid and a gas – of xenon, in order to better understand the phenomenon. This point is very difficult to attain in ground-based experiments, because the fluid becomes highly compressible or 'elastic'. In such cases, even if the critical point could be reached, its stability could not be maintained for long enough to achieve meaningful observations. In space, with a virtual absence of gravity, these complications were dramatically reduced. As it approached its critical point during USMP-3, the xenon turned milky-white and Zeno's team monitored more than 100 channels of data, updated every second, to plot statistical curves which measured the sample's temperature and pressure.

The USMP-3 operations ran so smoothly and generated such valuable data that on 4 March NASA decided to extend the STS-75 mission by 24 hours to almost 15 days. In the words of USMP-3 Mission Manager Sherwood Anderson, it represented a commitment by the space agency to the kind of research which would later be performed aboard the ISS. By 7 March, after two weeks of meticulously measuring the xenon sample, Zeno had reached its figurative peak and investigators hoped that its data would lead to applications such as liquid crystals and new semiconductors. At several points, the SAMS accelerometer supported the experiments in order to adjust the samples' growth rates in response to Shuttle manoeuvres. This knowledge could help to develop new furnaces which would automatically compensate for unavoidable acceleration changes induced by their carrier spacecraft. Very little of the USMP-3 research could have been done without the capacity for investigators to control their experiments from the ground. In total, more than 2,300 instructions were transmitted to Columbia during STS-75.

Not all of the mission's experiments were housed on the MPESS structure in the payload bay. A new facility was provided in the middeck. Known as the Middeck Glovebox, it was developed by MSFC and was essentially a scaled-down version of a glovebox which had for several years been used aboard the Spacelab module. Fitted with video and still cameras and foot-activated audio-recording equipment, it supported three combustion experiments. Activated on 22 February, its cameras and recorders were evaluated and on the 27th it began full operations. For each experiment, the crew removed a sample kit from the middeck stowage lockers and inserted it through the glovebox door, then tightly sealed it. Using gloves, they manipulated the samples, which focused on the effect of airflow on flame-spreading characteristics, explored the conditions leading to the ignition of ashless filter paper and examined the behaviour of soot in microgravity. Early results indicated that flames produced in space differed dramatically from those on Earth. In total, 65 samples, ranging from paper to Teflon, were burned and according to investigators provided more than 125 percent of the planned data. Sixteen flat and cylindrical paper specimens were burned, yielding significant variations in flame colour, size, growth rate, airflow speed and fuel temperature. New combustion phenomena were also observed, including 'tunnelling', in which flames followed a narrow path, rather than fanning out from the burn site. Two smoke detectors were also tested.

In terms of microgravity research, STS-75 had proven a superb success, hampered only by the tether breakage that lost almost a full day of electrodynamic measurements. Nevertheless, the reflight did demonstrate the concept of powering a spacecraft using a conducting tether system. Columbia's return to Earth, already extended until 8 March, was postponed by an additional 24 hours due to a forecast of low clouds and the chance of rain and gusty winds in Florida. Although weather conditions were acceptable at Edwards Air Force Base in California, NASA managers decided to hold out for an improvement on the East Coast. A cold front passed through KSC on 7 March and was expected to become stationary by the 9th, perhaps leading to an upper-level low-pressure system which could produce clouds and showers. Fortunately, the weather on the 9th proved acceptable and Allen and Horowitz guided Columbia to a smooth landing on Runway 33 at 8:58:21 am EST.

"All right!" yelled the entire crew, in unison, at the instant of touchdown.

"We copy your elation," came the reply from Mission Control.

With Allen and Nicollier both making their third Shuttle missions and Horowitz, Cheli and Guidoni on their first flights, it was Hoffman and Chang-Díaz – both of whom were on the fifth space voyages – who were the most experienced members of the STS-75 crew. In fact, both men accumulated a total of 1,000 hours of experience aboard the Shuttle during this mission; Hoffman was the first person to pass the milestone on 29 February 1996, with Chang-Díaz entering second place on 8 March, shortly before Columbia returned to Earth. "I was the first person to complete 1,000 hours on the Shuttle," Hoffman told the NASA oral historian, "which I hadn't really thought about." However, test pilots Allen and Horowitz reminded him that in their military community, becoming the first person to accrue 1,000 hours was highly commendable. "Now it turned out that Franklin wasn't going to get 1,000 hours if we had come back when we were supposed to land, but we had a weather delay, so we had one extra day in orbit," explained Hoffman. "Then he was the second person to get 1,000 hours, so we have a nice picture of the two of us floating together holding a big sign saying 2,000. That was quite nice."

Meanwhile, Kenneth Szalai's review panel charged with investigating the TSS-1R tether breakage had been established on 26 February. "Given the public investment in the tethered satellite, it is important that we find out what went wrong," explained Wil Trafton, NASA's associate administrator for the Office of Space Flight. "To do any less would be a disservice to the American and Italian people." By the time the board's 358-page report was published in June 1996, it blamed "arcing and burning of the tether, leading to a tensile failure after a significant portion of the tether had burned away". The arcing itself was caused by either penetration from a "foreign object" – though not orbital debris or micrometeoroids – or a tether defect breached its insulating material. (Certainly, it was stressed that "the degree of vulnerability of the tether insulation to damage was not fully appreciated.") This had apparently triggered a local electrical discharge from the copper wire in the tether to a nearby electrical ground. "The board found that the arcing burned away most of the tether material at that location," *Flight International* noted, "leading to separation of the tether from tensile or pulling force." In his concluding remarks, Szalai highlighted that the problem was "not indicative of any fundamental problem in using electrodynamic tethers", adding that "constructing a tether that was strong, lightweight and electrically conducting took the project into technical and engineering areas where they had never been before". To the STS-75 astronauts, a short circuit had been at the forefront of their minds from the outset. "I was able to hook up a very powerful train of optics [and] telephoto lenses and take a close look at the broken end of the tether," said Jeff Hoffman. "I could see that it was brown and charred, so we knew before we ever came home that it almost certainly had been a short circuit that had melted the tether."

RECORD RENDEZVOUS

Early on 20 May 1996, astronaut Mario Runco Jr grappled SPARTAN-207 – a small, free-flying spacecraft, equipped with a unique payload – and unberthed it from Endeavour's payload bay with the Canadian-built RMS mechanical arm. With his five crewmates, he had launched into orbit barely 24 hours earlier, to kick off the ten-day STS-77 mission. As was discussed earlier in this chapter, SPARTAN was a frequent flier aboard the Shuttle, having supported numerous astronomical and solar physics experiments, as well as space technology investigations. On STS-77, however, it undertook its most ambitious mission to date. Runco released the satellite, precisely on time at 7:29 am EDT, after which Commander John Casper manoeuvred the Shuttle to a distance of about 250 m. He held this position for about an hour, then conducted a partial fly-around, to a point directly 'above' the satellite, and began an 80-minute-long station-keeping exercise to observe a remarkable experiment; an experiment which carried potentially enormous benefits for a multitude of applications, from space radar to mobile communications, from astronomy to Earth observations and from environmental research to analyses of soil moisture and ocean salinity.

Two hours after Runco released SPARTAN-207, the Inflatable Antenna Experiment (IAE) got underway. Designed and built by L'Garde, Inc., a small aerospace company, based in Tustin, California, together with JPL, it sought to inflate a 14 m Mylar antenna dish, at the apex of three deployable struts, as part of an investigation into how large, expandable structures behaved and functioned in the microgravity environment. Since 1971, L'Garde had pioneered the construction of thin-skinned, multi-task balloons and its products included a decoy missile for the Department of Defense. It had long been recognised that the mass and stowed volume of inflatable space components was significantly less than an equivalent solid structure and that this carried the potential to reduce by 10 to 100 times the cost of future missions. In 1988, L'Garde began working on the IAE and, according to the company's founder and vice president, Alan Hirasuna, the $14 million cost of this inflatable antenna was a mere fraction of the $200 million to build a similar-sized antenna with more conventional materials. Moreover, its compactness and 60 kg weight meant that it could be carried aboard much smaller launch vehicles.

At 9:38 am EDT, as six pairs of astronaut eyes and a battery of still, video and motion-picture camera equipment aboard Endeavour looked on, SPARTAN-207 commanded the deployment of IAE's supporting tripod, each of whose neoprene-coated Kevlar limbs unfolded to a length of 28 m. At their apex, a canister of pressurised nitrogen gas inflated the antenna, in just five minutes, to its full 14 m diameter. With a silver reflective surface on its topside and a clear underside, the IAE was observed and photographed over the following 90 minutes by the STS-77 crew and illuminated by an array of lights aboard SPARTAN-207 to precisely measure its smoothness. The antenna was then jettisoned, steadily moving 'below' and 'ahead' of the satellite, as Casper executed an RCS thruster firing to manoeuvre 'above' and 'behind' it. They would maintain a distance of 90-110 km from SPARTAN-207 for the next two days.

It was a great success for the entire crew, and not least for Casper and Runco, who were making their second Shuttle mission together. John Howard Casper, an Air Force colonel, was making his fourth flight overall. He was born in Greenville, South Carolina, on 9 July 1943 and after high school entered the Air Force Academy

1-16: The Inflatable Antenna Experiment (IAE) is deployed from SPARTAN-207.

to study engineering science. He earned his degree in 1966 and completed a master's qualification in astronautics from Purdue University, early the following year. Following initial flight instruction, Casper received his wings at Reese Air Force Base in Texas, trained on the F-100 Super Sabre and was despatched to Vietnam, where he flew more than 200 combat missions. Upon his return to the United States, he continued to fly the F-100, as well as the F-4 Phantom, and was assigned as an exchange pilot to a tactical fighter wing at RAF Lakenheath in England. Casper then attended test pilot school at Edwards Air Force Base, graduated in 1974 and headed the F-4 Test Team, performing weapons separation and avionics testing. He later worked at the Pentagon as deputy chief of the Special Projects Office, developing Air Force positions on requirements, operational concepts, policy and force structure for tactical and strategic programmes. In February 1990, Casper flew his first mission as pilot of STS-36 and subsequently commanded STS-54 and STS-62.

Mario Runco Jr's quite remarkable career path had seen duty as a police state trooper, then a Navy oceanographer and meteorologist and ultimately an astronaut. Born in the Bronx, New York, on 26 January 1952, of Italian-American parentage, Runco's close resemblance to Mr Spock (Leonard Nimoy's character in *Star Trek*) proved the root of many jokes whilst at NASA. He studied Earth and planetary sciences at the City College of New York and, whilst there, played intercollegiate ice hockey. After graduation in 1974, he moved to Rutgers University for his master's degree in atmospheric physics, again played ice hockey, and spent two years as a research hydrologist, carrying out groundwater surveys for the US Geological Survey on Long Island. In 1977, Runco joined the New Jersey State Police and spent a year as a patrol trooper, then entered the Navy in June of the following year. His early military duties, not surprisingly, took full advantage of his academic and work training: he served as a research meteorologist at the Navy's Oceanographic and Atmospheric Research Lab in Monterey, California, then as a meteorological officer aboard the USS *Nassau* from 1981-83. During this latter assignment, Runco was designated a Naval Surface Warfare Officer. He then worked as a laboratory instructor at the Naval Postgraduate School and performed hydrographic and oceanographic surveys of the Java Sea and Indian Ocean aboard the naval survey vessel *Chauvenet*. Selected by NASA in June 1987, Runco's police background led to the nickname of 'Trooper'. He had flown twice aboard the Shuttle, aboard STS-44 and with Casper on STS-54.

Over the next two days, following the IAE deployment, Endeavour remained at a distance of 90-110 km from SPARTAN-207, which carried other technology experiments, including a new solid-state recorder and advanced integrated circuits in several of its electronics boxes. By the evening of 20 May, a few hours after deployment, the discarded IAE was being tracked at a distance of more than 185 km 'ahead' and 'below' the Shuttle and although its large size and relatively low weight made it difficult for trajectory specialists to predict exactly how long it would remain in orbit, they anticipated no longer than 17-24 hours. As circumstances transpired, IAE re-entered the upper atmosphere and burned up early on the 22nd, the same day that Casper and his men re-rendezvoused with SPARTAN-207 to begin retrieval activities.

Awakened from their slumbers that morning to the sound of Fifth Dimension's song 'Up, Up and Away', in honour of their completed experiment, the astronauts wasted no time in preparing their equipment for the rendezvous. In a similar manner to the approach procedures followed by the STS-69 and STS-72 crews, a Terminal Initiation burn of Endeavour's RCS thrusters kicked off the final approach and Casper guided his ship to just 10 m from SPARTAN-207, whereupon astronaut Marc Garneau – who in October 1984 had become the first Canadian in space – extended the Canadian-built RMS arm and grappled the satellite at 10:53 am EDT. As the satellite hung on the end of the arm, the crew performed a video and photographic survey before it was berthed in the payload bay.

All six astronauts were intimately involved in the SPARTAN-207 rendezvous, including pilot Curtis Lee Brown Jr and flight engineer Daniel Wheeler Bursch, both of whom were on their third Shuttle missions. Brown hailed from Elizabethtown, North Carolina, where he was born on 11 March 1956. "My dream as a small kid was to fly," he once said. "I fell in love with aircraft and flying movies and things about flight and built all the little airplanes that kids always did when I was growing up." After completing high school in his hometown, Brown entered the Air Force Academy to study electrical engineering and earned his degree in 1978. He was commissioned as a second lieutenant and underwent flight instruction at Laughlin Air Force Base in Texas, later flying A-10 Thunderbolt aircraft in South Carolina. Brown's work subsequently earned him a position as an instructor pilot on the A-10 at Davis-Monthan Air Force Base in Arizona. Graduation from the Air Force's Fighter Weapons School at Nellis Air Force Base in Nevada in 1983 was followed by duties as an A-10 weapons and tactics instructor. He was selected for Test Pilot School at Edwards Air Force Base in June 1985 and upon completion of this rigorous and demanding course in mid-1986 he worked as a test pilot for the A-10 and the F-16 Falcon until his selection as an astronaut candidate in June 1987. Almost a decade later, at the time of STS-77, Brown already had two space missions under his belt as a pilot and by the end of his NASA career in December 1999 he became the first person to have served six times as a Shuttle pilot or commander. It is a distinction that, even today, after the retirement of the orbiters, he shares with only one other American astronaut, Jim Wetherbee.

As Mission Specialist Two, Bursch supported Casper and Brown during ascent, re-entry and throughout STS-77's rendezvous activities. Earlier in his career, he had also earned the unenviable reputation of having endured *two* harrowing on-the-pad aborts for both his first and second space missions. Bursch was born on 25 July 1957 in Bristol, Pennsylvania, and studied physics at the Naval Academy. Upon graduation in 1979, he became a naval flight officer and trained as a bombardier/ navigator in the A-6E Intruder. Overseas deployments followed to the Mediterranean aboard the USS *John F. Kennedy* and to the North Atlantic and Indian Oceans aboard the USS *America* and upon his return to the United States he was selected for Naval Test Pilot School. Completion of the course in December 1984 was followed by A-6 test work and later duties as a flight instructor. He also served again overseas as Strike Operations Officer aboard the USS *Long Beach* and USS *Midway* and was in the process of completing his master's degree in engineering science at the Naval Postgraduate School when selected as an astronaut in January 1990. He flew aboard STS-51 in September 1993 and STS-68 in September 1994, both of whose inaugural launch attempts were stalled in the final seconds of the countdown, and *after* the ignition of the Shuttle's main engines. Fortunately, Bursch's run of cruel fortune appeared to end on his third mission and STS-77 ran like a charm.

With the successful retrieval of SPARTAN-207, the Shuttle's mission was barely a quarter complete, but its wide range of scientific and technological experiments were well underway. Flying for the fourth time as a research facility on STS-77, Spacehab was not dissimilar in appearance to the Spacelab module, but with several key differences: it was smaller, consuming just a quarter of the payload bay, and it was designed and built not by governments, but by private enterprise. It also differed from Spacelab in overall physical shape; it was cylindrical, but with a flat roof. When the module rose from Earth for the first time aboard STS-57 in June 1993, it marked the culmination and realisation of a decade-long dream for aerospace engineer Bob Citron, who founded the Spacehab company in 1983 and incorporated it the following year.

"People often ask me why I started Spacehab," Citron recalled on the website www.astrotech.com, "and my response usually goes something like this: It took a small group of wide-eyed dreamers and determined space enthusiasts who believed we could pull it off. We didn't have a clue about the enormous problems we would encounter and the nearly insurmountable technical, financial and institutional roadblocks that would stand in our way. Nobody had done anything like this before." The primary goal of Citron, who died from prostate cancer in January 2012,

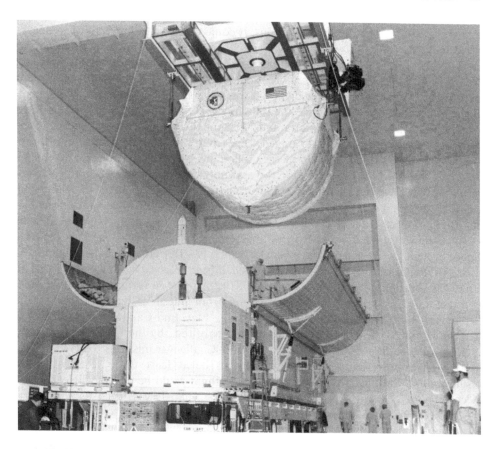

1-17: The Spacehab module is prepared for installation into its payload canister, ahead of delivery to Endeavour.

was to create the world's first privately funded company to support human space missions, using the cavernous payload bay of the Shuttle as a carrier of commercial pressurised research modules.

The need for such provision was self-evident. In the Shuttle's pre-Challenger heyday, many missions were planned each year and a primary thrust of the Reagan Administration's 1983 space policy was for the commercial exploitation of the microgravity environment. Already, middeck lockers were being used to carry out experiments in crystal growth and pharmaceutical development, but the limited volume meant that their commercial viability was restricted. The Spacehab module, accessed, like Spacelab, by a tunnel adaptor connected to the middeck airlock hatch, measured 2.8 m long, 3.4 m high and 4.1 m across the width of the payload bay and could increase the Shuttle's pressurised research volume by 31 m^3, effectively quadrupling the available working and storage space. The module, which weighed approximately 4,300 kg and could house a total payload of more than 1,300 kg, provided environmental provisions, electrical power, temperature control and experiment command and data functions. It could carry a maximum of 61 middeck-sized lockers and up to two large racks, in either 'single' or 'double' configurations.

After incorporating Spacehab in 1984, Citron admitted that the company was "on the verge of failure on a number of occasions during its first years", until he brought in critical professional management personnel and "things started to happen" through negotiations with NASA, the Italian Alenia Spazio and German MBB-Erno organisations and Martin Marietta and McDonnell Douglas. By the end of 1985, only weeks before the Challenger tragedy, Spacehab was tentatively scheduled for its first flight in 1987 and a lease of $5 million per mission was quoted by *Flight International*. Early plans called for the assembly of three modules, but centre-of-gravity issues gave NASA cause for concern and threatened to affect the placement of other cargoes in the payload bay. Ultimately, McDonnell Douglas was selected as the lead contractor and a decision was made to build two flight modules and an engineering test model. Other worries lingered, but in late 1986 Spacehab signed a Space System Development Agreement, in which NASA agreed to fly five inaugural missions. By October 1987, 42 firm requests had been received, with lockers priced at $300,000 for non-government users and $100,000 for government agencies and contractors. The concept was now growing from initial designs into reality. In September 1989, Patent No. 4,867,395 for a 'Flat End-Cap Module for Space Transportation System' was awarded to Spacehab, Inc., by the US Patent Office.

Hopes to fly the modules on very early post-Challenger missions received a rude awakening, however, and it was expected to be at least three years after the resumption of Shuttle operations before a first flight could be realistically expected. According to NASA's April 1988 schedule, Spacehab-1 was listed as a primary payload on STS-51 in June 1991, co-manifested with the retrieval of EURECA and the deployment of Italy's LAGEOS-II satellite. This schedule slipped and, at length, the first module was not unveiled until early 1992 at the custom-built Spacehab Payload Processing Facility (SPPF) at KSC. By this time, under the Commercial Middeck Augmentation Module procurement, initiated in February 1990 and formally signed the following December, NASA had agreed to lease a total of 200 Spacehab lockers at a cost of $184.2 million. Over the coming years, the SPPF would host more than a hundred astronauts and cosmonauts, enabling them to train on real experiment hardware.

However, all was not smooth sailing for Spacehab, which supported three dedicated science missions between June 1993 and February 1995 and was embarking on its fourth 'stand-alone' flight on STS-77. In the aftermath of the first flight, few contracts other than NASA's materialised for use of the laboratory, in spite of aggressive efforts to market its capabilities. This was due to a number of factors, chiefly high Spacehab locker prices in the region of $925,000 and anticipated commercial prospects failing to materialise. "No commercial companies are ready yet to make an independent commitment to research," said Rebecca Gray, Spacehab's manager of government and public relations, in a *Flight International* interview in late 1995, despite the company having delivered its services on time and on budget. NASA's contract was renegotiated to cover four missions, of which STS-77 – whose module, though full, was dominated almost exclusively by NASA-funded experiments – was to be the last. "If NASA requires more flights of this nature," *Flight International* continued, "it is likely only to be at a rate of about one flight a year and another contract will have to be negotiated." The company had already signed a $54 million lease contract with NASA to utilise its modules for logistics, rather than experiments, on several missions to Russia's Mir space station in 1996-1998. It also invested $15 million of its own capital to develop a 'double' logistics module to be used "as a laboratory and a cargo carrier", with the capacity to transport up to 2,720 kg of payloads to Mir, as well as soft-stowage canvas bags. Nevertheless, in the coming years, the option for more 'stand-alone' research missions would be renegotiated and two would be flown, including STS-107, which proved to be the final voyage of Columbia.

More than 90 percent of the Spacehab-4 payloads were directly sponsored by NASA's Office of Space Access and Technology, through its Commercial Space Centers and their industrial partners, as well as by several of the agency's field centres. Right from the outset, STS-77 was dedicated to opening the commercial frontier of space, with the Spacehab module hosting almost 1,400 kg of equipment in 28 lockers, four soft stowage bags and a pair of single racks to support around a dozen investigations in the fields of biotechnology, electronic materials, polymers and agriculture. These included the Advanced Separation Process for Organic Materials (ADSEP) to demonstrate separation and purification technologies for potential medical and pharmaceutical applications, the Commercial Generic Bioprocessing Apparatus (CGBA) and Plant Generic Bioprocessing Apparatus (PGBA) to investigate the influence of microgravity on molecular, cellular, tissue and small animal and plant systems, as well as studying compounds which might someday prove beneficial as chemotherapy and anti-malaria agents and the Fluids Generic Bioprocessing Apparatus (FGBA) to explore the management of liquids in space. The latter received corporate sponsorship from the Coca-Cola Company, which flew both Coke and Diet Coke aboard STS-77. The experiment, noted NASA, "will provide a testbed to determine if carbonated beverages can be produced from separately stored carbon dioxide, water and flavoured syrups and determine if the resulting fluids can be made available for consumption without bubble nucleation and resulting foam formation". If so, it was hoped that such experiments might lead to a better understanding of altered tastes in target populations, such as the elderly, and how future drinks could be tailored to increase hydration. International co-operation on STS-77 was underpinned by the presence of the Commercial Float Zone Furnace (CFZF), a joint effort between the respective space agencies of the United States (NASA), Canada (CSA) and Germany (DARA) to produce large, ultra-pure compound semiconductor and mixed-oxide crystals of gallium arsenide and gallium antimonide for electronic devices and infrared detectors. Another payload, the Space Experiment Facility (SEF), used a furnace to produce through the vapour-diffusion technique a series of crystals of mercurous chloride, which is considered a valuable electro-optical material of commercial interest, and to bond powdered metals through the liquid-phase sintering process, as part of efforts to design new composites for the machine tool industry.

With such a large number of payloads aboard, it was imperative for the STS-77 crew to begin activating as many experiments as possible on the first day of their planned ten-day flight. Launch was originally targeted for 16 May 1996, but was pushed back to the 19th, since the earlier date was not available to NASA on the Eastern Range schedule. Commander John Casper, pilot Curt Brown and mission specialists Andy Thomas, Dan Bursch, Mario Runco and Canada's Marc Garneau had been training for the mission for almost a year, having been assigned in June 1995, and took their seats aboard Endeavour for a rousing, early morning liftoff at 6:30 am EDT. Shortly after entering orbit and doffing their pressure suits, the six men set to work. In spite of a problem with a cooling device for one of the orbiter's hydraulic power units, Thomas and Garneau opened the hatch into the Spacehab module, floated inside and began powering up experiments. Later in the day, Thomas also checked out the Shuttle's RMS arm in preparation for the SPARTAN-207 deployment.

As the only first-time spacefarer on STS-77 crew, Andrew Sydney Withiel Thomas served as the payload commander, with overall responsibility for all of the mission's research objectives and science goals. Born in Adelaide, South Australia, on 18 December 1951, Thomas' father would later recount that his son became fascinated by space exploration as a child and built a steady stream of model rockets from cardboard and plastic. He received much of his early schooling in Adelaide, then earned a first-class degree in mechanical engineering from the University of Adelaide in 1973 and stayed on to pursue his PhD. "As a young kid, growing up in Australia, the likelihood of me being selected as an astronaut was pretty thin," Thomas told a NASA interviewer, "and I didn't really consider it a very realistic possibility." In his mind, the power of gaining a good education was the fundamental cornerstone, coupled with a conscious effort to steer his subsequent career goals towards the space programme. "I took steps in my career that would give me exposure to the right kinds of technical problems and technical experiences," he said, "which would make me a good candidate for an astronaut position." In 1977, shortly before receiving his doctorate, Thomas became a research scientist with the Lockheed Aeronautical Systems Company, with responsibility for investigations into the control of fluid-dynamic instabilities and aircraft drag. He was promoted to Principal Aerodynamic Scientist in 1980 and accepted the headship of the Advanced Flight Sciences Department in 1983, overseeing a group which explored problems in advanced aerodynamics and aircraft flight testing. With a keen eye on joining NASA as an astronaut, Thomas became a US citizen in December 1986, became the manager of Lockheed's Flight Sciences Division the following year and in 1989 moved to Pasadena, California, to join JPL, later rising to supervisor of the Microgravity Research Group. Three years later, in March 1992, he was selected by NASA as an astronaut candidate. His lifelong dream had been realised, and all due to the power of education. Thomas' attitude was that education opened a path to many of his life experiences. "In fact," he concluded, "it can open doors that you can't even imagine and that would remain forever closed."

Flying in space for the first time on STS-77 was a totally new path for Andy Thomas, even when placed alongside many of the challenges he had overcome on Earth. "And although the environment is totally alien and unnatural, your body starts to accept it as being natural and psychologically you accept it as being natural," he said. "You start to feel that being weightless is just the normal way to be; that everything *does* float and that *you* float. It's just the amazing adaptability of the human body. Of course, the down side to that is when you come back, you have to re-educate your body to working in gravity. You feel all your internal organs being pulled down, your arms being pulled down, your head being pulled down and you stand up and just feel this ponderous mass of your entire body. It gives you a perspective on gravity ... that people just generally never get unless you go through that. You realise just what a demanding force it actually is."

In Endeavour's middeck, the Immune-3 experiment tested the ability of insulin-like growth factor to prevent or reduce the detrimental effects of microgravity exposure on the immune and skeletal systems of rats, whilst three investigations sought to crystallise various protein crystals for investigations that addressed a range of diseases and the Gas Permeable Polymer Membrane (GPPM) pioneered the development of enhanced polymers for manufacturing rigid gas-permeable contact lenses. The National Institutes of Health provided a tissue culture incubator and its experiments focused on the influence of weightlessness on the muscle and bone cells of chicken embryos. Elsewhere, the metamorphosis of tobacco hornworm was examined, as part of efforts to understand the synthesis of muscle-forming proteins and the processes of fertilisation and embryonic development of small aquatic organisms, including starfish, mussels and sea urchins, were monitored on the first flight of the Canadian-built Aquatic Research Facility (ARF). In fact, the latter was one of the first experiments to be activated by Mario Runco, a few hours after liftoff.

With such substantial involvement from Canada, it seemed unsurprising that STS-77 featured Canadian astronaut Marc Garneau as one of its four mission specialists. In fact, Joseph Jean-Pierre Marc Garneau became Canada's first man in space when he served as a payload specialist aboard Shuttle mission STS-41G in October 1984. He came from Quebec City, where he was born on 23 February 1949. His father was an army officer "and we travelled quite a bit when I was growing up", he told a NASA interviewer, "and I thought that I would like to have a military career, although I was drawn more towards the Navy". Garneau would explain this decision in just three words: love of adventure. He was educated in Quebec and received his degree in engineering physics from the Royal Military College of Canada in Kingston in 1970. Garneau travelled to England in pursuit of his doctorate, studying electrical engineering at the Imperial College of Science and Technology in London. After gaining his PhD in 1973, he joined the Canadian Forces Maritime Command as a Navy engineer, serving aboard HMCS *Algonquin* and later as a forces fleet school instructor. During this period, Garneau, designed a simulator to train weapons officers to use the missile systems aboard Tribal-class destroyers. Whilst at staff college in 1982, he was promoted to the rank of commander in the Canadian Navy. He was transferred to Ottawa the following year to design naval communications and electronic warfare systems and equipment. In December 1983, he was selected as an astronaut candidate and in February was seconded from the Department of National Defence to commence full-time training. Following his first Shuttle mission, Garneau was promoted to captain in the Navy.

He then retired from active military service in 1989 to become deputy director of the Canadian Astronaut Programme (CAP). In mid-1992, alongside fellow Canadian Chris Hadfield, he was selected by NASA for mission specialist training.

If Endeavour's crew cabin and the Spacehab module both represented a hive of activity during STS-77, then the payload bay was similarly packed with experiments. The Brilliant Eyes Ten Kelvin Sorption Cryocooler Experiment (BETSCE) was carried as part of ongoing efforts to develop technologies to rapidly cool infrared and other sensors to near-absolute zero Kelvin (minus 273.15 degrees Celsius). The aim of BETSCE was to employ a highly reliable 'sorption cooler' that exhibited virtually no detrimental effects of vibration, to cool infrared sensors to just 10 degrees Kelvin (minus 263.1 degrees Celsius). "Sorption coolers work by using specialised metal alloy powders, called metal hydridges, that absorb the hydrogen refrigerant through means of a reversible chemical reaction," NASA explained in its pre-flight press kit. "In the sorption compressor, the metal powder is first heated to release and pressurise the hydrogen, and then cooled to room temperature to absorb hydrogen and reduce its pressure. By sequentially heating and cooling the powder, the hydrogen is circulated through the refrigeration cycle. Ten degrees Kelvin is achieved by expanding the pressurised hydrogen at the cold tip of the refrigerator. This expansion actually freezes the hydrogen to produce a solid ice cube. The heat load generated by the device being cooled then sublimates the ice. This closed-cycle operation is repeated over and over." Previous astronomy missions which conducted their studies in the infrared needed to carry large, heavy and expensive dewars of liquid helium or hydrogen to accomplish operating temperatures as low as 10 Kelvin, but the duration of their useful lives were restricted when the cryogens eventually boiled off and ran out. "The ability to achieve a lifetime of ten or more years, with no vibration," NASA added, "opens the door to a wide variety of future missions that could benefit from this novel technology."

Mounted atop a Hitchhiker structure in Endeavour's payload bay was GSFC's Technology Experiments for Advancing Missions in Space (TEAMS), which featured four investigations: the GPS Attitude and Navigation Experiment (GANE) to gauge the effectiveness of Global Positioning System technology, which was then in its infancy, the Vented Tank Resupply Experiment (VTRE) to evaluate improved methods for refuelling in space, the Liquid Metal Thermal Experiment (LMTE) to test potassium-filled liquid metal heat pipes in microgravity and the Passive Aerodynamically Stabilised Magnetically Damped Satellite-Satellite Test Unit (PAMS-STU). Developed by GSFC, PAMS-STU was a technology demonstrator for the principle of natural 'aerodynamic stabilisation', which it was hoped might increase the orbital lifetime of satellites by reducing or eliminating the need for large quantities of attitude-control propellants. "Aerodynamic stabilisation works the same way as a dart," NASA explained shortly before the STS-77 launch. "The front of the dart is weighted and once the dart is thrown, it will always right itself with the head facing forward. In the same manner, the PAMS-STU satellite will eventually be oriented with the heavy end facing forward in orbit. This principle can be used to partially control the attitude of small satellites."

On the fourth day of the mission, 22 May, the PAMS-STU operations got

1-18: The STS-77 crew pose on the runway, in front of Endeavour. From left to right are Curt Brown, Marc Garneau, Mario Runco, Andy Thomas, Dan Bursch and John Casper.

underway when Runco deployed the cylindrical, 60×90 cm satellite at 5:18 am EDT from a canister at the rear end of Endeavour's payload bay. As intended, it drifted away in a rotating, unstable attitude, in order to evaluate how quickly and effectively it could stabilise itself using natural aerodynamics. Casper and Brown manoeuvred the Shuttle to a distance of 14.6 km to begin the first of three scheduled rendezvous exercises. A little over 4.5 hours later, they drew closer to about 600 m to track PAMS-STU with the laser-based Attitude Measurement System (AMS) in the payload bay. However, it was noticed that the satellite had not yet stabilised itself and a strong 'lock' could not be obtained. With two further rendezvous sessions, each lasting around 6.5 hours, planned for 24 and 25 May, Endeavour withdrew to a maximum distance of about 103 km.

During their second rendezvous on 24 May, Casper and Brown reached a station-keeping point just 520 m from PAMS-STU and held their position for more than six hours, but a problem arose with the SEF in the Spacehab module and Thomas was called away to commence troubleshooting. It was clear from video imagery acquired by the crew that the satellite had begun to stabilise itself with natural aerodynamic forces, albeit somewhat more slowly than expected. The third and final rendezvous was postponed by 24 hours, until 26 May, in order for engineers to evaluate the

AMS, which provided high-accuracy data (to within one-tenth of a degree) on the behaviour and relative motions of PAMS-STU. Although it had proven its ability to track the small satellite, the laser system seemed to be locking onto an unknown target – perhaps a structure in the payload bay – and was subjected to intensive troubleshooting. During those 24 hours, Endeavour moved away from the satellite, reaching a maximum distance of about 185 km, before Casper executed an RCS thruster firing on the 26th to begin the third period of rendezvous. Closing on PAMS-STU at about 3.7 km per orbit, this time the operation ran by the book, with Casper and Brown guiding their ship to within 550 m of their quarry. As the orbiter's payload bay faced the satellite, GSFC flight controllers successfully commanded the AMS to calculate its attitude to within one-tenth of a degree. The pilots moved closer, to just 500 m, and held their position for seven hours and 45 minutes. This was about 70 minutes longer than planned, as controllers verified that the AMS laser was indeed impinging on PAMS-STU's reflectors. Throughout the third rendezvous, the satellite remained very stable and validated the aerodynamic stabilisation concept.

Departing PAMS-STU for the final time, the STS-77 crew entered the final days of their mission, heading for a landing at KSC on 29 May. It was expected that the small satellite would re-enter the upper atmosphere to destruction after a few weeks, although trajectory specialists noted that it might remain aloft until as late as January 1997. (As circumstances transpired, PAMS-STU ended its days on 26 October 1996.) With more than 21 cumulative hours of formation flying, Endeavour's crew had secured a new record for themselves by becoming the first mission to execute four discrete periods of rendezvous. STS-77 also marked the first Shuttle flight to be powered into orbit by a full set of three Block I main engines and the first to be fully controlled from the new Mission Control Center at JSC. Weather forecasts at KSC and Edwards Air Force Base in California were highly favourable for an on-time landing, with two opportunities available at each site on the 29th. The first opportunity to land in Florida was taken and Casper fired the OMS engines at 6:09 am EDT and guided his ship to a smooth touchdown on Runway 33, precisely an hour later, at 7:09 am. The ten-day mission was Endeavour's final flight before a major period of modification in Palmdale, California, to prepare her for ISS assembly. Her next mission, STS-88, then scheduled for launch in December 1997, would deliver the first elements into orbit for that multi-national outpost.

Already, earlier in 1996, NASA and the Russian Space Agency had announced that astronaut Bill Shepherd and cosmonauts Sergei Krikalev and Yuri Gidzenko would become the first three-person team to spend a period of several months aboard the ISS. Their five-month flight was scheduled to begin sometime in mid-1998. With the first Shuttle assembly crew also assigned to STS-88, the year 1996 was shaping up to be a significant one, as major steps were taken to make the eagerly awaited space station a reality. Yet the Shuttle provided a 'miniature' space station of its own, and on the next flight, STS-78, which launched several weeks after STS-77 landed, seven astronauts aboard Columbia would spend 17 days in orbit, with an international crew and support team, conducting the very kind of research that it was hoped would someday be done aboard the ISS, on a year-round basis.

A PRECURSOR TO SPACE STATION

A few days after Kenneth Szalai's investigative board had presented its report into the TSS-1R mishap, Columbia rocketed into orbit again on her 20th space mission. Perhaps more than any previous Shuttle flight, it was expected to virtually mirror the research which would be conducted by astronauts aboard the ISS. Designated the Life and Microgravity Spacelab (LMS), the flight – according to Mission Manager Marc Boudreaux of MSFC – had "the key ingredients to take us into the next era of space exploration". One of those ingredients was the relatively short period of time it took from the initial blueprints for the mission until its realisation: just 18 months. Historically, Spacelab missions were intricately complex affairs, involving hundreds of scientists and engineers from numerous institutions spread across the globe, and took up to four years to prepare from conception to launch. In the wake of the second International Microgravity Laboratory (IML-2) mission in July 1994, it was recognised that there would be a conspicuous lack of life and microgravity research flights until the space station became operational at the end of the decade.

With this in mind, in December 1994, NASA announced its intention to stage a one-off, 16-day mission in mid-1996. Partway through the following May, seven astronauts were assigned to the flight, which became known as STS-78: payload commander Susan Helms, mission specialists Rick Linnehan and Chuck Brady and a quartet of Frenchmen, Canadians, Spaniards and Italians to fill a pair of payload specialist slots. Of those four, Canadian physician Bob Thirsk and French physicist and metallurgist Jean-Jacques Favier were named to serve as prime payload specialists, with Italian Air Force physician Luca Urbani and Spanish aeronautical engineer Pedro Duque as their backups. Although non-flying on the LMS mission, Urbani and Duque would help to co-ordinate the experiments from MSFC.

In addition to her role as payload commander, Susan Jane Helms also served as the flight engineer for ascent and re-entry during STS-78. After her NASA career, Helms returned to active duty in the US Air Force, rising to become a lieutenant-general, and today holds the record for achieving the highest rank of any female military astronaut. Until her retirement from the Air Force in early 2014, she served as commander of the 14th Air Force (Air Forces Strategic) and the Joint Functional Component Command for Space at Vandenberg Air Force Base in California. On her first Shuttle mission, STS-54 in January 1993, she also became America's first military woman in space. Born on 26 February 1958 in Charlotte, North Carolina, Helms was the daughter of a career Air Force colonel and after high school in Portland, Oregon, she entered the Air Force Academy to study aeronautical engineering. Upon receipt of her degree in 1980, she received her military commission and trained as an F-16 weapons separation engineer with the Air Force Armament Laboratory at Eglin Air Force Base in Florida. Helms rose through the ranks and within two years was the lead engineer. "I'm not pilot-qualified," Helms told the NASA oral historian, "so my entrance into the Air Force didn't have anything to do with the flying bug. I have very bad eyesight, so I never had the opportunity to even become an Air Force pilot. I was in the Air Force about seven years before I realised that there were jobs available where engineers could fly."

In 1984, by now a captain, she was selected to pursue a master's degree in aeronautics and astronautics at Stanford University and in 1985-1987 accepted an assistant professorship at the Air Force Academy. Helms' next assignment was to Air Force Test Pilot School at Edwards Air Force Base in California, from which she qualified as a flight test engineer and later worked at Canadian Forces Base Cold Lake in Alberta, Canada, as project officer on the CF-18 Hornet. She was serving in Canada when she learned of her selection as an astronaut in January 1990. "I'd like to be able to say that I was born with this burning desire to become an astronaut," Helms reflected, "but it didn't really happen that way. What ended up happening was I sort of grew into the idea over several years."

Within months of entering NASA, she was persuaded to join the ranks of the all-astronaut rock band, 'Max Q', playing keyboards. Helms had taken lessons for 11 years, as well as having played concert drums and xylophone in marching bands and choirs and as part of a jazz combo. In an interview with Michael Cassutt, she described her tastes as "pop, Top 40, everything *but* country and western" and fellow Max Q member Kevin Chilton described her as "hugely talented" and capable of listening to a song on the radio and playing it. "She was able to teach us harmonies," said Chilton. On STS-54, Helms carried a mini-keyboard as part of her personal kit and managed to tap out a one-finger version of the Air Force anthem 'Wild Blue Yonder' whilst in orbit. She also flew STS-64 in September 1994 and in her later astronaut career would become the first woman to live for an extended period of several months aboard the ISS.

Richard Michael Linnehan was the first professional veterinarian-astronaut ever selected by NASA as a mission specialist. Born in Lowell, Massachusetts, on 19 September 1957, he was raised by his paternal grandparents, Henry and Mae Linnehan, in New Hampshire. "There was no time in my life," he once told a NASA interviewer, "that I can remember growing that I *didn't* want to be an astronaut." Although he pestered his grandparents to take him to Florida to see a launch, their residence in New Hampshire made it impossible, but the idea lingered in the young boy's mind. He attended high schools in his home state and later entered the University of New Hampshire, earning a degree in animal sciences and microbiology in 1980. "At the time, I was debating whether I wanted to be a pilot, a military pilot or a veterinarian," he admitted, "and I was lucky to get accepted into both programmes, but my advisor in undergraduate [school] was a veterinarian and he convinced me that I wanted to be a veterinarian and not, as he put it, a glorified bus driver!"

Consequently, Linnehan gained his doctorate from Ohio State University College of Veterinary Medicine in 1985 and was accepted onto a two-year internship programme in zoo animal medicine and comparative pathology at the Baltimore Zoo and Johns Hopkins University. Following his internship, Linnehan joined the US Army in 1989. He served as a captain in the Veterinary Corps and was assigned to the Naval Ocean Systems Center in San Diego, California, where he served as chief clinical veterinarian for the Navy's Marine Mammal Program. "I was really lucky to get into the military programme," he said, "because I did a lot of flying, a lot of military airlifts in different aircraft, helicopters and large jets. Although I

1-19: The STS-78 crew, including backup payload specialists, poses during the Terminal Countdown Demonstration Test.

wasn't a pilot, I got a lot of air time in some of those military aircraft." He had applied to NASA in 1985, immediately after leaving veterinary school, but was turned down. He tried again in 1991 and remembered being queried at interview – *What does a veterinarian want to work at NASA for?* – to which his response was simple. He wanted to work at NASA for exactly the *same* reason that a physicist, or a doctor, or a pilot wanted to work at NASA: in order to become part of the space programme. "I can get into this a lot," he told them. "A lot of the early stuff done in the space programme was with primates and that was done by veterinarians. In fact, they were heavily involved early on with a lot of the animal research." Linnehan figured at some point NASA would need a veterinarian again. As circumstances transpired, for his career as an astronaut, it was a shrewd move.

He was standing on a pier in the middle of San Diego harbour when he received the call from the space agency. "I had been busy all day in the water with a bunch of animals that I was taking care of, and I didn't know that NASA had called eight or nine times," he recalled. When he got back to his office, Linnehan found Post-Its all over the walls, telling him to call NASA, right away. He spoke to Don Puddy, the head of Flight Crew Operations, and after being told that he had been selected, by his own admission, Linnehan "took off, ran off the pier ... and jumped in the harbour!" In his mind, everything that had happened to him in his life had led to this moment. "A lot of people, early on in my undergraduate education, helped me a lot. If it wasn't for them, I would never have probably been accepted into veterinary school," he said, "and if it wasn't for veterinary school, I would probably never have been accepted into the Navy programme that I ended up working for in the military and if it wasn't for the Navy programme I probably would have never been accepted by NASA." *Everything* played its part for Linnehan, as if a long chain of people had guided his steps.

Selected in March 1992, Linnehan – along with STS-78 crewmate Chuck Brady – would jointly wait the longest from his astronaut class (nicknamed 'The Hogs') to fly into space. "I was a bit of a straggler," Linnehan once quipped. Training for STS-78 produced many surprises. "You don't know what to expect," he continued. "You do all this training and you're wondering. It's like being in vet school again. What should I study? What's the most important? What do I need to know more than this?" When those questions had been answered, the next series focused on which areas he needed to prioritise and how to prepare for them. "That first mission, you want to do so well. You *want* to perform and you're really hyped up. You don't have time to enjoy yourself at all until you're finally up in space and you realise: *Man, I'm here!*"

The mission was intended to be 'international' in its flavour, providing yet another ingredient in preparing for the ISS, and many of its payloads were carried over from IML-2 and the second Spacelab Life Sciences (SLS-2) flights. Principal investigators for the experiments were based at four institutions in Europe and four in the United States and a broad use of telescience would be employed to effect control from the ground. Already, the United States Microgravity Payload had demonstrated the ability of scientists to monitor their experiments remotely and on LMS they would do the same with research facilities inside the pressurised Spacelab module. However, unlike many earlier missions, the STS-78 crew would operate a single-shift system. The benefit of this approach was that life science experiments, which, by their very nature, would require a significant amount of crew time and video coverage, but only minimal power and energy, could be executed when all seven astronauts were available as test subjects and operators. The automated and remotely controlled materials science investigations, on the other hand, could then be run via telescience as the crew slept. Full telecommand and video facilities would still be available, but at the same time the absence of movement from the astronauts would not disturb the highly sensitive microgravity investigations. Developed at a cost of $138 million, LMS comprised 22 major experiments, supported by NASA and ESA, as well as the Canadian, French and Italian space agencies, together with adjunct research teams based in ten other nations. Specifically, the life science experiments explored the responses of living organisms to the weightless environment, with emphasis on musculoskeletal physiology. The second side of the LMS coin – microgravity research – probed the subtle influences at work whilst processing a variety of materials and examining the behaviour of fluids.

In May 1995, around the same time that the first portion of the STS-78 crew was announced, technicians in KSC's Operations & Checkout Building began integrating the LMS hardware into the Spacelab module. By the following April, when the fully laden facility was closed out and loaded aboard Columbia, the crew for the mission had expanded to seven members. In October 1995, STS-70 veterans Tom Henricks and Kevin Kregel were announced as the commander and pilot. Overall, Columbia's three-month processing flow after STS-75 ran with exceptional smoothness, further demonstrative of the efforts during her last modification period in Palmdale, California, to reduce turnaround times and enhance capabilities. In fact, the only problems were the need to replace a Tactical Air Navigation device and correct a software error with one of the vehicle's Master Events Controllers. Finally, on 18 June 1996, just two days before launch, engineers opened her aft compartment in order to X-ray the power drive units for the ET access doors. The latter were suspected of having loose screws in their circuitry boards, but the examinations proved that they were secure. Right on time, at 9:49 am EST on the 20th, at the opening of a 2.5-hour window for that day, STS-78 speared into orbit for what was already anticipated to be a lengthy Shuttle flight. Officially, it was scheduled to run for just under 16 days, but NASA managers confidently expected that conservation of electrical power reserves should be sufficient to extend the mission by 24 hours or more.

Commander Tom Henricks provided television viewers with a unique perspective of his fourth launch into orbit, thanks to the presence of a small video camera mounted in the forward flight deck. Its footage began just prior to the thunderous ignition of Columbia's main engines and boosters and proceeded through to the separation of their ET and the establishment of the vehicle into a preliminary orbit. The launch itself was certainly a bone-jarring affair, although a subsequent analysis of the recovered SRBs revealed worrying damage to their field joints, caused by hot gases. It was through such 'blow-by' of one of the boosters' seals that Challenger had been lost in January 1986. Although the STS-78 ascent did not compromise the

1-20: Experiment racks are loaded into the Life and Microgravity Spacelab (LMS).

astronauts' safety, it marked the first occasion that the Redesigned Solid Rocket Motor joints had experienced "combustion product penetration". NASA managers pointed out that there was a hot gas 'path' through the motor's field joint, but not through the 'capture' joint, and that, consequently, the boosters performed within their design requirements. Nevertheless, the incident did raise questions over a new adhesive and cleaning fluid which had recently been added to the SRBs to comply with Environmental Protection Agency (EPA) regulations.

Even as Columbia circled Earth, a problem with the adhesive was already expected to delay the next Shuttle mission, STS-79, previously scheduled for late July. Unfortunately, NASA was unable to revert to its old-style adhesive because the EPA forbade its methyl-based material. Developing a new adhesive was anticipated to take a lengthy period of time, but the severity of the anomaly led Space Shuttle Program Manager Tommy Holloway to tell journalists that it was "a serious situation until we determine it's not serious". He stressed that the damage did not affect the rubberised O-ring seals within the boosters, describing the joint as "an order of magnitude more robust than the joint on Challenger". In total, engineers found six joints where hot gas had apparently penetrated their heat shields. Ultimately, the discovery and the necessary repairs led to the postponement of STS-79 from 31 July until mid-September. Aside from the SRB issue, which did not come to light until the end of June, during their disassembly and inspection, Columbia's ascent was nominal.

Once inserted into a 280 km orbit, inclined 39 degrees to the equator, which enabled the crew to maintain their terrestrial sleep/wake circadian rhythms, the seven astronauts set to work transforming their vehicle into a home and laboratory for the next 16 days. By 11:15 am, less than 90 minutes into the flight, Columbia's payload bay doors were opened and the Spacelab module and its experiments were activated ahead of schedule. "Today was the busiest first shift of activities we've ever had for Spacelab," explained LMS Mission Scientist Patton Downey on the evening of 20 June. "Virtually every experiment on board either had its equipment activated or checked out." From their stations on the flight deck, Henricks and Kregel oriented the Shuttle into a 'gravity gradient' attitude, with her tail pointing Earthward, so that very few thruster firings would be required, thereby avoiding disrupting the sensitive microgravity experiments. In spite of his training, Rick Linnehan – seated alongside Susan Helms on the flight deck for ascent – was astounded by the view of the blues, browns, greens and whites of Earth through Columbia's overhead windows. "My first job was to take pictures of the External Tank as it separated with a huge camera lens," he remembered, "and I had to wedge myself in the window and take these pictures. As you looked down, there's the tail of the orbiter and just the Earth and the oceans. Everybody was around the aft windows ... and no one said *anything*. There was nothing to say."

Yet there was little time to observe their surroundings, for the activation of the LMS payloads took priority. One of its most important features was ESA's Advanced Protein Crystallisation Facility (APCF), which carried 11 experiments to study three different growth methods. This device employed vapour-diffusion, liquid-to-liquid diffusion and dialysis and, in total, more than 5,000 video images would be acquired during STS-78 of the development of key protein crystals. The APCF had previously flown aboard the IML-2 mission and was a relatively late arrival on LMS, only arriving in Florida from Europe on 13 June 1996 and then being installed into the Spacelab module, whilst in a vertical configuration on Pad 39B, less than 24 hours before liftoff. It was activated six hours into the mission, and since the facility had no space-to-ground telemetry capability the crew typically provided daily reports of the status of its displays.

Also provided by ESA was the Advanced Gradient Heating Furnace (AGHF), which successfully processed 13 samples for one semiconductor and five metallurgical experiments. The furnace was switched on by Jean-Jacques Favier on the afternoon of 20 June and ran near-continuously throughout the mission, performing better than in ground-based tests. Its objective was to solidify alloys and crystals in a number of investigations designed to understand the conditions in which the structures of freezing materials changed in the solidification process. Investigators hoped that AGHF research would increase scientists' knowledge of the physical processes involved in solidification and lead to enhancements in ground-based materials research. The first actual experiment in the furnace got underway when Helms inserted a sample cartridge to examine the transition in solidifying metal mixtures from ordered, column-like grains to unordered, round ones. Processing of AGHF samples was sequential and required the exchange of experiment cartridges and the activation and deactivation by a crew member. On 22 June, a sample of pure aluminium, reinforced with zirconia particles, was placed into the furnace as part of an investigation into the physics of liquid metals containing ceramic particles as they solidified. Other experiments included a polycrystalline sample in order to gather information on how to combine liquid metal alloy components into precise, well-ordered solid structures. It was anticipated that knowledge from melting and resolidifying such compounds could help manufacturers make higher-quality metal alloys and semiconductors. Later, an aluminium-copper mixture was solidified as part of a French experiment. Astronauts Brady and Helms also ran an experiment which sought to control the internal structures of aluminium and indium alloys during solidification, which was expected to have terrestrial benefits in producing new materials for engineering, chemical and electronics applications.

The third European microgravity facility was the Bubble, Drop and Particle Unit (BDPU), which was used on STS-78 to observe and record the behaviour of fluids under differing temperature levels and concentrations. Early in the mission, it was utilised for one fluid physics investigation to explore the processes that control evaporation and condensation. In its specially designed test cell within the BDPU, a small heater emitted an electrical charge into liquid freon, supersaturated with gas, which produced a single bubble through boiling. Although it required some minor in-flight maintenance by Kregel and Favier after suffering a blown fuse, the device functioned well and processed all nine of its test containers. One experiment examined surface tension and interactions of gases and liquids in microgravity. Various sizes of air bubbles were injected into a water and alcohol solution with temperature gradients ranging from 'hot' to 'cold'. Another study observed the behaviour of inert gas bubbles within silicone oil. From such experiments, it was hoped that new insights might be gained into controlling defects in many aspects of materials processing, perhaps leading to the production of stronger and more resilient metals, alloys, ceramics and glasses. A second glitch with the BDPU led to a remarkable repair on 28 June, the procedure for which was uplinked to the crew in the form of a video from Mission Control. Their efforts proved successful when they activated the Electrohydrodynamics of Liquid Bridges experiment, which focused on changes occurring in 'bridges' of fluid suspended between a pair of electrodes. Fluids under study included castor oil, olive oil, eugenol and silicone oil. Another, two-part, investigation was provided by the University of Naples and looked at the interaction of moving, pre-formed bubbles and the melting and solidifying 'edge' of a solid, whilst its second segment examined the ways in which droplets were captured by, or pushed away from, a moving solidification front. The experiment called for the injection of water droplets of differing diameters into a liquid alloy, in order to study their behaviour during the application of heat. Despite difficulties with the injector, which refused to retract from its deployed position, ground-based engineers devised a solution and the experiment ran successfully.

With such a large complement of European experiment facilities aboard STS-78, it made sense for an ESA astronaut to serve as a payload specialist. Jean-Jacques Favier was born on 13 April 1949 in Kehl, south-western Germany. He attended primary and secondary schools in the French city of Strasbourg, which lies directly opposite Kehl, across the River Rhine. In 1971, he received a degree in engineering from the National Polytechnical Institute of Grenoble, followed by doctorates in engineering from the *École des mines de Paris* (the Mining School of Paris) and in metallurgy and physics from the University of Grenoble, both in 1977. Favier's subsequent career saw him advising the head of the Material Science Research Centre at the French Atomic Energy Commission and proposing and serving as principal investigator for the MEPHISTO experiment, which flew aboard the United States Microgravity Payload on the Shuttle. Selected by France's *Centre National d'Études Spatiales* (CNES, the French National Centre for Space Studies) as a Shuttle payload specialist candidate in September 1985, Favier trained as a backup crew member for the IML-2 mission, before securing a prime assignment to LMS. Alongside his IML-2 activities, he was a visiting professor at the University of Alabama at Huntsville in 1994-1995.

Elsewhere in the LMS module, the crew – which included a pair of physicians, Brady and Thirsk, and veterinarian Linnehan – also concentrated on the second complement of activities: the life science investigations. These were further categorised under five disciplines: human physiology, musculoskeletal, metabolic, neuroscience and space biology. Two hours after reaching orbit, payload specialists Favier and Thirsk kicked off the human physiology research by donning electrodes and sensors to monitor their eye, head and torso movements for the Canadian Torso Rotation Experiment. It had been known for three decades that many astronauts suffer motion sickness in space – particularly during their first few days aloft – and the aim of the Canadian study was to identify and ultimately avoid movements which contributed to this sense of illness.

In recognition of the many Canadian experiments aboard LMS, the STS-78 crew included Canadian payload specialist Robert Brent Thirsk, whose subsequent career as an astronaut would also involve a six-month expedition to the ISS. In fact, at the time of writing, in early 2014, Thirsk stands as Canada's most flight-experienced spacefarer, having accrued more than 204 days in orbit. "Without a doubt, I was inspired by the early American space programme, in particular the Apollo missions," he once remarked. "I would go to the school library and I'd flip through those *National Geographic* [and] *Life* magazines, get up early in the morning to watch the launches on TV. A Canadian typically knows an awful lot about the hockey teams and the hockey players, but I was unusual in that I knew just as much about the early astronauts and cosmonauts."

Thirsk was born on 17 August 1953 in New Westminster, British Columbia, and attended local primary schools and graduated from high school in Calgary. "I grew up in western Canada," he told a NASA interviewer, shortly before his ISS mission. "It's a nice blend of cosmopolitan cities and also of wide open spaces, so cities like Vancouver, Calgary, Edmonton, Regina and Winnipeg play significant roles on the world's stage, and yet we have this awesome landscape that includes oceans, temporate rainforests, mountains and prairies. The people still have this spirit of exploration, the pioneering spirit, the 'can-do' attitude. If someone comes up to me or another Western Canadian and says that something can't be done, we're pretty sceptical of that. We think things *can* be done."

Thirsk received his bachelor's degree in mechanical engineering from the

University of Calgary in 1976. A few years later, he received the university's first Distinguished Alumni Award, and in 2009, whilst aboard the ISS, he became the first person ever to receive an honorary doctorate in orbit, also from the University of Calgary. His academic career would soon take an unexpected turn. "When I was in my third or fourth year of engineering," Thirsk recalled, "one of my professors took me aside and he noticed that I had this dream of doing something different, something challenging in my life, so he gave me some good advice to consider further engineering degrees and even a medical degree. That's what I did." He subsequently attended Massachusetts Institute of Technology in the United States, gaining a master's credential in mechanical engineering in 1978. Thirsk next entered medical school, earned his doctorate from McGill University in 1982 and worked on the family medicine residency programme at the Queen Elizabeth Hospital in Montreal. In December 1983, he was selected by the National Research Council of Canada, alongside engineer Marc Garneau, physicians Ken Money and Roberta Bondar, physicist Steve MacLean and meteorologist Bjarni Tryggvason, as one of Canada's first group of astronaut candidates, training for payload specialist opportunities on future Shuttle missions. Thirsk served as backup to Garneau when the latter became Canada's first man in space aboard STS-41G in October 1984. Five years later, in 1989, the Canadian Space Agency (CSA) was formed and the six-strong group transferred to the new entity. In 1993-1994, just before his assignment to STS-78, Thirsk served as chief of the CSA astronaut corps. During this period, he also participated in the CAPSULS simulated space mission and took a sabbatical year in Victoria, British Columbia, where he upgraded his clinical practice skills, space medicine research and Russian language training.

The second physician aboard STS-78 was Charles Eldon Brady Jr, nicknamed 'Chuck', who came from Pinehurst, North Carolina, where he was born on 12 August 1951. A search of Brady's name in an Internet search engine today will unfortunately raise the spectre of the fact that, in July 2006, he became the first NASA astronaut to take his own life, for reasons apparently associated with the severe pain and near-paralysis caused by rheumatoid arthritis. This is saddening, for Brady's real contribution to the space programme was as an accomplished military flight surgeon and astronaut. After completing high school, he took the pre-medical (or 'pre-med') educational track at the University of North Carolina at Chapel Hill and graduated in 1971. He completed his medical degree at Duke University in 1975, then proceeded to the University of Tennessee Medical Center in Knoxville for his internship. Brady worked as a team physician in sports medicine for Iowa State University, then moved into family practice for several years, before joining the Navy in 1986. He was trained as a flight surgeon at the Naval Aerospace Medical Institute at Naval Air Station Pensacola, Florida, receiving the Fox Flag for the highest academic achievement, and was assigned to Carrier Air Wing Two aboard the aircraft carrier USS *Ranger* in June of that year. Subsequent assignments to attack and electronic countermeasures squadrons were followed by service with the Blue Angels (the US Navy's Flight Demonstration Squadron and the second-oldest of its kind in the world, after the French *Patrouille de France*) in 1988-1990. One of Brady's contemporaries in the Blue Angels was Donnie Cochran, the squadron's first

African-American member and subsequent commander. At the time of his selection by NASA in March 1992, Brady was serving in Tactical Electronic Warfare Squadron 129 and had reached the rank of commander in the Navy.

Four years later, aboard LMS, Brady and Linnehan joined their colleagues in the torso rotation study and participated in initial tests to evaluate their muscle strength and control. It was already known that muscle fibres became smaller, or 'atrophied', in microgravity, which resulted in a steady loss of muscular mass. Many of these observations tend to be short-lived, generally vanishing when astronauts return to Earth, but the potential impact of longer, six-month tours aboard the ISS on human muscles remained largely unknown. Using an ESA-built device known as the Torque Velocity Dynamometer, the astronauts were able to take precise measurements and calculate their muscle performance and function, including strength, amounts of force produced and resultant fatigue. Blood samples were taken throughout the mission to enable ground-based physicians to better understand metabolic and biochemical changes in their bodies in space. Additionally, the Astronaut Lung Function Experiment was used to gauge the influence of microgravity exposure on lung performance and respiratory muscles during rest and periods of heavy exercise. On 22 June, the human physiology studies entered the first of two specialised three-day 'blocks' to probe changes in sleep and performance patterns. It marked the first-ever comprehensive study of sleep, circadian rhythms and task performance in space. As astronauts circle Earth every 90 minutes – experiencing 16 'sunrises' and 'sunsets' in each 24-hour period – their 'normal' timing cues are significantly affected. Investigators hoped that such work might be beneficial to workers on Earth, as well as sufferers of jetlag. Typically, during the sleep experiments, all four science crew members (Linnehan, Brady, Favier and Thirsk) wore electrode-laden skullcaps to observe their brainwaves, eye movements and muscular activity. A second block of time for the experiment began on 2 July when they filled in questionnaires at the start and end of their shifts and again wore the skullcaps whilst asleep. Following Columbia's landing, the data from both blocks was compared to create a picture of how the astronauts' sleep patterns had changed during the mission.

Extending this physiological research into the musculoskeletal arena were a range of experiments to explore the underlying causes of muscular and bone loss, which featured the first-ever collection of muscle tissue biopsy samples before and after the mission. Almost immediately after Columbia touched down at KSC on 7 July, the science crew underwent MRI scans and biopsies for comparison with pre-flight samples. It was expected that this might lead to improved countermeasures to reduce in-flight muscular atrophy. The astronauts routinely used a bicycle ergometer in the Shuttle's middeck, which was fitted with a large, weighted flywheel, surrounded by a braking band to resist imparting their pedalling to the hull, enabling the crew to exercise without disturbing the sensitive microgravity experiments. The Torque Velocity Dynamometer was used to measure calf-muscle performance during these exercise periods. Additionally, Rick Linnehan and Jean-Jacques Favier wore electrodes on their legs that applied precise electrical stimuli to cause involuntary muscular contractions. Data from these experiments was expected to yield new insights into why muscles lost mass in space. The dynamometer was also employed for musculoskeletal tests to measure the astronauts' arm and hand-grip strength. On 24 June, they strapped their arms into the machine, curling and extending them as it provided resistance. Overall, the device performed near-flawlessly, with the exception of a few mechanical set-up problems and software glitches, and operated for a total of 85 hours during STS-78.

In light-hearted reference to all the electrodes worn and blood and tissue samples taken, Linnehan, Brady, Favier and Thirsk jokingly called themselves "the rat crew". Even their food and drink intakes were carefully monitored, as part of ongoing investigations of metabolic changes and calcium loss during space missions. They typically took non-radioactive calcium isotopes at each meal, from ten days before launch until a week into the mission, and by tracking its relationship to food-and-drink intake, scientists were able to distinguish calcium intake and excretion and determine the total amount used by their bones. Their efforts, pain and discomfort did not go unnoticed, particularly by Tom Henricks. "I'm one of those non-scientist pilots who believes that doing medical experiments in orbit is part of an astronaut's job," he told a Smithsonian interviewer, years later. "When so few humans go to space, we'd better get as much information as we can out of every person who goes. In my opinion, people who don't want to participate in those experiments picked the wrong profession." For the STS-78 science crew, he had the utmost respect. "The most painful thing they had to endure was using an electric charge to contract their calf muscles to their maximum contraction. That's why I'm disappointed when I see astronauts who don't want to participate in minor experiments that cause some discomfort. I saw those guys on STS-78 being so brave and doing things that were extremely uncomfortable."

Many of the LMS research facilities were cross-disciplinary and Canada's Torso Rotation Experiment was also employed for some of the neuroscience studies, by providing for the first time an opportunity to bridge the gap between space motion sickness and the causes of disorientation and nausea on Earth. It was hoped that this could help scientists to learn more about the problems associated with postural disorders and vertigo leading to falls and broken bones. The device precisely tracked the positions of the astronauts' eyes, heads and upper bodies as they went about their everyday activities. The most important observations came soon after entering orbit and they reported symptoms associated with adaptation to the microgravity environment. Voluntarily fixing the head to the torso with a neck brace has acquired the name 'torso rotation' because the subject has to turn their entire body in order to move their head. On Earth, this gradually leads to motion sickness in an example of deliberate 'egocentric' motor strategy, during which the subject concentrates on a body frame of reference, rather than an external world reference. Similar motor strategies are often adopted by astronauts, thereby exacerbating the onset of space sickness symptoms. "The findings will make a contribution to a further under-standing, countermeasures and rehabilitative programmes for not only astronauts, but also for people in hospitals on Earth," Bob Thirsk told journalists in Toronto during a space-to-ground news conference on 30 June. "With this information, we can figure better ways to keep people in space healthier and fight off muscle and bone degeneration and also use the information on Earth." Another device used as part of the neuroscience research was a piece of headgear for the Canal and Otolith Integration Studies, which explored the impact of microgravity exposure on the vestibular system of the inner ear and resulting changes in eye-hand-head co-ordination. Throughout the experiment, astronauts wore high-tech modified ski goggles that carefully tracked their eye and head motions as they watched a series of illuminated targets. Typically, they remained either in a fixed position on the bicycle ergometer or 'free-floating' in the Spacelab module and the targets displayed themselves across the inner surface of the goggles.

Seven astronauts were not the Shuttle's only living passengers on STS-78. Also hitching a ride were embryos of the hardy Medaka fish, provided by Columbia University College of Physicians and Surgeons in New York, which were flown as part of experiments into gravity's role in the development of animals. At intervals, an on-board video microscope provided television viewers with pictures of the growth of the transparent embryos. It was recognised that understanding the impact of microgravity on vertebrate development would become increasingly important as long-duration space station missions got underway. A total of 36 embryos of the Medaka – which is known to be particularly tolerant of reduced temperatures – developed during the mission. Judging from the video images downlinked from Columbia, the specimens in orbit appeared to develop at a slower rate than the control specimens on Earth. Other living creatures included 20 loblolly pine seedlings in a special plant-growth facility. When trees growing on Earth 'bend', then right themselves, they form so-called 'reaction wood', which is structurally inferior. During LMS, biologists carefully examined the cellular structures of the pine seedlings as part of efforts to devise ways to prevent reaction wood formation on Earth, which would prove enormously beneficial to the paper and lumber industries. After several days of growth, Favier and Helms harvested the seedlings, applied a chemical fixative, photographed them and stored them for landing.

Although the bulk of their time was spent inside the Spacelab module, the astronauts had some free time, which first-time spacefarer Bob Thirsk used to look through the Shuttle's windows at the ever-changing Earth, 280 km below. "Every morning, within minutes of wake-up," he wrote in an article for the *Toronto Sun* newspaper on 29 June, "we pass over a virtually cloud-free Europe and Asia. I rush unshaven to the window with a camcorder to capture the view one more time. The orbital pass begins over Portugal and Spain, through the Mediterranean Sea, across the boot of Italy and the Peloponnesian peninsula. At 8 km/sec, we overfly Cyprus, Israel and the Persian Gulf. Clouds then begin to thicken as we near India and Sri Lanka, which are now in their monsoon season. It is a thrill to recognise features such as the Straits of Gibraltar, the Bay of Naples, Mount Vesuvius, Athens, the Jordan River and the Nile Delta. These are regions of Earth that were the cradle of civilisation millennia ago and even today play a major role in global affairs." Summing up, Thirsk called the view "a recap of history and current events in 15 minutes". Thirsk wrote two columns for the *Calgary Sun* during STS-78 and became the first astronaut to write and file a newspaper story – and have it published – whilst in orbit.

Columbia's marathon mission was expected to get within hours of the 16.5-day duration of Endeavour's STS-67 flight a year earlier. On 29 June, it became official that STS-78 would snare the new record, when NASA told the crew that they would remain aloft until 7 July, producing a 17-day mission. The announcement was accompanied by background music from the movie *Mission: Impossible*, to which Tom Henricks responded that his crew was "willing, able and eagerly anticipating" the extension to their flight. As the days wore on, the LMS science received nothing but praise from scientists on the ground. "We have 41 principal investigators involved and all but very few have 100 percent, if not 200 percent, of the data they hoped to collect," Patton Downey explained. Telescience had been spectacularly trialled and Kevin Kregel lauded the ability to conduct video conferences and uplink videotaped repair instructions. "It made fixes a lot easier," he said, "as opposed to sending up the message and trying to interpret the fix on paper."

Right on cue, at 6:37 am EST on 7 July, Henricks and Kregel fired the Shuttle's OMS engines – during their record-setting 271st circuit of the globe – to begin the hypersonic glide back to KSC. An hour later, at 7:37:30 am, she settled gracefully onto Runway 33 to conclude a mission of 16 days, 21 hours, 48 minutes and 30 seconds, eclipsing the STS-67 record by about six hours. Television viewers would later be treated to a unique perspective of landing, as seen from a tiny video camera, attached to a bracket on Kregel's glare shield in the pilot's window. The lipstick-sized device provided stunning coverage of the final seven minutes of Columbia's 20th descent from space. Immediately after disembarking, Henricks and Kregel participated in an Olympic Torch ceremony and told journalists that "it's been a pleasure to stay up this long and we know it will be a short-lived record". Their words proved prophetic. Not only would the STS-78 record be broken again that very same year, 1996, but it would be broken by Columbia herself, on her very next mission, STS-80. The difference on that flight was that it would set an endurance record which would remain unbroken until the very end of the 30-year Shuttle era.

RECORD-SETTERS

In many ways, STS-80 in November-December 1996 seemed a charmed mission: two satellite deployments and retrievals, a record-smashing flight of almost 18 days for the Shuttle, a wide range of life and microgravity science investigations and the oldest person at that point to have ventured into space. "And then, two flights later ... they *beat* us," said Rick Linnehan of the short-lived STS-78 record being broken by the STS-80 crew. "We were hoping to hold that record for a little bit longer." However, one thing that STS-80 did not have was an EVA, after a balky airlock handle prevented mission specialists Tammy Jernigan and Tom Jones from becoming the first-ever astronauts to conduct a spacewalk from Columbia. Still, as STS-80 commander Ken Cockrell put it, the mission offered "a little warm-up" in readiness for the ISS.

So it was that in the early hours of 23 November 1996, less than four days since launch, no fewer than three separate spacecraft trailed one another in a delicate orbital ballet. One was Columbia herself, whilst the others were the Wake Shield

Facility – flying its third mission overall – and a German-built payload known as the Shuttle Pallet Satellite (SPAS). The latter had flown multiple times since June 1983 and can be thanked for having taken the first photograph of a 'full' Shuttle in orbit when it captured Challenger on STS-7. Originally intended as a means of providing a standardised support structure and resources for research payloads, SPAS flew successfully on three occasions, carrying its first major scientific instrument (an experimental infrared background signature survey for the US Department of Defense) on STS-39 in April 1991. It was later superseded by an improved model, known as 'ASTRO-SPAS', which first flew on STS-51 in September 1993, transporting the Orbiting and Retrievable Far and Extreme Ultraviolet Spectro-meter (ORFEUS) into space. STS-80 marked ORFEUS' second mission.

Both ASTRO-SPAS and ORFEUS were the product of a co-operative venture between NASA and the *Deutsche Agentur für Raumfahrtangelegenheiten* (DARA), the German Aerospace Agency from 1989-1997, predecessor of today's *Deutsches Zentrum für Luft- und Raumfahrt* (DLR). Unlike its first-generation predecessor, ASTRO-SPAS had the capability to remain in autonomous free-flight for up to two weeks, commanded by the mobile German SPAS Payload Operations Centre (SPOC). The power for the satellite and its payloads came from a new lithium-sulphate battery pack and precise attitude-control was provided by a three-axis-stabilised cold-gas system, a star tracker and a space-borne Global Positioning System (GPS) receiver. Weighing 3,150 kg, more than half of which was accounted for by operational science payloads, the carbon-fibre-composite ASTRO-SPAS measured 4.5 m across the width of the Shuttle's payload bay and was 2.5 m in depth from its front to rear faces. The satellite's precise attitude-control capabilities enabled it to support sensitive astronomical and Earth-observation sensors, with several missions planned. Two of these would carry a set of infrared telescopes and spectrometers to examine the upper atmosphere, one was scheduled (but never flown) to demonstrate advanced automated rendezvous and capture technologies and two others carried ORFEUS.

This instrument, with a large 1.2 m-diameter telescope, was designed to investigate 'very hot' and 'very cold' matter in the Universe, combined with an Interstellar Medium Absorption Profile Spectrograph (IMAPS). Of these, ORFEUS, which extended to a length of 2.4 m through the middle of the ASTRO-SPAS satellite, was by far the largest instrument. It was to observe the far ultraviolet (90-125 nm) and extreme ultraviolet (40-90 nm), a region of the electromagnetic spectrum obscured from ground-based astronomers by the atmo-sphere. Yet it bears the highest density of spectral lines – especially from various states of hydrogen and helium – which are emitted or absorbed by matter of very different temperatures. ORFEUS was expected to add a great deal to scientific understanding of the life-cycles of celestial objects by studying hot stellar atmospheres and white dwarfs, together with supernova remnants, active galactic nuclei and star-forming clouds of gas and dust. Operating alongside the telescope, IMAPS continued an earlier series of experiments aboard high-altitude sounding rockets to observe bright galactic objects at 95-114 nm and examine the fine structure of interstellar gas lines. It was capable of very precisely measuring the

1-21: The Wake Shield Facility (WSF) is prepared for launch.

motions of interstellar gas clouds with an accuracy of 1.6 km/sec. During orbital operations, ORFEUS' far ultraviolet (nicknamed 'Echelle', indicative of its diffraction grating) and extreme ultraviolet spectrometers were operated alternately, by 'flipping' a mirror into the beam reflected off the instrument's primary mirror. Two reflection gratings then dispersed the incoming light from celestial sources into a spectrum, which was projected onto a two-dimensional microchannel plate detector. Meanwhile, the Surface Effects Sample Monitor (SESAM) provided a passive carrier to evaluate state-of-the-art optics and potential future astronomical detectors, whilst high school students from Ottobrunn in Germany supplied an electrolysis experiment to investigate various salt solutions. It was planned that on STS-80 ORFEUS-SPAS-2 would spend 14 days in quasi-autonomous flight alongside the Shuttle.

When Columbia reached orbit on 19 November 1996, the ORFEUS-SPAS-2 deployment was the first major objective on the crew's agenda, scheduled to occur about seven hours after launch. The launch itself was three weeks overdue, having been originally targeted for 31 October. Firstly, problems had arisen earlier in the year with a new-specification adhesive and cleaning fluid in the SRBs, which NASA engineers suspected was chiefly to blame for the field-joint damage experienced during STS-78 ascent. That problem had led to the postponement of Atlantis' STS-79 mission to Mir from 31 July until mid-September, as her set of SRBs were replaced with those previously earmarked for STS-80. Moreover, preparations at KSC in anticipation of Hurricane Fran had forced managers to delay Columbia's launch until no earlier than 8 November, a date that quickly slipped to the 11th. Otherwise, readying the orbiter for her 21st mission ran smoothly, with the exception of having to replace two of her flight deck windows, following an engineering analysis which determined that windows with a high number of flights tended to fracture more easily during the period of maximum aerodynamic turbulence (known as 'Max Q'), experienced by the ascending Shuttle stack about 60 seconds into flight. One of the windows had flown eight previous missions, the other seven.

By mid-October, with the STS-80 stack on the launch pad, the ORFEUS-SPAS-2 and WSF-3 payloads had been installed, but on the 28th NASA elected to postpone the mission again by at least a week beyond 11 November. This was in order to investigate an issue surrounding erosion of the SRB nozzles during the STS-79 launch. The concern surrounded insulating material, which had experienced higher than normal 'grooving' erosion in the nozzle's throat. Although it presented no safety of flight issue, it was decided to evaluate and clear the problem before committing to launch. After agreeing to a new date of 15 November, Tommy Holloway told journalists that "the effort is nearing a point where it will provide us with a good understanding of the phenomenon", at the same time ensuring "that the final portions of the investigation are not rushed". In conclusion, engineers noted that the most likely cause for the 'out-of-family' erosion pattern seen on the STS-79 boosters was due to a 'pocketing' erosion effect, triggered by slight ply distortions in the ablative material of the nozzle's throat ring and normal variations in other material properties. Ordinarily, the manufacturing of the throat rings was accomplished by wrapping the ablative material in a 'criss-cross' fashion and curing it at elevated temperatures and pressures. However, it was suspected that during the curing process, the material near the surface of the insulation shifted slightly, producing the distortions. When hot gas flowed through the SRBs, the distortions significantly raised stresses in the material, which could result in the pocketing effect and the ablator wearing away unevenly. Nonetheless, analysis showed that safety margins could be maintained even with the ply distortion condition in the worst possible configuration.

Columbia's liftoff was delayed a further four days until 19 November by the postponement of an Atlas expendable rocket launch from nearby Cape Canaveral Air Force Station and by predicted poor weather at KSC. Instead of halting, and then restarting, the countdown, the clock was brought back from T-11 hours to T-19 hours and held at that point until the evening of the 18th. Liftoff the following afternoon was delayed slightly at T-31 seconds, when a minor hydrogen leak was detected in the Shuttle's aft compartment. During a two-minute hold, engineers carefully monitored the hydrogen concentrations and concluded that they were at "acceptable levels" and the countdown proceeded without further incident. Among the five-member STS-80 crew was mission specialist Franklin Story Musgrave, who not only became the first person to embark on a sixth Shuttle flight, but also became the oldest man in space, aged 61. "I'm hugely blessed, just hugely blessed," he said before boarding Columbia. Musgrave also became the only person to fly into space aboard all five orbiters: two missions on Challenger and one each on Columbia, Discovery, Atlantis and Endeavour.

Scientist, physician, engineer, pilot, mechanic, poet and literary critic, Franklin Story Musgrave had amassed extensive flying time in the Marine Corps and by 1987 had secured no fewer than *six* academic degrees. He was never called by his first name – not even by his parents – and his middle name, 'Story', honoured an old family surname from several generations back. "I got into this business to be on the intellectual and physical frontier," he explained of his decision to pursue a career with NASA. "I wanted a *transcendental* experience – an existential reaction to the environment. I'm not talking about an illusion, or seeing something that wasn't there, but a magical, emotional reaction to the environment. *That's* what I've been after all my life: to experience and feel new sensations."

That life, certainly, began badly for Musgrave, and in *NASA's Scientist-Astronauts*, historians Dave Shayler and Colin Burgess would characterise his formative years as "a childhood filled with despair". Born in Stockbridge, Massachusetts, on 19 August 1935, his parents were alcoholics – his mother meek and acquiescent, his father malicious and brutal – and the family's isolation on a farm meant that visitors were rare. Two things 'saved' him: one was that his ancestry on his father's side had boasted *nine* straight generations of doctors and the other was the ability to escape into the natural environment. "The unhappy situation," wrote Shayler and Burgess, "would often cause young Story to flee his home by night, making his way into the embrace of a nearby forest, where he would lie on his back, look up and marvel at the stars. He recalls doing this when aged *only three*, but the darkened forest held no fears for him." Even at this age, Musgrave considered nature to be his home – "my solace," he said, "a place where there was beauty, in which there was order" – and he was soon building rafts and becoming more self-reliant.

His salvation came in 1945, when his abused mother finally decided that she could no longer bear a violent existence with her husband and fled, taking Story with her. Yet tragedy was never far away: both of his parents *and* his younger brother would eventually commit suicide, whilst his older brother died in an aircraft accident. Psychologically, these calamities helped to improve Musgrave's self-reliance and mould him into the man that he would become. Surprisingly, in view of his later academic accomplishments, he did not shine at high school. "He hated school and all that was associated with it," wrote his biographer, Anne Lenehan, on Musgrave's website, www.storymusgrave.com. "He was constantly in trouble with the school authorities and was subjected to almost continual disciplinary action." A motorcycle accident caused him to miss out on his final exams, but Musgrave's perspective was that home life and school life offered a "fantastically narrow" window on the world. "I felt the urge to expand my horizons," he told Lenehan, "and to see other worlds." He joined the Marines in 1953 and trained as an aviation and instrument technician, completing active-duty assignments in Korea, Japan and Hawaii and serving aboard the aircraft carrier USS *Wasp* in the Far East. These years also rekindled an earlier interest in aviation and he resumed his studies to gain his pilot's licence.

Clutching a National Defense Service Medal and an Outstanding Unit Citation from his Marine Corps squadron, Musgrave left the military and enrolled at Syracuse University in New York to study mathematics and statistics. Shortly before gaining his bachelor's degree, he was employed as a mathematician and operations analyst by the Eastman Kodak Company in Rochester, New York. He then followed up with a master's in business administration from the University of California in 1959, a bachelor's credential in chemistry from Marietta College in 1960 and a doctorate in medicine from Columbia University in 1964. With his medical degree under his belt, Musgrave began his own research into the human nervous system, with a one-year surgical residency at the University of Kentucky Medical Center in Lexington. His achievements enabled him to win post-doctoral fellowships from both the Air Force and the National Heart Institute. During this period, Musgrave's interests broadened to encompass aerospace physiology, temperature regulation and clinical surgery ... and *another* master's degree (this time in physiology and biophysics) in 1966. When NASA announced its intention to select a second group of scientist-astronauts, Musgrave was convinced that his experiences had led him to the door of space. At first, NASA considered him to be *over-qualified* – five degrees so far, an active laboratory, a surgical practice and a licenced commercial pilot, flight instructor and accomplished parachutist (he would ultimately log over 500 jumps) – but his potential was noticed and he was selected in August 1967.

Musgrave's advantage over the other members of his class of scientist-astronauts was that he was alone amongst them in being a qualified pilot ... although even he had never flown high-performance jets. Jungle survival training and flight instruction were therefore mandatory for all group members. Musgrave went to Reece Air Force Base in Lubbock, Texas, completing his 53 weeks of training with the highest scores ever recorded at the base and a commendation. In April 1969, he was detailed to the Apollo Applications Program, subsequently renamed 'Skylab'. Less than two years later, his name was announced as the backup science pilot for the first Skylab mission. At the time, plans existed for a second space station – 'Skylab B' – and Shayler and Burgess have speculated that had this actually flown, in around 1975-1977, Musgrave would have been a primary candidate for a seat on one of its missions. Sadly, budget cuts and the emphasis on getting the Shuttle ready to fly ultimately sounded the death knell for Skylab B. In 1974, he was assigned to the life sciences branch of the astronaut office and in October of that year participated in a medical development test of Spacelab with Dennis Morrison of JSC's Bioscience Payloads Office. The two men spent seven days inside a Spacelab mock-up and conducted a series of biomedical demonstrations in order to perfect operational procedures for real missions. Musgrave participated in a second such test for five days in January 1976, together with nuclear chemist Robert Clark and cardiopulmonary physiologist Charles Sawin; on this occasion, they lived and worked in a full-scale mock-up of not only the Spacelab facility, but also the middeck and flight deck of the orbiter itself.

During this period, Musgrave continued clinical and scientific training as a part-time surgeon at Denver General Hospital and as a part-time professor of physiology and biophysics at the University of Kentucky Medical Center. In his NASA capacity, he participated in the development of the Shuttle space suit – together with the airlock, life-support systems and a new Manned Manoeuvring Unit – as well as working in the Shuttle Avionics Integration Laboratory. Moreover, for an astronaut who was selected with *no* jet experience, Musgrave would eventually amass 17,700 hours of experience; more than a third of this time was in NASA's T-38 Talon jets, but in all he flew over 160 different types of aircraft. By comparison, no other astronaut has come close to this total, with most former chiefs of the office averaging 7,000 hours. In fact, even the most flight-experienced chief astronaut, John Young, logged 15,200 hours in his career. "From the beginning of his NASA days," wrote Anne Lenehan, "Story flew just about every day, sometimes *twice* a day, and around a hundred hours a month … an extraordinary amount, given his other commitments." Nor was Musgrave a cautious aviator. He was nicknamed 'Dr Details' by his fellow astronauts. He exuded self-confidence both in the missions he flew and in the tasks he fulfilled. "It was sheer play for me," said the man whose life had begun under such a cloud of menace, "to be able to so completely interact with my environment."

Following a spectacular launch at 2:55:47 pm EST, NASA's long-range tracking cameras spotted something unusual on STS-80: an apparent *fire* between the two SRBs. However, it was considered so minor that managers did not even mention it during the post-launch press conference, with NASA spokesman Doug Ward remarking that it was nothing unusual. "When you get close to booster separation and you get out of Earth's atmosphere, there's no strong airstream to keep the fire … pointing down," he said, adding that the ET was heavily insulated to safeguard against a potential explosion. Fortunately, when the STS-80 astronauts photographed the ET as it tumbled away, 8.5 minutes into the mission, they reported no obvious damage. After establishing themselves in their correct orbit and stowing their pressure suits and seats, they readied ORFEUS-SPAS-2 for deployment. With mission specialist Thomas David Jones at the controls of the RMS mechanical arm, the satellite was released – slightly later than planned, due to an extended checkout phase – at 11:11 pm EST. It was the first RMS assignment for Jones, who, although making his third Shuttle flight, had previously flown two Space Radar Laboratory (SRL) missions in 1994.

Born on 22 January 1955 in Baltimore, Maryland, Jones grew up with the early astronauts of Mercury and Gemini as his heroes. "I knew all of these guys by what missions they flew," he later told a NASA interviewer, "and I was really hoping to follow directly in their footsteps." Although no one in his family had served in the military, or had been a scientist, Jones was guided by his teacher father and by the influence of the early astronauts, all of whom showed him "how important it was to excel in academics and school". One day, as a Cub Scout, Jones visited the Martin Aircraft production plant at one of their 'Open House' days, where the Titan II boosters for NASA's Gemini VII and VIII missions were being prepared. Gazing upon those two enormous machines, Jones was entranced, enthralled … and hooked for life. After leaving high school, he entered the Air Force Academy and earned a degree in basic sciences in 1977, then began his military career. He underwent flight training and spent six years flying strategic bombers at Carswell Air Force Base in Texas and served as a pilot and commander of the B-52D Stratofortress, leading a six-strong combat crew. By the time he resigned from active duty in 1983, Jones had attained the rank of captain and was recipient of the Air Force's Commendation

1-22: The STS-80 crew included Story Musgrave (centre), who became the oldest spacefarer to date and the first person to record six Shuttle missions.

Medal. As a civilian, he was admitted into the University of Arizona in Tucson and began working towards his doctorate in planetary sciences, which he completed in 1988. Jones' research focuses included the remote sensing of asteroids, meteorite spectroscopy and the applications of space resources.

During this period, he tried twice for admission into NASA's astronaut corps, but was unsuccessful. Until he had received his PhD, Jones wrote in his memoir, *Skywalking*, "I just wasn't competitive". Shortly thereafter, he was hired by the CIA as a programme management engineer at the Office of Development and Engineering in Washington, DC, and sent off another application for NASA's 1990 selection cycle. "I tried to keep my expectations low," Jones wrote. "About five percent of all qualified applicants would be interviewed; fewer than one percent would be hired." He completed a week-long series of interviews and medical and psychological evaluations in October 1989. At around the same time, he was selected as a senior scientist with Science Applications International Corporation (SAIC) in Washington, DC, supporting advanced mission planning for NASA's Solar System Exploration Division. Jones was to assume his new role in January 1990 and took a vacation with his family before his start date. During the vacation, he received a telephone call, inviting him to become an astronaut, and that led to an awkward, but gracious, conversation with his SAIC boss, Harvey Feingold. Jones explained that he *could* accept the post with SAIC ... but he could *only* accept it for six months, as NASA required him to be in Houston in July. "For someone who had just been told his company had, in effect, finished in second place," Jones wrote, "Harvey was gracious to a fault. He said that although he was sorry to lose me, he was happy that my dream was moving closer to reality." In April and September 1994, Jones flew aboard both SRL missions, STS-59 and STS-68.

Three hours after the deployment of ORFEUS-SPAS-2 in the early hours of 20 November 1996, to the relief of ground controllers, the telescope's aperture door opened satisfactorily. Within 24 hours of deployment, Columbia led the satellite by about 53 km and, following the first of several station-keeping manoeuvres, closed to enable radar data to provide a more precise 'fix'. It was hoped that this would enable Mission Control to better compute future manoeuvres. Over the next few days, until ORFEUS-SPAS-2's retrieval on 3 December, STS-80 commander Ken Cockrell and pilot Kent Rominger performed a series of thruster firings to guide the Shuttle closer, then further away, from the satellite. This task was complicated yet more by the deployment of their second payload, WSF-3, on 23 November; in fact, coming just a year after STS-69 (which also featured Cockrell on its crew), STS-80 was only the second Shuttle mission to deploy and retrieve two major payloads and execute intricate rendezvous manoeuvres with them. However, unlike STS-69, whose first deployable payload was retrieved before the second was deployed, Cockrell's crew would contend with *both* satellites in free flight at the same time.

Since even a few atoms could have detrimental effects on WSF-3's ability to grow ultra-pure thin films, its 'wake' side had to be very carefully cleaned before it was installed into Columbia's payload bay on the ground. In a similar manner to that employed by the STS-69 crew, further cleaning took place in space, prior to deployment. On the afternoon of 22 November, Tom Jones powered-up the RMS arm for the second time and grappled the satellite at 2:25 pm EST. Shortly afterwards, fellow mission specialist Tammy Jernigan activated the Space Vision System (SVS) to carefully track WSF-3's position and Cockrell oriented the Shuttle in a gravity gradient attitude to minimise disturbances. The Canadian-built SVS sought to improve astronauts' perception of large structures under unfavourable viewing conditions and would play a key role in the construction of the ISS. It was already known that, in space, visual acuity is compromised by frequent periods of darkness and lighting and by a paucity of reliable points of spatial reference. This had led many astronauts to express concerns about the difficulty of accurately judging distances and range rates. Meanwhile, Jones' next task after unberthing the satellite was to hang it over the port-side payload bay wall, with its underside facing into the Columbia's direction of travel, to 'cleanse' its wake side for 2.5 hours using the harsh atomic oxygen prevalent in low-Earth orbit. Later, at 6:45 pm, he manoeuvred the payload across the bay and positioned it over the starboard wall, this time with its underside facing away from the direction of travel, in order to check out its Automatic Data Acquisition and Control System and thrusters. Finally, Jones set it up, high above the payload bay, again with its underside facing away from the direction of travel, to await the opening of the first, 41-minute deployment 'window', at 8:06 pm.

Following a slight delay, he released the payload at 8:38 pm, as the Shuttle flew high above the western Pacific Ocean. Shortly afterwards, WSF-3 ground controllers commanded the payload to fire its tiny nitrogen-gas thruster, which produced just 45 grams of thrust, for about 19 minutes to push it away from Columbia. This positioned the satellite at a distance of 30-50 km and set it up for three days of autonomous operations. At this stage, ORFEUS-SPAS-2 was about 93 km 'behind' them. By mid-morning on the 23rd, the first of seven planned semiconductor processing runs got underway. However, a potential problem would soon rear its head. Early on 24 November, the astronauts were awakened to the news that ORFEUS-SPAS-2 seemed to be closing on WSF-3 at a faster than anticipated rate. Original plans called for Cockrell and Rominger to perform up to two manoeuvres per day to maintain proper separation distances between themselves and the two satellites, complemented by a daily firing by WSF-3's cold-gas thruster. It was expected that these procedures would keep them both 30-40 km away from the Shuttle and about the same distance from each other. After a lengthy period of tracking analysis and predictions of the satellites' positions, Mission Control decided that their separation distances were within limits, but Cockrell was instructed to retrieve WSF-3 three hours early.

The early retrieval did not impair its semiconductor growth operations, with all seven planned films having been successfully processed by the early evening of 25 November, and WSF-3 was grappled by the RMS at 7:01 pm. Throughout the entire rendezvous process, Columbia drew no closer than 10 km to ORFEUS-SPAS-2, which was exactly as predicted by trajectory planners and well within safety margins. Yet it was not the end of WSF-3 operations. At 7:06 pm on the 26th, Jones again grappled the satellite for what should have been 3.5 hours of experiments. The work was carried out in support of the Atomic Oxygen Processing investigation, in which it was angled 45 degrees into the Shuttle's direction of travel to explore the usefulness of atomic oxygen in low-Earth orbit to grow aluminium oxide films. So successful were these experiments that scientists were granted an additional three hours of work, before Jones finally reberthed WSF-3 onto its cross-bay carrier at 1:53 am on the 27th. Throughout the procedure, Jernigan was also able to gather data on the satellite's position in the payload bay using the SVS.

Applications of the Wake Shield research were envisaged to extend much further than the semiconductor and electronics industries, with the possibility of building replacement rods and cones into human retinas to convert light into electrical impulses and aid sufferers of macular degeneration and *ritinitis pigmentosa*. Originally, four WSF missions were planned, three of which were NASA-funded and the final flight was expected to be financed by industrial partners. At the time of STS-80, it was anticipated that WSF-4 would benefit from solar arrays, additional on-board computing capabilities and robotic substrate sample manipulation for extended operations in low-Earth orbit. It was also planned that the final mission would produce as many as 300 thin films. The integration and test schedule for WSF-4 was approved, but by the end of 1997 NASA funding ran out. Several months later, in May 1998, exclusive licence was granted to Spacehab, Inc., to market, manage and operate the WSF and there were hopes that a five-year Mark II free-flyer might be launched. Ultimately, with the emphasis of the Shuttle upon building the ISS – which, in 2003, was tragically punctuated by the loss of Columbia – this did not come to pass and the WSF never flew again.

In spite of the success of STS-80, the mission would long be remembered as the Shuttle flight in which a pair of planned EVAs were cancelled due to a faulty airlock hatch. When Tammy Jernigan and Tom Jones, and their three crewmates, were named to the crew in January 1996, they eagerly anticipated the opportunity to perform two spacewalks, each lasting six hours, to prepare for the procedures and tasks which would someday prove vital in the construction of the ISS. Like the excursions performed on STS-69 and STS-72, they were part of the EDFT effort to build spacewalking expertise among NASA's astronaut corps. "Of all the space station assembly missions coming up, probably more than 80 percent of them" require spacewalks, noted Jones. "They're going to depend on these concepts that we think we've gotten right, but we've got to prove." Whilst Jernigan and Jones were outside in the payload bay, becoming the first astronauts ever to perform a spacewalk from Columbia, their colleagues would have choreographed their every move and manipulated the RMS arm for several tests.

It was the culmination of the astronaut career of Tamara Elizabeth Jernigan, who would go on to record five Shuttle missions by the time she was 40 years old. When she flew her first flight in June 1991, she became the youngest American woman ever to journey into space, surpassing the previous record-holder, Sally Ride, by a matter of days. It is a record that Jernigan still holds for the United States. Yet Jernigan exhibited a no-nonsense competence. "In meetings," wrote Tom Jones in *Sky-walking*, "she would quickly get to the crux of any issue on the table." (In her later career with NASA, she would serve as deputy chief of the astronaut office.) Jernigan came from Chattanooga, Tennessee, and was born on 7 May 1959, but much of her early education took place on the West Coast, in California. Aged only ten, she vividly recalled stepping out of her front door to look at the Moon, one summer evening in 1969, knowing that a pair of astronauts had just landed there. For Jernigan, *that* was her first memory of aspiring to become an astronaut herself, someday. The dream took on a more serious turn after high school in Santa Fe, when she moved to Stanford University to study physics. "I was a physics major," she told a NASA interviewer in the weeks before her final Shuttle mission. "In 1978, they started taking astronauts from the scientific community in earnest and also taking women. It was in 1978 that I thought I might have a chance to be selected."

A keen intercollegiate athlete and varsity volleyball team member whilst at Stanford, Jernigan earned her degree in 1981, followed by a master's credential in engineering science, two years later. Her career with NASA began immediately after graduation, as a research scientist in the Theoretical Studies Branch of the agency's Ames Research Center in Moffett Field, California. A second master's degree, in astronomy, from the University of California at Berkeley, followed in 1985. In June of that same year, she was selected as an astronaut, aged only 26. Alongside the demanding programme of astronaut candidate training, Jernigan pursued her PhD in space physics and astronomy and achieved her doctorate in 1988. Aged only 29 when she was assigned to her first mission, STS-40, in April 1989, Jernigan would serve as the flight engineer, seated behind and between the commander and pilot to offer a crucial additional set of eyes to monitor the instruments and displays and respond to emergency situations. Five years after STS-40, at the end of 1996, Jernigan was on her fourth Shuttle flight on STS-80, with one more mission still ahead of her.

During her first EVA with Jones, planned for 28 November, it was intended that they would conduct an 'end-to-end' demonstration to replace an ISS battery, using a 1.8 m, 70 kg crane. This was expected to require three hours. Then they would test the crane's ability to move a small cable caddy, previously used during by the STS-72 spacewalkers in January 1996. Jernigan and Jones' second EVA on 30 November would have involved each of them spending two hours working on the battery, whilst standing on a mobile work platform at the end of the RMS. Other activities included evaluating the Body Restraint Tether, which offered a stabilising bar for spacewalkers, and a Multi-Use Tether (MUT), capable of fitting square Russian-built handrails and round US ones. The crane, meanwhile, included a boom which telescoped outwards as much as 5 m in length and could manoeuvre payloads as massive as 270 kg to various locations on the space station's truss structure.

In anticipation of EVA-1, on 21 November, Jernigan and Jones, with assistance from Musgrave, began checking out their suits in Columbia's middeck. Then, late on the 27th, the cabin pressure was lowered from 101.3 kPal to 70.3 kPal, thus reducing the amount of time that the spacewalkers would need to pre-breathe pure oxygen before venturing outside. It also served to prepare them for the 29.6 kPal pure oxygen atmospheres of their suits. Ironically, as it turned out, the STS-80 crew was awakened to the sound of Robert Palmer's 'Some Guys Have All The Luck', although Jernigan and Jones' fortunes were bad for that day when the outermost airlock hatch refused to open.

"Initially, I thought we just had a sticky hatch and the fact that Tammy's initial rotation wasn't able to free it up was just an indication that we'd have to put a little more elbow grease into it," Jones told CNN in a space-to-ground news conference on 2 December. "Certainly, we are feeling some combination of disappointment at the failure of the hatch, but pleasure in being part of this mission that's been in every other way very successful," added Jernigan. Both astronauts commented that space exploration was a complex business, but remained upbeat, adding that lessons would be learned from the experiment. The handle of the hatch had apparently stopped after about 30 degrees of rotation, insufficient to release a series of latches around its circumference. An engineering team was promptly assembled to determine the most likely cause of the mishap and Mission Control adjusted the STS-80 crew's schedule in anticipation of a second attempt on 29 November. A minor problem was also noted with a signal conditioner in Jones' space suit and it was decided to replace this unit, should the EVA go ahead.

By the afternoon of the 29th, however, Mission Operations representative Jeff Bantle told journalists that analysis suggested a misalignment of the hatch against the airlock seal. Meanwhile, engineers worked to assess emergency procedures to open the hatch – which might still be needed in the event of a contingency EVA to close the payload bay doors – including warming it by orienting the orbiter's topside towards the Sun and having Jernigan and Jones apply pressure from inside the airlock. According to veteran spacewalker Jerry Ross, under normal conditions only "fairly light forces" were needed to open the hatch, "certainly not as high as a lug wrench on a bolt". Awakened by David Bowie's song 'Changes' on the evening of 29 November, the crew underwent another day in limbo, working on their experiments, until the Mission Management Team concluded that a definitive cause for the problem could not be identified. Capcom Dom Gorie radioed the disappointing news early on the 30th, telling them that both EVAs would be cancelled. Still, Jernigan and Jones conducted an evaluation on the middeck of a pistol-grip power tool that would have been used during their excursions. They successfully loosened and tightened bolts and screws on floor panels. Ultimately, both astronauts would fly again and perform their 'missed' spacewalks; albeit on missions to the 'real' ISS. After Columbia had returned to Earth, inspections revealed that a small screw had worked its way loose from an internal assembly and lodged in a gearbox-like actuator. This made all of Jernigan and Jones' efforts to open the hatch fruitless. When the actuator was replaced, the hatch opened normally.

A glimmer of good news emerged on the horizon on 2 December, when STS-80 was officially extended to 17 days, bringing it within hours of the duration record set by Tom Henricks' crew in July. Awakened that day by Jackson Browne's song 'Stay', the decision gave ORFEUS-SPAS-2 an additional day of observations and allowed the crew to press on with their experiments. One of these was the Space Experiment Module, provided by GSFC and making its first flight. It was carried as part of efforts to increase educational access to space and targeted students from kindergarten to undergraduate level, providing research volume within a GAS canister in the payload bay. The investigations encompassed studies of gravity and acceleration and observations of bacteria and crystal growth and carried algae, bones, yeast, photographic film and even a variety of children's play items, including crayons, chalk and Silly Putty. Other experiments in the middeck included a pair of medical investigations, studying blood pressure regulation in rats and the impact of microgravity on bones at the cellular level. As the mission entered its final days, Jernigan successfully recaptured ORFEUS-SPAS-2 with the RMS at 3:26 am EST on 3 December, manoeuvring it through a series of exercises to gather positioning data with the SVS, then reberthing it. In total, the ultraviolet telescope made 420 observations of around 150 astronomical objects, including the Moon, nearby stars, distant sources in the Milky Way, active galaxies and a quasar.

With a chance of less than favourable weather in Florida on 6 December, NASA managers revised their decision to allow the astronauts to remain aloft for 17 days and instructed them to prepare for a homecoming on the 5th instead. However, both landing opportunities at KSC for that day were called off, due to unacceptable cloud cover moving towards the SLF runway. With a more optimistic outlook on the 6th, Entry Flight Director Wayne Hale opted not to call up Edwards Air Force Base in California and try again for Florida. Unfortunately, both landing opportunities on the 6th were also scrubbed due to low-level fog. Eventually, following a cold front which moved through KSC on the night of the 6th, Cockrell and Rominger received the go-ahead to execute the irreversible de-orbit burn.

From her point of atmospheric entry to landfall in Florida, Columbia streaked hypersonically across half of the planet, creating a spectacular light show for observers on the ground and for her crew. Story Musgrave, making his sixth Shuttle re-entry and the last of his career, had planned to videotape the entire hour, through the overhead flight deck windows. "I was able to observe it from Mach 25 all the way to the ground, which no one had ever done before," he told a Smithsonian interviewer. "The video is lousy, because I'm standing up with 80 pounds of gear on – helmet, gloves, parachute and so on – ready for bailout. I'm taking all the Gs after an 18-day Shuttle flight. I have no cooling in my suit, because I'm supposed to be downstairs, plugged into the cooling lines, instead of looking out the windows up on the flight deck. The video is a mess, but for whatever it's worth, I have it." As he gazed on the ever-changing plasma beyond Columbia's windows, Musgrave was convinced that he was seeing a cosmic-scale painting, organised by shock waves and the Shuttle's own thrusters. An hour after beginning her descent from orbit, Columbia touched down smoothly on Runway 33 at 6:49:05 am EST on 7 December, smashing the record set by the STS-78 crew by about 18 hours. At the instant of touchdown, Ken Cockrell's mission had lasted 17 days, 15 hours, 53 minutes and 18 seconds and had covered 11.2 million km. Until the end of the Shuttle era in July 2011, no other single mission would surpass STS-80 in terms of flight duration, offering a fitting tribute to Columbia and her multitude of crews.

2

A new partnership

When the crew of STS-63 was announced by NASA in September 1993, they could hardly have anticipated that theirs would be a voyage of destiny. Scheduled for launch in May of the following year, the five astronauts – commander Jim Wetherbee, pilot Eileen Collins and mission specialists Bernard Harris, Mike Foale, Janice Voss and Russian cosmonaut Vladimir Titov – were tasked with supporting a range of scientific and technological experiments aboard the Spacehab-3 research module and deploying and retrieving the SPARTAN-201 free-flying solar physics satellite. The eight-day mission would showcase many of the Shuttle's capabilities, it was true, and serve as a pathfinder for the future space station, but it was hardly a voyage of destiny.

The presence of Eileen Marie Collins on the crew added flavour to STS-63, for she was the first female Shuttle pilot ever selected into NASA's astronaut corps. In fact, when one counted women astronauts and cosmonauts, and excluding the Soviet Union's Valentina Tereshkova, Collins was the first female to serve in a command position aboard a space mission. In July 1999, after two missions as a pilot, she went on to become the first female commander of a Shuttle flight. She had been chosen as an astronaut candidate by NASA in January 1990, to great fanfare. Accompanied by US Army aviator Nancy Sherlock (later Currie) and US Air Force test engineer Susan Helms, she was one of the first three military women astronauts ever selected by the space agency and alone among them in her selection as a pilot. By the end of her astronaut career in 2006, Collins would have flown four missions, including command of STS-114, the flight which returned the Shuttle fleet to operational status in the aftermath of the Columbia disaster.

In spite of her historic position in the annals of human space exploration, Collins had no pervasive childhood urge to someday become an astronaut. The daughter of Irish immigrants from County Cork, she was born in Elmira, New York, on 19 November 1956. "The desire to be an astronaut came later in life, but I was always interested in flying as a child," she told a NASA interviewer. Elmira was (and remains) the home of the National Soaring Museum, atop Harris Hill, and Collins grew up visiting summer camps and watching gliders flying overhead. "My family didn't have the money to give me flying lessons when I was a child," she admitted, "so I had to wait until I was 16, so I could get a job and save up money." Three years later, with $1,000 in savings to her name, she visited her local airport and asked for flying lessons. Collins began her aviation adventure in Cessnas, then gained her glider licence, and gravitated towards military service in the Air Force.

"I never really had a real talent for languages," she admitted, "but I found it very interesting because I enjoyed learning about different cultures and their history. I found that I was good at math and science and I think that's just because I was born with genes that favoured those subjects. That's where I decided to go with my career." After high school, Collins entered Corning Community College and earned an associate degree in mathematics and science in 1976, then graduated from Syracuse University in 1978 and entered the Air Force as one of four women chosen for pilot training at Vance Air Force Base in Oklahoma. Her instructor was an F-4 fighter pilot and Vietnam War veteran and he inspired the young Collins to succeed. Upon receipt of her wings, she remained at Vance for three years as a T-38 Talon instructor pilot, then moved to Travis Air Force Base in California to fly Lockheed's C-141 Starlifter. Collins gained a master's degree in operations research from Stanford University in 1986 and served for three years as an assistant professor of mathematics at the Air Force Academy and as a T-41 Mescalero instructor pilot. In 1989, following receipt of a second master's degree in space systems management from Webster University, she became the second woman ever chosen to attend Air Force Test Pilot School and was selected into NASA's astronaut corps as the first female Shuttle pilot in January 1990.

Astronaut was an attractive profession to Collins, who first took notice when NASA selected its first female Shuttle mission specialists in January 1978. "That's when the career of being an astronaut first became a reality to me," she recalled, "and when I really started looking forward to doing this someday." Coupled with her love of aviation, she harboured a fascination for history, astronomy and geology and considered NASA the perfect place to marry each of these interests. In 1989, she applied for *both* the Shuttle pilot *and* mission specialist categories, acknowledging that both options appealed to her. "I think the reason NASA chose me as a pilot was due to all of my flying experience," she said. "Being chosen as a mission specialist would have been great, too. I would have had the opportunity to do spacewalks or to work the robot arm. I would have learned more about the actual science of it. Both jobs really do have a lot to offer."

Looking back to her childhood, she remembered an incident when she was in the fourth grade. Her school subscribed to a magazine which once featured an article about the pros and cons of spending money on the space programme. Collins could not understand why anyone would oppose such spending. "It was obvious to me, as a fourth-grader, that we needed to learn about space," she said. "I was more fascinated with what we *didn't* know, rather than what we *did* know. That was when I started learning about the astronauts and their backgrounds and figuring out what you needed to do to become an astronaut." By the beginning of February 1995, after

38 years of life and five years since being selected by NASA into the world's most elite flying fraternity, Collins was poised on the cusp of realising her ultimate dream.

By contrast, her five crewmates had all flown into space before. James Donald Wetherbee, the commander, came from Flushing, New York, where he was born on 27 November 1952. He grew up *tall* in more ways than one … in fact, even today, at 1.93 m – six feet and four inches – he remains one of the tallest spacefarers ever launched. Wetherbee attended high schools in New York and earned a degree in aerospace engineering from the University of Notre Dame in 1974, then entered the Navy, hoping to become a pilot and land on aircraft carriers. Certainly, aviation was important to him, since his father had served as an Army Air Corps aviator in the Second World War and later became an American Airlines captain and chief pilot. In his youth, becoming an astronaut was a dream, but Wetherbee recognised that the chances of bringing that dream to fruition were very low. He received his aviator's wings in December 1976 and trained initially in the A-7E Corsair, flying for three years aboard the USS *John F. Kennedy* and accumulating more than a hundred carrier landings. Wetherbee's next step was test pilot school, from which he graduated in 1981, and he was serving as a project officer for the weapons delivery systems and avionics integration of the new F/A-18 Hornet fighter. He later flew the Hornet as an operational pilot. It was his wife, Robin, who spotted NASA's call for astronauts and encouraged Wetherbee to apply. Selected in May 1984, with the rank of a lieutenant, Wetherbee was the youngest and most junior of the pilots … yet he would fly more often than any of them. He had drummed for the marching band at Notre Dame as an undergraduate and had put his kit away when he entered the Navy – "I *couldn't* take the drums on an aircraft carrier!" – but little did he know that in the summer of 1987 he would be recruited by astronauts Robert 'Hoot' Gibson and Brewster Shaw to join the all-astronaut 'Max Q' rock band. By the time of his assignment to STS-63, Wetherbee had two Shuttle missions under his belt, having served as pilot on STS-32 in January 1990 and as commander of STS-52 in October 1992.

Leading the quartet of mission specialists was Bernard Anthony Harris Jr, who became only the sixth African-American astronaut when he launched on his first space flight, STS-55, in April 1993. Born in Temple, Texas, on 26 June 1955, Harris completed high school in San Antonio, from which he emerged with a strong interest in the sciences. He entered the University of Houston and earned a degree in biology in 1978, followed by a medical doctorate from the School of Medicine of the Texas Tech University Health Sciences Center in 1982. After an internal medicine residency at the Mayo Clinic, he completed a National Research Council Fellowship at NASA's Ames Research Center in 1987 – with an emphasis upon musculature physiology and disuse – and trained as a flight surgeon at Brooks Air Force Base's Aerospace School of Medicine in San Antonio the following year. His fellowship at Ames led to a position at the Johnson Space Center (JSC) in Houston, Texas, as a clinical scientist and flight surgeon, in which post Harris worked on space adaptation syndrome countermeasures. Selected by NASA in January 1990, Harris was one of the earliest members of his class to draw a flight assignment. On STS-63, his second mission, in addition to fulfilling the role of payload commander, with key

2-01: The STS-63 crew, including Russian cosmonaut Vladimir Titov (far right), are pictured during pre-flight training.

responsibility for the mission's scientific objectives, he would become the world's first African-American spacewalker.

STS-63's flight engineer was British-born Colin Michael Foale, who went on to fly six space missions, including two long-duration expeditions aboard Russia's Mir space station and the International Space Station (ISS), and a Hubble Space Telescope (HST) servicing. Foale, who was awarded a CBE in the Queen's New Year's Honours List in December 2004, came from Louth, in Lincolnshire, England. He was born on 6 January 1957, the son of a Royal Air Force pilot father and an American mother. "I grew up with the sound of jets," Foale later told the NASA oral historian, "and I lived in exotic places and I developed a taste for not so much adventure, but new vistas, new places, new things. I quickly decided that I wanted to fly." Frequent visits to his mother's family in Minnesota introduced him to the space programme as a child, when he saw John Glenn's Friendship 7 capsule in the state fair. "My father didn't discourage in any means my interest in being a pilot or an astronaut," he reflected. "I'm not sure he credited very realistic my dreams and aspirations there, especially since Britain did not have – and *still* does not have – a human spaceflight programme. However, there was quiet support."

A test piloting and military aviation career beckoned, but in his mid-teens Foale was misdiagnosed with a vision issue and it was this turning point which altered his focus toward the sciences. He entered Queen's College, Cambridge, and would earn a first-class degree in physics in 1978. By the time the misdiagnosis was uncovered and he realised that his vision was perfectly fine for military aviation, NASA's astronaut requirements had changed. Test piloting credentials were no longer mandatory and he could apply as a scientist. After his degree, he completed a doctorate in laboratory astrophysics in 1982 and made his first move towards NASA the following year. In June 1983, Foale entered the Johnson Space Center in Houston, Texas, as a payload officer in Mission Control. He unsuccessfully applied for admission into the astronaut corps in 1984 and 1985, but in the aftermath of the Challenger disaster he altered the focus of his application essay from describing his own dreams to considering the managerial realities faced by NASA. On his third attempt, he was accepted as an astronaut candidate in June 1987. By the time of his assignment to STS-63, Foale had two Shuttle missions under his belt.

Janice Elaine Voss came from South Bend, Indiana, where she was born on 8 October 1956. Space exploration as a real possibility first entered her head whilst in the sixth grade, when she read Madeleine l'Engle's novel *A Wrinkle in Time*; this guided her deeper into science fiction and from thence into science generally. "I just found the whole thing so fascinating," she told the NASA oral historian before her last Shuttle mission, "that I've never been interested in doing anything else." Her love of learning was such that, in 1998, with four Shuttle missions and a handful of degrees to her credit, she once told a group of pre-teen girls that her next personal goal was to learn the piano. In high school, she pursued mathematics and science. Later, she earned a degree in engineering from Purdue, whilst working on a NASA co-operative programme in computer simulations with the Engineering and Development Directorate at the Johnson Space Center. Reflecting on Purdue, Voss was well aware of the fact that the exalted college had produced more astronauts than any other US educational institution, outside of the military academies. After gaining her degree in 1975, she completed a master's qualification in electrical engineering and returned to JSC in 1977 as a crew trainer, teaching re-entry guidance and navigation. Ten years later, she gained her doctorate in aeronautics and astronautics from Massachusetts Institute of Technology and worked for a time for Orbital Sciences. Finally, on her fifth application to NASA, she was accepted as an astronaut candidate in January 1990 and flew her first mission on STS-57 in the summer of 1993.

The final member of the STS-63 crew, Vladimir Georgyevich Titov, spoke with a distinctly different accent. He was only the second Russian cosmonaut ever to fly aboard the Shuttle as part of an agreement between the two former superpowers to merge their respective space programmes and co-operate on the construction of the ISS. When he joined STS-63, Titov had more than six times as much space flight experience as the rest of the crew, put together. Yet the notion of being a cosmonaut, he said, was 'just' his job, and it was a job that he loved with every fibre of his being. Even as he grew older, commanding the first year-long Mir mission in 1987-1988 and flying the Shuttle twice with American crewmates, he would yearn for more. It was an enthusiasm that he would pass on to his son, although not only a love for space, but a love for the Home Planet, too. "He doesn't understand about borders," Titov told a NASA oral historian in the summer of 1998. "For him, it's the same: America or Russia. He doesn't want to understand ... different countries ... because, for him, it's his house here, house there. It's the same planet, same home." In 1983, Titov was almost granted the uncommon opportunity to view that planet from space, not once, but *twice*. Both opportunities would meet with intense disappointment: on the first occasion, in the late spring, aboard Soyuz T-8, he and fellow cosmonauts Gennadi Strekalov and Alexander Serebrov failed to dock with the Salyut 7 space station and, on the second, in the autumn, again teamed with Strekalov, Titov came within a hair's breadth of death when their rocket caught fire and exploded on the launch pad. He was born on 1 January 1947 in Sretensk, in the Zabaykalsky Krai region of southern Siberia, which shares lengthy borders with China and Mongolia. After completing his secondary education, he entered the Higher Air Force College at Chernihiv in the Ukraine in 1970 and remained for several years at the college, serving as a pilot-instructor, responsible for up to a dozen students. Titov later worked as a flight commander with the air regiment, flew ten different types of aircraft – including the MiG-21 supersonic fighter – and accrued 1,400 hours in his logbook and the qualifications of Military Pilot First Class and Test Pilot Third Class. He was selected as a cosmonaut in August 1976 and by the time of his assignment to STS-63 had accrued 368 days in space on two missions.

With Russian cosmonauts flying aboard the Shuttle, the human space programmes of the United States and the former Soviet Union were accelerating towards a long-term partnership. Of course, this co-operation did not come from nowhere. A one-off joint manned mission, the Apollo-Soyuz Test Project (ASTP), had been conducted in July 1975, but simmering mistrust between the two nations in the years which followed cooled their attitudes towards each other substantially. A little more than a decade later, efforts steadily resumed to foster closer ties. The reasons were chiefly political. As the Soviet bloc teetered on the road to collapse, the risk of Russia haemorrhaging its technology to undesirable destinations, such as Iran, the United States took a keen interest in keeping the vast country together. It would help, in the words of US President George H.W. Bush, to preserve "a new world order". As early as July 1990, NASA contracted to fly one of its Total Ozone Mapping Spectrometer (TOMS) instruments aboard a Soviet Meteor-3 satellite for joint environmental monitoring and in July of the following year the countries agreed to fly a Russian cosmonaut aboard the Shuttle and a US astronaut for several months aboard Mir. The deal was inked between President Bush and Soviet General Secretary Mikhail Gorbachev at a two-day summit in Moscow. It followed initial discussions, a year earlier, led by Vice President Dan Quayle, in his role as chair of the National Space Council. "The purpose of the exchange of flights," explained a NASA news release, "is to conduct life sciences research of mutual interest. It would advance current efforts to standardise in-flight medical procedures, which would improve comparability of data taken by each side."

By the summer of 1992, the Soviet Union had collapsed. A bloodless coup the previous August – only three weeks after Bush and Gorbachev's summit – by hard line Communists had failed and in December Gorbachev stepped down from power. In his place at the head of the new 'Russian Federation' came President Boris Yeltsin and in June 1992 his fragile nation and the United States issued a 'Joint Statement on Co-operation in Space'. For the first time, it laid out the plans for cosmonauts to be flown aboard the Shuttle, a docking mission between the Shuttle and Mir in 1994-1995 and at least one long-duration flight (of at least 90 days) by a US astronaut. After several other meetings during the summer, in October NASA Administrator

Dan Goldin and his counterpart Yuri Koptev, Director-General of the newly established Russian Federal Space Agency, met in Moscow to sign the 'Implementing Agreement on Human Space Flight Co-operation', which explored the plan in further depth. Titov and fellow cosmonaut Sergei Krikalev were named to train for a Shuttle mission and a US astronaut would spend several months aboard Mir. Moreover, the Shuttle would physically dock with Mir and exchange crew members. At around the same time as Goldin and Koptev's meeting, in November-December 1992, US and Russian working groups began exploring the possibility of utilising the Soyuz-TM spacecraft as an 'interim' Assured Crew Return Vehicle (ACRV) for the future space station.

Aside from human space missions, this burgeoning spirit of co-operation permeated into other areas, too. In May 1992, for example, NASA and Russian scientists participated in a joint research expedition to the Bunger Hills Oasis of eastern Antarctica, which enabled them to practice working in extreme environments and demonstrate advanced robotics and telescience applications. Several weeks later, in early July, NASA appointed Sam Keller as its first Associate Administrator for Russian Programs to oversee the dawn of a new era of collaboration between two former foes. In the months which followed, the countries partnered on the development of joint scientific instruments for the Mars '94 mission to the Red Planet, engaged in the geological exploration of the volcanic Kamchatka Peninsula and in October 1993 NASA physicians supported their Russian counterparts through a unique satellite-based telemedicine programme, known as 'Spacebridge to Moscow'.

In April 1993, Bill Clinton and Boris Yeltsin met in Vancouver, Canada, and one of the points of discussion included an expansion of joint human missions to feature not one, but *several*, long-duration visits to Mir by US astronauts. Initially, the talk was of a pair of three-month missions and as many as four others, lasting maybe six months apiece, and it was agreed that in return Mir's new Spektr ('Spectrum') and Priroda ('Nature') scientific modules would be subsidised and utilised for US research equipment.

Also that April, Sergei Krikalev was formally assigned as a mission specialist on STS-60, which finally flew in February 1994, with Vladimir Titov serving as his backup. It came as little surprise when Titov later joined the STS-63 crew, which, for the first few months, appeared to be little more than another Russian cosmonaut on a Shuttle science mission. The situation changed markedly in September 1993, when US Vice President Al Gore and Russian Prime Minister Viktor Chernomyrdin chaired the first meeting of the US-Russian Joint Commission on Energy and Space. The two leaders agreed to begin 'Phase 1' of International Space Station co-operation with the so-called 'Shuttle-Mir' project, encompassing 21 months of total US astronaut time aboard the Russian space station. Within weeks, in early November, an addendum had merged the American, European, Japanese and Canadian components from the former Space Station Freedom with Russia's planned Mir-2. Under its provisions, 'Phase 2' would then realise the construction of the ISS and 'Phase 3' would see full operations as an international scientific research facility, to be permanently staffed by a six-strong crew of astronauts and cosmonauts by 2001.

2-02: Discovery roars into orbit on 3 February 1995.

The first Shuttle-Mir docking mission was scheduled for STS-71 in June 1995, but the language of the documents produced at the Gore-Chernomyrdin Commission in December 1993 added flesh to the bones of up to *nine* follow-on flights. It was highlighted that American financial support was the only available lifeline which might save the increasingly decrepit Mir from an ignominious end. On 16 December 1993, Dan Goldin and Yuri Koptev signed the 'Contract for Human Space Flight Activities' and Shuttle-Mir was open for business. Specifically, it was highlighted that STS-74 in October-November 1995 "may include Russian cosmonauts" and would carry an array of equipment to Mir, "including power supply and life-support system elements" and would "include activities on Mir and possible extravehicular activities to upgrade solar arrays". Further downstream, it was expected that the United States and Russia would jointly develop a solar dynamic power system, "with a test flight on the Space Shuttle and Mir in 1996", together with "the joint development of spacecraft environmental control and life-support systems and studies on potential development of a common space suit, starting with the compatibility of respective space suits". These ambitious Shuttle-Mir flights would deliver and exchange long-duration, "NASA-designated" astronauts, who would focus their attention upon "a specific programme of technological and scientific research" aboard Spektr and Priroda to "expand ongoing research in biotechnology, materials sciences, biomedical sciences, Earth observations and technology".

The first mission under the Shuttle-Mir umbrella would be STS-63, tasked with performing a close-range rendezvous with Mir and testing VHF radio communications and navigation equipment, ahead of the first planned docking flight by Space Shuttle Atlantis on STS-71 in June 1995. Under the new plan, STS-63 commander Jim Wetherbee would guide the Shuttle to a distance of just 200 m from Mir. The idea of performing a rendezvous had been under discussion for several months. As early as April 1993, *Flight International* explained that NASA was "considering" such a mission as a key preparatory step, ahead of the first docking mission. "The complexity of the Atlantis mission, particularly the close proximity flying with the Mir – the largest object with which the Shuttle has made rendezvous and one with several large solar arrays – has made NASA consider making STS-63 a rehearsal mission," it was pointed out.

By assigning the Mir rendezvous commitment to STS-63, NASA was also faced with the challenge of redesigning the mission's trajectory plan. Instead of launching into a standard 28.5-degree-inclination orbit, the Shuttle would fly into a 51.6-degree path, in order that it could enter the same orbital plane as the Russian station. However, STS-63 had already been delayed from May 1994 until no earlier than January 1995, due to a paucity of bookings for experiments aboard its Spacehab-3 module. In the months to come, experiments already assigned to the third and fourth Spacehab flights were combined and produced a full plate of mainly NASA payloads on STS-63. "Spacehab's position reflects the lack of interest in commercial microgravity processing," explained *Flight International*, "which has been caused mainly by the high cost of experiments and the long-term nature of the research to develop viable processes."

According to Jim Wetherbee, the original plan for the Mir rendezvous was to approach no closer than about 330 m, although US and Russian trajectory planners and managers eventually reduced this to 120 m and finally 33 m. However, for Wetherbee, 33 m seemed impractical, since one of their key goals was to test procedures and the tracking radar and hand-held laser range-finders. "All of these systems give range and range-rate information to the target – Mir, in this case – and the camera system that was going to optically give us information on the attitude misalignment by looking at the target," he told a NASA interviewer. "We were going to check out the procedures and the flight profile and the directions that we were approaching. The first thing that I noticed was that ... the visual target that we used *couldn't* be used accurately until we got to about 30 feet [10 m]." Although Wetherbee was able to convince the NASA trajectory planning team to accept a closure distance of just 10 m, gaining Russian approval was a far harder nut to crack. "It resulted in a pretty interesting series of meetings and discussions about why we wanted to get so close to their station," said Wetherbee, "that is, of course, a national asset, like our Shuttle. It was risky enough for them, for little benefit, since we weren't actually going to dock and transfer any people." Psychologically, it was hard for the Russians, having the Shuttle approach their station and having no effective control. At one meeting, Wetherbee was in discussion with Viktor Blagov, co-chair of the Russian Flight Operations and Systems Integration Working Group, who probed him on the issue. Wetherbee responded that, on STS-71, the Shuttle commander, Robert 'Hoot' Gibson, needed to halt at 10 m to make visual corrections, before moving in for the final approach. "It would be very good," Wetherbee told Blagov, "if we could get in there and make that kind of an assessment, to see whether or not the target was going to work, and if it *didn't*, we'd still have time to change things. I don't know that we convinced him that day, but in the next couple of weeks they finally said, 'Yes, it's okay. You can go to 10 m.' I was happy at that point, because we were going to get to do a pretty valuable piece of test flying."

It was made clear to Wetherbee, by representatives on the US and Russian sides, that he was not to approach Mir closer than 10 m. "I was *not* going to violate that limit, by even a single centimetre," he said. "I was going to go to 10 m and I wasn't going to be one centimetre closer, because I knew *both* space agencies in *both* countries were going to be watching as we approached on this flight." With Spacehab payload problems having already postponed STS-63 from May 1994 until at least January 1995, there would be a gap of just a few months before the rendezvous mission and the first docking mission and this correspondingly shortened the period of time for any lessons learned to be implemented into training simulations for the STS-71 crew.

As the STS-63 crew dug in with their training, on 3 February 1994 NASA announced the selection of veteran Shuttle astronauts Norm Thagard and Bonnie Dunbar to train for the United States' first long-duration space station mission since the Skylab era. Lasting three months, the flight would begin with Thagard's launch aboard Soyuz TM-21, accompanied by two Russian crewmates, in March 1995, and would end with the arrival of the first Shuttle-Mir docking mission in June. STS-71 would carry a crew of two replacement cosmonauts for Mir and would bring

Thagard and his Russian crewmates back to Earth. It was noted that Dunbar's training in a backup capacity would position her appropriately for a long-duration mission at a later date. Two weeks after the assignment of Thagard and Dunbar, veteran astronaut Ken Cameron was named as NASA's first director of operations in Russia, based at the Star City cosmonaut training centre, near Moscow. "Cameron's responsibilities will include supervising NASA astronaut training at Star City, developing a training syllabus for Shuttle crew members for Mir rendezvous missions and co-ordinating training for scientific experimenters," it was explained. Other tasks for the role included "establishing and maintaining operations, operational relationships, plans and procedures to support flight operations between NASA and the Russian Space Agency in joint Shuttle-Mir flights and space station development, assembly and operations". It was noted that Cameron was tipped to command one of the early Shuttle-Mir docking missions and, indeed, he went on to lead the STS-74 flight in November 1995. Cameron's assignment began a directorship which endures to this day in the ISS era. In support of these early astronaut appointments, NASA's Office of Space Communications installed a video teleconferencing facility at Brown University in Providence, Rhode Island, to expand its coverage to Russia.

In May 1994, negotiating teams for NASA and the Russian Space Agency met for the first time and so too did the Task Force on the Shuttle-Mir Rendezvous and Docking Missions, chaired by former astronaut Tom Stafford, in order to review planning, management, training, operations and to make recommendations. On 23 June, Gore and Chernomyrdin met in Washington, DC, to sign a definitive contract between the two space agencies, under which the United States would pay $400 million for up to 21 months of astronaut time aboard Mir, up to nine Shuttle dockings after STS-71, a specialised docking module and the use of the station's Spektr and Priroda modules for research.

At around the same time, the General Accounting Office in the United States was highly critical of the increased costs associated with collaborating with Russia, including the need to upgrade the Shuttle with expensive rendezvous and docking hardware. It was noted that about $746 million from additional Shuttle missions and the $400 million contract to fly astronauts for long-duration missions to Mir should be scored against the projected $2 billion in savings to the US taxpayer by bringing Russia into the ISS partnership. NASA Administrator Dan Goldin responded that accessing Mir – a long-duration resource in orbit – would be enormously beneficial as the United States strove to develop its own space station. Yet there was something more fundamental in Goldin's mind. "While there are tangible benefits to Russian co-operation, which the GAO report fairly and accurately points out," he said, "auditors cannot put a price tag on the intangible benefits of international co-operation. It's good foreign policy and it's good space policy. The Cold War is over and co-operation with the Russians demonstrates that former adversaries can join forces in a peaceful pursuit which will generate tremendous benefits for both nations."

By the dawn of 1995, the STS-63 launch had slipped until no earlier than 2 February. Russian mission managers had insisted that both the rendezvous, nicknamed 'Near-Mir', and subsequent dockings had to occur whilst Mir was in contact with Russian ground stations, requiring the Shuttle's final approach to occur during orbital darkness. "To make a daylight rendezvous would mean delaying [STS-63] for at least two weeks," *Flight International* told its readers on 25 January, "which cannot be tolerated within the NASA Shuttle schedule for 1995." Although the Mir rendezvous on Flight Day Four was arguably the most publicly visible aspect of the eight-day mission, the primary payload aboard Space Shuttle Discovery was the Spacehab-3 module, loaded with over 20 experiments, sponsored chiefly by NASA's Offices of Space Access and Technology and Life and Microgravity Sciences and Applications, together with the Department of Defense. It continued the capabilities demonstrated on two previous flights of the commercial middeck expansion module in June 1993 and February 1994, but had been modified with a number of features to minimise the demands on crew time. These included a video switch to reduce the need for the astronauts to work on video operations and an interface between the experiments and the Spacehab module's telemetry to cut crew time on data downlink. On earlier Spacehab missions, the astronauts had been required to set up camcorders and manually switch from one to the next, which proved extremely time-consuming, but on STS-63 up to eight cameras could be cabled into the module's video switch. Mission Control could then switch one of the camcorders into the Shuttle system for downlink capability. In support of the Near-Mir rendezvous, a pair of 30 cm windows were installed in the module's flat roof. One of these windows housed a NASA docking camera to assist in proximity operations.

Launch on 2 February was postponed by 24 hours, due to the failure of one of Discovery's three Inertial Measurement Units (IMUs), which was promptly replaced and tested. With a 'window' barely five minutes in length, the precise liftoff time was calculated relatively late in the countdown, based upon new Mir state vectors for the Shuttle's rendezvous phasing requirements. These were updated about 60 minutes ahead of launch and Discovery and her crew rocketed away from Pad 39B at the Kennedy Space Center (KSC) in Florida at 12:22:03 am EST on 3 February, turning night into day along much of the United States' east coast. Although Jim Wetherbee was embarking on his third Shuttle flight, he made sure to check on the progress of his rookie pilot, Eileen Collins, but need not have worried, for her performance was that of a veteran. "As a commander flying with any rookie, you always want to see how they're going to do," he told an interviewer, years later, "and you can tell pretty much when they get in the vehicle that they're going to be okay. During the liftoff, I could hear the things she was saying were just like we say in the sim[ulator] and so I knew she was performing like a veteran, instead of a rookie, and that was good."

Within eight minutes, Discovery had reached her preliminary orbit, although not without problems, for one of the aft-mounted Reaction Control System (RCS) thrusters failed and another exhibited a minor leak. As soon as the External Tank (ET) was jettisoned, Wetherbee and Collins were alerted by a 'Jet Leak' and two 'Jet Off' RCS messages on their instrument panel. The Shuttle carried a total of 44 primary and backup RCS thrusters on its aft compartment and on its nose, which were employed for manoeuvring in orbit. Two aft-mounted thrusters had failed, one on the port side of the vehicle (L2D) and one on the starboard (R1U). The former was 'deselected' and remained off for the remainder of the mission, but R1U was a primary thruster and critical to the Mir rendezvous. Wetherbee was dismayed: he had trained for more than a year to test critical procedures and equipment in support of STS-71 and his mission was potentially hamstrung by an RCS malfunction, only minutes after achieving orbit. "As the day wore on, I was pretty much convinced that we were not going to be able to rendezvous with Mir," he said, "and it was pretty disappointing." Although it was stressed by NASA that the thrusters were fully redundant, with plenty of backup capability, and were not expected to cause any violations to the Near-Mir rendezvous, it did not bode well for the mission. Flight rules dictated that all aft-firing RCS needed to be operational for the orbiter to be permitted to advance closer than 330 m from Mir. Later that afternoon, Mission Control asked Wetherbee to position the Shuttle in such a manner that the Sun would shine onto the vehicle's top side in an effort to warm up the leaking thruster, which was losing up to 0.9 kg of oxidiser every hour. This was somewhat less than originally feared and mission managers declared that it was within limits and the thruster's temperature of 12.2 degrees Celsius remained stable and well above its minimum redline of 4.4 degrees Celsius. It was recognised that if the thruster's temperature dipped below this redline, it would be shut down, necessitating the use of another RCS thruster ... and eliminating any chance of approaching to within 330 m of Mir. In the meantime, about nine hours after launch, the crew executed the first of a series of RCS phasing burns to adjust their altitude and begin the rendezvous procedure, which would conclude with a flyby of Mir on 6 February. By the end of the first day of the mission, Discovery was trailing the Russian space station by about 11,260 km and closing the distance by about 330 km with each 90-minute orbit of Earth. Over the next three days, Wetherbee and Collins would perform five manoeuvres to position Discovery at a point about 75 km 'behind' Mir at the same altitude as the Russian space station, preparatory to the final manual rendezvous.

Late on 3 February, the Orbital Debris and Radar Calibration Spheres (ODERACS) was deployed from a Getaway Special (GAS) canister in Discovery's payload bay. This experiment, which had flown on several previous Shuttle missions, provided a means of observing small targets, injected into low-Earth orbit, for the purposes of calibrating ground-based radar and optical systems in the ongoing effort to characterise the presence of potentially hazardous debris. Since the launch of the world's first artificial satellite, Sputnik 1, in October 1957, more than 3,200 rockets had placed over 6,500 payloads, weighing in excess of two million kilograms, into orbit. Although the Space Surveillance Network of US Space Command (USSPACECOM) catalogued all satellites, only a tiny percentage represented fully functional spacecraft, with about 94 percent classified as debris. Moreover, USSPACECOM only had the capability to track objects greater than 10 cm in diameter and it was known that far smaller objects – including paint flecks – carried the potential to cause significant damage to spacecraft, with the future ISS particularly at risk. Following their deployment from STS-63, the six ODERACS spheres, which varied in diameter from 5 cm to 15.2 cm and were of polished, blackened or whitened stainless steel or aluminium, remained in orbit for between 20 and 280 days, before burning up in the atmosphere. During their short lives, they were observed by tracking stations in Florida, North Dakota, Massachusetts, Germany and the Marshall Islands of the South Pacific.

About six hours after launch, Vladimir Titov powered up Discovery's Canadian-built Remote Manipulator System (RMS) mechanical arm and used it to grapple the Shuttle-Pointed Autonomous Research Tool for Astronomy (SPARTAN)-204 payload for 4.5 hours to examine the effects of the mysterious 'Shuttle glow' phenomenon. Like the SPARTAN-201 payload flown aboard other missions, SPARTAN-204 was tasked with performing astronomical research during two days in autonomous free flight. However, for its early operations at the end of the RMS, it was focused on the Shuttle herself, acquiring far ultraviolet imagery of day and night primary RCS thruster firings. Its primary instrument was the Naval Research Laboratory's Far Ultraviolet Imaging Spectrograph (FUVIS), which was employed to determine the ultraviolet spectral intensity distributions in Shuttle glow and thruster plumes and the chemical species making the emissions. The experiment was funded exclusively by the Department of Defense. After these initial activities were completed, Titov returned SPARTAN-204 to its berth in the payload bay. It would be deployed for its core astronomical research programme later in the mission.

In the meantime, under the direction of Bernard Harris, the activation of more than 20 experimental payloads aboard the Spacehab-3 module got underway. These included 'Charlotte', an experimental robot, built by McDonnell Douglas Aerospace, which was designed to demonstrate automated servicing of experiments and allow remote video observations within the module. Charlotte had no gantries, jointed arms or complex systems, but was instead suspended on cables, which were easy to install and remove. It was used to operate knobs, switches and buttons inside Spacehab-3, and it also carried the potential to change experiment samples and data cartridges and perform inspections and manipulation tasks, thereby freeing up valuable crew time.

Other experiments included the Astroculture plant growth system. Although making its fourth Shuttle mission on STS-63, this was only the first time that it actually carried plants. On its three previous flights, it was employed to test water and nutrient delivery systems, light-emitting diodes and temperature and humidity controls, but aboard Spacehab-3 it was loaded with wheat seedlings and special fast-growing plants, developed at the University of Wisconsin-Madison's College of Agriculture and Life Sciences. It tested the NASA-built Zeoponics nutrient composition control system. Astroculture also evaluated carbon dioxide controls, contaminant removal facilities and a video and data acquisition feature, ahead of a long-duration, 16-day mission on STS-73 in late 1995 to study plant starch metabolism and carbohydrate translocation in potato leaves. Located in Discovery's middeck was the Biological Research in Canisters (BRIC) payload, which investigated carbohydrate production in the microgravity environment and its effect upon the development of soybeans, which are an ideal food source for future long-duration space missions. The CHROMEX experiment explored the relevance of changes within plant cell walls to the development and appearance of abnormalities

2-03: Vladimir Titov works aboard the Spacehab-3 module.

in cell shape and structure. Investigators used superdwarf wheat, which had been planted 48 hours before launch, and which developed under laboratory conditions until the specimens were loaded aboard the Shuttle for flight.

Elsewhere, the Bioserve Pilot Laboratory (BPL) was used to study the behaviour of *Rhizobium trifolii*, a bacterium which infests plants at an early stage in their seedling development and forms nodules on their roots. The bacteria in these nodules derive nutrient support from the plants and, in turn, provide the plants with nitrogen fixed from the air, in what is known as a 'symbiotic relationship'. Other plants which form such relationships include alfalfa, clover and soybean and they do not require synthetic fertilisers into order to grow. By understanding the process of infection, it was hoped to develop techniques for manipulating the process which causes infection in other crop plants, potentially making enormous savings in fertiliser production. The BPL also supported research into the bacterium *E. coli*, usually found in mammalian gastrointestinal tracts, and was under investigation in terms of developing techniques for controlling bacterial infection within closed environments and exploiting microorganisms in future ecological life-support systems and for waste management. BioServe, a NASA Center for the Commercial Development of Space at the University of Colorado at Boulder, together with Kansas State University, also provided IMMUNE-2, which investigated the ability of polyethylene glycol-interleukin-2 (PEG-IL2) to prevent or reduce the detrimental effects of space flight exposure on the immune-system responses of laboratory rats. This carried potential future applications for studying the responses of the human immune system to the microgravity environment, because PEG-IL2 (and recombinant IL2) was known to be an immunoregulatory agent to treat microbial infections, particularly in the elderly. The six white male rats on STS-63 were housed in a pair of Animal Enclosure Modules (AEMs) in Discovery's middeck. The National Institutes of Health supported three experiments in a Space Tissue Loss (STL) module to study animal and human bone-forming cells (known as 'osteoblasts'), the muscle cells of rats and the growth rate, development and mineralisation of chicken osteoblasts. It was anticipated that such research would contribute to the diagnosis and treatment of prolonged skeletal immobilisation and mineral abnormalities, which carried great weight in planning future space missions.

Meanwhile, the Commercial Generic Bioprocessing Apparatus (CGBA) supported a range of investigations in pharmaceutical testing and development, small agricultural and environmental systems development and biotechnology and biomaterials systems development. These experiments focused upon immune system disorders, bond and development disorders, wound healing and cancer and cellular disorders and their results carried important applications for the treatment of cancer, osteoporosis and AIDS. Other CGBA research involved seed germination, the development of brine shrimp and the growth of large protein and ribonucleic acid (RNA) crystals for use in the commercial drug development industries. A series of commercial protein crystal growth experiments were also carried aboard Discovery, employing the Vapour Diffusion Apparatus and the Protein Crystallisation Facility in order to produce large, well-ordered specimens with X-ray diffraction yields of superior quality to Earth-grown samples. Alpha interferon, which has found applications in the treatment of human viral hepatitis B and C, was used for the experiments and analysis of crystals grown on earlier Shuttle missions had revealed that in 20 percent of cases they were superior to their Earth-grown counterparts. A Hand-Held Diffusion Test Cell (HH-DTC) demonstrated experiment chambers for a future Observable Protein Crystal Growth Apparatus, which sought to understand why space-grown crystals were of greater quality than those produced on the ground. It was hoped that if the mechanisms of the growth process were better understood, scientists might be able to devise improved models for crystal production on Earth and in space. Candidate proteins in the study included lysozyme, haemoglobin, satellite tobacco mosaic virus, concanavalin-B and canavalin. The Protein Crystallisation Apparatus for Microgravity (PCAM) supported the temperature-controlled vapour diffusion of human cytomegalovirus assemblin, the HIV inhibitor pseudoknot-26 and the blood-clotting agent human antithrombin-III. With 378 samples housed in a single experiment, PCAM could handle more than six times the volume of any protein crystal growth apparatus on previous missions.

The University of Alabama in Huntsville provided the Equipment for Controlled Liquid Phase Sintering (ECLIPSE) payload, whose furnace sought to process specimens of copper for far longer periods (up to an hour at a time) than had been previously achievable aboard high-altitude sounding rockets. 'Sintering' is a process whereby metallic powders are consolidated into a metal at temperatures of only 50 percent of those required to melt all of their constituent phases, but on Earth the influence of gravitational factors, such as sedimentation, produced materials which lacked homogeneity and dimensional stability. In space, however, such gravitational factors were largely removed, opening up the possibility of developing metal composites within tough metal matrices with excellent strength and wear properties. On Earth, such composites were of interest in the construction of bearings, cutting tools, electric brushes and high-stress mechanical parts.

Other payloads included BioServe's Fluids Generic Bioprocessing Apparatus (FGBA), flying as part of efforts to develop advanced fluids management technologies in space, with particular interest from Coca-Cola, which wished to create hardware that would carbonate water on demand and mix and dispense its beverages with minimal loss of carbonation. During STS-63, the crew's taste perception changes were tracked as they sampled Coca-Cola and Diet Coke. The Gas Permeable Polymer Materials (GPPM) experiment investigated the similarities and differences between space-grown and Earth-grown polymers, whose applications included improved long-term-wear contact lenses and for technologies such as dialysis and blood-gas monitoring. The Solid Surface Combustion Experiment (SSCE) explored flame spreading across a Plexiglas sample, in a closed environment of 50 percent oxygen and 50 percent nitrogen at one atmospheric pressure. Measurements of the influence of acceleration on sensitive experiments were provided by the Space Acceleration Measurement System (SAMS) in Discovery's middeck and by the Three-Dimensional Microgravity Accelerometer (3-DMA), whose sensor heads were located at four different positions within the Spacehab-3 module. Unfortunately, 3-DMA collected data during the Shuttle's ascent, but was shut down after orbital insertion because its hard drives failed to cycle properly and caused it to overheat. It was switched to descent mode and repowered in time for re-entry on 11 February.

Other research facilities aboard Discovery included the Cryo Systems Experiment (CSE) and Shuttle Glow (GLO-2) experiments, which were mounted on a Hitchhiker truss structure in the payload bay. Together with ODERACS, the combined suite was known as CGP/ODERACS-2. The CSE payload validated and characterised the performance of a pair of thermal management technologies as part of a hybrid cryogenic system: a new-generation, long-life, low-vibration cryocooler and an oxygen diode heat pipe. It was expected to provide baseline data for the design of future cryogenic cooling systems for NASA and military spacecraft. Meanwhile, the GLO-2 experiment utilised a battery of imagers and boresighted spectrographs to examine the mysterious 'Shuttle glow' phenomenon, as well as the effects of the ambient magnetic field, the orbital altitude, the mission elapsed time and the impact of thruster plumes. In spite of difficulties with a problematic data recording unit, the experiment also successfully acquired spectra of Jupiter, the Moon and terrestrial aurorae.

Despite the successful conduct of the mission's scientific research, the RCS problems continued. The L2D aft thruster had been deselected, but the critical R1U thruster continued to leak. The problem was exacerbated when one of the forward RCS thrusters in Discovery's nose (F1F) began leaking during a two-phase 'hot fire' test on 4 February. During the test, the thruster's oxidiser injector temperature dropped to 73.8 degrees Celsius, and later fell still further to less than 4.4 degrees Celsius. This marked the third of the orbiter's 44 thrusters to experience a problem during STS-63. F1F's oxidiser supply line was closed and the Shuttle was manoeuvred into a nose-to-Sun attitude in an effort to warm the thruster. Jim Wetherbee and Eileen Collins also closed and reopened the manifold of F1F on several occasions, to allow the force of pressurising the manifold to clear any contaminants. They were ultimately successful in clearing the F1F leak, but were unsuccessful in their efforts with the R1U aft thruster. A total of four attempts were made to bring R1U back online, closing and reopening the manifold valves to allow the pressure to decrease and repressurise, and although the leak decreased it did not stop. "By the third day," reflected Wetherbee in his NASA oral history, "it was still leaking and you could look out the back window and you could see the propellant going up for *miles*. It kind of goes in a cone-shaped pattern, because there's no atmosphere to attenuate its motion, and it just goes up pretty straight and it just continues, like a snowstorm, for five miles up in space." Contamination of Mir was of paramount concern, not only in terms of the station's fragile solar arrays, but also the Soyuz TM-20 spacecraft, which would ferry the crew of Alexander Viktorenko, Yelena Kondakova and Valeri Polyakov back to Earth in late March. Its delicate optical sensors were needed to align the navigation platform for re-entry into the atmosphere and the risk of damaging them did not bear thinking about. On the evening of 5 February, the night before the rendezvous and closest approach between Discovery and Mir, it seemed to Wetherbee that Mission Control would give them a 'Go' to proceed to 10 m. Yet he remained uneasy.

He pulled Vladimir Titov aside.

"You know, if this leak doesn't get any smaller, I will *not* bring our vehicle close to the Mir," he told the cosmonaut, "even if they give us a 'Go', because I don't want to cause any problems for the cosmonauts when they're coming back."

By the time the astronauts bedded down for their third night's sleep in space, Discovery was trailing Mir by about 1,850 km, closing at less than 130 km with each orbit. Early on the morning of the 6th, only hours before arriving in the vicinity of Mir, they awoke to encouraging news: the R1U leak had slowed down and stabilised, but had not stopped. Looking out of the aft flight deck windows, they could *see* that its plume was less pronounced. In the meantime, NASA and Russian mission controllers settled upon two possible rendezvous scenarios. The first called for Discovery to manoeuvre, as planned, to a distance of 10 m, whilst the second called for a far more conservative approach to no closer than 120 m. The early stages of both scenarios were similar, calling for Wetherbee to fire the thrusters at 9:16 am EST and again at 10:02 am to decrease his rate of closure along the so-called 'V-Bar', approaching Mir along its velocity vector, or direction of travel, towards the space station's Kristall ('Crystal') module. Later, at 11:37 am, at a distance of 14.8 km, he would perform the so-called 'Terminal Initiation' (TI) burn to kick off the final stages of the rendezvous, coming to a halt about 120 m directly 'ahead' of Mir, just one orbit later, at 1:16 pm. "At this point, the Shuttle's rendezvous radar system begins providing range and closing rate information to the crew," explained NASA's STS-63 press kit. "The manual phase of the operation begins just after Discovery passes about [0.8 km] below Mir, when Commander Jim Wetherbee takes the controls at a distance of about [610 m]."

In their seminal work *Shuttle-Mir: The United States and Russia Share History's Highest Stage*, Clay Morgan noted that STS-63 Flight Director Bill Reeves was finding it increasingly difficult to explain the effect of the RCS malfunction to his Russian counterparts. At one stage, he attended a meeting which included all of the Mir systems experts and safety people. After Reeves completed his presentation, they asked him an odd question.

"We understand what you're saying, but what about the 180-gram snowball?"

Reeves looked blankly at them. "What are you talking about?"

"The 180-gram snowball in your fax."

It turned out that a group of NASA engineers had worked on an extremely remote, worst-case scenario of an RCS thruster freezing and becoming packed with ice. Since communications between Moscow and Houston were poor at the time, when the fax was sent it went directly to the Russians, bypassing Reeves. It left him in the rather awkward position of having to explain that the issue was not a major problem. By this point, it was expected that the decision would have been made as to whether to perform a flyaround of the station at 120 m or continue to 10 m. That decision and its transmission to the STS-63 came in a rather unusual fashion; not from Capcom Story Musgrave, seated in Mission Control in Houston, but from Russian cosmonauts Alexander Viktorenko, Yelena Kondakova and Valeri Polyakov ... aboard *Mir*. "We were testing out a new VHF radio that ... all the crews used to talk directly with the Mir," said Wetherbee. "You need to have, in the

2-04: Mir emerges from the gloom, as seen from Discovery.

two vehicles, an ability for the crew members to talk to each other, to co-ordinate things separately from the air-to-ground voice loop, from the ground control centre in Moscow and from the orbiter to the control centre in Houston. We had a separate communication, directly with Mir. The two mission control centres were co-ordinating the approach and they finally had agreed that they were going to let us make the final approach, but Houston didn't tell us right away. We heard from Mir, from the Russian cosmonauts, that they were going to give us a 'Go'."

Unfortunately, the actions of the STS-63 crew on the flight deck were being recorded and televised live to Houston, although without audio, and there was some surprise when mission controllers saw the astronauts and Titov cheering and clapping. The Mir crew had inadvertently stolen Mission Control's thunder and Wetherbee had to tell his crew to calm down until Houston gave them a definitive 'Go' to initiate the final approach. A few moments later, at 10:25 am EST, Story Musgrave radioed Wetherbee, with the slightest hint of humour in his voice: "We think you *know* this, because of the reaction we saw, but you *do* have a final 'Go' to approach to 10 m!" As if to echo Musgrave's words, the STS-63 crew 'simulated' their euphoria again.

"In hindsight," Eileen Collins told a Smithsonian interviewer, several years later, "this was one of the major successes of the Shuttle-Mir programme, because we had a real-time failure [involving the RCS], and in three days were able to come to an agreement and do the mission as planned. It showed that we can work together in space and, when it really came down to it, we had the same goals." Collins' own flight as the first female Shuttle pilot also did not go unnoticed and in the aftermath of STS-63 she received the Harmon Trophy, one of three awards which are given annually to recognise the world's outstanding aviator, female aviator and balloon or dirigible aeronaut.

Flying the orbiter from instruments on the aft flight deck, Wetherbee intersected the V-Bar and after passing within 330 m began executing RCS firings in a so-called 'Low-Z' mode, whereby he employed thrusters which were slightly offset to the space station, rather than those which pointed directly at Mir. This helped to eliminate the possibility of contamination. Additional laser ranging data was provided by the Trajectory Control System (TCS), mounted in Discovery's payload bay, and Wetherbee made use of the centreline camera in the roof window of the Spacehab-3 module. At 120 m, he halted Discovery to await permission to advance closer, which was granted. At 60 m, VHF ship-to-ship radio communications was initiated between the Shuttle and Mir crews and, by 2:23:20 pm, Discovery reached a distance of 12 m, or about 37 feet, from the station.

At least, that was the detail in the issued press releases. "The closest point of approach between structure was 33 feet [about 11 m]," said Wetherbee in his NASA oral history. "The number that was reported in the press was 37 feet, because that was the distance between the laser system and the target on Mir. You had to triangulate and that's the longer hypotenuse of the triangle. Not that *that* matters; it was just interesting to see that when we came back it was reported, and is *still* reported, as 37 feet, but it really was 33 feet and not one centimetre closer. We did not get a single *inch* closer to Mir." True to his word, Wetherbee had flown a perfect, precise rendezvous.

In the hours before the event, he had another issue on his mind: the right words to say over the space-to-ground communications loop when US and Russian piloted spacecraft reached their closest point in orbit for almost 20 years. Wetherbee knew that the rendezvous would be watched by both a US audience and a Russian audience and he was determined to deliver his message in both languages. That proved easier said than done, for although he had studied a little Russian, his powers of translation were poorer, and he approached Vladimir Titov for help. He wrote down what he wanted to say, showed none of his other crewmates, and passed it to Titov, who read it and translated it on his kneeboard. Armed with the translation, Wetherbee practiced a handful of times until he was satisfied that he could deliver the message and properly pronounce the words. As the point of closest approach neared, the time to speak those words finally came.

"As we are bringing our space ships closer together, we are bringing our nations closer together," Wetherbee told Mir's commander, cosmonaut Alexander Viktorenko, via VHF radio link. "The next time we approach, we will shake your hand and together we will lead our world into the next millennium."

"We are one. We are human," came Viktorenko's heartfelt reply.

For the three cosmonauts aboard Mir, there was no noticeable vibration or movement in the station's solar arrays and its attitude-control system performed normally. Wetherbee held Discovery at 12 m for about 15 minutes, before withdrawing to a distance of 120 m by 3:00 pm. He then completed a slow, quarter-loop flyaround of Mir, maintaining a relative distance of about 122 m from the station, and conducted a final separation manoeuvre at shortly after 4:00 pm. This positioned the orbiter 'ahead' of Mir and began their departure, opening up the distance between them with each successive orbit. Fortunately, no other RCS issues arose during the rendezvous. During a subsequent ship-to-ship communications session, relayed through an interpreter in Houston, Wetherbee and Viktorenko shared their personal perspectives of the historic operation. For Wetherbee, the high point was when Mir manoeuvred into a new attitude as Discovery performed the quarter-loop flyaround of the station. "It was like dancing in the cosmos," he told Viktorenko. "It was great." The two commanders pledged to meet on Earth after their respective missions were over.

With the Near-Mir operations behind them, the STS-63 crew turned their attention to the deployment of SPARTAN-204 on 7 February. At 7:26 am EST, Vladimir Titov released the satellite from the RMS arm and reported that it had successfully performed a pirouette to confirm the health of its attitude-control system. Over the next two days in free flight, SPARTAN-204 employed its FUVIS instrument to investigate astronomical and artificially induced sources of diffuse far ultraviolet radiation, including distant nebulae, celestial background radiation and galaxies. Of specific interest was the Orion nebula, a cloud of interstellar material which was excited to glow by the far ultraviolet light emitted by very hot stars embedded within it. Other targets included the Cygnus Loop, which is the remnant of a 50,000-year-old supernova expanding in space, and FUVIS also supported observations of nearby galaxies, such as the Large and Small Magellanic Clouds and the Andromeda Spiral. Early on 9 February, about four hours ahead of the scheduled retrieval, Wetherbee and Collins manoeuvred Discovery to within range of SPARTAN-204. Assisted by Titov, astronaut Janice Voss grappled the satellite with the RMS at 6:33 am. It was then lowered onto its support structure in the payload bay and deactivated.

Although its primary goal was based upon astronomical research, SPARTAN-204 also supported a series of detailed test objectives for the ISS programme. It was fitted with a series of six laser retroreflectors to support the Tracking Control System and these were used during its unberthing, deployment and retrieval processes for tracking purposes. Nor did its task end with berthing in the payload bay on 9 February, for on that very same day STS-63 crewmen Mike Foale (designated 'EV1', with red stripes on the legs of his suit) and Bernard Harris ('EV2', wearing a pure white suit) ventured outside on a five-hour spacewalk that would be the first in a series of Extravehicular Development Flight Tests (EDFTs) over a number of missions to increase the spacewalking experience base within NASA's astronaut corps, in readiness for the anticipated 'wall' of EVAs that would be required in order to construct the new space station. Since the airlock was connected to the Spacehab-3 module, Foale and Harris had to pre-breathe on masks for about four hours, because they could not lower the cabin pressure for fear of compromising the experiments. With Eileen Collins monitoring their progress and acting as their primary communicator from inside Discovery's cabin, and Titov operating the RMS, the two spacewalkers left the airlock at 6:56 am EST, less than 30 minutes after the SPARTAN-204 retrieval, and immediately set to work establishing their tools and tethers for working in the vacuum of space.

"Having been in space for about five days," Harris told a Smithsonian interviewer, years later, "I considered myself pretty acclimated to the environment. But when I opened up the hatch to come out of the airlock, I got a *faceful* of Earth! We were flying upside down, so I immediately saw the grandeur of Earth down below and I had two sensations. The sense of movement was striking; it made me feel dizzy. I got a little uneasy ... and I didn't get that from looking out the Shuttle window. Then, as I was getting ready to step out of the spaceship, it felt for an instant like gravity was going to grab hold of me and pull me down toward Earth. Being a scientist and an astronaut, you *know* that's not going to happen, but within a split second I had those two sensations. Your natural response is to hesitate and grab on harder." At length, after convincing himself that he was *not* going to fall, Harris composed himself and began his work.

In making this EVA, Harris became the first African-American to perform a spacewalk, whilst Foale – a naturalised US citizen – became the first person of British ancestry to venture outside the confines of a spacecraft in a pressurised space suit. Previous spacewalkers had reported a tendency to become extremely cold when working in open or shaded areas and it was recognised that working to build and maintain the ISS would place future astronauts in similar positions for extended periods. Consequently, on STS-63, Foale and Harris would test a number of improvements. On their undergarments, the cooling tubes running down the length of their arms had been bypassed, in order that their arms would not be chilled, whilst additional layers of insulation had been added to the suit and to the exterior of their gloves for warmth. Shortly after entering the payload bay, Titov raised Foale and Harris on the RMS to a position about 9.1 m above the payload bay, away from the Shuttle's radiated heat. Wetherbee had oriented the orbiter with its belly towards the Sun and the bay was thus shadowed, producing one of the coldest environments ever experienced by spacewalkers. At this vantage point, the two men did nothing for 15 minutes, save reporting their comfort levels as sensors in their gloves and a 'thermal cube' sensor on the mobile foot restraint collected data for subsequent comparison to local ambient temperatures.

Operating in a high-inclination orbit of 51.6 degrees, in pitch darkness, Harris anticipated that the coldness would be acute. "For years, astronauts who'd done spacewalks on the night side of Earth at this higher latitude had come back and told the NASA engineers that they felt cold inside the suit," he recalled. However, it was far colder than either Harris or Foale could have expected. Within ten minutes, the temperature of their suits plummeted to below freezing. "Even with the suit temperature control in full *Hot* position," Harris recalled, "it got so cold that our hands felt like they were in a vat of liquid nitrogen." When they returned inside the orbiter, they would learn that the exterior temperatures had reached minus 148 degrees Celsius, as recorded by the thermal cube. This sapped the heat from their suits. "You felt it mostly in your feet and hands," he continued. "The gloves are the thinnest part of the suit, because you want dexterity. I pulled my fingers away from the edges of the gloves and began to do calisthenics in the suit to increase my body temperature. I think Mike did the same thing."

In the later stages of the EVA, the 1,200 kg SPARTAN-204 satellite again entered the picture, supporting a range of 'mass-handling' tasks. Foale positioned himself in a portable foot restraint on the end of the RMS and Harris perched at the port side of the satellite's support structure. From inside Discovery's cabin, Janice Voss unlatched SPARTAN-204 and Titov manoeuvred Foale, on the RMS, until he was directly 'above' the satellite. Foale grabbed SPARTAN, using one of its trio of EVA handling attachment points, and passed it to Harris. As EV2 finished up his portion of the activity, both astronauts reported that their hands and feet were getting cold and Mission Control decided to cancel Foale's mass handling task and end the EVA slightly earlier than planned. Harris had not used the thermal over-gloves during the SPARTAN mass-handing task and when he subsequently put them on, he noted that conditions were a little warmer, but that the cold was still apparent. On a scale of 1 to 8, with '1' being unbearable, the spacewalkers rated their situation as a '3', which was classified as "unacceptably cold". The frigidity was so intense that even their suits' Electronic Cuff Checklists went blank and the spacewalkers were obliged to use their printed checklists. Since the retrieval of SPARTAN-204 and the beginning of the EVA were timed to occur so closely together, it was always recognised that a late or delayed retrieval might mean the spacewalk – planned for four hours and 50 minutes – might be shortened. As circumstances transpired, it ran to almost exactly its scheduled duration and Foale and Harris returned to the airlock at 11:34 am EST, after four hours and 38 minutes. The only point of note during repressurisation was when Harris experienced a burning sensation in his eyes. After removing his suit, he flushed his eyes with cold water, but post-flight inspections would reveal no evidence of any contamination. It was speculated that Harris may have fallen victim to contact with the anti-fogging soap solution on his helmet visor.

With the completion of the EVA, the bulk of STS-63 was complete and on 10 February, the day before landing, Wetherbee and Collins performed the standard tests of Discovery's flight systems ahead of re-entry. Following a flawless passage through the upper atmosphere, the orbiter alighted on Runway 15 at the Shuttle Landing Facility (SLF) runway at 6:51 am EST on the 11th, ending an eight-day flight. The Near-Mir rendezvous had inaugurated the joint programme between the United States and Russia in spectacular fashion. A few weeks later, in mid-March, NASA astronaut Norm Thagard would join two Russian cosmonauts aboard Soyuz TM-21, launching from the Baikonur Cosmodrome in Kazakhstan, for a planned three-month stay aboard Mir. Their return to Earth, in June, would be aboard STS-71, the first of at least seven Shuttle-Mir docking missions.

In fact, the delivery and retrieval of *cosmonauts*, as well as astronauts, to and from the space station, had given Russian officials some cause for concern. As early as August 1993, *Flight International* had pointed to controversy over the actions of former astronaut Guy Gardner, then serving as NASA's Associate Administrator for Russia, who had recommended using the first Shuttle-Mir docking flight as a means of transporting two new cosmonauts to the station and bringing home the old crew. The Russians were concerned that such a move might "interfere with training regimes", even though the potential benefits of a Shuttle visit were enormous: including the capacity to transport 500 kg of much-needed equipment to the space station and return 1,000 kg of failed gyrodynes for engineering inspections. Joint EVAs were also considered and it was explained by Clay Morgan in *Shuttle-Mir: The United States and Russia Share History's Highest Stage* that such collaborative work would serve "to extend Mir's useful lifetime through the end of 1997". In the end, Gardner's plan ran through to completion and in June 1994 a cadre of Russian cosmonauts arrived at JSC in Houston to begin training on US hardware. They included Vladimir Dezhurov and Gennadi Strekalov, who would accompany Thagard on the United States' first long-duration mission in over two decades.

AN AMERICAN 'LIVING' IN SPACE

Not since the end of the Skylab era had an American astronaut spent longer than 18 days in orbit on a single mission. In fact, the single longest US human space flight of all time had been 84 days – just shy of three months – which Gerry Carr, Ed Gibson and Bill Pogue accomplished on 8 February 1974. Since that time, the space endurance records had been dominated by the Soviet Union, and later Russia, and even in the second decade of the 21st century the top 20 places of the most experienced spacefarers of all time are exclusively held by cosmonauts. That is expected to change, for at the time of writing, in 2014, the first US-Russian year-long expedition aboard the ISS is planned, but the prevalence of Russian names on the experience list illustrates that their technology has enabled them to transform space from a place to not simply 'visit', but to 'live'.

And at the dawn of 1995, the 17th crew of cosmonauts was in residence aboard Mir, the latest of Russia's orbital space stations. Cosmonauts Alexander Viktorenko and Yelena Kondakova – the first woman ever to participate in a long-duration mission – had been aloft for three months, having been launched aboard Soyuz TM-20 in October 1994. Their crewmate Valeri Polyakov, on the other hand, eclipsed the world space endurance record on 8 January 1995 as he passed the 366th day of his own flight. He had flown to Mir with another crew aboard Soyuz TM-18 in January 1994 and, although his original comrades had returned to Earth after six months, Polyakov had trained for a far longer mission of 14 months.

For almost two decades, Valeri Vladimirovich Polyakov has held the world record for the longest single space mission, spending a total of 437 days in orbit aboard Mir from January 1994 until March 1995. When added to the 241 days he also spent aboard the space station between August 1988 and April 1989, Polyakov also ranks within the top five on the list of the world's most experienced astronauts and cosmonauts, with a personal achievement of 678 days (more than 22 months). Yet the name which today continues to grace the Guinness Book of Records is not even his birth name. He was actually born Valeri Ivanovich Korshunov on 27 April 1942 in the industrial city of Tula, a couple of hundred kilometres south of Moscow, but changed his middle patronymic and his family name to 'Vladimirovich' and 'Polyakov' at the age of 15, when he was adopted by his stepfather. Polyakov completed his secondary schooling in Tula in 1959 and entered the I.M. Sechenov First Moscow Medical Institute, from where he gained his doctorate. He then specialised in astronautical medicine at the Institute of Medical and Biological Problems at Moscow's Ministry of Public Health, his fascination with how the human organism adapts to the microgravity environment having been inspired by the flight of the first medical specialist, Boris Yegorov, aboard Voskhod-1 in October 1964. A little over seven years later, in March 1972, Polyakov was selected, along with two other physicians, Georgi Machinsky and Lev Smirenny. However, due to his ongoing postgraduate research towards a candidate of medical sciences degree, Polyakov did not begin his cosmonaut training until October 1972. He received his degree in 1976 and joined another cosmonaut selection, two years later, to complete the remainder of his training. He was an early candidate for a dedicated medical research mission to the Salyut 6 space station in late 1980 and served as physician Oleg Atkov's backup on the long-duration Soyuz T-10B flight to the Salyut 7 station in 1984. He might also have flown aboard the Soyuz T-13 mission, but was replaced by military engineer Alexander Volkov.

If Polyakov achieved the world record for the longest single space mission in 1994-1995, then his crewmate Yelena Vladimirovna Kondakova, born on 30 March 1957 in Mitischi, within the Moscow region, actually became only the third Russian woman to venture into orbit, after Valentina Tereshkova and Svetlana Savitskaya. At the time of writing, she also remains the most recent, although Yelena Serova is scheduled for a six-month mission to the International Space Station in 2014-2015. Kondakova graduated from Moscow Bauman High Technical College in 1980 and worked for several years for the Energia design bureau, before being selected into the cosmonaut corps in January 1989. When Kondakova was launched aboard Soyuz

TM-20 in October 1994, she became the first woman in history to attempt a long-duration space mission; no one from either the United States or Russia had embarked on a single flight in excess of 14 days. Kondakova thus snatched a personal and world record for women within the first quarter of her mission to Mir.

Commanding Mir at the dawn of 1995 was veteran cosmonaut Alexander Stepanovich Viktorenko. He came from Olginka, in northern Kazakhstan, where he was born on 29 March 1947, and after a stellar career in the Soviet Air Force was selected as a cosmonaut candidate in May 1978. Together with fellow selectee Nikolai Grekov, they joined a larger detachment of civilian engineers and physicians the following December to commence training in earnest. However, whereas the remainder of the group graduated in October 1980, Viktorenko was delayed by a serious training accident and did not complete his final examinations until February 1982. Nevertheless, he quickly established a reputation for himself as a dedicated professional. Of the eight members of his 16-strong group who went on to actually fly in space, Viktorenko was fifth.

In time, he would undertake four space missions, spend a cumulative 489 days in orbit and would be in command of Mir in March 1995 when the first American visitor, NASA astronaut Norm Thagard, arrived. "A pleasant fellow, but also extremely competent", was how Thagard later described him. "Sasha Viktorenko was one of those who was probably more like you would expect to find in an American commander. Although we've got some American commanders who are fairly authoritarian ... I think if you looked at the scale of things, you'd find the Russians ... a little bit further down towards the less authoritarian end." After completing his high school education, Viktorenko entered the Advanced School of Military Aviation in Orenburg, graduated in 1969 and entered the Soviet Air Force, where he rose rapidly through the ranks. He embarked on his first voyage into orbit aboard the joint Soviet-Syrian Soyuz TM-3 mission to Mir in July 1987 and commanded two long-duration flights in September 1989-February 1990 and March-July 1992.

The space station aboard which Viktorenko, Kondakova, Polyakov and, very soon, Norm Thagard, would fly, and which the STS-63 astronauts had seen from afar in February 1995, had been progressively assembled in orbit for over a period of nine years. Starting with the name, 'Mir' is roughly translatable as either 'World' or 'Peace' and seems to have formed part of a propagandist Soviet effort to present their space programme as one of 'Star Peace', diametrically opposed to US President Ronald Reagan's belligerent 'Star Wars'. This theme was reinforced by former cosmonaut Alexei Leonov, speaking shortly after Mir's February 1986 launch, that "by naming the space station in this way, we want to emphasise once again that the Soviet programme for space research and for the use of outer space is intended solely for peaceful purposes". This notion was supported by Mir's first commander, Leonid Kizim, who responded to a question about whether the new station would be used for military purposes. "The programme for our work," he replied, "does *not* contain any experiments for military purposes. As for the statements by US officials, it seems to us that they are being made in order to justify their *own* plans for transferring the arms race to space." Relations between the two nations were improving, it is true, and the attempts of General Secretary Mikhail Gorbachev – elected in March 1985 – to reform the Soviet Union were bearing fruit, but the intrinsic mistrust between East and West remained strong.

The roles and capabilities of the Mir space station were at least an order of magnitude higher than anything previously attempted. Unlike the 'monolithic' Salyut stations, launched between 1971 and 1982, Mir represented something entirely novel: a first effort to create a 'modular' complex in space. After the insertion into low-Earth orbit of its 'core' module, equipped with a unique, multiple-port docking adaptor, the ambition was for it to be rapidly expanded into a fully-fledged research facility by the addition of large laboratory modules, each weighing around 20,000 kg, devoted to astrophysics, remote sensing and the life and microgravity sciences. For a decade and a half, counting from the launch of its core in February 1986 to the end of its life in March 2001, Mir would prove itself to be one of the most successful and influential space stations ever launched. Record after record would be secured by its cosmonauts, many of which still stand to this day, including records for the longest time spent in orbit, and it would be occupied continuously for more than ten years. In total, more than a hundred individuals from a dozen nations – Russia, Syria, Bulgaria, Afghanistan, France, Japan, the United Kingdom, Austria, Germany, the United States, Canada and Slovakia, to say nothing of the many Soviet and Russian cosmonauts who traced their racial and cultural ancestry to Azerbaijan, Kazakhstan, Kyrgyzstan and elsewhere – would work aboard Mir, would gaze at the Home Planet through its windows, would share meals at its dinner table, would exercise on its treadmill and would share in its successes, trials and near-disasters. From its high vantage point, typically around 350 km above Earth, Mir would not only eclipse the achievements of every space station which preceded it, but would also lay the critical cornerstones and smooth the way for today's ISS.

Mir's story began in February 1976, when a Soviet decree identified it as quite distinct from earlier Salyuts: it would possess docking ports at each end of its core module, just like them, but its spherical multiple adaptor would carry two radial ports for additional research laboratories. By the summer of 1978, this plan had changed to a final configuration of a single aft port and no fewer than *five* ports on the multiple adaptor: one at the front and another four radial ports, spaced at 90-degree intervals. It was intended that these radial ports would accommodate small laboratories, weighing in the order of 7,000 kg apiece, and probably derived from the Soyuz spacecraft but with the orbital and descent modules replaced by a pressurised research compartment. Plans shifted yet further in the spring of 1979, when a government resolution called for the project to be merged with a cancelled military space station project, known as 'Almaz' ('Diamond'), and Mir's ports were reinforced to handle larger modules based on the TKS (*Transportniy Korabl Snabzheniya*, known for a time in the West as the 'Star' and 'Heavy Cosmos') spacecraft in size and shape, each weighing around 20,000 kg.

Mir would benefit from digital flight computers and gyrodyne flywheels for attitude control, an automated rendezvous system, a dedicated satellite communications network and improved oxygen generators and carbon dioxide scrubbers.

Unfortunately, financial resources during this period were very much directed into the Soviet Union's Buran shuttle programme and it was not until the spring of 1984 that Valentin Glushko – a legendary rocket engine designer and head of the Kaliningrad-based Energia design bureau – was ordered to place the new station into orbit by February-March 1986, to coincide with the 27th Communist Party Congress. (It is probably not entirely coincidental that Glushko received his order only weeks after Ronald Reagan announced plans for the United States to build its own space station.) It would be difficult to get Mir ready for launch in such a short span; in fact, original plans had not envisaged the core module being placed into orbit before late 1986. Static and dynamic testing of the core was completed in December 1984, but the remainder of the processing was far from complete.

There were other headaches, too. The weight of electrical cabling pushed the core module a couple of thousand kilograms outside its Proton launch vehicle's performance envelope. In response, a sizeable chunk of the experiment hardware had to be removed and remanifested onto subsequent Progress supply missions, but the weight overspill was still unacceptable. At length, in January 1985, the planned 65-degree orbital inclination for Mir was relaxed to 51.6 degrees, like the Salyuts, although this would reduce the photographic coverage of the Soviet Union. Difficulties with the new Strela ('Arrow') flight computers forced engineers to install older-specification units, as a short-term measure, in order to meet the politically mandated launch target. In April 1985 it was decided to ship Mir's core module to the launch site in what is today the independent nation of Kazakhstan to complete systems testing and integration. When it arrived at the desolate launch site in early May, more than *eleven hundred* wires – almost half of the total – required substantial rework, but by October it had completed its final clean room inspections and the focus changed to communications checks (using the Cosmos 1700 satellite which had recently been inserted into geostationary orbit).

Mir left Earth from one of the most desolate locations on the planet, Tyuratam, as it is properly known, some 200 km east of the Aral Sea, in Kazakhstan. Today, it is still variously called 'Tyuratam', after the tiny railhead, or, more often, 'Baikonur', which covers a broader and different geographical location to the north-east. The first launch attempt for Mir's core module came on 16 February 1986, less than three weeks after America had reeled from the loss of Space Shuttle Challenger. It was scrubbed when Mir failed a communications test, but the launch occurred successfully at 12:28 am Moscow Time on the 20th, with the Proton booster inserting the 20,400 kg infant station perfectly into a 170×300 km orbit, inclined 51.6 degrees to the equator. By early March, Mir was established in its operational orbit. Physically, the core measured 13.1 m long and 4.2 m in diameter at its widest point and, with its twin solar arrays fully deployed, had a wingspan of 20.7 m. These arrays covered 76 m^2 and utilised high-performance gallium arsenide cells, with an initial electrical power yield of around nine kilowatts. (A third, 'dorsal' solar panel, presumably omitted at launch, due to weight considerations, would be installed during a pair of spacewalks in June 1987.)

From 'front' to 'back', Mir's most visible feature was the spherical multiple docking adaptor, which measured 2.2 m in diameter and 2.8 m long. "It serves as a lobby for spacecraft docking," noted a 1987 US Senate report on Soviet space activities, "and houses the airlock for egress into space." It was intended that an arriving module would firstly dock at the 'forward' port, and then the arm-like 'Ljappa' manipulator device on the module would robotically separate it and swing it around to one of the radial ports. A pair of 'sockets' for these devices were situated on the multiple adaptor. Moving out of the adaptor, the cosmonaut entered Mir's cylindrical working and living compartment, 7.7 m in length and comprising two sections, which measured 2.9 m and 4.2 m in diameter, connected by a short conical frustum. Although this dual-section compartment took the form of a stepped cylinder, it provided a single internal volume. At the rear was a service propulsion area and transfer adaptor, the same diameter as the main compartment and measuring 2.3 m long, which carried communications and rendezvous equipment, and four propellant tanks for the 32 thrusters of the attitude control system. Lastly, it housed a pair of liquid-fuelled main engines, each with a thrust of 300 kg. These could not be used after the docking at the rear port of the first scientific module, the astrophysics laboratory 'Kvant-1' ('Quantum').

Inside the core, Mir differed greatly from the earlier Salyuts. With the exception of the station's control consoles, it was largely devoid of experimental apparatus. It had instead been designed as a 'habitat' or living area, which the cosmonauts came to affectionately nickname 'the base block', and could comfortably accommodate a crew of two. Each man was provided with his own private cabin, equipped with a window, a sleeping bag and a desk. It also housed a toilet, an entertainment system for movies and music, exercise equipment, medical supplies, and a galley and fold-down dinner table with food warmers and movable 'chairs'. "The crews cook on a hot plate," noted the US Senate report, "and are allowed to select their own food, as long as they consume the required number of calories." *Spatial awareness* was provided by a dark green 'carpet' on the 'floor', light green 'walls' and a white 'ceiling' on which were installed fluorescent lamps, and the installation of equipment was deliberately arranged to mimic this Earth-like orientation. Communications were to be provided via a network of command and control satellites, known as 'Luch' ('Ray'), which could trace their genesis to the very same February 1976 decree which had brought Mir into existence, although at that time there was only one satellite in operation. Similar in function to NASA's Tracking and Data Relay Satellites, the Luch network provided a high-rate space-to-ground link for the station and its visiting Soyuz spacecraft, together with communications support for the Soviet Navy. In total, a trio of three-axis-stabilised Luch satellites were launched between October 1985 and December 1989, under the cover names of Cosmos 1700, Cosmos 1897 and Cosmos 2054. Although much of the hardware destined for Mir would be launched by Progress or aboard subsequent modules, the core did arrive in orbit with some of its research facilities in place: crystal growth apparatus, melting and solidification facilities, electrophoresis equipment, photometers and spectro-meters, metallurgy hardware and cameras.

A little over a year after the launch of Mir's core, in April 1987, the first additional module, Kvant-1, arrived and was successfully docked onto the rear port as a permanent fixture. The next pair of laboratories, Kvant-2 and Kristall

('Crystal'), were to be positioned on opposing radial ports ('zenith' and 'nadir') of the multiple docking adaptor and it was considered desirable that they should be installed in rapid succession in order to preserve the 'balanced' configuration of the complex. Their arrival was expected early in 1989, but this was extensively delayed and for about five months in the summer of that year Mir was left unoccupied. Cosmonauts returned and switched on the lights in September and from then until the end of August 1999, just a few days shy of ten full years, Mir would be permanently inhabited by a succession of rotating crews. The thought that at least two humans were always circling Earth for an entire decade should not be underestimated, for in the 1990s the pioneering single-orbit flight of Yuri Gagarin remained very much in living memory.

In that decade, Mir would seesaw violently between triumph and near-disaster: it would expand, just as the Soviets originally intended, into an impressive 'dragonfly' (as Bryan Burrough called it) with modules, arrays, cranes and antennas sprouting in a myriad directions, with no fewer than *five* research laboratories, and would host dozens of crews from a dozen nations. On the other side of the coin, however, it would fall victim to the kind of funding issues which were an integral feature of post-Soviet Russia and from 1989 onwards the country's space budget established itself on a slippery slope which would see it plummet by 80 percent. That lack of funding would delay modules, scrub missions and oblige cosmonauts to live aboard an outpost which was no longer fit for purpose. As early as April 1989, half of Mir's equipment was effectively inoperable and the lack of additional modules made the interior cramped, as Sergei Krikalev observed. Batteries would not properly hold charge, creating chronic power supply difficulties.

Even as Russia prepared to launch NASA's first long-duration astronaut, Norm Thagard, in March 1995, the omens for the future of Mir were bad. The civilian space programme had been described in several quarters as "doomed", with its budgets declining to barely ten percent of their 1989 levels, and the financing of military and rocket-production efforts had fallen by a factor of ten. Only half of Russia's planned launches in 1994 were accomplished, the new Russian Space Agency – still less than three years into its existence – had received a meagre 12 percent of its required fiscal allocation, barely enough to keep it alive, and around half of all planned long-term programmes had been either cancelled or severely curtailed. Senior managers felt that, even with $400 million of American cash to fly NASA astronauts to the station, Mir could probably not survive for much longer and might have to be evacuated. Even the booster for an unmanned Progress cargo ship, scheduled for April 1995 to carry some of Thagard's research equipment into orbit, had to be drawn from the Russian military arsenal. As NASA was faced with the stark reality of cutting $5 billion from its budget between 1995-2000, so the Russian Space Agency suffered a deep, 40-percent cut to 1.2 billion roubles (about $180 million), which placed it on a funding par with India and prompted fears of up to 360,000 redundancies in 1995 alone. Indeed, thousands of skilled workers were being laid off or were departing the space industry in droves, to such an extent than in February 1995 the Russian parliament adopted a new resolution to finance the space programme at a level of no lower than one percent of the gross national

2-05: Norm Thagard (left) and Bonnie Dunbar, clad in Soyuz-TM space suits, prepare for their prime and backup roles on the first long-duration US mission to Mir.

product and to implement a new national space policy. The early post-Soviet years were among the most desperate ever experienced by Russia's once-proud human space effort.

Yet an operational space station was acutely needed by NASA as a means of preparing its own astronauts for the ISS and, ahead of the arrival of the first US long-duration resident, the Russians had already begun supporting US experiments. In October 1994, they activated the Space Acceleration Measurement System (SAMS), which had been delivered to Mir a month earlier aboard an unmanned Progress supply craft. It was part of preparations for US research, which would begin in 1995, and sought to gather acceleration and low-frequency vibration data with three sensor 'heads', positioned throughout the Mir complex at the future locations for NASA's Protein Crystal Growth experiment and Russia's Gallar materials processing furnace. A few weeks later, in early November, the Soyuz TM-19 crew of cosmonauts Yuri Malenchenko and Talgat Musabayev and German astronaut Ulf Merbold returned to Earth, bringing the SAMS data with them. "Flying SAMS on Mir," noted a NASA news release, "meets a US National Research Council recommendation that NASA measure and characterise the vibration environment on all spacecraft carrying chemistry and physics experiments and to pursue strategies for producing the lowest possible gravity conditions." Other early experiments included the Tissue Equivalent Proportional Counter (TEPC), a dosimetry monitor to measure the radiation environment inside the station, and the Mir Interface to Payload Systems (MIPS)-1, a data-support payload for the upcoming US life sciences research to be performed by Norm Thagard.

For Thagard, a veteran of four previous Shuttle missions, selection to fly the first NASA 'increment' to Mir was a prize he had sought for several years. In the summer of 1991, as he was preparing to fly as payload commander aboard STS-42, scheduled for launch the following January, he heard about the US-Russian agreement that would involve a long-duration mission to Mir. About a year later, Thagard was chatting with fellow astronaut Dave Hilmers about what would entice them to remain at NASA; Hilmers had already been accepted into medical school, and intended to retire from the space agency, but Thagard was fascinated by the possibility of flying with the Russians. His mind was made up when Chief Astronaut Dan Brandenstein asked him, point-blank, if he would fly the mission. Thagard's response was immediate: *"Absolutely!"*

At first, Brandenstein told him, he would not have a backup crew member for his increment and Thagard began teaching himself Russian, before the astronaut office brought in an instructor to give weekly, hour-long language classes in October 1992. This allowed Thagard to steadily improve his language skills before work each morning. At around the same time, he accompanied Don Puddy, the head of Flight Crew Operations, on a fact-finding mission to Russia to discuss using the Soyuz-TM spacecraft as an Assured Crew Return Vehicle (ACRV) for Space Station Freedom. Several months later, in the summer of 1993, Dave Leestma, who had replaced Puddy as head of Flight Crew Operations, sent Thagard to the Defense Language Institute in Monterey, California, for more in-depth instruction. "Unfortunately, the funds were limited," Thagard told a NASA oral historian, "and I actually wound up signing shared cost orders, meaning that while the per-diem rate out at Monterey ... was $34 per day, plus transportation, I would up with $10 per day and *no* transportation. So I wound up driving my own car out there." Knowing that he was heading to Russia in the near future, and with a car that was already ten years old, Thagard was cautious about the move, and rightly so. "True to form, going into Death Valley," he continued, "on my way out to Monterey from Houston, one of the wheels actually came off the car. There was a wheel bearing failure and the whole wheel just separated and, of course, the car skidded, but we managed to get it safely over to the side." Several hours later, a flatbed truck picked Thagard up and he ultimately traded the old car in for a newer model, which he drove to Monterey. It seemed a bad omen, for misfortune had befallen him in his homeland, before he even set foot in Russia.

From July until December 1993, Thagard and NASA flight surgeon Dave Ward worked on the Russian language at Monterey. Although it was not 'total immersion' in the language, the classes were "pretty intensive", eight hours per day, for five days each week. Thagard's name had not been formally announced and, in August, *Flight International* reported that his seat was "being challenged" by veteran astronaut Bill

Shepherd, who was then deeply involved in the Space Station Freedom redesign process as Assistant Deputy Administrator (Technical) at NASA's Washington headquarters. Another name which *Flight International* linked to the flight, although likely in a backup capacity, was physician Drew Gaffney, who had served as a payload specialist on the first Spacelab Life Sciences (SLS-1) mission. On 3 February 1994, NASA formally announced Thagard's name as the prime candidate for the first long-duration increment to Mir. The Russians had insisted that a backup astronaut should also be named and fellow Shuttle veteran Bonnie Dunbar was named. In her NASA oral history, Dunbar recalled being approached by Dave Leestma in December 1993. Despite her reservations about the Russian language, he knew that she had learned German during training for an earlier mission, and was convinced that she was the right person to take the backup position. Still, her assignment was initially perplexing to Thagard, for unlike himself and a handful of others, including Shannon Lucid and Story Musgrave, Dunbar had not volunteered for the Russian language classes at JSC. "It's bad policy to send people over to Russia," said Thagard, "who don't have some experience in Russian before they get there." Nonetheless, in late February 1994, Thagard, Dunbar, fellow astronaut Ken Cameron and NASA flight surgeons Dave Ward and Mike Barrett travelled to a country which had remained largely closed to outsiders for many years.

On 4 February 1994, the day after Thagard and Dunbar's assignment to the first long-duration Mir flight, 'Team Zero', the first of ten US-Russian working groups, met to begin the process of mission planning, cargo and scheduling, public affairs, safety, operations, science, training, integration, EVA and medicine. (Russia's Shuttle-Mir Program Manager, former cosmonaut Valeri Ryumin, was overheard to joke that the group was called 'Zero' because "managers don't do *anything!*") Within weeks, however, NASA's Office of Life and Microgravity Sciences and Applications began soliciting the scientific community for experiment proposals for the long-duration Mir missions. It was expected that the NASA astronauts would utilise Russia's 'Spektr' ('Spectrum') module for their living quarters and for their research. The 20,000 kg module would house 1,130 kg of scientific equipment for Thagard's mission, but had been extensively delayed.

Spektr was originally conceived in the Soviet era as a military space station module, carrying surveillance equipment, including an experimental optical telescope, known as 'Pion' ('Peony'). Working from a command post at the forward end of the module, it was envisaged that cosmonauts would operate Oktava experimental interceptor rockets, whose progress would be identified and tracked using a battery of optical equipment, as well as demonstrating a multi-spectral device to discriminate between missiles and decoys. Other instruments would have performed spectral observations of Earth's surface, upper atmosphere and charged particle environment. Following the collapse of the Soviet Union, in 1992 the military and political leadership ordered all such efforts to cease and the 'military' Spektr never left the assembly building. It was subsequently retasked for remote observations of Earth's atmosphere and surface. Its military payload was replaced with a new, conical section to house a second, V-shaped pair of solar panels – giving it a total complement of four arrays, capable of providing half of the station's power supply – and hopes were high that it would be launched in 1993. However, as Russia's post-Soviet economic crisis deepened, it was delayed yet further and expectations of launching it in February 1995 proved hopelessly optimistic. In October 1994, the rouble collapsed and the Russians were forced to advise NASA that launch dates for Spektr and the next Mir module, Priroda, could not be met. A new launch date of 10 May 1995 was established for Spektr, which meant that Norm Thagard would spend more than half of his three-month mission *without* much of the equipment that he needed for his expansive scientific research programme.

For Thagard, training in Russia was more akin to training in the United States than he had anticipated. Having been selected by NASA in January 1978, he was one of the agency's most experienced astronauts, with four flights and a cumulative 25 days in orbit to his credit. Born Norman Earl Thagard on 3 July 1943, his birthplace is given as Marianna, Florida, and in his NASA oral history he said that his father was a Greyhound bus driver and Marianna was the bus changeover point, "so I was born at a time when he was doing that". Thagard attended high school in Jacksonville and studied engineering science at Florida State University, gaining bachelor's and master's qualifications in 1965 and 1966. (As a high school senior, Thagard told his classmates that his aspiration was to become an electrical engineer, a jet pilot, a medical doctor and an astronaut, in *that* order ... which, incidentally, is precisely the path that his career ultimately took.) Shortly after receiving his master's degree, he entered active duty as a reservist with the Marine Corps, quickly attained the rank of captain and was designated a naval aviator in 1968. Under normal circumstances, as a designate for the Platoon Leaders Class-Aviation (PLC-A), Thagard would have been put through Marine basic school, but "the Vietnam War was on by that time and they needed pilots desperately, so they basically eliminated the requirement ... for PLC-A designates". After initial flight instruction at Naval Air Station Pensacola in Florida, Thagard flew the F-4 Phantom at Marine Air Corps Station Beaufort in South Carolina, and then, in January 1969, went to Vietnam. Thagard flew a total of 163 combat missions, primarily supporting the fleet in the Tonkin Gulf, providing fighter cover for B-52 Stratofortress bombers and close aerial support for ground units in the South; he never flew over North Vietnam.

Returning to the United States, Thagard served as an aviation weapons division officer back at Beaufort and resumed his academic studies in March 1971, pursuing additional work in electrical engineering and a medical degree. At first, he attended Florida State University, working for a PhD in electrical engineering, but it was decided soon afterwards to terminate the whole school – "the Apollo programme was winding to an end," Thagard recalled, and "engineering was kind of on a down cycle that year" – and medicine seemed the most likely alternative. Here, he found severe age discrimination: Baylor College of Medicine in Waco, Texas, warned him that applicants over 25 were rarely given serious consideration and Duke University in North Carolina told him that would take the same view. However, a trump card appeared from the most unlikely of places: Thagard's *mother-in-law*! "She knew the dean of the School of Medicine, University of Texas Health Science Center [in San Antonio, Texas]," he explained, "because she was working there at the time." She asked the dean on Thagard's behalf and was told that if he moved to Texas and

2-06: Norm Thagard works with Vladimir Dezhurov, who would command his four-month mission to Mir.

established residency there, he would be admitted. More trouble was afoot after the Thagards' arrival in Texas in June 1972. "We got there on a Friday evening at about five o'clock," he continued, "and got up the next morning, Saturday, and the headlines in the *San Antonio Light* [newspaper] read 'Medical School Dean Fired'. *This* was the fellow that said if I'd move there, they'd admit me to his school!"

Thagard worked initially as an engineer, whilst awaiting responses from other medical schools. He eventually accepted a place at the University of Texas Southwestern Medical School in Dallas and received his doctorate in 1977, only to learn that NASA was taking applications for its astronaut corps. His wife actually sent off for the application pack, "before I ever even *knew* about it", and Thagard applied for *both* the pilot and mission specialist categories, although his 804 hours of jet experience fell short of the minimum 1,000 hours needed for a pilot. "Realistically," he recalled, "the pilot [way] wasn't really going to happen. I didn't meet 1,000, but ... it wasn't *just* 1,000: you also needed to be a graduate of a military test pilot school, which I also was not." The mission specialist application, though, bore fruit and a month after starting his medical internship, in the summer of 1977, paperwork for security clearances, background checks and requests to attend his local police station for fingerprint profiles began to arrive at the Thagards' home. Late in October, he received the call from NASA to come to Houston for a week of screening. He heard nothing more until January 1978. "I actually had one of those rare Saturday evenings off," he told the oral historian, "and we had some friends over to the house. The news was on and there was an announcement ... that NASA on Monday was going to announce the first new group of astronauts. I was a little embarrassed, because to me that meant that I *wasn't* taken, because it seemed to me that this is *Saturday* and they're going to announce *Monday*, they've already told the people who are selected." Two days later, on Monday morning, Thagard was at the Veterans Administration Hospital, about to start work, when he received *the* call which would change the direction of his life.

The assignment of Bonnie Jeanne Dunbar as Thagard's backup for the three-month Mir mission brought with it an expectation in many areas of the media that she would also fly a long-duration expedition, perhaps in 1996. She was born in Sunnyside, Washington, on 3 March 1949. Her grandparents had arrived from Scotland and homesteaded in Oregon and the young girl grew up on a ranch, raising Hereford cattle in eastern Washington State. She often listened as her elderly grandfather told her why he came to the United States: to own his own soil. His perspective was that life was an adventure and the key to unlocking that adventure was education. Her father had similar, yet also different, ideals. "He was a Marine in World War II in the South Pacific," Dunbar recalled, and had "fought for [his] sons *and* daughters to be able to become what they want to become, if they wanted to work hard enough to do it." Her parents instilled this ethic into her and she remembered that the first books she received as a child were a set of encyclopaedias. If anything appeared on the television which the young girl did not understand, her parents would direct her to the shelf and ask her to read the encyclopaedia entry to the rest of the family. "It was fun," Dunbar recalled. "Reading was mandatory to be able to participate in this game."

The Soviet Union's launch of Sputnik-1 in October 1957 switched Dunbar on to the idea of space travel for the first time and she became engrossed in the work of H.G. Wells and Jules Verne. In eighth grade at school, her principal asked her what she wanted to do in life and the girl, embarrassed to admit that she wanted to be an astronaut, told him instead that she wanted to design and build spaceships. He did not laugh. Nor did he belittle her dream. "You'll have to know algebra," he told her. She had no idea what *algebra* actually was, but in later years would come to appreciate that algebra, geometry, trigonometry and calculus were effectively the 'keys' for an engineering career. Dunbar had little idea how challenging it would be to *enter* such a career. After high school, she entered the University of Washington to study engineering and was advised by Dr James Mueller, the dean of ceramic engineering, to change her major from aeronautics to either materials or ceramic engineering. It would prove a pivotal moment in her future career and it was Mueller who introduced her to many key engineering minds who were working on research for the Shuttle's thermal protection system. "That then put me on a course of becoming more involved in the real parts of space," Dunbar recalled. "They had summer positions open and so one of the summer positions at the University of Washington was doing X-ray diffraction research on some of the fibres that NASA was considering." Very few women studied engineering and Dunbar was received with difficulty by some students and professors. "My response to any negative attitudes," she concluded, "was to *ace* the class!" Much of her work was not subjective; it involved work whose results were either *right* or *wrong* and, as one supportive professor told her: If you've got it *right*, you've got it *right*! Gender made no difference.

Graduation in 1971 was followed by two years as a systems analyst with Boeing and a call from Mueller to invite her back for a master's degree in ceramic engineering. From 1973 until 1975, Dunbar worked on the mechanisms and kinetics of ionic diffusion in sodium beta-alumina, part of a project to develop a high-energy-density battery, primarily for the automotive industry. In fact, Dunbar would later give a lecture to the Ford Motor Company, who were performing similar research. By the summer of 1975, she had completed the requirements of her master's degree and was made a quite astonishing offer: the rigorous quality and extent of her research meant that she had effectively completed most of the requirements for a PhD, awardable in just *six months*. However, Dunbar was broke and opted instead for a six-month visiting scientist position in England at the Atomic Energy Research Establishment in Oxford, studying the wetting behaviour of liquids on solid substrates. Years later, she would profoundly regret not going ahead and completing her doctorate. Back in the United States, she joined Rockwell International to work on the Shuttle's thermal protection system and later received the Engineer of the Year Award for it. She applied, unsuccessfully, for admission into the astronaut corps, but, in 1978, entered NASA's Johnson Space Center as a payload officer and flight controller, providing guidance and navigation for the Skylab re-entry. She was finally selected as an astronaut in May 1980 and three years later completed her doctorate in mechanical and biomedical engineering at the University of Houston. Her PhD focus was an evaluation of the effects of simulated orbital flight on bone strength and fracture toughness. In October 1985 she flew her first Shuttle mission, a joint Spacelab flight with West Germany, and followed this with two more flights in 1990 and 1992.

Having trained and flown aboard the Shuttle, Thagard and Dunbar were rapidly indoctrinated into the inner workings of both the Mir station and the Soyuz-TM spacecraft, which was Russia's means of delivering cosmonauts into orbit. The Soyuz was the latest in a long line of piloted vehicles flown since the mid-1960s and had been the brainchild of Chief Designer Sergei Korolev, with the original intention of supporting both Earth-orbiting missions and lunar voyages to rival Project Apollo. As early as 1964, the design and definition of Soyuz was well underway and technical documentation and a mockup revealed it as a craft capable of lofting two or even three cosmonauts. Even its *name* was no accident: for the Soviet Union's official moniker – *Soyuz Sovietskikh Sotsialisticheskikh Respublik*, the Union of Soviet Socialist Republics – was often popularly known amongst its citizenry as 'Soyuz' (the 'Union'). Therefore, the name of the spacecraft not only reflected its role in supporting rendezvous and orbital stations, but was also a highly symbolic and political statement. In his seminal work *Challenge to Apollo*, historian Asif Siddiqi noted that when Korolev first saw the mockup, he proudly declared that Soyuz was "the machine of the future". Indeed, Soyuz has had the most long-lasting impact on the world. Since its first manned flight in April 1967, Soyuz and a modified version of its original R-7 launch vehicle continue to be used operationally today; a fitting legacy to an enduring talent.

By the end of the decade, seven manned Soyuz spacecraft had rocketed into orbit. They employed a pair of large rectangular solar panels, mounted on the instrument module, to generate electrical power. The total surface area of these wing-like appendages was 14 m^2, with each wing measuring 3.6 m long and 1.9 m wide. Beginning at its base, the cylindrical instrument module , 2.3 m long and 2.3 m wide, carried chemical batteries and two large solar panels to charge them, together with a thermo-regulation radiator and an integrated propulsion and attitude-control system. The latter, designated 'KTDU-35', comprised a pair of engines, one primary and one backup, sharing the same oxidiser and fuel supply. Propellants took the form of unsymmetrical dimethyl hydrazine and an oxidiser of nitric acid, housed in spherical tanks within the instrument module. Attitude control came from hydrogen peroxide thrusters. Guidance, rendezvous, communications and environmental gear filled the remainder of the cylindrical compartment. The bell-shaped descent module, which measured 2.2 m long and 2.3 m wide at its base, sat directly above the instrument module and housed the crew during ascent and re-entry. It had a habitable volume of some 2.5 m^3. The commander's seat was located in the centre, flanked by positions for a flight engineer and a research cosmonaut or 'test' engineer. Many of Soyuz' flight regimes were pre-programmed from the ground. Consequently, the main instrument panel presented the crew with readouts and visual displays of the performance of on-board systems, together with a monitor for the external television camera, an optical periscope called 'Vzor' ('Visor') for attitude manoeuvres and the 'Globus' ('Globe') device to show the spacecraft's position above Earth. In the event of a failure of the automatic systems, and to facilitate rendezvous and docking, it was expected that the commander would assume manual control. As a result, two hand controllers (one for velocity, the other for attitude) were located directly underneath the instrument panel. Rendezvous and docking were supported by the Vzor, together with a system of gyroscopes, attitude-control sensors and thrusters and the 'Igla' ('Needle') radar. The latter would automatically navigate the spacecraft to its target and draw to a halt at a range of 200-300 m, after which the commander would take charge and accomplish the final approach and docking. Atop the descent module, the spheroidal 'orbital module' measured 2.65 m long and 2.25 m wide. It housed bunks for the cosmonauts, a cupboard for food and water, life-support equipment, a rudimentary toilet, controls for experiments, cameras and other gear for the mission.

The descent module would be the only component capable of surviving the intense heat of atmospheric re-entry and bringing the cosmonauts back to Earth. At the end of each mission, the instrument and orbital modules would be jettisoned and the descent module would employ hydrogen peroxide engines to provide roll, pitch and yaw controllability during the early stages of re-entry. To protect its occupants, it was coated with a heat-resistant ablator, together with a thermal shield at its base that would detach shortly before touchdown to expose the four solid-propellant landing rockets. A 14 m^2 drogue parachute would deploy 9.5 km above the ground in order to stabilise the craft, prior to deploying the main canopy. If a problem occurred, a secondary canopy could be deployed. Seconds before touchdown, an altimeter would command the landing rockets to fire to cushion the impact.

Soyuz promised to be one of the safest manned spacecraft ever built, possessing as it did the Soviets' first 'true' launch escape system. This consisted of a tower atop the R-7's payload shroud and a multiple-nozzle, solid-fuelled rocket engine. In the event of an emergency from 20 minutes before launch until 160 seconds into the ascent, the shroud would split at the base of the descent module and the escape tower's engine would lift the descent and orbital modules to safety. At the top of the arc, the descent module would be released to parachute back to Earth, landing a couple of kilometres from the pad, in the case of a pad abort. Early predictions estimated that the crew could be exposed to gravitational loads as high as 10 G during such a scenario.

As their training got underway, it became clear to Thagard that Dunbar's lack of Russian language preparation had been problematic, particularly when she was assigned an interpreter. "You're never going to learn a language if you're using an interpreter," Thagard told the NASA oral historian, "and Bonnie struggled mightily virtually throughout the thing because of no Russian language before she went over there." In the months and years to come, US astronauts would be specifically given language tuition at the Defense Language Institute and elsewhere, before heading to Russia. Shortly after Thagard's mission, in July 1995, JSC would contract with The Corporate Word of Pittsburgh, Pennsylvania, to provide Russian language interpretation, translation and training. "I didn't feel totally prepared, especially in the area of language," Dunbar admitted to a NASA interviewer. "I took Russian classes in 1992, so I could read it, sound out the words; I knew the letters, but I wasn't able to really *speak* it." For Dunbar, a veteran of three previous Shuttle flights, the difficulty was not the engineering or the science or the Russian technical systems, but being able to attend lectures and take oral exams, which were *only* delivered in Russian. It was, she said, like being a first-grader in graduate school; she *knew* the answers to questions, but without the foundation of vocabulary could not easily communicate replies. Working with a language instructor from Moscow State University, for four hours daily, she focused on grammar, vocabulary and everyday talk. After about six months, by the late summer of 1994, she began to feel more comfortable with the Russian language, to such an extent that when she was assessed at the Defense Language Institute she was graded at 3.5 out of 5.

Shortly after their arrival, Thagard and Dunbar went through Soyuz winter survival training in the woods outside the cosmonauts' training centre in Zvezdny Gorodok ('Star City'), on the forested outskirts of Moscow. "Star City has always been kind of felt as the forbidden city or the hidden city," said flight surgeon Mike Barratt in a NASA oral history. "It wasn't on any maps; it was a secret cosmonaut training base. Of course, everyone knew where it was, but it was considered a closed and secure city." During Thagard and Dunbar's survival training, a simulated Soyuz spacecraft was immersed in hot water, then plopped into the snow in the middle of the forest and they were left alone for 48 hours to build a temporary shelter, don survival gear, chop wood and construct fires, just as they would be expected to do in the dire event that the Soyuz landed far from civilisation. Dunbar, who had grown up in the north-western United States, the daughter of a US Marine Corps father, felt quite at home with the outdoors and the training came very naturally to her. The NASA astronauts were joined for the exercise by cosmonaut Vladimir Nikolayevich

2-07: The Mir mockup at the Star City cosmonauts' training centre.

Dezhurov, who was named in March 1994 to serve as the commander of Thagard's mission. Dezhurov came from the settlement of Yavas, in the Zubovo-Polyansky district of Russia's western republic of Moldovia, where he was born on 30 July 1962. He grew up with a love of aviation. "First of all, I decided to be a pilot and a fighter pilot," he told a NASA interviewer, many years later, "and I finished Air Force military school and then Air Force military academy." Dezhurov graduated from the S.I. Gritsevits Kharkov Higher Military Aviation School in 1983 with a pilot's diploma and flew operational missions in the MiG-21 and MiG-23 fighters. Four years later, in December 1987, he was selected as a cosmonaut candidate. "I received an invitation to go to the Cosmonaut Office," he said, "and check the health. After that, if everything is okay, I can start my space preparation." After 18 months of initial training, Dezhurov became a fully-fledged cosmonaut and later was a correspondence student at the Gagarin Air Force Academy.

Not long after the winter survival exercise, Thagard was approached by General Yuri Kargapolov, the training officer at Star City, who asked him for his opinion of Dezhurov. It was at that point that Thagard knew that the 'rookie' cosmonaut would be the commander of Soyuz TM-21. However, the name of the second Russian cosmonaut, the 'flight engineer', had yet to be revealed. In the weeks which followed, something unusual happened. The original plan for 1994 was that cosmonauts Yuri Malenchenko, Gennadi Strekalov and Talgat Musabayev would fly Soyuz TM-19 on a long-duration mission to Mir in May-October. However, Strekalov – a veteran cosmonaut, with four previous flights to his credit – was removed from the crew, on the basis that his seat was needed for additional cargo. (According to *Flight International*, Strekalov would have remained aboard Mir for just 14 days, then returned to Earth with the outgoing Soyuz TM-18 crew, leaving Malenchenko and Musabayev for a long-duration mission.) At some point after his removal from Soyuz TM-19, Strekalov joined Dezhurov and Thagard to commence training for Soyuz TM-21. Backing them up would be cosmonauts Anatoli Solovyov and Nikolai Budarin, together with Dunbar. Training for a long-duration mission was quite unlike anything the two Americans had done before. "Shuttle flights are short, so you can intensively train for virtually every aspect of them," said Thagard, "and that's not true for a three-month flight. You simply cannot do that. In fact, you're going to wind up having things happen during the course of the flight that you never anticipated at all." Initially, Thagard and Dunbar worked together, but by about November 1994 they began to function as members of their respective crews, learning and mastering Mir and Soyuz-TM systems. By her own admission, Dunbar felt that Solovyov and Budarin were "very protective" of her, which she regarded as a positive symbol of crew bonding. When the backup crew attended their final exams, they were evaluated by a psychologist, who described them as having operated "like a symphony".

Living in Russia was also quite different, and far more spartan, than the United States. Thagard and Dunbar were given accommodation in a high-rise block within Star City. "It was a three-bedroom apartment," Thagard recalled. "It had new furniture. They had gone to the trouble of doing that and I thought the apartment was fine, even by US standards. It wasn't a luxury apartment, but by Russian standards it certainly was." Having said that, the upkeep of the buildings were poor. The astronauts were assigned drivers, but frequently had to request them a day or two in advance, which left Thagard making the journey into Moscow only about once per week. Dunbar, meanwhile, was embraced by the women of Star City, who invited her into their homes for tea and showed her around the military complex. The middle of the 1990s were an exceptionally difficult time in Russia, as the fragmented nation struggled to turn away from the old Soviet Union and move towards a more pluralistic system of democracy. "The economy was extremely poor," recalled Dunbar. At Star City, there were no fresh vegetables on site, which necessitated journeys into Moscow every Saturday to visit the US Embassy to buy a newspaper or to visit one of the handful of department stores which were springing up in the capital in order to buy food and essentials. In their apartments, they had Russian televisions with just one channel, no heating for weeks at a time, the telephones did not call long-distance and were difficult to hear and there were no tumble dryers, so they had to hang laundry over the bathtubs. Funding took two months to arrive between JSC in Houston and Star City, since it had to be funnelled through NASA Headquarters, the US State Department, the US embassies in Paris and Moscow, before it ended up (in cash) in the hands of Ken Cameron for distribution to his team. Shipments of the astronauts' own items from home, including clothes, irons and ironing boards, were notoriously slow to arrive, and since the arrangement between Russia and the United States was strictly *quid pro quo*, the astronauts were told in no uncertain terms that they should expect nothing more and should learn to "live like Russians". On one occasion, Ken Cameron got hold of some margarita mix and the three astronauts watched videos and ate popcorn as Thagard's noisy washer breakdanced its way across the bathroom and dislodged the sink from the wall. "*That* was our entertainment," said Dunbar, "for several weeks."

Late in February 1995, with more than a year of this technical, language and cultural training thus completed, it was formally announced that Dezhurov, Strekalov and Thagard had passed their required tests and were declared ready for launch. They flew from Star City to Tyuratam to be quarantined. The prime crew threw a party for their support staff and bought supplies, including a couple of cases of cognac. Not all of the cognac was consumed and the cosmonauts decanted it into litre-sized plastic bottles, labelled them with 'SOK', the Russian word for 'juice', and had them loaded aboard Soyuz TM-21. "So we launched with quite a lot of cognac on the Soyuz," Thagard recalled, then added that *none* of it was brought back from Mir . . .

The delays to the Spektr module had already pushed the Soyuz TM-21 launch date back from its original target of 3 March, but NASA and the Russians eventually settled on the 14th. Early that morning, the three men suited up for launch and departed their crew quarters into the bitterly cold Kazakh morning. "It was below freezing," said Thagard, "and there was quite a strong wind blowing and it was the only time in my life when I was actually glad I had a pressure suit on, because those things are usually hot and uncomfortable, especially if you start moving around in them. Yet it was just perfect for that day." Having launched previously from subtropical Florida, Thagard was worried that the cold weather would scrub the launch. Gennadi Strekalov told him not to worry. "The colder the better," he said. Thagard next pointed out that gale-force winds were whipping across Tyuratam. Strekalov grinned. All would be fine, he said, "as long as it's not a hurricane!" (Incidentally, 'Hurricane', or the Russian word *Uragan*, was Soyuz TM-21's radio callsign . . .)

In traditional fashion, Dezhurov, the commander, symbolically requested permission to conduct the Soyuz TM-21 mission from the commanding general. Arriving at the base of the launch pad, the crew was faced with the rocket which would deliver them into orbit. Its basic design had remained the same, having been conceived as the R-7 intercontinental ballistic missile under the direction of Sergei Korolev in the late 1950s: a three-stage behemoth, fed by liquid oxygen and a highly refined form of kerosene, known as 'Rocket Propellant-1' (RP-1). Strapped around its lower stage were four tapering boosters, each measuring 19.6 m in length. With the Soyuz spacecraft and escape tower in place, the rocket stood 49.3 m tall and was transported to the launch pad in a horizontal configuration, atop a railcar. It is a method still used to this day. Four cradling arms, nicknamed 'The Tulip', supported the booster and a pair of towering gantries provided pre-launch access. The process of loading liquid oxygen and RP-1 aboard the rocket got underway about five hours before liftoff, after which it transitioned into a topping mode, whereby all cryogenic boil-off of propellants were rapidly replenished. This ensured that the liquid oxygen tanks were maintained at Flight Ready levels, ahead of the ignition of the rocket's single first-stage engine and the engines of the four strap-on boosters.

Unlike the United States, where all non-essential personnel were kept away from the launch pad area in the hours preceding a flight, it surprised Thagard that he, Dezhurov and Strekalov had to wade through a gaggle of people to reach the elevator which took them to the top of the rocket. "You're literally brushing by them as you go through," he told the NASA oral historian. As he ascended the steps, someone called his name in an American ascent; Thagard turned to respond and almost lost his grip. Composing himself, he waved back, then walked to the top of the steps and strode across to the elevator, which took the crew up to boarding level. Thagard was first to clamber through the side hatch of Soyuz TM-21's orbital module, followed by Gennadi Strekalov, who quickly moved into his couch on the left-hand side of the descent module. "He had to turn on electrical power," recalled Thagard, "so Gennadi went in, and then I went in and then Veloga [Dezhurov] went in, and then, of course, you shut the hatch." By this stage, about two hours remained before liftoff. Although Thagard and Strekalov both had windows on their respective sides of the cabin, they were covered by the rocket's aerodynamic shroud, offering them no view of the titanic events which would soon engulf them as they left Earth.

In those two hours, however, there was little time to do anything other than the final tests for which they had trained. As Dezhurov and Strekalov busied themselves with running through their spacecraft's myriad systems, Thagard took care of activating the radios and periodically switching the views of two on-board television cameras, one of which focused on himself and the second upon the two cosmonauts. In the final moments, internal avionics within the rocket were initiated and the on-board flight recorders were spooled-up to monitor the vehicle's systems. At T-10 seconds, turbopumps on the core and strap-on boosters came to life. After confirmation that the engines were running at full power, the fuelling tower was retracted and at 9:11 am Moscow Time on 14 March 1995, the joint US-Russian crew of Soyuz TM-21 speared into the cold Tyuratam sky.

"It's very similar to what the Shuttle feels like," remembered Thagard of the adrenaline-charged ride into orbit, "not as much noise, not as much vibration, but similar." In making the flight, he became the first American in history to ride a Russian rocket and spacecraft into orbit. From his perspective, the staccato crackle of the Solid Rocket Boosters (SRBs), which characterised first-stage flight on the Shuttle, were not present with the Soyuz, which was far less noisy. On the other hand, second-stage flight on the Shuttle, powered only by the three liquid-fuelled main engines, felt to Thagard like a smooth electric ride, whereas aboard the Soyuz the noise and vibration remained. Rising rapidly, the vehicle exceeded 1,770 km/h within a minute of liftoff, during which time the maximum amount of aerodynamic stress, known as "Max Q", impacted its airframe. At T + 118 seconds, at an altitude of about 45 km, the four depleted strap-on boosters were jettisoned, leaving the central core and its single engine to continue the ascent. By two minutes into the flight, the rocket was travelling at over 5,390 km/h. The payload shroud and escape tower were jettisoned shortly afterward, and, some four minutes and 50 seconds after leaving Tyuratam, the core stage separated at an altitude of 170 km and the single engine of the third stage ignited to boost Soyuz TM-21 to a velocity in excess of 21,600 km/h. By the point of third stage separation, about nine minutes into the flight, the vehicle was in space, describing an orbit of 190×220 km, inclined 51.65 degrees to the equator.

"The main difference at the end was that when the Shuttle main engines cut off, they just cut off," said Thagard. "It's not a huge, emphatic thing, but when the main engines cut off on the Soyuz it *was* very emphatic, almost like a 'clang'. One possible explanation I've been told is that the Shuttle throttles back, so it's at 65 percent when the main engines cut off, whereas the Soyuz third-stage engine is at full bore when it cuts off." With the aerodynamic shroud gone, Thagard was greeted by his first glimpse of Earth.

On the ground, at Tyuratam, the launch had indeed been memorable, but for other reasons. It had been a cold day at the cosmodrome and gale-force winds had directed the rocket's exhaust across the launch pad, though not into the flame trench, and had started a fire which caused damage to some ground support equipment and cables. No injuries were reported, but the windy conditions intensified the damage by tearing up slabs of melted asphalt in the vicinity of the launch complex.

Meanwhile, in orbit, Soyuz TM-21 began the complex process of deploying its solar arrays and communications and navigation antennas, ahead of the docking at Mir on 16 March, two days hence. Three hours after reaching orbit, the crew performed the first of two manoeuvring system thruster firings to raise their orbit slightly to 231×306 km and, early on 15 March, executed a small 'phasing' burn and two further rendezvous burns to adjust their altitude to 390×396 km, matching that of Mir itself. During those two days, Dezhurov, Strekalov and Thagard were confined to the cramped descent and orbital modules of the Soyuz and, although there were rudimentary toilet facilities, the complete lack of privacy had led many cosmonauts to avoid responding to the need 'to go' for as long as possible. The toilet, said Thagard, was "a funnel-like affair that's attached with a flexible tube to the structure", but stressed that "folks usually tried to get themselves in a position so they don't have to defecate while they're on the Soyuz". Most spacefarers took enemas before launch, as Thagard did, and he noticed that the probability of suffering space sickness – properly termed 'Space Adaptation Syndrome' – was reduced aboard the Russian craft. He had suffered from the nausea-like malaise on all four of his Shuttle missions, but not aboard Soyuz TM-21, and the explanation was simple: it was too cramped for him to move around. "There's not that big volume on the Soyuz", compared to the Shuttle, he said, "so you just don't move around and you certainly don't have as many head movements." On the orbiters, shortly after arrival in space, he had entered the relatively large volume of the middeck, moving his head around, opening stowage lockers and retrieving equipment . . . and very quickly making himself disorientated and nauseous. "You've got so much to do and so little time in which to do it," he said, "that you're up, just darting here, and throw big head movements, lots of movements, and there is absolutely no question that's what causes and exacerbates space motion sickness." To a physician like Thagard, it therefore made sense for astronauts and cosmonauts to travel to the space station in a small, cramped vehicle for two days, because by the time they arrived and entered the large, open volume of Mir, the susceptibility period for space sickness would have passed.

Under Vladimir Dezhurov's deft control, Soyuz TM-21 docked perfectly at the aft port of Mir's Kvant-1 module at 10:45:26 am Moscow Time on 16 March, a little over 49 hours into the mission. From his position, Thagard described the "non-violent" docking as carrying the same minor punch as backing a car into a loading bay and impacting a set of rubberised cushions; "kind of a little bump, but nothing awesome, nothing scary." Less than two hours later, following customary pressurisation and leak checks, the hatches between Soyuz TM-21 and Mir were opened and the newcomers were engulfed in bear hugs from the resident crew of Alexander Viktorenko, Yelena Kondakova and Valeri Polyakov. In a traditional Russian welcome, Kondakova carried a tray with several packages of bread and salt affixed in place with Velcro. A new record had been set with this mission, for in addition to the six people on Mir, another seven souls orbited independently aboard Space Shuttle Endeavour, flying STS-67, marking the first time in history that as many as 13 individuals were in space at the same time. In fact, Thagard and STS-67 commander Steve Oswald had flown together three years previously, and on 16 March, shortly after arriving at Mir, NASA's first 'cosmonaut' spoke to his former crewmate via radio link.

Permanently occupied since September 1989, it was standard procedure aboard Mir for the incoming and outgoing crews to overlap duties for about a week. "The old crew can hand over to the new crew," said Thagard, "because it turns out that when something's been up there for years and years, the ground never really knows the full state of everything. The only way the new crew can get all of the up-to-date information is by talking to the old crew, this handover period, and that's exactly what went on." Although the station boasted its base block and three other modules (Kvant-1, Kvant-2 and Kristall) and was relatively spacious, it was not designed to support a crew of six for a lengthy period of time.

By 16 March, Valeri Polyakov had been aboard Mir for over 14 months, securing a single-flight record which still stood strong in 2014. Keenly excited to be heading home to see his family in a few days' time, Polyakov was in jubilant mood. Together with Kondakova, he showed Thagard around the station and pointed out where his research equipment was situated. Although Mir did have a logistics tracking system, it was an imperfect one and ground controllers' knowledge of where everything was kept in the cluttered complex was woefully incomplete. "There were things I probably never would have found if they hadn't physically led me by the hand," Thagard recalled, "and said 'Okay, this is here and that's there.'"

There were awkward points, too. Vladimir Dezhurov was the commander, but he was also the youngest and, on his first mission, the least flight-experienced of the Soyuz TM-21 crew; for Thagard had four prior Shuttle missions to his credit and Strekalov had four space station missions to his credit. In fact, Gennadi Mikhailovich Strekalov became only the second Russian to complete a fifth space mission and he remains one of only six Russians to have flown on so many occasions. He came from Mytishchi, a major industrial hub, situated to the north-east of Moscow, where he was born on 26 October 1940. Judging from the nature of his birthplace, it is hardly surprising that he should have gravitated towards a career in science and engineering. His father was killed in 1945, during the Red Army's liberation of Poland, only weeks before the end of the bloody conflict with Nazi Germany. The young Strekalov completed his schooling and became an apprentice coppersmith, before enrolling at the prestigious Bauman Moscow Higher Technical School. He received his engineer's diploma in 1965 and moved directly to work for the Moscow-based organisation which evolved from the OKB-1 design bureau into TsKBEM and eventually Energia, helping with the design of Soyuz. Strekalov was chosen as a civilian cosmonaut in March 1973 and within months began formal training. In his first four space missions, he accomplished virtually everything that a cosmonaut could expect to accomplish: his first flight conducted lengthy repairs of the Salyut 6 space station, his second mission narrowly missed docking with Salyut 7, his third mission was a co-operative voyage with an Indian cosmonaut and his fourth mission offered his first taste of long-duration flight and spacewalking. He was also a member of the ill-fated Soyuz T-10A crew in September 1983, who narrowly escaped death when their launch vehicle exploded on the pad. In the words of Norm Thagard, this last unenviable achievement meant that Strekalov "actually had four-and-a-half flights, before this one … because he was on the one that blew up, so he got a rocket launch, but he only went up three miles, instead of getting into space."

According to Thagard, Vladimir Dezhurov had been "just the nicest guy" throughout their year of training together and the astronauts and cosmonauts frequently invited each other to their apartments and entertained each other's families. That changed shortly after Dezhurov boarded Mir. A distinct chill in his character first became apparent when the six crew members were gathered around the dining table in the station's base block, one evening, and Dezhurov curtly ordered Strekalov to do something.

Yelena Kondakova seemed surprised. "My, you're bossy, aren't you?" she said, then suggested that Dezhurov do the task himself.

Looking on with a mixture of bemusement and dismay, Thagard shared Kondakova's thought process. He started wondering if he could spend three months in orbit with an authoritarian commander. "I don't think that sort of approach is ever going to appeal greatly to Americans," Thagard told the NASA oral historian, "and I kind of wondered how Gennadi was going to take it, because Gennadi had all the experience and [Dezhurov] didn't." Certainly, the commander's curtness did not seem to extend in Thagard's direction at first. From a psychological perspective, Thagard was convinced that Dezhurov felt ill at ease with being the commander, and yet also the flight's only rookie, and needed to assert his authority in some manner. For his own part, Strekalov took it on the chin for a few days, then "all of a sudden he'd just level a blast" at his commander and Dezhurov would back off. Shortly afterwards, a similar situation arose with Thagard, who also retorted in kind, and after that point the relationship of the crew steadily began to improve. By the end of the mission, Dezhurov came to refer to Thagard as "my friend" whenever he spoke to him. Years later, Thagard wondered if Dezhurov felt some sort of nationalistic urge to exhibit his command and authority at an early stage. In fact, Thagard was the first American to serve under the direct command of a Russian military officer. "I thought it was extremely ironic," he said, "because when I was flying missions in Vietnam in 1969 as an F-4 pilot, I thought there was an excellent chance that at some point in time I'd have interactions with the Russians, but I thought they would be of a very different nature." If anyone had told Thagard in 1969 that he would someday end up in space with a Russian Air Force lieutenant-colonel as his commander, he would have declared them crazy.

After six days, late on the evening of 21 March, the time came for Viktorenko, Kondakova and Polyakov to return to Earth. Although Thagard liked them, he felt that three residents was optimum. "It stretches the resources on the Mir station," he said of the effect of having a crew of six. "You've only got one toilet, for instance, so all of a sudden it's like being in the coach cabin on the airplane." As Soyuz TM-20 undocked from the forward port of the multiple docking adaptor at 3:40 am Moscow Time on the 22nd and departed from the vicinity of the space station, there remained a great deal of chatter on the radio, particularly from Polyakov, who was returning to *terra firma* after more than 437 days. "He's a big guy, an extroverted guy," said Thagard, "but can be sort of like a bull in a china shop. He was apparently being real rambunctious, just listening to him on the radio as they were undocking and flying around before re-entering." So excited was Polyakov at the thought of going home that, on several occasions, Viktorenko had to calm him down. Soyuz TM-20 touched down safely at 7:04 am Moscow Time, about 50 km from the Kazakh city of Arkalyk. In completing their mission, Viktorenko and Kondakova had spent 169 days in orbit, whereas Polyakov had amassed 437 days, 17 hours and 58 minutes (and 6,927 orbits of Earth) since his 8 January 1994 launch. It was a remarkable accomplishment and provided aerospace physicians and physiologists with their closest analogue for the effects of microgravity upon the human body during a journey to Mars. After landing, Polyakov did not appear to have suffered significantly from his lengthy spell in orbit and was able to walk from the Soyuz TM-20 descent module over to the chairs provided for transportation to a field hospital.

With the old crew gone, Dezhurov, Strekalov and Thagard were now on their own until June, as Mir's 18th long-duration crew. Yet, as already noted, the arrival of Spektr and the bulk of the NASA-financed research payload was still weeks away. They very quickly settled into a routine: they awoke at 8:00 am Moscow Time, spent two hours washing, eating breakfast and preparing for their daily tasks, with the official working day running from 10:00 am until 7:00 pm, punctuated by breaks for lunch and dinner. After their evening meal, they prepared reports on the day's activities and reviewed plans for the next day; from 10:00-11:00 pm they had personal time, followed by a nine-hour sleep period. Dezhurov, Strekalov and Thagard's first robotic visitor was the unmanned Progress M-27 cargo spacecraft, which launched on 9 April and docked two days later. It carried foodstuffs, propellant for Mir's thrusters, water, spare parts, experiments and personal items for the crew, including a *New York Times* crossword puzzle book, sent by Thagard's wife, Kirby. Also aboard was a group of 48 fertilised Japanese quail eggs, which would be preserved (or 'fixed') whilst on the station, then returned to Earth aboard STS-71 in June. "The primary objective is to answer key questions about the effects of microgravity on avian development," explained Shuttle/Mir Payload Manager Gary Jahns of NASA's Ames Research Center in Moffett Field, California. "From

2-08: Cosmonauts Dezhurov and Strekalov at work outside Mir.

the beginning, this was developed as a joint US-Russian experiment, with investigators from both countries sharing all samples." Also aboard was a spherical satellite, about 21 cm in diameter, built by Germany, called GFZ-1, which would make measurements of Earth's gravitational field. Weighing 20 kg, it was deployed from the scientific airlock of the station's base block on the 19th and departed the vicinity of Mir at about 1.2 m/sec. An interesting facet of Progress deliveries was that, as well as providing items from home, they also provided *smells* from home. "One of the things you notice is that the air smells different inside the vehicle, but it's not any special air supply or anything," said Thagard. He guessed it was just the ambient air from the Tyuratam launch site, but would later notice a similar odour when the Spektr module arrived and its hatch was opened for the first time.

Of course, Spektr would not arrive until well past the midpoint of his mission, but Thagard was able to occupy himself with around 28 experiments, which encompassed seven scientific disciplines, ranging from fundamental biology and microgravity studies to human metabolic, neurosensory, motor performance and cardiovascular responses to the characterisation of Mir's environment. Much of his work centred on a freezer, left aboard the station following a past European Space Agency (ESA) mission, which Thagard used to store biological samples, including blood and urine specimens. Unfortunately, from the beginning of April onwards, the unit caused problems and ran frosted up. By the end of the month, it had failed completely, which led to the cancellation of many of the medical and biological experiments scheduled for May.

For Thagard, this was unthinkable. "The way I debriefed it when I came back is the most important thing from a psychological standpoint is to be reasonably busy, with meaningful work," he explained. His Russian crewmates were chronically overworked, which led to tension, whilst he was chronically underworked, which fostered boredom. Even watching Earth drift past the windows lost its lustre after a few weeks. At one stage, he watched winter come to the southern Andes Mountains and Tierra del Fuego, the southernmost tip of South America. "You could watch the mountain lakes start to ice over and some of the passageways between the Atlantic and Pacific," he said. "I was also really intrigued by a road down in Argentina or southern Chile that seemed to run from a port city down to the south-west, as though it were heading down to Tierra del Fuego." In those quiet moments, Thagard made a mental note to himself to someday travel there and rent a car to find that particular road.

With the first rendezvous and docking between Atlantis and the space station planned for June 1995, it was recognised that the Kristall module had to be relocated from its previous position on the 'nadir' port of the multiple adaptor (known as the -Y axis), firstly to the front longitudinal port (known as the -X axis) and later to the port-side port (known as the -Z axis). Moreover, since it solar arrays would interfere with Atlantis' arrival, it was mandatory that they be retracted as a preliminary to being removed and transferred to the Kvant-1 module at the opposite end of the Mir complex. This would then allow Atlantis to dock with sufficient clearance from solar arrays and other appendages. Original plans called for Viktorenko and Polyakov to perform this EVA transfer in late 1994, but the delays to Spektr pushed it back into the spring of 1995. At the end of April, Dezhurov and Strekalov were advised that their first EVA, scheduled for the 28th, had been postponed, due to a further delay to the Spektr launch until at least 20 May, caused by the need to add equipment to counter a possible Kurs rendezvous system malfunction. Moreover, the Russians said that Spektr would need at least a month of on-orbit checkout before Atlantis could fly the STS-71 docking mission. Early in May, it was decided to postpone STS-71 until no earlier than 19-24 June. As pointed out in Chapter 1, this had implications for the STS-70 mission that was scheduled for 8 June, and NASA stressed that it would wait until Spektr's launch before making a decision on which Shuttle mission would fly first.

In the meantime, on 12 May, Dezhurov and Strekalov ventured outside Mir on the first EVA of their mission. It was Strekalov's second career spacewalk and Dezhurov's first, although it might not have happened at all. A few days earlier, on the 5th, Strekalov had scratched his hand during cleaning work aboard Mir and the injury had quickly become inflamed. Thagard administered medication and there was some talk that the American might replace him on the EVA. This claim was quickly denied by Russian officials, who declared that Thagard had insufficient training in the space suit. That suit was the 'Orlan' ('Sea Eagle') and was the third generation in a series of suits, first trialled aboard the Salyut 6 station in December 1977. Built by the Zvezda design bureau, the Orlan consisted of flexible limbs and a one-piece, rigid body-helmet unit. The cosmonaut entered the suit through a hatch in the rear of the torso and, for Salyut 6 operations, it had a maximum operating time of around three hours. For Mir, this increased to nine hours. In its earliest guise, the Orlan's integrated design meant that it did not need external oxygen hoses and its operating pressure required a relatively short pre-breathing period of only 30 minutes. Electrical power came from umbilicals connected to the station and the cosmonauts controlled the function of the suit from a chest panel. The distance to which a cosmonaut could venture beyond the hatch was therefore limited by the length of his umbilical.

The Orlan received significant modification over the years. By the Salyut 7 era, in the early 1980s, its 'DM' configuration had a generally sturdier construction, with rubberised fabric shoulder belts, and afforded greater flexion for the wearer. It also featured bright lights at the headset temples for illuminating suit controls. The 105 kg 'DMA' version, used aboard Mir, was first trialled by Soyuz TM-4 cosmonauts Vladimir Titov and Musa Manarov in October 1988. Its improvements included lighter, more flexible and tougher composite fabrics in the arms and legs (which could be removed from the suit for repair or replacement) as well as having more durable electrical motors and more dexterous gloves. The Orlan-DMA also carried an inflatable forearm cuff, which used air from the suit's backup oxygen tank to respond to a glove puncture; in essence, this would seal off the glove until the cosmonaut was able to return to and pressurise the airlock. Its integral backpack measured 1.2 m long and 48 cm wide and operated at pressures of 26.2-40 kPal. Like its predecessors, the Orlan-D of the Salyut 6 era and the Orlan-DM of the Salyut 7 era, the DMA possessed dual polyurethane rubber pressure bladders, an integrated 'Korona' voice communication system with dual microphones and earphones and primary and backup transceivers and amplifiers. The Korona antenna was embedded within the DMA's outer fabric layer.

In anticipation of their first EVA together, Dezhurov and Strekalov spent several days performing an inventory of cables, reviewing a training video, preparing their Orlan-DMA suits and familiarising themselves with their tools and equipment. During this familiarisation session, which took place on 10 May in Mir's depressurised multiple docking adaptor, they discovered a problem with a radio transmitter in one of their suits. Nevertheless, the EVA went ahead as planned on the 12th. The cosmonauts were expected to change wiring on the Kvant-1 module, in readiness for moving the 12.2 m solar arrays from the Kristall module. Dezhurov and Strekalov spent six hours and eight minutes outside Mir, changing the Kvant-1 wiring and practicing how to fold up three small panels on Kristall's 28-panel solar array. They were assigned to remove a US-built space exposure experiment, in a scheduled 20-minute task, but this was deferred to a later EVA when they ran about 15 minutes over their suits' safety limits. Upon returning inside Mir, both cosmonauts were reportedly "very tired" and rested all of the next day. In fact, at 54 years old, Strekalov became the oldest Russian spacewalker to date.

Several days later, on the 17th, after replenishing consumables and replacing batteries in their suits, the duo were back outside to get started on the removal and reconfiguration of one of the 500 kg solar arrays on Kristall. As a crew, they successfully retracted the array, with Thagard cycling closure servo motor switches from inside Mir, as Strekalov monitored the closure process from his position atop a 45 kg telescoping crane, known as 'Strela' ('Arrow'). Meanwhile, Dezhurov manoeuvred his crewmate on the Strela, using a hand-cranking mechanism. Strekalov had to manually close one panel of the solar array, but physically detaching it from Kristall required no tools, as it had been designed for easy removal by spacewalking cosmonauts. Dezhurov then cranked the Strela over to the Kvant-1 module, which occupied the aft port on Mir's longitudinal axis, then joined Strekalov for the next stage of the installation. Since the batteries for the array resided inside Kristall, it was necessary for electrical cabling to run along the entire length of Mir. As time ticked onwards, the two men had insufficient time to install the array and open it, so they lashed it to its mounting point with tool tethers. This created a temporary power deficit aboard the station, which was partly supplemented by the arrays on the Progress M-27 cargo ship, which was kept attached to Mir for an additional two days for the purpose. Dezhurov and Strekalov returned inside the station after six hours and 52 minutes.

In the meantime, Spektr finally rose into orbit atop a Proton-K booster from Tyuratam at 6:33 am Moscow Time on 20 May, a full ten weeks into Thagard's mission. By this time, the sole remnant of its previous military life was the Phaza spectrometer for observations of Earth's surface at resolutions of about 200 km. It also carried the Balkan-1 lidar to monitor upper cloud altitudes, the Astra-2

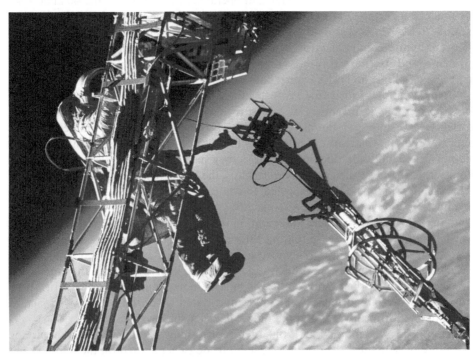

2-09: Vladimir Dezhurov is shadowed against Earth during one of his EVAs.

atmospheric environment experiment, the Taurus/Grif detector to analyse Mir's induced X-ray and gamma ray 'background', Switzerland's KOMZA interstellar gas instrument, a binocular radiometer, an ultraviolet spectroradiometer and a photometer. Another instrument, to be installed in July 1995, was the joint French-Belgian Mir Infrared Atmospheric Spectrometer (MIRAS) to measure neutral atmospheric composition data. Measuring 2.5 m in length and weighing 220 kg, MIRAS was originally intended for attachment to Russia's follow-on Mir-2 station, but after this was combined with the ISS in late 1993 the instrument was modified to be launched inside Spektr and installed on the module's exterior during an EVA. To fit through Mir's internal passageways, MIRAS was split into two halves and its diameter was reduced.

Spektr measured 14.4 m in length and 4.35 m in diameter at its widest point and weighed approximately 19,600 kg. It consisted of an unpressurised compartment, about 5.6 m long, primarily for its Earth resources instrumentation, and a pressurised segment, about 8.8 m in length, with a habitable volume of 62 m^3. It had a pair of main engines and attitude control thrusters for its approach and docking to Mir. Its four solar arrays (two measuring 38 m^2 and two measuring 27 m^2) were expected to provide half of the entire space station's electricity supply. Its facilities included a small airlock and manipulator arm to attach external payloads and deploy miniaturised satellites.

With Spektr on its way, and docking anticipated in less than two weeks, the clock was ticking for the completion of the solar array movement and the transfer of Kristall to its new location. On 22 May, the third EVA got underway and was planned for a little over six hours in duration to complete the installation of the solar array onto Kvant-1 and to stow Kristall's second array, in preparation for its own detachment and movement. During the course of the spacewalk, an electrical shortage interfered with communications between Mir and Russian mission controllers and insufficient power existed to support television coverage through the Altair communications satellite in geostationary orbit. In spite of this difficulty, Dezhurov and Strekalov installed the first Kristall array onto Kvant-1 and, from inside Mir, Thagard commanded it to unfurl, which restored the station's electrical capabilities. Towards the end of the EVA, the cosmonauts retracted 13 of the 28 segments of the second array on Kristall, allowing this to continue producing electricity whilst also leaving enough clearance for the module itself to be repositioned. Dezhurov and Strekalov returned inside Mir after five hours and 15 minutes.

Later that evening, its task of assisting with power generation completed, Progress M-27 was packed with unneeded equipment and trash, ready to be undocked. Interestingly, one of the items to be disposed of was the broken shower in the Kvant-1 module, which Dezhurov and Strekalov had to chop into small pieces with a machete in order to fit inside the Progress. The shower's place aboard the station was taken by a new set of gyrodynes. Progress M-27 undocked from Mir late on 22 May and burned up in the atmosphere in the early hours of the following morning. This freed the forward port (along the station's -X axis) for the 'temporary parking' of Kristall. On 26 May, Dezhurov oversaw the first robotic movement of Kristall from the -Y axis to the -X location. Using the stubby 'Ljappa' mechanical manipulator arm, he grappled the module, pivoted it by 90 degrees and repositioned it onto its new location. "The arm," wrote Tim Furniss in *Flight International*, "is attached to the module and rotates and grasps an attachment ... The module then undocks, moves away a very short distance and is rotated around to a side port for hard docking – not by power from the arm, but by very careful thrusting by Mir to pitch the module correctly." Several days later, on the 29th, Dezhurov and Strekalov embarked on a short EVA of just 21 minutes in order to prepare the station's spherical multiple docking adaptor for the second transfer of Kristall to the -Z docking port. Although they remained in the multiple docking adaptor, and did not venture into open space, they were working in an unpressurised environment and so wore their Orlan-DMA suits. Upon entering the adaptor, they sealed the hatches which linked it to Kristall, the base block and Kvant-2 and evacuated its atmosphere. Next, the cosmonauts removed the Konus docking drogue and cone mechanism from the -Y port, closed it with a hinged flat-plate door, opened an identical door on the -Z port and installed the Konus. They then repressurised the multiple docking adaptor and re-entered the base block to join Thagard. Next day, Dezhurov used the Ljappa arm to robotically move Kristall again, this time from the -X axis to the port-side -Z adaptor. In spite of a temporary failure in a hydraulic connector, the transfer was successfully concluded on his third attempt.

On 1 June, at 3:56 am Moscow Time, Spektr finally docked at Mir's forward longitudinal port, along the -X axis. Next day, Dezhurov and Strekalov performed another internal EVA, lasting just 23 minutes, to prepare the multiple docking adaptor for the transfer of Spektr to its final position on the -Y axis radial port of the adaptor. In a similar fashion to the 29 May internal EVA, they isolated and depressurised the multiple adaptor and moved the Konus to the nadir-facing -Y port to support the relocation of Spektr. Next day, the 3rd, Spektr was transferred by means of the Ljappa manipulator arm onto the -Y axis. The procedure took about an hour to complete. On 5 June, two of Spektr's four solar arrays were commanded to deploy, although one failed to open fully and left the module with 20 percent less electrical output than anticipated. The cause was traced to a launch restraint which had failed to release and the crew tried, without success, to extend it by sending pulses to the motor or firing Mir's thrusters. Plans were formulated for an unrehearsed EVA on 15 June, lasting about five hours, to fully open the stubborn array, ahead of the arrival of Atlantis, but it was postponed by a day and eventually cancelled when Gennadi Strekalov refused to participate. His consensus was that the spacewalk was unnecessary and its inadequate level of preparation added further risk to what was already a delicate procedure. Although Dezhurov reportedly "argued" with him over his actions, Strekalov was adamant. After returning to Earth in July 1995, both cosmonauts were fined the equivalent of $9,000, which represented about 15 percent of their contracted fee to conduct the mission. Officially, the Russians noted that the cosmonauts lacked the correct tools for the operation and that the EVA would be performed by Mir's next crew, Anatoli Solovyov and Nikolai Budarin, who would carry the proper equipment.

This incident led to a remarkable example of international co-operation. With STS-71 scheduled to launch in late June, the opportunity presented itself to design, build, test and certify and appropriate tool in enough time to load aboard Atlantis. In Moscow, Russian engineers set to work on the tool, whilst a 'tiger team' at JSC in Houston was assembled and a machinist suggested a car steering wheel cutter. He found one such cutter at a local fire department. Its arms were lengthened, in order to function from a distance, in a manner not dissimilar to a tree-limb lopper, and it was tested with bulky space suit gloves. The arms were then segmented, in order to be dismantled and boxed for launch, and within six days the task was complete. In Moscow, the Russian team had also developed a tool which resembled a scissors-jack, but the JSC tool was ultimately selected, packaged and loaded aboard the Shuttle. In those final days, Solovyov and Budarin received training whilst in quarantine at KSC in how to use the tool on a simulated solar array.

On 10 June, in the final robotic move of this mission, Kristall was detached via Ljappa from the -Z port and installed onto the -X port to await Atlantis. The transfer was originally scheduled for 6 June, but was postponed, according to the Russians, "due to activities with a higher priority". It was suggested that the delay was associated with an air seal problem experienced after the redocking of Kristall on 30 May.

With Spektr finally open for business, Thagard set to work activating two US-built experiment freezers for the storage of biomedical samples. He also set up the exercise bicycle, the laptops, the CD recorder for data, the centrifuge and an array of metabolic gas equipment. In the light of the delays, around 270 kg of US research equipment had been delivered aboard unmanned Progress cargo ships and Spektr arrived at Mir with 755 kg of NASA experiments and 45 kg of ESA experiments for use by German astronaut Thomas Reiter on a subsequent long-duration mission. The Priroda module, which was at that time expected to launch in November 1995, would transport a further 935 kg of payloads to the space station. On 6 June, Thagard also quietly exceeded the 84-day US endurance record, set by the final Skylab crew, more than two decades earlier.

Among the experiments was an investigation into Thagard's metabolism and diet. His food supply for the three-month mission consisted of a basic, repeating, six-day menu, four of whose items were canned fish, which he hated. Unlike the Shuttle, where crew members could choose their food items, the element of choice was not provided by the Russian dieticians. On a regular basis, Thagard found himself swapping the fish dishes that he so loathed with the asparagus dishes hated by Dezhurov and Strekalov. Each food package was carefully barcoded and they were obliged to record with a scanner exactly what they consumed, with about 2,600 calories scheduled for each day. Although 'supplementary' foods *were* available, with about 400-600 'extra' calories, the diet was wholly inadequate; Dezhurov and Strekalov gave it up shortly after reaching orbit, admitting that somewhere between a quarter to a half of their food came from dipping into the supplementary supplies, but Thagard stuck religiously to his regimen. The obvious consequence was that he was constantly hungry and lost weight drastically. At one stage, Mission Control realised that he had shed 7.9 kg and Russian physicians warned him that he was not just losing fat, but also muscle mass. "They told me that I was free to eat anything on-board," Thagard recalled, "*other* than my crewmates!"

In spite of the large amount of work, it was difficult for the three men to be unaffected by the profound sense of isolation from Earth. They were allowed to talk to their families, but in Thagard's words the quality of communication with the ground was far from perfect. "There were days when we had as little as 42 minutes of communication time for the *whole* 24-hour period," he said. "That's for *everything*! Obviously, the stuff I was doing can't have priority over stuff that you need to do to keep the Mir station running. I think there were four times during the flight when I went 72 hours without talking to anybody in the mission control centre." After the mission, Thagard would tell NASA Administrator Dan Goldin about the sense of extreme cultural isolation aboard Mir and his worry about the "psychological stress of longer flights". Going for several days without talking English or speaking to any of his US colleagues or hearing any news of the world was profoundly troubling. "Dr Thagard made it very clear to us," said Goldin, "that we need to take a look at the psychological wellbeing." As the mission wore on, family ties became increasingly important for the Russians, too, particularly when Dezhurov was given the heartbreaking news on 18 June that his mother had died. He was given two days off to allow him some respite to grieve.

At the opposite extreme, Gennadi Strekalov grew frustrated that the launch of STS-71 – his ride home – had been delayed by several days past 24 June. His daughter was getting married in September 1995 and he was ready to go home. "I'm not sure [he] ever wanted to fly the mission in the first place," Thagard told a Smithsonian interviewer, years later. "They had called him out of semi-retirement for the flight. If the Shuttle didn't pick us up, we might have to had to stay as long as six months, when we'd come back in the Russian Soyuz spacecraft." As they flew over Cape Canaveral, and beheld solid cloud cover across KSC and the whole south-eastern portion of the United States, Strekalov wondered what all the fuss was about. *Why* did they need to delay Atlantis? Dezhurov and Thagard exchanged glances, mentally wondering *'What is he thinking?'* because the weather situation was obviously poor in Florida. Normally calm and unruffled – "a sweetheart," according to Thagard – Strekalov suddenly allowed his emotions to boil over.

"There's absolutely *no* excuse for a crew not to launch for *anything* other than a problem with a rocket," he fumed. "We'll just stay up here and go back on the Soyuz." He then launched into a tirade about the problems with the Shuttle.

Thagard understood Strekalov's anger, but stopped him dead in his tracks. "That's fine with me, Gennadi," he replied. "I'd like to go home on the Soyuz. I've never done a re-entry on a Soyuz. You, on the other hand, are going to miss your daughter's wedding." In truth, Thagard was ambivalent about coming home; he was happy to be returning to his family, but would have been equally happy to remain aboard Mir for longer. He knew that even if the Shuttle was delayed for a long period of time, with Soyuz TM-21 docked, and taking into account the spacecraft's six-month operational lifetime, the latest that he would be back on Earth would be September. Moreover, for every day that STS-71 was postponed, Thagard could add an extra day to the US space endurance record. "Isn't that the way it is?" he would later joke with STS-71's commander, Robert 'Hoot' Gibson. "You call for a taxi ... and it takes *weeks* to get here!"

In spite of their eager anticipation for Atlantis to arrive, Dezhurov, Strekalov and Thagard's final days aboard Mir were exceptionally busy ones. "We had to do a lot of stowage," Thagard recalled, "getting things ready to come home, because some items were going to be brought back on the Shuttle side." Three days later, on 27 June, the weather at KSC co-operated and Atlantis was finally ready to fly one of the most ambitious and remarkable missions of the entire Shuttle programme, heralding the dawn of a new era.

"ATLANTIS IN FREE DRIFT"

In terms of international co-operation between two former foes, it would be no understatement to assert that STS-71 was the singular most important Space Shuttle mission of the 1990s. For the first time, the reusable orbiter accomplished what it had been designed to do: it docked with an Earth-circling space station and exchanged crew members of different nationalities. It supported shared research and it marked the largest target with which the Shuttle had ever performed a rendezvous. The technical and human success of STS-71 enabled each of the incremental steps which followed and guided the United States and Russia onto a new path in space, from which even political differences have thus far not obliged either side to deviate. The first docking between the orbiter Atlantis and the Mir space station in June 1995 was a direct stepping stone to STS-88, the first mission to begin construction of the ISS, and its ramifications can still be heard in the successful partnership which endures in the second decade of the 21st century. In the words of NASA Administrator Dan Goldin, the mission of STS-71 heralded "a new era of friendship and co-operation between our two countries".

Yet the idea of a joint US-Russian Shuttle-station docking mission was far older and could trace its ancestry back to the era of the Apollo-Soyuz Test Project. As early as May 1975, NASA made its first proposal to the Soviets to fly a cosmonaut aboard one of the reusable orbiters, which were then scheduled to begin flying later in the decade. A rendezvous and docking with a Salyut station and participation by a cosmonaut on a Spacelab flight were also discussed, as noted by *Time* magazine on 4 August 1975. A little over a year later, a series of US-Soviet talks at NASA Headquarters in Washington, DC, established a "meeting of minds" of the two nations on future manned space co-operation, with two principal foci: a scientific venture involving Shuttle-Salyut or the development of "a space platform ... bilaterally or multilaterally". In May 1977, NASA Acting Administrator Alan Lovelace and Anatoli Alexandrov of the Soviet Academy of Sciences explored the Shuttle-Salyut option in greater depth and produced a ponderous document, entitled 'Objectives, Feasibility and Means of Accomplishing Joint Experimental Flights of a Long-Duration Station of the Salyut Type and a Reusable Shuttle Spacecraft'. Ongoing meetings over the next year or two raised a glimmer of hope that it might happen and in April 1978 *Flight International* suggested that a rendezvous mission might take place as soon as 1981, followed by a docking at some point thereafter. Unfortunately, these plans came to nought when the deteriorating geopolitical

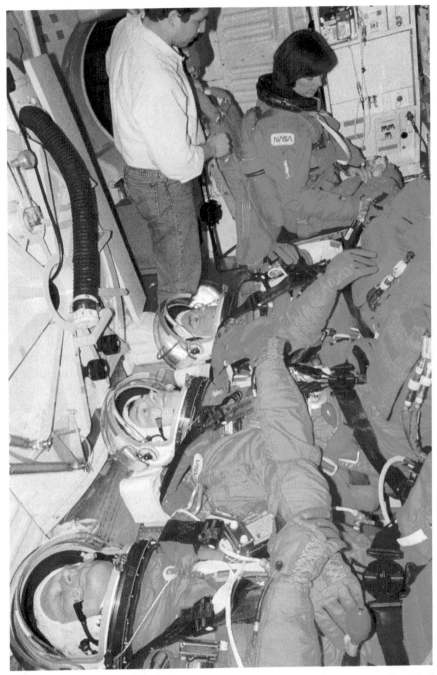

2-10: Norm Thagard, Vladimir Dezhurov and Gennadi Strekalov occupy their recumbent seats during pre-mission training for their Shuttle landing. Bonnie Dunbar is seated in the background.

situation in the late 1970s and early 1980s (particularly the Soviet Union's invasion of Afghanistan) led to a rapid cooling of relations.

It was more than a decade later, as pointed out earlier in this chapter, before the next serious inroads were made in terms of collaboration in human space exploration. The first formal planning for a Shuttle-Mir mission began in July 1991, when US President George H.W. Bush and Soviet General Secretary Mikhail Gorbachev met for two days in Moscow and agreed to fly a cosmonaut aboard the orbiter and an astronaut aboard the space station for three months. A year later, planning expanded to include a single Shuttle-Mir docking mission, but by the time US Vice President Al Gore and Russian Prime Minister Viktor Chernomyrdin met in December 1993, flesh was added to the bones of up to *nine* similar flights in 1995-1997. These flights would deliver equipment and supplies and exchange crew members, including a succession of long-duration NASA astronauts.

In addition to bringing Vladimir Dezhurov, Gennadi Strekalov and Norm Thagard – the 18th long-duration Mir crew, dubbed 'Mir-18' – back to Earth, STS-71 was also tasked with delivering their replacements, the Mir-19 crew of cosmonauts Anatoli Solovyov and Nikolai Budarin. When combined with five US astronauts on STS-71, that meant that Atlantis would launch with seven crew members and, uniquely, would return to Earth with *eight*. Although a crew of eight had flown aboard the Shuttle in October-November 1985, STS-71 would be the first occasion on which an orbiter had returned from space with different crew members *and* with a larger number of crew members. It was a situation about which the Russians had been decidedly unhappy at first. NASA Associate Administrator for Russia, former astronaut Guy Gardner, had recommended using STS-71 as a crew-exchange flight, but was initially met with concern from Moscow that such a move might "interfere with training regimes". At length, agreement was reached and on 3 June 1994 the core of the STS-71 crew was formally announced. In command was Robert 'Hoot' Gibson, a veteran of four previous Shuttle missions, who was at the time also serving as chief of NASA's astronaut corps. He would be joined by pilot Charlie Precourt and mission specialists Ellen Baker, Greg Harbaugh and Bonnie Dunbar, the latter of whom was training to serve as Thagard's backup. She would join the STS-71 crew on a full-time basis in March 1995, following Thagard's launch.

Gibson's assignment to STS-71, at first glance, seemed unsurprising, for the Russians had specifically requested the chief astronaut to command the important flight. Yet according to Bryan Burrough in his controversial 1998 book *Dragonfly*, the machinations behind the assignment were somewhat more complex. When Gibson was named as chief astronaut in December 1992, part of his mandate was to assign Shuttle crews and he was reluctant to name himself to a 'plum' mission, as several of his predecessors had done. Gibson's choice to lead STS-71 was four-time Shuttle commander Steve Nagel and it was his name that he proposed to George Abbey, the powerful deputy director (and, from January 1996, the director) of JSC. Abbey rejected it and it was left to Dave Leestma, in his role as head of Flight Crew Operations, to formally ask Gibson to command STS-71. Although Nagel said little of the assignment in his NASA oral history, it is not difficult to read the disappointment in his words. "I was vying for another flight and my name had been submitted for STS-71," Nagel admitted, "but that wasn't going anywhere. I wasn't going to get on that flight." He left the astronaut office a few months later, in March 1995.

Robert Lee Gibson grew up with the moniker 'Hoot', as had his father. "I always tell people that it comes from 'not worth a hoot'," he once told an interviewer, but in reality it originated from Edmund Richard Gibson (not a relation), a famous rodeo champion who turned into a cowboy film star in the 1920s and 1930s; *his* nickname of 'Hoot-Owl' came from co-workers and, later, evolved simply into 'Hoot'. "So after that," his astronaut namesake continued, "*everybody* whose name is Gibson usually picked up the name 'Hoot'." In fact, when he progressed into the Navy Fighter Weapons School – the famous 'Top Gun' – Gibson chose the nickname for his radio callsign. Born in Cooperstown, New York, on 30 October 1946, he completed high school in Huntington and entered Suffolk County Community College on Long Island to study for an associate degree in engineering science. He later earned a bachelor's degree in aeronautical engineering from California Polytechnic State University in 1969.

Gibson's father was a test pilot and inspector for the Civil Aeronautics Administration and built his own private aircraft in the garage, whilst his mother was one of the few women to fly general aviation aircraft in her day; in her youth, she and two friends bought a J-2 Taylor Cub. With such an impressive pedigree, it is hardly surprising that their son should have charted his own course for the skies and beyond. As a boy, Gibson travelled frequently with his father on CAA business and, on one occasion, the pair were at an airport in Phoenix, sitting in a Beechcraft Bonanza with just a single yoke, and Paul Gibson handed the controls to his son to perform the takeoff. The boy was just ten years old. "I was so proud that he trusted me," Gibson recalled years later. "He was my inspiration." That was just the start. Gibson soloed in a Piper Colt on the "windy, rainy, solid overcast" day of his 16th birthday and gained his private pilot's licence at 17.

After completing his bachelor's degree, Gibson entered the Navy and received basic and primary flight instruction in Florida and Mississippi, then advanced training in Kingsville, Texas, and eventually moved to Naval Air Station Miramar in California for assignment to the F-4 Phantom fighter. "I was in awe of the F-4," he told Robin White in an interview for *Air & Space* magazine. "It looked so big and heavy and the wings seemed so small. I was reluctant to slow it down. I was sure it would fall out of the sky, but it was just totally rock-solid on approach to the carrier." From April 1972 until September 1975 he served aboard the USS *Coral Sea* and the USS *Enterprise*, flying the F-4 over Vietnam during two tours of duty. When his commanding officer asked him if he wanted a *third* tour, Gibson was not enthusiastic, until he learned that the tour would involve operational deployment of the new F-14 Tomcat. If Gibson was in awe of the old F-4, then this new fighter functioned on a totally different level. On one occasion, with just 30 hours' experience in the Tomcat, he faced a thousand-hour F-4 veteran for a training dogfight. "We called *Fight's On*," Gibson recalled, "and 30 seconds later I was sitting in his six [behind him]. We ran the engagement three times. The results were *always* the same. An F-14 with a *nugget* at the stick could out-manoeuvre, out-turn and out-fight a Phantom flown by an old hand!" Completion of the Navy's Top Gun course was followed by assignment as an F-14 instructor pilot and graduation as a test pilot in June 1977. The following January, he was selected as one of ten Shuttle pilot candidates and flew four times between 1984 and 1992.

Pure serendipity and circumstance led to STS-71 also becoming America's 100th human space mission since the inaugural voyage of Al Shepard in May 1961. Originally scheduled for the end of May 1995, ahead of the STS-70 mission, delays to the launch of the Spektr module and the need for additional on-orbit checkout time pushed STS-71 until late June. This should have seen STS-70 flying in the second week of June, but as described in Chapter 1 of this volume, the antics of woodpeckers on Discovery's ET forced extensive delays and the two missions ended up taking place in their original sequence, although numerically in the wrong order. Nor was Atlantis herself immune from problems during the final weeks of processing for STS-71. In mid-May, engineers replaced the High-Pressure Fuel Turbopump (HPFT) on her No. 3 main engine and, a month later, on 21 June, whilst the Shuttle sat on Pad 39A, a repair was also necessary on a leaking RCS thruster. NASA and Russian managers had settled on a five-day block between 19-24 June to launch Atlantis and, according to the pre-flight press kit, the final date was the 23rd, within a tight 'launch window' of between five and ten minutes to create the optimum conditions to reach Mir on the third day of the ten-day mission.

Unfortunately, it became clear that the often unpredictable Florida weather simply would not co-operate on 23 June and the launch was scrubbed, ahead of fuelling the ET, due to severe weather and lightning strikes a mere 5 km from the pad. Heavy cloud cover and thunderstorms put paid to a second attempt on the 24th – much to Gennadi Strekalov's annoyance, watching as he was from aboard Mir – and the launch was rescheduled for 3:32 pm EDT on the 27th, at the opening of a carefully timed window, which extended for ten minutes and 19 seconds.

With only a 60-percent chance of acceptable weather for the third launch attempt, the seven-member crew suited-up in their orange pressure suits and departed KSC's Operations & Checkout Building at 11:20 am, arriving at Pad 39A about 15 minutes later. The weather situation steadily improved, however, and the skies cleared to permit a spectacular, on-time liftoff of STS-71. Nine minutes later, Atlantis was inserted into an orbit with an apogee of 292 km and a perigee of 157 km, "the lowest altitude ever flown by a Space Shuttle," according to NASA, which allowed her to close the 12,960 km distance to Mir at an initial rate of about 1,630 km with each 90-minute circuit of Earth. A little under four hours into the mission, Hoot Gibson and Charlie Precourt fired Atlantis' twin Orbital Manoeuvring System (OMS) engines for two minutes in the so-called 'NC-1' burn to raise their altitude to 389×292 km and slow their rate of closure on the space station. By the time the crew bedded down on the evening of the 27th, Atlantis was trailing Mir by 10,000 km and gaining on her quarry by about 520 km per orbit.

STS-71 was the second space mission for Charles Joseph Precourt, who would go on to command two subsequent Shuttle-Mir flights and later serve as chief of the astronaut office. He was born in Waltham, Massachusetts, on 29 June 1955, and completed his schooling in his hometown of Hudson, then attended the Air Force

Academy, from which he graduated with distinction in aeronautical engineering. During the course of his studies, Precourt participated as an exchange student with the French Air Force Academy in 1976. Undergraduate pilot training at Reese Air Force Base in Texas was followed by roles as a T-37 and T-38 instructor and maintenance test pilot. Precourt flew operationally in the F-15 Eagle at Bitburg Air Base in Germany in 1982-1984 and completed test pilot school at Edwards Air Force Base in California a year later. He received awards as both the most outstanding undergraduate pilot and the most outstanding test pilot instructor. Precourt flew test missions in the F-15E Strike Eagle, F-4 Phantom, A-7 Corsair and A-37 Dragonfly. In 1988 he gained a master's degree in engineering management from Golden Gate University and had just completed another master's credential in national security affairs and strategic studies at the Naval War College when he was selected as an astronaut candidate in January 1990. In his NASA oral history, Precourt admitted that he was "quite anxious through the whole ten to 12 months" of training for STS-71, as there existed an awful lot of unknowns, including language barriers and the idea of bringing two enormous spacecraft together for the very first time with an untried docking mechanism.

Next morning, 28 June 1995, Gibson pulsed the OMS engines for 14 seconds to further reduce Atlantis' rate of closure with Mir, adjusting their altitude to 390 × 300 km and narrowing the distance to the station by about 460 km per orbit. By this time, the Spacelab long module, which was being carried for the first and only time aboard Atlantis, had been opened and activated. Also deployed and tested was the $95.2 million Orbiter Docking System (ODS), which would permit the physical connection between the Shuttle and Mir. The development of the ODS got underway in July 1992, shortly after the Shuttle-Mir docking mission was first defined, and within 19 months it had progressed through preliminary and critical design reviews on both the US and Russian sides. The 286 kg docking mechanism, known as the Androgynous Peripheral Docking System (APDS), was manufactured by Russia's Energia design bureau in Kaliningrad, under an $18 million subcontract signed in June 1993, and was based on an original design that would have been used by the Soviet shuttle, Buran. Although a docking between US and Russian spacecraft had been achieved two decades earlier, the Shuttle-Mir system was far more complex in terms of its capacity to handle larger structures and accommodate axial loads of up to 1,000 kg with a greater centre-of-gravity offset. Physically, the ODS measured 4.6 m wide, 1.9 m long and 4.1 m high and weighed 1,590 kg. It was mounted in Atlantis' forward payload bay and featured a docking mechanism, attached to a docking base, atop the orbiter's external airlock and supporting truss structure, which were themselves linked by tunnel to the Shuttle's middeck. A camera was situated on the ODS centreline in order to transmit television images of the docking port of the Kristall module during the final approach as a visual cue for Gibson. Equipped with a capture ring and three inboard-canted guide 'petals', the APDS mated with a similar mechanism on the end of Kristall. After docking, the mechanisms provided an internal pressurised tunnel some 80 cm in diameter to allow astronauts and cosmonauts to pass in shirt sleeves between the two vehicles. Rockwell International, NASA's prime contractor for the Shuttle, accepted delivery of the Russian-built

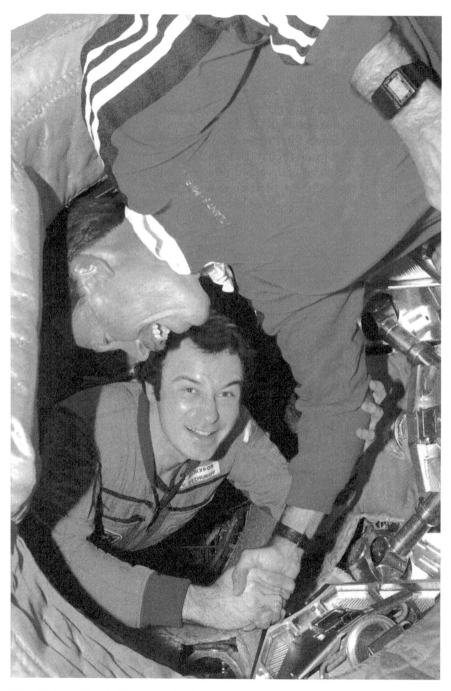

2-11: Robert 'Hoot' Gibson shakes hands with Vladimir Dezhurov inside the Kristall module after the historic Shuttle-Mir docking.

hardware from Energia in September 1994 and set to work integrating it into the ODS. Three months later, it was delivered from Rockwell's Space Systems Division in Downey, California, to KSC for final processing and loaded aboard Atlantis in March 1995. The orbiter herself had received extensive wiring upgrades to accept the new system.

Putting the ODS to the ultimate test came on the morning of 29 June, as the STS-71 crew awoke in anticipation of the final rendezvous and docking with Mir. The day began in celebratory mood for Precourt, who turned 40 years old. Shortly before 4:00 am EDT, he and Gibson fired Atlantis' OMS engines for 45 seconds in the NC-4 burn to raise the orbiter's perigee and position them about 14.8 km 'behind' the station. One orbit later, Gibson performed the Terminal Initiation (TI) manoeuvre to begin the final phases of the two-day orbital ballet. By 6:15 am, the Shuttle had reached a point a few hundred metres 'below' the space station, flying along the 'R-Bar' (or 'Earth Radius Vector'), which described an imaginary line, extending 'upwards' from the centre of Earth towards Mir. The profile differed fundamentally from the 'V-Bar' (or 'Velocity Vector') technique employed on earlier Shuttle rendezvous missions, including STS-63. By approaching its quarry from 'beneath', Atlantis could exploit the gravitational gradient to naturally brake her final approach. In fact, because she would have to thrust *against* gravity to maintain the approach, this would also provide a margin of safety in the event of an RCS failure. Moreover, it was recognised that plume impingement would have to be minimised during proximity operations with Mir, lest the thrusters cause damage or deposit contaminants upon the station's surfaces and, in particular, its solar panels. If such firings *were* necessary, using the R-Bar approach, they would be executed in 'Low-Z' mode, whereby Atlantis would use slightly-offset thrusters on her nose and tail, such that the plumes were not aimed toward the station. In November 1994, the R-Bar profile had been trialled during a satellite rendezvous on STS-66, with favourable results.

As the morning wore on, Atlantis drew steadily closer to Mir and Gibson took manual control of his ship from the aft flight deck at a distance of about 800 m. By this time, he was receiving timely inputs from the Shuttle's rendezvous radar, which supplied range and rate-of-closure data, whilst the Trajectory Control Sensor in the payload bay provided additional laser tracking support. At about 75 m, he entered a 90-minute period of station-keeping to await authorisation from NASA Flight Director Bob Castle and Russian Flight Director Viktor Blagov to proceed. Upon receipt of their permission, Gibson advanced at a rate of about 3.3 cm/sec until he was just 10 m from Kristall, using the ODS centreline camera for guidance. Throughout these tense minutes, he was able to call Dezhurov, Strekalov and Thagard via VHF radio to keep them updated with the Shuttle's major events, including confirmation of contact and capture.

Shortly before docking, the STS-71 crew placed the ODS into its 'active' mode, extending the capture ring 'outward' to its full 28 cm extent and disengaging its five locking devices. Atlantis remained the 'active' partner, with Mir's docking mechanism in a 'passive', non-operational, stowed configuration. Docking began as Gibson manoeuvred the orbiter to bring the interfaces of Atlantis' active mechanism into contact with Mir's passive mechanism, at which point the maximum allowable axial rate of approach was just 6.1 cm/sec. "Minor misalignments of the two mechanisms of up to [20 cm] and five degrees are corrected as the orbiter interface is displaced and rotated, so the capture ring can latch onto the opposing androgynous interface ring," explained NASA's STS-71 press kit. "This rotation is produced by the relative velocity of the two vehicles. If the alignments are exceeded, the passive half will not be captured and the two vehicles will simply separate. Further docking attempts can then be made." This proved unnecessary, for Gibson had accomplished a smooth docking on his first attempt and precisely on schedule at 9:00 am EDT as the two spacecraft flew 400 km above Russia's Lake Baikal. At the point of capture, the two spacecraft had a total mass of almost 230,000 kg, making it the single largest vehicle ever assembled into orbit.

Five seconds after the initial capture, dampers were activated for about 30 seconds to reduce the relative motion between Atlantis and Mir and, after this movement had subsided, the 'retraction' phase got underway, whereby the latched capture ring and the passive station mechanism were pulled into the Shuttle's mechanism. 'Structural latching' occurred at the completion of this retraction process, providing for the shirt-sleeves transfer of astronauts and cosmonauts between the two spacecraft.

"Houston, Atlantis," Gibson radioed, triumphantly, "we have capture."

"Copy, capture," came the reply from Capcom Dave Wolf in Houston. "Congratulations Space Shuttle Atlantis and Space Station Mir. After 20 years, our spacecraft are docked in orbit again. Our new era has begun."

After docking, a lengthy period of almost two hours ensued as pressurisation and leak checks were conducted. In his oral history, Charlie Precourt remembered floating into the ODS and spotting the Mir crew through the porthole. "Their hatch is already open and we're doing pressure checks with our hatch," he recalled, "and our hatch opens last, to physically give us access, so you can look through this little porthole and wave to the guys on the other side and you can see that they're really antsy for us to open our hatch." In the aftermath of the leak checks between the two vehicles, the hatch was finally opened and in a highly symbolic gesture – and with a nod to the Apollo-Soyuz mission, two decades earlier – the commanders of Atlantis and Mir, Hoot Gibson and Vladimir Dezhurov, shook hands and exchanged greetings at the interface between the ODS and the Kristall module.

Precourt's first view of Norm Thagard was a comical one. "Norm," the newcomers shouted, "you guys are *upside down!*"

"Naw," replied Thagard. "*You* guys are upside down!"

After all seven STS-71 crew members had boarded Mir, the ten-strong group assembled for a welcoming ceremony. On that same day, the responsibilities of the Mir crews changed. After transferring their personal gear and their specially moulded seat liners over to Soyuz TM-21, Anatoli Solovyov and Nikolai Budarin immediately became the 19th long-duration crew of the station and slept aboard it on the night of 29 June, whilst Vladimir Dezhurov, Gennadi Strekalov and Norm Thagard moved over to the Shuttle.

For the next five days, more than twice as long as ASTP, the Atlantis-Mir combination orbited Earth in a tight mechanical embrace. During that period, medical samples from Thagard's research, including disks and cassettes, over 100 urine and saliva samples, some 30 blood samples, 20 surface samples, 12 air samples, numerous water samples and even specimens of the astronaut's breath, were transferred to the Shuttle for return to Earth and post-flight analysis. A broken computer from Mir was also removed and about 450 kg of water, generated by the orbiter for waste system flushing and electrolysis, was loaded into Russian tanks and moved over to the station for use by Solovyov and Budarin. The spacewalking tools which would be used during the repair of Spektr's solar array were also transferred and oxygen and nitrogen from Atlantis' environmental control system was used to raise air pressure on the station in order to improve Mir's consumables margins.

In the rear of the Shuttle's payload bay, connected to the middeck and ODS by means of a tunnel, was the 7-m-long Spacelab module, which was flying on STS-71, almost exclusively for research on Dezhurov, Strekalov and Thagard. The research work was led by Ellen Louise Baker, STS-71's payload commander, who was making her third space mission. She came from Fayetteville, North Carolina, the daughter of Dr Mel Shulman and politician Claire Shulman, where she was born on 27 April 1953. Raised in New York City, she completed high school in Queens in 1970 and earned a degree in geology from the University of Buffalo at the State University of New York in 1974. She then entered medicine and, upon receipt of her doctorate in 1978, trained in internal medicine at the University of Texas Health Science Center in San Antonio and was board-certified in 1981. Baker's career at NASA began that same year, as a medical officer at JSC, and she later served as a physician in the Flight Medicine Clinic, before selection into the astronaut corps in May 1984.

Aboard the Spacelab module, a battery of research hardware supported 15 of the 28 Shuttle-Mir experiments conducted under Norm Thagard's programme. These included six metabolic experiments, focusing on a range of physiological responses in the bodies of Dezhurov, Strekalov and Thagard, which expanded on studies already begun aboard Mir to examine human metabolism and endocrinology and determine how fluids redistributed themselves around the human body. The long-duration crewmen participated in efforts to understand whether prolonged exposure to the weightless environment affected their ability to mount an antibody response and if their immune cells were altered in any way. Other experiments involved the use of both Russian and American lower-body negative pressure devices to assess their usefulness as countermeasures to the kind of 'orthostatic intolerance' frequently reported by spacefarers upon their return to terrestrial gravity; the tendency to black out upon standing upright. As part of this research, Dezhurov, Strekalov and Thagard would return to Earth in a reclining position, aboard custom-moulded recumbent seats in Atlantis' middeck, with changes in their heart rates, blood pressure, voices and posture being continuously monitored throughout re-entry and landing. As part of ongoing neurosensory investigations, the crew measured muscle tone, strength and endurance by electromyography and utilisation of oxygen during 1-2 hours of daily walking or running on a treadmill, together with other exercise sessions. Microbial samples were taken from both Mir and the Shuttle, as well as specimens from the crew themselves, in order to determine if the closed environment of a spacecraft or space station affected microbial physiology and its interaction with humans in orbit. Muscle co-ordination and mental agility were also monitored. The Spacelab module also supported a variety of other experiments. It provided a means to return the pre-fertilised Japanese quail eggs to Earth, following their launch aboard Progress M-27 in April, and to deliver new sensors to Mir for the station's greenhouse. A series of several hundred protein crystal growth investigations, frozen in a thermos-bottle-like vacuum dewar, were delivered to Mir for the next four months; it was intended that they would be retrieved and brought back to the ground by the STS-74 crew in November 1995.

Early on 30 June, with the Russian and US flags as a backdrop in the Spacelab module, the crews exchanged gifts, including the ceremonial joining of a halved pewter medallion, which bore a relief image of the docked Shuttle-Mir combination. A 1/200-scale model of the two spacecraft was also joined, with the intention that both gifts would be presented to US and Russian heads of state after the mission. Furthermore, a proclamation was signed by all ten astronauts and cosmonauts, certifying the date and time of docking, which declared that "The success of this endeavour demonstrates the desire of these two nations to work co-operatively to achieve the goal of providing tangible scientific and technical rewards that will have far-reaching effects to all people of the planet Earth."

With the exception of a General Purpose Computer (GPC) alarm on 30 June, caused by a failure to synchronise with one of its siblings, only the most minor of issues troubled Atlantis herself during the docked phase of the mission. The need to reset a troublesome hydrogen valve also disturbed Gibson's sleep and, early on 1 July, the temperature on one of the nose-mounted RCS thrusters (F5R) fell below its temperature limits, requiring the crew to adjust Atlantis' attitude to provide solar warming. "The temperature drop was not unexpected," NASA explained, "due to the inertial attitude the Atlantis/Mir spacecraft has been flying." Gibson and Precourt also performed a thruster firing to test the integrity of the docking mechanism and found the hardware to be very secure.

It was already intended that Solovyov and Budarin would board and undock Soyuz TM-21 from the aft longitudinal port of the Kvant-1 module about 15 minutes prior to Atlantis' own separation on 4 July, in order to capture still and video imagery from a station-keeping distance of about a hundred metres. In preparation for this task, on 2 July, they checked out their pressure suits and performed leak checks. Finally, on the afternoon of 3 July, the time came for the two crews to part and the STS-71 astronauts gave Solovyov and Budarin gifts of flight pins, watches, fresh fruit and tortillas. The tortillas were Bonnie Dunbar's idea. Having trained with Solovyov, she knew that he loved them; in fact, both he and Budarin enjoyed American food. There was plenty left over aboard Atlantis, so she took some tortillas – "great big, soft, Mexican tortillas" – over to him, shortly before the hatches were closed. At this point, Solovyov pulled her to him. Which side of the *hatch* would she like to stay on, he asked. Dunbar grinned. As much as she would have loved to remain aboard Mir for a long-duration mission, she told Solovyov that she had to return on the Shuttle. Solovyov secured the hatch on the Mir side at 3:32 pm EDT, whilst Greg Harbaugh did likewise in the ODS a few minutes at 3:48 pm.

STS-71 was Gregory Jordan Harbaugh's third Shuttle flight. He was born on 15 April 1956 in Cleveland, Ohio. After high school, he gained a degree in aeronautical and astronautical engineering from Purdue University in 1978 and was immediately employed by NASA as an engineer at JSC in Houston. During the pre-Challenger era, Harbaugh supported Shuttle flight operations in Mission Control as a data processing systems officer and earned his master's degree in physical science from the University of Houston at Clear Lake in 1986. A year later, in June 1987, he was selected as an astronaut candidate and he flew two missions in the early 1990s, as well as serving as a backup spacewalker for the first Hubble Space Telescope (HST) servicing mission.

Fittingly, the undocking between Atlantis and Mir occurred in the early hours of US Independence Day, 4 July. At the same time, Solovyov and Budarin deactivated several station systems, in anticipation of their own undocking and flyaround. At 6:55 am EDT, Soyuz TM-21 separated from Mir and soon reached a station-keeping position of about 100 m, from which Budarin acquired stunning imagery as Atlantis herself undocked at 7:09:45 am. The undocking procedure required Harbaugh to depressurise the ODS docking base and command the unhooking of latches, after which pre-loaded separation springs pushed the two spacecraft apart at low velocity. At a distance of about 60 cm, after clearing the respective docking mechanisms, Gibson reactivated Atlantis' thrusters and pulsed them in a Low-Z mode to begin the relative separation. At 120 m, he began a steady flyaround inspection of Mir, during which time the redocking of Soyuz TM-21 was captured in still and video imagery at 7:39 am. The redocking occurred a minute earlier than planned when Mir's on-board computer malfunctioned and crashed. The station had been left in free drift during the brief flight, but was about ten degrees off its correct attitude and was becoming unstable and starting to drift. Solovyov and Budarin restored the situation to normal.

From his perspective, Gibson described the manoeuvres of Soyuz TM-21, Mir and the Shuttle as nothing less than "a cosmic ballet". With eight crew members now aboard Atlantis, this was the joint largest crew ever carried aboard the Shuttle in orbit during independent flight. Whilst the medical research continued aboard the Spacelab module, the pilots continued to maintain a steady separation distance from Mir. By the end of 5 July, they were 370 km 'behind' the station and increasing their separation distance by about 16 km with each orbit, although Gibson was still able to see Mir as a far-off point of light.

With landing scheduled for early on the 7th, Ellen Baker set to work assembling three recumbent seats in Atlantis' middeck for Dezhurov, Strekalov and Thagard. Having the Russians aboard was a pleasant experience for the Shuttle crew ... and presented the opportunity for Dezhurov to prank at Precourt's expense. One day, Precourt was working with a volt-ohm meter on the aft flight deck, repairing a broken circuit, when the cosmonaut floated up behind him and whispered *"Pfftt"* in his ear. "You can't *jump* in space, but you can *sure* go reeling in zero-gravity," Precourt told the NASA oral historian, "and I could've *choked* him, but he had this

2-12: As seen from Soyuz TM-21, the Atlantis-Mir combination prepares for undocking on 4 July 1995.

big grin on his face! It was just a neat experience to bring him on the Shuttle, show him around and let him feel at home."

In the meantime, most of the Spacelab hardware was deactivated and Gibson, Precourt and Harbaugh prepared the Shuttle's systems for re-entry. Two opportunities existed to land at KSC, the first occurring a few seconds before 10:55 am EDT and the second at 12:31 pm. In contrast to the difficulties in getting Atlantis off the ground, her return to Earth was charmed and at 9:45 am the OMS engines were fired for the irreversible deorbit to begin an hour-long hypersonic plunge into the 'sensible' atmosphere. Seventy minutes later, at 10:54:34, Gibson executed a perfect landing on Runway 15, closing out a ten-day voyage and one of the most spectacular human space flights ever attempted.

A normal homecoming came as a great pleasure for Dezhurov and Strekalov, for two reasons. Obviously, they were relieved that they had returned to Earth safely, but secondly they were pleased that they were not challenged by the US authorities for lacking passports and visas upon landing in Florida. During their final days aboard Mir, Strekalov pulled Thagard to one side and, in all seriousness, expressed concern that he had no passport or travel visa. Would he be arrested?

"I kept trying to allay Gennadi," said Thagard in his oral history. "Of course, he comes from a different culture, but knowing what I know of bureaucracy, I should have been worried a little bit for him, but I couldn't believe that in a million years they were going to arrest Veloga or Gennadi because they arrived in the United States with no passport. I hadn't even thought about it, but Gennadi obviously had been thinking about it quite a lot."

After 115 days, eight hours and 43 minutes in flight, Dezhurov, Strekalov and Thagard could have been forgiven for being a little unsteady on their feet. However, Thagard was one of the first of them to unstrap and stand up after landing. His recumbent couch was on the starboard side of Atlantis' middeck, furthest from the hatch, so he had to wait until Dezhurov, Strekalov and (in a normal, upright seat) Dunbar had departed before he could leave the orbiter. "I walked off with no assistance," Thagard recalled. "I didn't have that much of a problem." In fact, the biggest issue was the sheer volume of monitoring equipment, including an electrocardiograph and an irritating blood pressure cuff, which rhythmically pumped itself up every few minutes, leaving his arm bruised and little feeling in his hand.

Within an hour or two of landing, after medical tests, he no longer felt 'heavy', but it took a few days for him to return to his normal self. Nevertheless, the three men were flown back to Ellington Field in Houston, aboard an Air Force C-9 Medevac aircraft for several weeks of medical tests and readaptation to normal terrestrial gravity. The remainder of the STS-71 crew returned to Houston later on 7 July. "I still didn't feel totally gainly," he said. "I felt a little awkward; more so than on my shorter Shuttle flights. I had a real sensation that if I were to bend forward, if I weren't careful, I'd continue to go forward, and if I bent back, if I weren't careful, I'd continue to go back." Walking down hallways and turning, he felt the tendency to overshoot the corner, brushing his shoulder against a wall. "You just don't turn sharply enough," he said, "and that's all because of the gains that change in the vestibular system while you're there." By his own admission, after each of his Shuttle flights, Thagard had felt back to normal within about 24 hours of landing; after his Mir mission, it took around five days, which still represented a remarkably rapid readaptation to gravity. On 12 July, he went jogging with Charlie Precourt and Ellen Baker. "It was the *hardest* three miles I ever did," he said, "but I *did it*."

AN 'INTERNATIONAL' SPACE STATION

With the departure of Atlantis, cosmonauts Anatoli Solovyov and Nikolai Budarin took over Mir for the next two months. Since they would be utilising the Soyuz TM-21 spacecraft of Dezhurov, Strekalov and Thagard, and since that vehicle had an on-orbit lifetime of six months, theirs would be a relatively short expedition of just 75 days, ending in mid-September 1995. One of their earliest tasks was performing an EVA on 14 July to deploy Spektr's stubborn solar array and check the problematic air seal in the -Z docking mechanism on the port side of the multiple adaptor. An unexpected slow pressure loss had caused some difficulty and consternation during the initial plan to move Kristall over from the -Z radial port to the -X frontal port at the end of May and the beginning of June.

In support of its role during the STS-71 mission, Kristall had remained attached to the -X port, along the station's longitudinal axis, but with Atlantis gone it would be manoeuvred to its permanent location at the -Z port, thereby making room for future Soyuz-TM and Progress visiting vehicles. At 6:56 am Moscow Time on the 14th, Solovyov and Budarin entered the vacuum of space and set to work. The original plan was for them to spend about five hours and 15 minutes outside Mir, although the EVA ran a little longer than intended. The two men found nothing amiss with the -Z port, with no evidence of damage or contamination, which cleared the way to robotically transfer Kristall to its new location. They also used the Strela cargo crane to reach the jammed Spektr array, which they succeeded in unfurling with the NASA-built tool. Although small sections at the edge of the array remained oriented about 90 degrees from their intended final position, the electricity loss was judged insignificant and the repair was declared complete. Next, Solovyov and Budarin moved to the Kvant-2 module, located on the space-facing ('upper', or 'zenith') port of the multiple adaptor for inspections of an antenna and a malfunctioning solar array drive motor.

The EVA ended successfully after five hours and 34 minutes, concluding the first career spacewalk for Nikolai Mikhailovich Budarin. Born on 29 April 1953 in Kirya, Chuvashia, in the central area of European Russia, he grew up with the dream of someday becoming a cosmonaut. "I was eight years old when Yuri Gagarin made his first flight into space," Budarin told a NASA interviewer, years later. "It was a turning point in not only my life, but in the lives of all other boys and girls, and that was when the dream was born for me." He entered the Soviet Army in his teens and served in Czechoslovakia in 1971-1973 and later attended the night-time education department of Ordzhonikidze Moscow Aviation Institute in 1979, earning a diploma in mechanical engineering. During this period, he occupied various positions within the Energia spacecraft design bureau, working as an electrician, an electrical foreman, a test engineer and, by 1986, was the head of his group. Two years later, he was appointed as the lead specialist of Energia's checkout and testing facility, before being selected to join the cosmonaut corps in January 1989. Upon completion of his training, he served with Solovyov and Bonnie Dunbar on the Soyuz TM-21 backup crew and in March 1995 the trio began training for their role aboard the STS-71 mission. Budarin became the first Russian cosmonaut to fly his first mission aboard a foreign spacecraft.

With the pressure integrity of the -Z port confirmed to be uncompromised after Solovyov and Budarin's first EVA, the Kristall module was relocated, via the Ljappa manipulator arm, to the port side of the multiple adaptor on 17 July. The operation lasted about 90 minutes and marked Kristall's final move and its permanent location until the end of Mir's life. It also positioned the module in its correct configuration for the planned docking of the second Shuttle-Mir mission, STS-74, in early November 1995. After the installation of Kristall, Solovyov again checked the air seal in the -Z port and found it to be intact and undamaged. Two days later, on the 19th, the cosmonauts were back outside on their second EVA, tasked with the deployment of the joint Belgian-French MIRAS experiment on the aft end of the Spektr module. However, within minutes of venturing into space, the cooling system of Solovyov's suit malfunctioned and he was ordered to remain tethered by an umbilical to Kvant-2, close to the airlock hatch at the 'top end' of the module. Meanwhile, Budarin, whose suit was unaffected, was able to perform some preparation work, but the MIRAS installation had to be postponed. He was, however, able to retrieve a US cosmic ray detector, which had been outside Mir for four years, and exchange cassettes of sample construction materials. The cosmonauts returned to the airlock after three hours and eight minutes, but even after closing Kvant-2's hatch, they discovered a 2 mm gap in the seal, through which air was escaping. They had to work with the hatch for some time in order to get it fully closed and secured. The installation of MIRAS was still undone and on 21 July they returned outside for five hours and 35 minutes and, with Budarin operating the Strela crane, Solovyov successfully secured the experiment by three clamps. Shortly afterwards, mission controllers noticed a lack of data from MIRAS and traced the problem to Spektr's data transmission system, which the cosmonauts corrected.

At the close of the third and final EVA of their mission, Anatoli Yakovlevich Solovyov completed his ninth career spacewalk. By the time he completed his final flight in February 1998, he would have completed 16 EVAs and, to this day, Solovyov holds the record as the world's most experienced spacewalker, with a cumulative 82 hours. Born on 16 January 1948 in Riga, today's capital of Latvia, he completed the Lenin Komsomol Chernigov Higher Military Aviation School in 1972 and served until 1976 as a senior pilot and group commander in the Far Eastern Military District. Selected as a cosmonaut candidate in August 1976, Solovyov commenced more than two years of evaluation and training and served as backup commander for the Soyuz TM-3 Syrian mission. His military experience and upbringing had imbued him with a staunch support for Communism – he had been a party member since 1971 – and, during the Shuttle-Mir effort with the United States in the 1990s, it has been remarked that he was "not especially friendly" towards the Americans and gave NASA reason to suspect that he was "a stern critic of the two countries' collaboration". Bryan Burrough described him as "the Chuck Yeager of the Russian programme ... there is a bit of the Old Soviet in Solovyov". If Alexander Viktorenko was one of the most 'Americanised' of the Soviet commanders, in terms of his personable nature, then Solovyov was the reverse. "He gave *orders*," wrote Burrough, "harsh orders, often shouted." Having said this, NASA astronaut Dave Wolf, who flew aboard Mir with him in 1997-1998, praised

Solovyov's stance and felt that a more 'patronising' tone would not have made him feel like an integrated member of the team. Yet Solovyov's credentials stand for themselves and do not just extend to EVA: at the time of writing, he holds sixth place on the list of most experienced spacefarers in history and is one of only half a dozen cosmonauts to have completed five missions.

Following their EVAs, Solovyov and Budarin settled down to a full plate of interior work aboard Mir throughout the remainder of the summer. Their first visitor was the Progress M-28 resupply craft, which docked at the -X frontal port of the base block's multiple adaptor on 22 July, delivering food, water, fuel and oxidiser, together with about 335 kg of scientific equipment for ESA's EuroMir '95 mission, scheduled to begin in late August. Interestingly, the cargo ship also carried a pair of icons of the third-century Christian martyr St Anastasia, which spent the next several months aboard Mir. Following their return to Earth aboard Soyuz TM-22 in February 1996, the icons would be displayed in shrines around the world. During August, Solovyov and Budarin worked on a range of life sciences and astrophysics experiments, together with smelting investigations in the Gallar furnace. They also installed a set of new gyrodynes, carried aloft aboard Progress M-28, into the Kvant-2 module, as well as repairing the seals on other gyrodyne cases.

In the meantime, Soyuz TM-22 had been postponed from its original target launch date of 18 August until no earlier than 3 September. Its crew consisted of cosmonauts Yuri Gidzenko and Sergei Avdeyev, together with German astronaut Thomas Reiter, who would be embarking on a mission of 135 days, the longest ever attempted by a non-Russian. The two EuroMir missions evolved from ESA's frustration that the US-led Space Station Freedom programme, and, by extension, the European Columbus laboratory module, was teetering on the brink of collapse. To hedge its chances of maintaining a viable space flight agenda, in May 1993 ESA opted to proceed with the two EuroMir missions at a cost of $53 million. The first, EuroMir '94, took place in October-November 1994 and saw German astronaut Ulf Merbold spend 30 days aboard Mir, whilst the second, EuroMir '95, scheduled to begin in the summer of 1995, promised to be far more ambitious: its crewman would spend more than four months in orbit. This would be substantially longer than the week-long missions previously flown by European astronauts. "Space station crews are not going to be changed each week," said Friedrich Engstrom, ESA's head of space station and microgravity applications, in an interview for *Flight International*. "Europe is preparing astronauts for the global space station era and this is an excellent opportunity to perform scientific experiments, as well as prepare the user community for that era."

The EuroMir missions were therefore considered 'Columbus Precursor' flights, supporting ESA's plans for scientific research aboard Columbus, although funding limitations required their cosmonauts to rely heavily upon equipment left aboard the station by previous European occupants. Candidates for EuroMir '95 were Germany's Thomas Reiter and Sweden's Christer Fuglesang and it became obvious at an early stage that an EVA would be manifested onto the mission, making either Germany or Sweden only the fourth discrete nation (after Russia, the United States and France) to have its own spacewalker. In August 1993, together with EuroMir '94 candidates Ulf Merbold and Spaniard Pedro Duque, the four men arrived at Star City to begin a gruelling syllabus of technical preparation, biomedical activities and more than 530 hours of mission-specific training. "Studying the Mir systems is interesting," admitted Merbold, "but I would have liked to spend more time preparing for the experiments which we shall be carrying out on the mission." His three comrades added that the cultural differences were a challenge – "The Russian way of thinking," explained Duque, "is not the same as ours" – as well as the need to learn the complex Cyrillic alphabet and language.

As described in the previous volume of this series, Merbold's EuroMir '94 mission ran relatively smoothly, although the weight constraints of the Soyuz spacecraft meant that he could return only about 16 kg of scientific samples to Earth, with many other experiments remaining aboard Mir in anticipation of EuroMir '95. More than 450 hours of experiments were planned for the second mission, with 18 investigations in life sciences, five in astrophysics, eight in materials science and a further ten in technology and applications. In readiness for the long mission, a 135-day ground-based simulation began in September 1994, under the auspices of Russia's Institute of Biomedical Problems in Moscow. Known as Human Behaviour in Extended Spaceflights (HUBES), the simulation was managed by the Norwegian Underwater Technology Centre and included three Russian research volunteers to compare and validate psychological methods and tools for use in crew training, monitoring and support.

At first, it was widely expected that Christer Fuglesang would be the prime candidate for EuroMir '95, but in March 1995 Thomas Reiter received the coveted assignment instead. Thomas Arthur Reiter was born on 23 May 1958 in Frankfurt and is currently the most experienced non-Russian and non-US spacefarer in the world, with a cumulative 350 days from EuroMir '95 and a five-month mission to the ISS under his belt. "When I was a boy, I was very closely following all the space activities," he told a NASA interviewer. "I remember the first flights into Earth orbit and when I was eleven I very well remember the first moment when Neil Armstrong put his feet on the Moon. *That* was actually the point in time when I thought, hey, this would be a good profession!" Years later, Reiter would stress that Armstrong became his hero at that point. As a European in the late 1960s, becoming an astronaut seemed a remote possibility and Reiter pursued a career in aviation and the German military. He completed Goethe-High School in Neu-Isenburg in 1977 and entered the Bundeswehr University in Munich, graduating with a master's degree in aerospace technology in 1984. He underwent military jet training at Sheppard Air Force Base, Texas, flew the Alpha-Jet in a fighter-bomber squadron, based in Oldenburg, Germany, and was involved in the development of computerised mission planning systems and became a flight operations officer and deputy squadron commander. Reiter completed Class 2 test pilot training at Germany's Flight Test Centre in Manching in 1990, he underwent conversion training on the Tornado aircraft and was assigned to the Empire Test Pilots School in Boscombe Down, England, for Class 1 test pilot training. Upon graduation from the school in early 1992, he was selected as one of six members – and the only German – in ESA's Group 2 astronaut intake. He vividly recalled the invitation from

2-13: Thomas Reiter plays the guitar aboard Mir.

a senior officer to apply for ESA. "One day, I was coming back from a mission. I was flying at this time in a fighter-bomber wing and in the evening I was called to report to the ops group commander. I couldn't imagine what the issue was. I thought maybe something with the flight today. Who knows? And he actually asked me if I would be interested in taking part in a selection for an astronaut programme." By his own admission, it was a simple question for Reiter to answer. A year later, in August 1993, he began training in Star City to support the EuroMir missions. By the time that Reiter was formally announced as the prime candidate for EuroMir '95, his two Russian comrades had already begun mission-specific training. In fact, commander Yuri Gidzenko and flight engineer Sergei Avdeyev had previously served with Pedro Duque as the EuroMir '94 backup crew on Soyuz TM-20.

For Sergei Vasilyevich Avdeyev, the flight of EuroMir '95 aboard Soyuz TM-22 was the second flight in a career which would establish him from 1999-2005 as the acknowledged record-holder for the longest cumulative time spent in space. In his three missions to Mir, Avdeyev spent a total of 749 days – a little more than two full years – away from the Home Planet and this accomplishment received much popular attention when it was announced that he had also secured the record for the greatest 'time dilation' experienced by a human being. Travelling at an orbital velocity of 27,360 km/h, Avdeyev 'aged' roughly 20 milliseconds (0.02 seconds) less than an Earthbound person, due to the special relativistic effect of time dilation. Avdeyev came from Chapayevsk in Samara Oblast, where he was born on 1 January 1956, and attended the Moscow Physics-Engineering Institute. Upon graduation in 1979, he worked as an engineer-physicist for eight years and was selected as a cosmonaut candidate in March 1987. He first flew aboard Soyuz TM-15 to Mir between July 1992 and January 1993.

His commander, Yuri Pavlovich Gidzenko, was embarking on his first flight on Soyuz TM-22. Born on 26 March 1962 in Elanets, within the Mykolaiv Oblast of southern Ukraine, he grew up with a love of aviation. "When I was a child, my dream was to be a pilot," he told a NASA interviewer. "Maybe not a civil pilot, but a military pilot, because my father was a military man and maybe he advised me to be a military man, too. When I was maybe 12 years old, I decided to be a military pilot." After high school, Gidzenko entered the Higher Military Pilot School in Kharkiv and after graduation in 1983 he served as a 3rd Class pilot and senior pilot in the Soviet Air Force, operating within the Odessa military district. His cosmonaut career began in the same group as Sergei Avdeyev, with selection in March 1987. It was not an over-arching desire to fly into space, but one which brought tremendous satisfaction. "I knew that it's very difficult to pass exams," he said, "because only one person from one hundred managed to be cosmonauts. That's why I thought about it in my dreams." Those dreams steadily turned to reality when Gidzenko began formal training in December 1987 and achieved qualified status in June 1989. An accomplished parachutist and parachute-landing instructor, Gidzenko was also a veteran of more than 170 jumps. During his training for the EuroMir '95 mission, he gained a degree in geodesy and cartography from Moscow State University of Geodesy and Cartography.

Originally scheduled for launch in late August 1995, Soyuz TM-22 rocketed into orbit at precisely 1:00 pm Moscow Time on 3 September, kicking off EuroMir '95. In the words of Jorg Feustel-Büechl, ESA's head of Manned Spaceflight and Microgravity, the ambitious mission would "provide European scientists with unprecedented data on long-duration space flight and further strengthen ESA's relationship with the Russian space programme". In readiness for the arrival of the new crew, Progress M-28 was undocked from the -X port on 4 September and intentionally destroyed in the upper atmosphere a few hours later. Next day, at 2:30 pm, Gidzenko guided his ship to a smooth docking with Mir and the hatches between the two spacecraft were opened shortly afterwards. Following the customary week of hand-over briefings and joint experiments, Anatoli Solovyov and Nikolai Budarin boarded Soyuz TM-21 in the early hours of 11 September, undocked from the aft longitudinal port of the Kvant-1 module and touched down about 300 km north-east of Arkalyk in Kazakhstan at 9:52 am Moscow Time. Their landing was described as "far away from the aiming point", but rescue forces found the two men to be in excellent condition.

Gidzenko, Avdeyev and Reiter were alone for their 135-day mission, which was originally scheduled to end in mid-January 1995. However, on 6 October, it was decided to extend EuroMir '95 by an additional 44 days, with an anticipated landing on 29 February 1996, after 179 days in orbit. The formal announcement was made a few days later, although the option of extending the mission had been discussed before launch, and the decision was principally based upon the desire to postpone Soyuz launch vehicle processing expenditure into the next fiscal year to relieve the stretched budget of the Russian Space Agency. Instead of launching in January, the next crew of Yuri Onufrienko and Yuri Usachev, aboard Soyuz TM-23, would fly on 21 February. It was also stressed that keeping Soyuz TM-22 docked until the end of February would optimise the spacecraft's capabilities, which, after all, were designed for a 180-day operational lifetime. A Progress resupply mission in December 1995 would provide additional EuroMir '95 experiment payloads. The extended mission was welcomed by ESA, not least by the fact that Reiter was offered the opportunity to perform a second spacewalk. In fact, also in October 1995, *Flight International* noted that ESA intended to propose a third EuroMir mission at its Council of Ministers meeting in Toulouse, later that same month. It was hoped to fly the mission in 1998.

The crew's first visitor was Progress M-29, which arrived at the Kvant-1 aft port on 10 October, laden with supplies, including 80 kg of EuroMir '95 research equipment. By this stage, Reiter's research had begun in earnest, with experiments focused upon the effect of microgravity on the human body, the state of his cardiovascular health through a network of blood pressure sensors, studies of his loss of bone mass density and observations of the functioning of his kidneys, lungs and muscles. He worked with the European Space Exposure Facility (ESEF) aboard the Spektr module, which sought to capture and analyse natural and man-made particles from low-Earth orbit, and supported a range of technological investigations, monitoring radiation levels on electronic components, measuring microbial contaminants within Mir, observing disturbances produced by a small robotic arm and processing alloys, glasses and semiconductor samples in the Titus six-zone tubular furnace. On average the crew worked about 4.5 hours per day on their research experiments, with the remainder devoted to exercise and maintenance tasks on the station itself. This maintenance work included a focus upon a troublesome coolant loop in early November, which leaked about 1.8 litres of ethylene glycol solution into Mir's interior. The leak forced the crew to shut down the coolant loop, which effectively disabled the primary carbon dioxide removal system and the oxygen replenishment system. Although the leak was repaired with putty, a backup air scrubber was employed to remove carbon dioxide and Kvant-2's alternate oxygen system was used for oxygen production until the next Progress arrived in December. However, it was decided to manifest additional lithium hydroxide regeneration canisters, together with a NASA-devised connection to the Mir system, aboard STS-74, the second Shuttle docking mission, scheduled for mid-November.

Notwithstanding the fact that this was ESA's precursor for ISS activities, the high point for Europe came on the 20th, when Avdeyev and Reiter ventured outside Mir on the long-awaited and much-anticipated first EVA of a German astronaut. Since Reiter was the first non-Russian to qualify as a Mir flight engineer, the spacewalk marked the first occasion that an EVA had been performed by two flight engineers; previous excursions had always involved the mission commander. Their primary task was to install exposure cassettes into the ESEF on the exterior of Spektr, for which they spent five days in preparation, assembling and checking the hardware and placing it into an EVA bag to prevent it from floating away into space. Eventually, the two men clambered out of the Kvant-2 airlock and set to work. First out was Reiter, who quickly tethered himself and his bag of equipment to the Strela boom. Meanwhile, Avdeyev – a veteran of four previous spacewalks – moved to the control crank of the Strela on the base block and manoeuvred Reiter over to the forward section of Spektr. He then climbed the length of the boom himself to join his German crewmate. Reiter threaded a tether through wire loops affixed to pins on ESEF, then pulled the tether to release covers, which exposed four attachment sites. During the EVA, which ran to five hours and 11 minutes, the men installed a pair of clam-shaped exposure cassettes, a spacecraft environment monitoring package and a control electronics box onto the ESEF. Of the cassettes, one would be opened to sample space debris, whilst the other would be used to gather particles from the Draconids meteor storm as Earth passed through the tail of Comet Giacobini-Zinner. Avdeyev and Reiter also exchanged cartridges on the joint Russian-Swiss KOMZA interstellar gas instrument.

In addition to his research workload, Reiter was granted the opportunity to attempt to pinpoint his homeland, although finding his specific birthplace proved more difficult. "In the centre of Germany are rural areas and huge urban areas," he explained, "so when I was flying over Germany I really had to try to find my home town, Neu Isenburg, close to Frankfurt, by following the River Rhine and then the River Main and then I usually could identify the Rhine-Main Airport and, close to the airport, is my hometown." He described the sight as "really a very, very nice feeling" from an altitude of close to 400 km and remembered gazing northward towards the North Sea. "It's magnificent!" he recalled.

Equally magnificent would be the arrival of the second Shuttle-Mir mission, STS-

2-14: The orange-coloured Docking Module, pictured atop Atlantis' external airlock, in the early stages of the STS-74 mission.

74, whose launch was planned for 11 November. Unlike its predecessor, STS-71, this mission would dock with the Kristall module in its permanent configuration on the radial -Z axis of the multiple adaptor, rather than the -X forward longitudinal port. Kristall had been specifically moved to the -X location in readiness for STS-71, in order to provide sufficient clearance for the Shuttle, but since this frontal port was normally a location for Soyuz-TM or Progress spacecraft it was not considered suitable to be occupied by a permanent module. Additionally, it was undesirable to constantly move Kristall from port to port in anticipation of each Shuttle docking, which was anticipated to take place every four months, not least because it would exceed the design lifetime of the Ljappa manipulator arm. As a result, in November 1993, at around the time that plans for as many as ten Shuttle-Mir missions were becoming a reality, concept discussions for a specialised Docking Module (DM) got underway between NASA and Russia's Energia design bureau. By the summer of 1994, this work had been finalised, creating the blueprint for a component which the Shuttle would install onto the end of Kristall, in order that the module could remain at its normal -Z port, but which also provided enough clearance for the Shuttle and for Mir's expansive solar arrays. In February 1995, Energia began the final assembly of the DM and its functional testing was completed in May.

On 7 June, the DM was delivered to KSC aboard an Antonov An-124 cargo aircraft and became the first flight-ready article to be prepared in NASA's Space Station Processing Facility (SSPF). This new facility had been opened just three months earlier. "This is a major operational hardware exchange between the United States and Russia," said Tommy Holloway, then serving as NASA's manager of the Phase 1 Program Office. "As we move into the Space Station era, these equipment exchanges will become almost commonplace. This particular hardware is also very important to the reconfiguration of Mir for future joint endeavours." Within the SSPF, the new module underwent a thorough systems checkout and two solar arrays (delivered to KSC in November-December 1994), together with a trunnion assembly were installed. The arrays were part of the Mir Co-operative Solar Array (MCSA) project, which brought together the advanced photovoltaic technology of NASA's Lewis Research Center of Cleveland, Ohio, with Russia's proven structures and mechanisms and formed part of efforts to extend the station's operational lifetime and support US scientific research. It was intended that 84 array modules would be integrated into Russian-built frames in 42 hinged pairs, providing up to six kilowatts of electrical power when deployed to their maximum extent of 2.7×18.2 m. In addition to the DM itself, a training mockup was delivered to JSC in Houston to enable the STS-74 crew to practice contingency EVAs. Although no spacewalk was planned for the mission, it was necessary to rehearse backup procedures in the event that problems arose with the DM in flight.

Measuring 4.7 m in length and 2.2 m in width, the cylindrical module weighed about 4,090 kg and was equipped with identical Androgynous Peripheral Docking System (APDS) mechanisms at either end, compatible with both Kristall and the Shuttle. Its exterior was protected from the brutal thermal and radiation environment of low-Earth orbit by an orange micrometeoroid shield and a layer of screen vacuum thermal insulation. "The colour was not chosen for the purpose of blaring to the Americans, rather insultingly, that *this is where you park*," explained astronaut Jerry Linenger in his memoir, *Off the Planet*, "but because the orange solar blanket was found in some Russian warehouse and, in order to cut costs, used instead of manufacturing a new white blanket. In any case, the glowing orange could *not* be missed!" In September 1995, the DM was installed into Atlantis' payload bay in the Orbiter Processing Facility (OPF), secured by means of three side latches and one keel latch. The task of the STS-74 crew was to unberth the DM, by means of the Shuttle's RMS mechanical arm, and attach it to the top of the ODS, ahead of docking with Kristall. When the time came for Atlantis to depart Mir, it would detach at the ODS-DM interface, leaving the module in place on Kristall for future missions. A Remotely Operated Electrical Umbilical (ROEU) provided power to the module whilst in the payload bay, but would be released shortly before unberthing and installation atop the ODS for docking onto Mir.

When the STS-74 crew was announced in September 1994, the selection of veteran astronaut Kenneth Donald Cameron to command the mission came as little surprise. As has already been explained, earlier in this chapter, Cameron had served as

NASA's first manager of operations in Star City from February until August of that year. Cameron was born in Cleveland, Ohio, on 29 November 1949, and enlisted in the Marine Corps at Parris Island, South Carolina. He completed Officer Candidate School at Marine Corps Base Quantico in Virginia in 1970 and later graduated from the Infantry Officer's Course and Vietnamese Language School and served as a platoon commander in the latter days of the bitter and highly divisive conflict in south-east Asia. After this year-long tour of duty, Cameron was also part of a company of Marine Security Guards at the US embassy in Saigon. In 1972, he reported to Naval Air Station Pensacola in Florida for flight training and became a naval aviator the following year, flying A-4M Skyhawks. Selected to attend Massachusetts Institute of Technology, Cameron commenced undergraduate studies in aeronautics and astronautics and received a bachelor's degree in 1978 and a master's degree in 1979. After graduation, he flew for a year out of Marine Corps Air Station Iwakuni in Japan and worked at the Pacific Missile Test Center, prior to admission into the Naval Test Pilot School at Patuxent River, Maryland, in 1982. He was then assigned as project officer and test pilot for the new F/A-18 Hornet fighter, a position he held at the time of his selection as an astronaut by NASA in May 1984. He flew two Shuttle missions in the early 1990s, the second as a commander.

Joining Cameron on Atlantis' flight deck for STS-74 was pilot James Donald Halsell Jr, making his second space mission. Born in West Monroe, Louisiana, on 29 September 1956, Halsell grew up with the strong ideals of setting ambitious personal goals and pursuing his interests. "I also had an uncle, Tommy Thompson, who was an airline pilot," he once told a NASA interviewer. "When I was growing up, I considered him one of my heroes, because he was flying airplanes and they were *paying* him to do it. *That* seemed like a good way to spend your life!" By the time Halsell finished high school, he had decided on the Air Force as a career path and that began with the prestigious Air Force Academy, from which he earned an engineering degree in 1978. Undergraduate pilot training followed and Halsell flew the F-4 Phantom aircraft from Nellis Air Force Base in Nevada and Moody Air Force Base in Georgia, gaining qualifications in conventional and nuclear weapons delivery. Two master's degrees followed – one in management from Troy University in 1983 and a second in space operations from the Air Force Institute of Technology in 1985 – and Halsell was selected to attend test pilot school. *Air Force Academy, fighter pilot, test pilot* had long been his goals and in 1986 he graduated first in his Test Pilot School class and was awarded the coveted Liethen/Tittle Trophy for performance and academic excellence. He served as a test pilot in the F-4, the F-16 Falcon and the SR-71 Blackbird, before entering NASA's astronaut corps, alongside Leroy Chiao and Don Thomas, in January 1990. "I think it might be wrong to paint myself as a super-charged 16-year-old, thinking I was going to be an astronaut," Halsell said later, reflecting upon his youth, "but certainly I had it in the back of my mind that could be a path that I would like to explore." Halsell first flew the Shuttle in July 1994.

Three mission specialists – Chris Hadfield, Jerry Ross and Bill McArthur – rounded out the five-man STS-74 crew. Jerry Lynn Ross is today one of only two people in history to have chalked up as many as seven discrete space missions. STS-

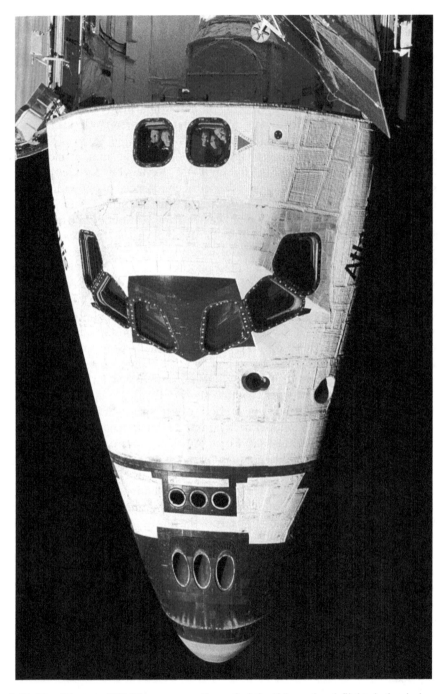

2-15: The five-man STS-74 crew peers through Atlantis' overhead flight deck windows
during rendezvous operations with Mir.

74 was his fifth. Ross was born on 20 January 1948 in Crown Point, Indiana, and represents one of only a handful of individuals to have dedicated virtually his adult working life to the astronaut business. He grew up at a time when the Cold War was at its peak and the idea of rockets, whether for carrying explosives or men, was steadily entering the popular consciousness as something more than a facet of science fiction. As a child, he watched television shows about space stations, read articles in *Life* magazine, created scrapbooks of space-related events and watched in awestruck astonishment when the Soviets launched Sputnik and America responded with Explorer-1. Even at this young age, he was introduced to the word 'engineer'. "I truly didn't fully understand what an engineer *was*," Ross told the NASA oral historian, "but I knew that they had to use a lot of math and science. I *liked* math and science, so I thought that's what I wanted to do. I wanted to become an engineer." By his own confession, this gave him a one-track mind, working on farms to earn money for a bank account which would someday pay his way through the prestigious Purdue University.

His three-step plan was relatively straightforward: (1) be an engineer, (2) go to Purdue and (3) get into the space programme. In Ross' mind, it was as cut-and-dried as that. Unlike so many others, he stuck doggedly to his plan and not only achieved it, but surpassed it. After completing high school in 1966, he entered Purdue to study mechanical engineering and received his bachelor's degree in 1970 and a master's in 1972. He entered active duty with the Air Force and worked on computer-aided design of ramjet engines and captives tests of supersonic ramjet missiles at Wright-Patterson Air Force Base in Ohio. Ross graduated at the top of his class from the Air Force Test Pilot School's flight test engineer course in 1976 and later served as project engineer for the flying qualities of the B-1 Lancer bomber. His role included both the training and supervision of all B-1 flight test engineers and also mission planning for the bomber's offensive avionics. "The B-1 at the time was the Air Force's highest-priority programme," he remembered, "and I was given the opportunity to come on-board as a B-1 flight test engineer and to work in the stability and flight controls areas of the B-1." It was shortly after the B-1 effort that Ross learned about NASA's plans to hire astronauts; he was one of thousands of hopefuls who submitted applications in 1977 and, though summoned to JSC in Houston for interview, was not successful. Still, he persevered and George Abbey offered him a position as a payload officer at JSC, working on the integration of military payloads into the Shuttle, with the hint that it might stand him in good stead for possible future selection. Two years later, in May 1980, after reviewing six thousand applications and interviewing 120 of those, NASA selected 19 new astronauts ... including Ross!

William Surles McArthur Jr was born in Laurinburg, North Carolina, on 26 July 1951, the son of an Army brigadier-general and Second World War veteran, and entered the Military Academy at West Point to study applied science and engineering. "In college, I'd really gotten fascinated with engineering," he told a NASA interviewer, years later. "And I liked aircraft; liked rockets. Engineering kind of put it all together." Upon receipt of his degree in 1973, McArthur was commissioned as a second lieutenant in the Army and after a tour with the 82nd

Airborne Division at Fort Bragg, North Carolina, he entered the Army Aviation School in 1975 and graduated at the top of his class the following year. He served as an aeroscout team leader and brigade aviation section commander in the Republic of Korea and later assumed duties as a company commander, platoon leader and operations officer with the 24th Combat Aviation Battalion in Savannah, Georgia. In 1983, he gained a master's degree in aerospace engineering from Georgia Institute of Technology and returned to West Point as an assistant professor in the Department of Mechanics. By this time, McArthur was aware that a fellow Army aviator, Bob Stewart, had been selected into NASA's astronaut corps. "And a little light came on," he remembered, "and I looked at it and, lo and behold, the goal all of a sudden became attainable." To McArthur, applying for NASA was like buying a lottery ticket and he submitted his application, knowing that "the chances might not be very good that you'll win, but they're a whole lot better than if you *never* buy the ticket!" The ticket failed him – in a sense – in 1987, when he was unsuccessful in his bid to join NASA's 12th class of astronauts. Yet the cloud had a silver lining. McArthur completed Naval Test Pilot School that year, was designated as an experimental test pilot ... and was accepted by NASA as an engineer on the Shuttle Vehicle Integration Test Team. A little more than two years later, he was selected for the astronaut corps. By the time he boarded Columbia for his first launch into orbit in October 1993, McArthur was 42 years old, one of the oldest members of his class. "Fortunately," he told an interviewer much later, "I haven't been forced to grow up just yet!"

An initial launch attempt on 11 November was postponed by 24 hours, due to poor weather at one of the Transoceanic Abort Landing (TAL) sites, which proved disappointing for the crew, who had lain uncomfortably on their backs for more than two hours before the scrub was called at T-5 minutes. Next morning, however, all went well. At 7:30:43 am EST, right on the opening of the launch window that would last for ten minutes and nine seconds, Atlantis rocketed away from Pad 39A into the Florida sky to begin her eight-day mission. In the next few hours, the Shuttle's payload bay doors were opened and Cameron and Halsell executed the first of several manoeuvres to prepare for a docking with Mir, about 65 hours into the mission, on 15 November.

Shortly after the manoeuvre came the first major task for Canada's first fully-fledged Shuttle mission specialist, Chris Austin Hadfield. Today, Hadfield is best remembered for having written and recorded his own songs during his five-month expedition to the ISS in 2012-2013, as well as having accrued over a million Twitter followers in perhaps the largest social-media following of any astronaut in history. Born in Sarnia, an industrial 'oil town' in south-western Ontario, on 29 August 1959, Hadfield grew up on a corn farm and loved aviation from an early age. "My philosophy has always been to try and take away a little bit from everyone that I meet," he said, "because everyone does something better than I do. You just have to find out what." In his youth, he drew inspiration from Charles Lindbergh, the first man to fly solo across the Atlantic, and from Neil Armstrong and Buzz Aldrin. He also read *Carrying the Fire*, the autobiography of Apollo 11 astronaut Mike Collins. On the eve of humanity's first Moonwalk, he had a *National Geographic* picture of the Moon over his bed and watched, fascinated, as Armstrong and Aldrin took their historic first steps. "To me," he said later, "it really crystallised what it was that I wanted to do when I grew up."

As an air cadet he won a glider pilot scholarship at the age of 15 and a powered pilot scholarship a year later. He graduated as an Ontario Scholar from high school in 1977 and joined the Canadian military in May of the following year. He initially studied at Royal Roads Military College in Victoria, British Columbia, and in 1980 entered the Royal Military College in Kingston, Ontario, from where he received a degree in mechanical engineering in 1982. During the course of his academic life, Hadfield underwent basic training at Portage la Prairie, Manitoba, and was named top pilot. Following receipt of his degree, he was honoured as the overall top graduate from Basic Jet Training and in 1984-1985 trained as a fighter pilot in Cold Lake, Alberta, flying the Canadian Air Force's CF-5 Freedom Fighter and CF-18 Hornet. During the course of the next three years, Hadfield flew CF-18s operationally for the North American Aerospace Defense Command (NORAD) with 425 Squadron, during which time he piloted the first CF-18 intercept of a Soviet Tu-95 'Bear' aircraft. Selected to attend the US Air Force Test Pilot School at Edwards Air Force Base in California, he won the Liethen-Tittle Award for top graduate in 1988 and later served as an exchange officer with the US Navy and the Strike Test Directorate at Patuxent River in Maryland. The US Navy named him as their Test Pilot of the Year in 1991, after which he worked on the testing of the F/A-18 Hornet and A-7 Corsair aircraft. He gained a master's degree in aviation systems from the University of Tennessee in 1992 and in June of that year was selected from 5,330 applicants to join Canada's second group of astronaut candidates. Years later, Hadfield remembered lucidly the advertisement in the newspapers: *Wanted – Astronauts Coast to Coast*. It was the beginning of a new and exciting era in his life.

Now, on his first space mission, Hadfield was responsible for bringing the DM and its myriad systems to life. Assisted by Bill McArthur, he also powered up the Orbiter Space Vision System (OSVS) to precisely align the RMS motions. This device formed one of STS-74's detailed test objectives and consisted of a series of large dots on the exterior of the DM and the ODS. Digitised television camera views of the dots allowed the OSVS to generate a display on a laptop computer to indicate alignments between the vehicles with great precision. The astronauts also installed the centreline camera in the ODS to support Cameron during proximity operations and docking with Mir. Meanwhile, Ross and McArthur checked out their space suits in case a contingency EVA became necessary. The pressure of Atlantis' cabin atmosphere was lowered from its normal 101.3 kPal to around 70.3 kPal to prepare Ross and McArthur for the 29.6 kPal pure oxygen of their suits. By the late afternoon of the 13th, Atlantis was about 3,220 km from Mir and closing at a rate of about 220 km with each 90-minute orbit of Earth. Early on the 14th, Hadfield successfully manoeuvred the DM out of the payload bay, pivoted it by 90 degrees into a vertical position, rotated it by almost 180 degrees and moved it to within 12.7 cm of the top of the ODS. He then placed the RMS into a 'limp' position, which effectively deactivated the brakes on its joints. At that point, Cameron gave Atlantis' RCS thrusters a brief burst and pushed the Shuttle towards the DM. When the two spacecraft were rigidly locked into a metallic embrace, the ODS docking ring retracted and a series of hooks and latches engaged to provide an airtight seal. Mating was confirmed at 2:17 am EST. The DM was then pressurised and, at 3:00 am, the RMS was ungrappled and moved to an overnight 'extended park' position. The crew entered the module to relocate the centreline camera to the docking mechanism at its far end, then raised the cabin atmosphere back to its normal 101.3 kPal pressure. By this point, the Shuttle was 2,330 km 'behind' Mir, closing at 290 km per orbit.

Back on Earth, a crisis was brewing. Conflict between President Bill Clinton and Speaker of the House of Representatives Newt Gingrich over funding allocations for Medicare, education, environmental monitoring and public health forced a government shutdown when the president vetoed the spending bill. Federal employees were furloughed from 14-19 November 1995 and again from 16 December until early in January 1996. Since a budget for Fiscal Year 1996 had not been approved, from 1 October 1995 the entire federal government operated on a

2-16: Mir seems to tower 'above' Atlantis and its orange-coloured Docking Module in this perspective of the two spacecraft linked together.

continuing resolution, which authorised interim funding until new budgets were approved. This resolution expired at midnight on 13 November, at which point all non-essential government departments ceased operations in order to prevent them from spending funds which they had not yet been allocated. Aboard STS-74, and within NASA, the effect was that much of the space agency's non-critical services, including its newsroom, were shut down throughout the furlough. However, the crew maintained a sense of humour. On a private Internet site, dubbed 'The Utterly Unofficial STS-74 Mission Guide', they posted updates and blamed "budget idiocy in Washington".

The consequence was that very little real-time detail of the events of the second Shuttle-Mir docking mission and the joint activities with Gidzenko, Avdeyev and Reiter were forthcoming until after STS-74 had returned to Earth. Rendezvous and proximity operations along the R-Bar broadly followed the same parameters as those pursued by the STS-71 crew, with the exception that in the final moments before docking Ken Cameron was reliant upon the RMS elbow camera for visual cues, due to the length of the DM, which partially obstructed his view. It was "like looking at the top of a building from the ground floor", according to Cameron. "You can see that it's up there, but you really can't ... accurately judge position or orientation." Docking with Mir was accomplished at 1:27:38 am EST on 15 November 1995, a little under three days into the mission. After standard checks of pressure and other parameters, the hatches were opened, Gidzenko and Cameron shook hands in the DM-Kristall tunnel and the eight-strong population of the two crews gathered ceremonially inside Mir. In doing so, Chris Hadfield became the first and only Canadian ever to board the Russian space station in orbit, and for those few days in November 1995 Mir cemented its credentials as the first 'international' space station as it played host to citizens from no fewer than four discrete sovereign nations (Russia, the United States, Germany and Canada) for the first time. "Mir was an amazing accomplishment," Hadfield said later. "It's very much a child of the Cold War, but it developed in the later years of its life into becoming a real crucible for international space operations." Moreover, in Hadfield's mind, Mir allowed the United States and its international partners to learn how to operate the Shuttle in conjunction with a space station and laid a vital cornerstone for the ISS. Flowers and chocolates, interspersed with handshakes and hugs, characterised those euphoric first hours, before the astronauts and cosmonauts settled down to three days of joint work and scientific research.

Atlantis' crew transferred over 900 kg of water and supplies to Mir and brought biomedical and microgravity science experiment samples and faulty equipment aboard the orbiter for return to Earth. One of the experiments coming home was the University of California at Berkeley's Trek investigation, which had been aboard the station since 1991. Key research areas included Earth observations, protein crystal growth, human life sciences and fundamental biology, with samples of dwarf wheat returned from Mir's joint Russian-Slovakian greenhouse. In the Shuttle's payload bay, the GLO-4 and Photogrammetric Appendage Structural Dynamics Experiment (PASDE) formed the GPP payload. Managed by NASA's Goddard Space Flight Center of Greenbelt, Maryland, GLO-4 consisted of a battery of experimental hardware to examine Earth's thermosphere, ionosphere and mesosphere energetics and dynamics via broadband spectroscopy, together with Shuttle and Mir 'glow' patterns, engine firings, fuel cell purges and waste water dumps. Meanwhile, PASDE comprised three canisters to 'photogrammetrically' record structural response data on the dynamic behaviour of Mir's solar arrays during the docked phase of the STS-74 mission. Photogrammetry was seen as a low-cost alternative to dedicated accelerometer-based structural response systems, especially when measurements were required for articulating or rotating spacecraft components, such as ISS thermal radiators and solar arrays.

At length, on 18 November 1995, the time came for Atlantis and Mir to part company. The initial separation at 3:55:44 am EST was performed by springs which pushed the Shuttle away from the docking mechanism, with the manoeuvring assets of both spacecraft shut off to avoid inadvertent firings. When their docking mechanisms were clear of each other, Cameron reactivated Atlantis' RCS thrusters and executed a Low-Z manoeuvre to begin a slow separation from Mir. At a distance of 120 m, he handed over control of the orbiter to Jim Halsell, who executed a two-circuit flyaround of the space station, during which time the rest of the crew performed a photographic survey, before departing for good. Two days later, at 12:01:27 pm EST on the 20th, Cameron guided Atlantis smoothly onto Runway 33 at KSC, on the first landing opportunity of the day.

In the aftermath of STS-74, the Mir crew settled down to their routine of scientific research and maintenance of the old space station, which was to mark the tenth anniversary of the launch of its base block in February 1996. Thomas Reiter worked with an Austrian-built 'Optovert' apparatus, which he employed to study the effects of weightlessness on his motor system performance and the interactions of the vestibular system and visual organs. During this period, the cosmonauts also conducted preventative maintenance on gyrodynes that had been installed inside the Kvant-2 module, using Mir's attitude control thrusters to maintain orientation whilst the devices were rendered inactive.

On 8 December, Avdeyev and Gidzenko departed the station for the second planned EVA of their mission. This brief 'internal' spacewalk – it lasted just 37 minutes – was to move the off-axis Konus docking drogue from the -Z (port-side) port of the multiple docking adaptor over to the +Z (starboard-side) port. This was done in anticipation of the launch of Mir's final scientific module, Priroda ('Nature'), which, like Spektr, had been extensively delayed due to the harsh post-Soviet economic downturn in Russia. Assembled between 1989-1991, Priroda measured 12 m in length and 4.35 m in diameter at its widest point, with a pressurised volume of about 66 m^3, and weighed about 19,700 kg. In addition to its pressurised area, it also included an unpressurised instrument compartment, covered by a payload shroud to protect it from aerodynamic effects during ascent. This compartment had an inner habitation and working area and an outer instrumentation area, separated by aluminium-magnesium plastic panels to form a fire break and contribute to Priroda's environmental control system by permitting the passage of conditioned air. In July 1993, NASA agreed to provide funding for the completion of both Priroda and Spektr and pledged to supply up to 700 kg of US equipment for the module. Due to the increasing weight growth of Priroda, a forward retractable solar array was deleted from the module, although it was intended for it to be delivered during a subsequent Progress cargo mission. Original plans called for Priroda to spend a full month in free-flight before docking at Mir, for which reason it was equipped with 186 expendable aluminium-lithium batteries.

After several years in limbo, Priroda was removed from storage in early 1994 and underwent its final systems testing in November of the following year. By this point, almost six years later than intended, Russia had other ideas for both Priroda and the already-launched Spektr, suggesting to NASA that they should ultimately be detached from Mir and incorporated into the ISS. This proposal was soundly rejected by NASA in December 1995, which noted that it was reluctant to deviate from the launch schedule or accept anything which increased costs or risk. "The bottom line," NASA was quoted by *Flight International*, "is that we will not redesign the space station." The two modules would remain part of Mir. In January 1996 Priroda arrived at Tyuratam, primed for its launch on 10 March. Primarily designed to support Earth sciences and remote-sensing applications, Priroda carried the Travers-III synthetic aperture radar, to be deployed after its arrival at Mir, and a suite of 'passive' microwave instruments: the IKAR-N nadir-facing radiometer, the IKAR-D four-channel scanning radiometer and the IKAR-P three- and five-channel radiometers. Other instruments included the joint Russian-Czech ISTOK-1 infrared radiometer, the French ALISA lidar to examine cloud heights, structures and optical properties, a surveying television camera, the OZONE-M atmospheric profiling sensor and the German MOS-A and MOS-B radiometers. Germany also provided an updated version of its Modular Optoelectronic Multispectral Scanner (MOMS-2P), which had flown several short-duration Shuttle missions in the 1980s and 1990s.

With Priroda scheduled to arrive in early 1996, Gidzenko, Avdeyev and Reiter spent the quiet time before Christmas packing up Progress M-29, which undocked from Mir on 19 December, to be replaced a day or so later by the fresh Progress M-30. This included among its 2,300 kg cargo additional supplies and equipment for the expanded EuroMir '95 mission, allowing Reiter to set to work on biomedical sampling and measurements, as well as testing the capacity of uncooled melts in the Titus furnace. Meanwhile, Gidzenko and Avdeyev focused on the effects of the microgravity environment upon hydrodynamics with the Volda-2 device, which utilised models of spacecraft fuel system elements, and worked with the Maria magnetic spectrometer to explore links between terrestrial seismic activity and high-energy charged particle fluxes. Early in January, they resumed work on a troublesome cooling system leak in the Kvant-1 module, hermetically sealing a manifold on the coolant line and refilling the loop with ethylene glycol.

The next major objective, and one which ESA had been promised in partial recompense for extending EuroMir '95 by six weeks, was a second EVA for Reiter. On 8 February, he and Gidzenko ventured outside Mir on a spacewalk for whose tasks they had actually trained via radio communication from ground controllers. Planned for 5.5 hours, the EVA comprised a number of tasks. The first required them to move a peculiar piece of equipment known as the *Sredstvo Peredvizheniy Kosmonavtov* (SPK, the 'Cosmonaut Manoeuvring Equipment'). This was of similar purpose to NASA's Manned Manoeuvring Unit (MMU), a jet-propelled space suit backpack, and had been tested in early 1990, then left inside the Kvant-2 airlock. Its sheer size posed an obstruction within the airlock and Gidzenko and Reiter moved it to a permanent location outside Kvant-2. Their next task was for Gidzenko to manoeuvre Reiter on the Strela crane over to Spektr to begin an intricate activity of retrieving a pair of 2 kg dust collectors from the ESEF payload, deployed the previous October. Gidzenko joined Reiter at the worksite and the two men laboured behind schedule to remove the collectors, eventually making up time by working through nighttime orbital passes. Their final task was to remove a malfunctioning antenna on one of Kvant-2's solar arrays, but this had to be abandoned when it proved impossible to loosen the bolts on one of the joints using the tools at their disposal. Gidzenko and Reiter returned inside Mir after just three hours and six minutes and the task was deferred to a subsequent EVA.

Entering the final stages of their six-month mission, on 20 February Gidzenko, Avdeyev and Reiter marked the tenth anniversary of the launch of Mir's base block, with Russian news sources reporting a "holiday atmosphere" aboard the space station for the occasion. Twenty-four hours later, at 3:34 pm Moscow Time on the 21st, their replacement crew was launched from Tyuratam aboard Soyuz TM-23. At least, that is, *part* of their replacement crew, for Russian cosmonauts Yuri Onufrienko and Yuri Usachev would soon be joined by NASA astronaut Shannon Lucid, to be launched in March aboard STS-76, to kick off an anticipated two years of continuous US presence aboard Mir. In preparation for the arrival of Soyuz TM-23, the Progress M-30 craft departed on 22 February and the quiet Onufrienko and the talkative Usachev docked at the aft port of the station's Kvant-1 module at 5:20:35 pm Moscow Time on the 23rd. For the next few days, the outgoing crew carried out a handover of tasks to the incoming crew and the five men performed joint experiments, including crop breeding and Earth spectrometry. Then, early on 29 February, Gidzenko, Avdeyev and Reiter donned their Sokol suits, boarded Soyuz TM-22 and undocked from the forward port of the multiple docking adaptor. They touched down safely at 1:42 pm Moscow Time, about 105 km to the north-east of Arkalyk, after 179 days in orbit.

For the next few weeks, Mir was reduced to a crew of just two men. Yuri Ivanovich Onufrienko, the commander, was making his first mission. Born on 6 February 1961 in Ryasne, within the Zolochiv Raion (district) of Kharkiv Oblast in north-eastern Ukraine, Onufrienko graduated from V.M. Komarov Eisk Higher Military Aviation School for Pilots in 1982 with the diploma of pilot-engineer. He later served as a pilot and senior pilot in the Soviet and later Russian Air Force, flying the L-29 and L-39 military jet trainers, the Sukhoi Su-7 fighter and the Sukhoi Su-17 attack aircraft and later gained a degree in cartography from Moscow State University. Onufrienko was selected for cosmonaut training in January 1989, although by his own admission it marked a natural progression from flying in the military. "It was maybe a dream since I was a child" to fly into space, he later told a NASA interviewer. "I read several books about future flights, about future life here and other planets ... but I think I [was] lucky when I was selected as a cosmonaut." Incidentally, Onufrienko was chosen for cosmonaut training in the very same group as his future Soyuz TM-23 crewmate, civilian flight engineer Yuri Vladimirovich Usachev. The two men completed their initial training together in early 1991. Like Onufrienko, Usachev was raised in eastern Ukraine. Born in Donetsk on 9 October 1957, he grew up wanting to become a pilot and later studied mechanical engineering at the Moscow Aviation Institute. Following his graduation in 1985, Usachev worked for the Energia design bureau and it was whilst there that he first came into contact with a group of real cosmonauts. "I thought this could be a good idea to try to work like them, like they did," he told a NASA interviewer, "and I tried to pass some exams, some medical tests, and there you go." Of course, his route into the cosmonaut corps was arduous, but he was selected as a civilian engineer. Following two years of general training, he served aboard Soyuz TM-18 for his first mission to Mir in January-July 1994. Asked years later to reflect upon the most important attributes that a cosmonaut requires for long-duration flight, Usachev was philosophical and paid tribute to his mother. "I think my mother may have more influence," he said, "because she is very patient. It's very useful for me to have enough patience to live six months in space!"

Within days of the departure of Gidzenko, Avdeyev and Reiter, the two men ventured outside Mir for their first EVA on 15 March to install a second Strela crane onto the base block, on the opposite side of the station to the point at which the existing Strela was mounted. The second crane was necessary because the original Strela could not reach the repositioned Kristall module on the port side of Mir. They also prepared cables and electrical connectors on the surface of Kvant-1 for another EVA in May 1996 to install the Mir Co-operative Solar Array (MCSA), delivered aboard STS-74. After extending the new Strela to its full 12 m length, Onufrienko and Usachev used it to make their way back towards the Kvant-2 airlock, concluding a spacewalk of five hours and 51 minutes. Upon returning to Mir, the cosmonauts settled down for a week of independent work, ahead of the arrival of Space Shuttle Atlantis in late March on what would herald the start of more than two years of continuous US presence in space.

A 'PERMANENT' AMERICAN PRESENCE?

It was always clear from the time of Norm Thagard's assignment that he would not be the only American astronaut to spend a period of several months aboard Mir. As early as 1992, NASA and the Russian Space Agency were deep into negotiations to send several other astronauts to Mir, 'rotating' them via Shuttle-Mir docking missions, to maintain a continuous US presence aboard the station for more than a year. In the early stages, it seemed likely that Thagard's backup, Bonnie Dunbar, might remain on Mir with Anatoli Solovyov and Nikolai Budarin after the departure of STS-71, but a combination of factors, including ESA's plan to fly Thomas Reiter, meant that the aging station would be maxed-out at three long-duration residents for the latter part of 1995. Another early plan was to fly veteran Shuttle pilot Bill Readdy to Mir on STS-71, after which he would remain aboard the station for a month, then return to Earth with Solovyov and Budarin on Soyuz TM-21. In fact, in his NASA oral history, Readdy explained that the original idea was that *he*, and not Dunbar, would have been Thagard's backup for the three-month flight. He recalled that in September 1993, he was sent to the Defense Language School in Monterey, California, to begin Russian classes. "The plan was to train as Norm Thagard's backup on the Mir-18 flight," Readdy told the NASA oral historian. "At that time, there was a Mir-18A and a Mir-18B, and the Mir-18B part was to fly up on 71 and then come back on the Soyuz." The plan made practical sense, for since late 1992 NASA had been investigating the use of Soyuz as a potential Assured Crew Return Vehicle (ACRV) for Space Station Freedom. Since Thagard would see the 'ascent' phase of the Soyuz TM-21 mission, Readdy would see its 'descent' phase, several months later. Readdy himself made little further reference to Mir-18B, although the unrealised mission was explored at length in Bryan Burrough's book, *Dragonfly*, although primarily as a means of uncovering the alleged political machinations of NASA's senior leadership.

These potential plans highlight the central focus of Mir as a proving ground for what would become today's ISS. In November 1994, veteran astronauts Shannon Lucid and John Blaha were formally named to train for the second of "at least four" further long-duration missions to the Russian outpost. The pair had both flown four Shuttle missions and knew each other well. They had been selected by NASA just two years apart, their children were about the same age and they were good friends.

The life of Shannon Matilda Wells Lucid is the story of triumph over terrible odds. She had been born in war-torn Shanghai on 14 January 1943 and her experiences during her formative years make it unsurprising that she grew up with a "zest for life, steely determination and resourcefulness", according to writer Peggy Mihelich. Her parents, Oscar and Myrtle Wells, were Baptist missionaries and they, together with Shannon, her younger brother, Joe, her aunts and an uncle and her grandparents were taken captive by the Japanese army and held in Shanghai's Chapei Civil Assembly Centre prison camp. She learned to walk in early 1944, whilst aboard the Swedish ship *Gripsholm*, which returned her family to the United States as part of an exchange of non-combatant citizens of the warring nations. It was a long and arduous voyage and, during a stopover at Johannesburg in South Africa, she received her first pair of shoes. After the war, the family returned to China – living at times in Shanghai, Nanking and Anking – and Shannon found herself the centre of attention, due to her blonde hair and blue eyes. Her fierce desire to learn to read prompted her parents to place her in a Chinese elementary school. Aged five, she took her first flight. As the DC-3 flew over mountainous terrain and landed on a gravel runway, the young girl was convinced that *flying* was the most remarkable thing for a human being to do ... and steeled herself to do the same when she grew up. Her family was expelled from China in 1949, after the Communist Revolution, and the young girl received her schooling in Bethany, Oklahoma.

She entered the University of Oklahoma to study chemistry and received her degree in 1963. By now, the first teams of astronauts had already been selected and Lucid was astonished that *all* of them were *male*; in fact, she had written to *Time* magazine in 1960, criticising NASA for choosing only men. Space exploration had fascinated her, ever since she read about the rocket experiments of Robert Goddard

... but there was another motive. "The Baptists wouldn't let women preach," she once said, "so I *had* to become an astronaut to get closer to God than my father!" During her undergraduate studies, she took flying lessons, gained her licence and encountered another cruel and harsh reality of life. One day, in her final year of study, she sat down with her professor to discuss her options for getting a job. The professor looked at her blankly. "A job?" he asked. "You plan on *working*? But you're a *girl!*" It underlined the reality that women were not taken seriously in many professional careers. Despite having a private licence, her efforts to become a commercial pilot led nowhere, for the same reasons. Fortunately, the Kennedy and Johnson administrations, with their incessant civil rights campaigning, smoothed the road over the next few years and she found work in academia as a teaching assistant and research chemist, firstly at her *alma mater*, then at the Oklahoma Medical Research Foundation and finally at the Kerr-McGee oil and gas corporation. By now married to Michael Lucid, she returned to study at the University of Oklahoma, earning a master's degree in biochemistry in 1970 and a PhD in 1973. With her doctorate, Lucid gained a job as a research associate with the Oklahoma Medical Research Foundation in Oklahoma City and remained in this position until NASA called for astronaut applicants. Lucid "scrambled" to complete and submit her application. In late August 1977, she was invited to Houston as part of the third group of finalists to be interviewed ... a 20-strong group which included a subset of individuals whose presence, a decade earlier, would have been inconceivable: *eight women.*

Three of those eight women – Lucid, Anna Fisher and Rhea Seddon – would form half of the female component of the astronaut class announced in January 1978. "It's a remarkable story," fellow astronaut John Fabian said of Lucid's life. "It's a story of the human spirit and I love to tell it ... because kids don't realise what opportunities really lie ahead of them. Some are very quick to worry about the disadvantages that they have in their own lives, or as they perceive in their own lives, and I think the Shannon Lucid story is just a great story about overcoming obstacles and blasting through ceilings and knocking down doors and never letting anything get in the way of doing the things that you believe are right."

As for John Elmer Blaha, he came from San Antonio, Texas, where he was born on 26 August 1942, the son of an Air Force pilot, although he received his high school education in Virginia. Blaha entered the Air Force Academy and received a degree in engineering science in 1965 and a master's in astronautical engineering from Purdue University in the following year. He earned his pilot's wings at Williams Air Force Base in Arizona in 1967 and undertook 361 combat missions in Vietnam, flying the F-4 Phantom, the F-102 Delta Dagger, the F-106 Delta Dart and the A-37 Dragonfly. By the time he returned from Vietnam and was accepted into the famed test pilot school at Edwards Air Force Base, Blaha had his sights on becoming an astronaut. "I had read the biographies of some of the early astronauts," he explained, "and realised many flew different types of airplanes and attended the test pilot school." Apollo 11 veteran Edwin 'Buzz' Aldrin happened to be commandant of the school at this time and Blaha got to know him and shared his goal. Aldrin recommended that he stay at Edwards and teach in the NF-104

2-17: Shannon Lucid, pictured during water survival training for her Mir mission.

Starfighter research aircraft, which, with rocket augmentation, achieved altitudes of over 100,000 feet. After test pilot school, Blaha followed Aldrin's advice and flew the NF-104 to an altitude of more than 104,400 feet – around *thirty kilometres*. Blaha next served as an F-104 instructor and taught his students to fly low lift-to-drag approaches, stability and control and spin flight test techniques. In 1973, he was assigned as an exchange pilot with the Royal Air Force at Boscombe Down in Wiltshire and flew a variety of aircraft, including the Jaguar, the Hawk, the Harrier and the Buccaneer. "Our time in England," he said, "was the best three years of our lives. We could eat breakfast together as a family and then I went to work at nine o'clock in the morning. By five o'clock in the evening, I was home." Graduation from the Air Command and Staff College at Maxwell Air Force Base in Alabama followed in 1976 and Blaha was assistant chief of staff for studies and analysis at the Pentagon when he answered NASA's first call for astronaut candidates. He was unsuccessful on his first attempt, but was selected two years later. In *Dragonfly*, Bryan Burrough summarised perceptions of Blaha from engineers, managers and fellow astronauts. Blaha often presented himself as "a dolt", who needed things repeated to him. "But John's far from stupid," engineer Mark Bowman said of Blaha. "He just wants to make sure he *gets it*. He wants to start from scratch and go over and over and over."

Three months after their formal announcement by NASA, in February 1995, Lucid and Blaha arrived at the cosmonauts' training centre of Star City, on the forested outskirts of Moscow, to begin formal training. At the end of March, NASA revealed that Lucid would train specifically for a five-month stay aboard Mir in March-August 1996, with Blaha serving as her backup. At the same time, two other veteran astronauts entered training and in late May 1995 arrived in Star City to join the steadily burgeoning ranks of the 'NASA Set'. Jerry Linenger would fly the third long-duration mission from August-December 1996, backed-up by Scott Parazynski. Flesh was added to the bones of these assignments by *Flight International* in April 1995; following Lucid's return to Earth, Blaha would fly his own long-duration mission from December 1996 until May 1997, whereupon Parazynski would launch for a further five months. He would return to Earth aboard STS-86, the seventh Shuttle-Mir docking mission, in early October 1997. It was stressed at the time that STS-86 was "unlikely" to carry a further NASA resident for Mir, because construction of the ISS was scheduled to get underway in December 1997 and permanent occupation of the new station was anticipated for mid-1998.

The assignments of Linenger and Parazynski, who were both physicians, made particular sense, for NASA sought medically qualified astronauts for the Mir missions, one of whose primary focuses would be upon life sciences research. Moreover, the Russians were adamant that all of the Americans *must* have had previous Shuttle experience, which had led NASA to fly Jerry Michael Linenger on STS-64 in September 1994, just a few months before sending him to Star City. Linenger was born on 16 January 1955 and raised in Eastpointe, Michigan. After leaving high school, he entered the Naval Academy to study bioscience and earned his degree in 1977. Additional academic credentials followed rapidly: a doctorate in medicine from Wayne State University School of Medicine in 1981, a master's degree in systems management from the University of Southern California in 1988, a master's in public health policy from the University of North Carolina at Chapel Hill and a PhD in epidemiology from the University of North Carolina, both in 1989. "Whenever anyone teases me about the number of academic degrees I have accrued," Linenger wrote in his memoir, *Off the Planet*, "I generally respond that I was a rather slow learner – I had to keep going to school until I got it right!" As a physician, he completed his surgical internship, his aerospace medicine training and his flight surgeon training, then served as medical adviser to Admiral Jim Service, commander of the Navy's air assets in the Pacific and Indian Oceans.

Fellow physician Scott Edward Parazynski had also flown one previous Shuttle mission and, in time, would also fly with John Glenn, would help to build the ISS, would volunteer at the Houston Astrodome in 2004 to support the victims of Hurricane Katrina and would also become the first veteran spacefarer to summit Mount Everest in May 2009. Born on 28 July 1961 in Little Rock, Arkansas, Parazynski was well-travelled as a youth, since his father was a Boeing engineer during the Apollo era. "I sort of had a nomadic upbringing," Parazynski admitted to a NASA interviewer. "There are no camels involved, but a *lot* of travel." An understatement indeed: he received his early schooling in Dakar, Senegal, and in Beirut, Lebanon, and attended high schools in Tehran, Iran, and in Athens, Greece. "I think the fundamental lesson that I learned through all that experience," he continued, "is that some of life's greatest lessons come *outside* of the classroom. Travel opened up for me new experiences. Being adventurous, meeting new people, travelling to new places, really broadens your horizons and it gave me the motivation to do some of the things I did later in life." Returning to the United States, he entered Stanford University, graduated in biology in 1983, and remained to study for his medical degree, which he received in 1989. At approximately the same time, he competed on the US Development Luge Team and was ranked among the Top Ten competitors in the nation during the 1988 Olympic Trials. Parazynski served his internship at the Brigham and Women's Hospital in Boston, Massachusetts, and was two years into a residency programme in emergency medicine when he was selected by NASA in March 1992, in the same class as Jerry Linenger, as an astronaut candidate.

More astronauts were scheduled to support subsequent Mir and ISS assignments. In June 1995, Wendy Barrien Lawrence began Russian language training in the United States, with the expectation that she would travel to Star City to train as John Blaha's backup. Lawrence, a Navy helicopter pilot and veteran of one previous Shuttle mission, was born in Jacksonville, Florida, on 2 July 1959. Her father was Vice-Admiral William P. Lawrence, a finalist for Project Mercury, former Vietnam prisoner of war and former Superintendent of the US Naval Academy. Seeing Neil Armstrong walk on the Moon, and watching *Star Trek*, imbued her with excitement and her father guided her steps. "His very wise advice to me was follow in the footsteps of the first several groups of astronauts," she told a NASA interviewer. The first stage was attending the Naval Academy in Annapolis, Maryland. Lawrence left the academy in 1981 with a degree in ocean engineering and when she flew on STS-67 in March 1995 she became its first female graduate ever to journey into space. She completed flight training with distinction and became a naval aviator in July 1982, ultimately flying six different types of helicopter, accruing 1,500 hours and more than 800 shipboard landings. At one stage, attached to Helicopter Combat Support Squadron Six, she was one of the first two female helicopter pilots to make a long deployment to the Indian Ocean as part of a carrier battle group. In 1988, she gained a master's degree in ocean engineering from the Massachusetts Institute of Technology and Woods Hole Oceanographic Institute and later worked as a physics instructor at the Naval Academy from October 1990 until her selection as an astronaut in March 1992.

The arrival of Lawrence and Parazynski in the Shuttle-Mir programme led to one of the strangest and most awkward situations of the entire US-Russian co-operative effort. In early October 1995, NASA announced that Parazynski would have to discontinue his training as Jerry Linenger's backup, "due to concerns over his ability to safely fit in a Soyuz descent vehicle for landing". Standing 188 cm (six feet and two inches), Parazynski was one of the tallest astronauts in NASA's corps at that time and exceeded the maximum height allowable for the Soyuz. According to Charlie Brown, co-chair of NASA's Crew Training and Exchange Working Group, astronauts who were taller than allowable could be more susceptible to injuries during a Soyuz re-entry, particularly leg injuries. In her NASA oral history, Wendy Lawrence recalled "hearing rumblings" about Parazynski's removal in the August-September timeframe, but continued her preparations regardless. At the end of September, as part of her training, she was obliged to try a Russian Orlan-DMA space suit, when a Hamilton Standard engineer pulled her to one side. He told her that he had seen a Russian memo, in which the *minimum* allowable height for a cosmonaut to sit safely aboard the Soyuz had been increased from 160 cm (five feet and three inches) to 164 cm (five feet and *four* inches).

Lawrence, one of the shortest astronauts, stood exactly 160 cm tall ...

As a result, NASA was forced to the embarrassing conclusion that two of its astronauts had to be dropped from long-duration Mir missions, because one was too tall and the other too short. (In fact, Parazynski and Lawrence would earn the nicknames of *Too Tall* and *Too Short* and would even wear self-deprecating name tags to this effect on their flight jackets.) During one evaluation in the Orlan-DMA, it became obvious that Lawrence's hands did not even reach the space suit's gloves when it was fully pressurised. To be fair, it *was* known to both NASA and the Russians at the time of Parazynski's assignment that he was "slightly outside the nominal height to fly on the Soyuz", but that although a "preliminary evaluation cleared him for training", follow-on discussions over deceleration loads during re-entry meant that "safety margins against injury would be unacceptably reduced". Veteran astronaut Frank Culbertson, then serving as Acting Director of the Phase 1 Shuttle-Mir Program Office, pointed out that it was "understood that there were modifications that could be made that would allow [Parazynski] to use the [Soyuz] descent vehicle". However, Culbertson also added that the Russians "do not have the latitude or sufficient modification capability on the Soyuz to allow Scott to return to Earth in the vehicle with a level of risk that we would be comfortable with".

It was against this backdrop, throughout the latter part of 1995 that Shannon Lucid trained for what was expected to be the longest space mission ever undertaken by a citizen of the United States. Unlike Thagard, she would not launch or land aboard a Soyuz, but would fly into orbit aboard Atlantis on STS-76 in March 1996 and return to Earth aboard the very same orbiter on its next mission, STS-79, in early August, after about 143 days. Although both Lucid and Blaha would be assigned to their own respective Mir crews, they actually spent much of their early training together. "We sat in a classroom together," Lucid told the NASA oral historian. "It was just the two of us and an instructor for whatever classroom it was. We didn't interface with anybody else. Only toward the end did we do just a very few sims with the Russian crew; we got in the Soyuz and went through a sim, but it was all very minimal." The primary focus of their training was upon learning conversational and technical Russian and mastering the US scientific research payloads for their long-duration missions. Key to this research, from Lucid's perspective, was the arrival of equipment aboard the Priroda module. Already delayed from November 1995, the launch of Priroda had slipped further to March 1996 and then no earlier than 15 April. The fear, of course, was that Priroda would fall victim to a Spektr-type delay and Lucid would be forced to spend the majority of her mission with insufficient equipment to complete her scientific research programme. It was therefore under a cloud of both excitement and pessimism that Lucid joined her five STS-76 crewmates to be launched to Mir. Although it was the third Shuttle-Mir docking flight, STS-76 was notable in that it would be the first to deliver a US long-duration crew member and would feature the first EVA by NASA astronauts outside a space station.

Also flying aboard Atlantis was the first Spacehab logistics module. The company's expectations of great commercial interest in its middeck augmentation laboratory had been severely disappointed in the early 1990s and many payloads aboard the first four Spacehab missions were exclusively NASA-financed. In 1993, Spacehab signed a $54 million contract with NASA to change its mode of operations to logistics and resupply for Shuttle-Mir, employing the modules to transport experiments, food, water, equipment and provisions. The company invested an additional $15 million into the construction of a 'double' Spacehab module, using one of two single modules and a Structural Test Article, linked by an adaptor ring. This could carry up to 2,720 kg of cargo, whilst a system of canvas soft stowage bags increased the maximum capacity to 4,080 kg. Quoted by *Flight International* in December 1995, a Spacehab spokesperson explained that the Shuttle-Mir contract was "an example of NASA's continuing emphasis on cost savings through commercialisation to achieve certain aspects of its agency mission in a fiscally constrained environment". STS-76 carried a single module, although the subsequent STS-79, STS-81 and STS-84 docking missions were expected to carry double modules.

In order to accommodate Spacehab on STS-76, Atlantis carried a short tunnel and airlock, connected to her middeck and ODS, and a longer, 10 m tunnel extending to the aft section of the payload bay and the pressurised module itself. It housed a mixture of Russian logistics, EVA tools, ISS risk mitigation experiments with a focus upon astronauts' health and safety, US logistics for Shannon Lucid and scientific and technological investigations. The Russian items included a replacement gyrodyne and three replacement batteries for Mir, a Soyuz seat liner for Lucid to use in the event of an emergency return to Earth and various other items of food, water, clothing and personal hygiene supplies in soft stowage bags. Other equipment included EVA tethers and tools, as well as brackets and hardware for the Mir Environmental Effects Payload (MEEP), which was to be installed on the exterior of the Docking Module during a spacewalk. In addition to the logistics, the Spacehab module was also loaded with several research facilities, including a double rack for ESA's Biorack and the Life Sciences Laboratory Equipment Refrigerator/Freezer (LSLE). Biorack provided an incubator and scientific glovebox facility for a total of 11 radiation experiments on plants, tissues, cells, bacteria and insects and its research was supported by dozens of investigators from the United States, France, Germany, Switzerland and the Netherlands. The LSLE was a vapour compression refrigerator that operated at minus 22 degrees Celsius. It was to store processed Biorack samples, as well as blood, urine and saliva specimens for post-flight analysis. Other payloads included parts for the Mir Glovebox, the Canadian-provided Queens University Experiment in Liquid Diffusion (QUELD) and the High Temperature Liquid Phase Sintering for materials research.

With Lucid embarking on her fifth space mission, her five STS-76 crewmates were also veterans of at least one previous Shuttle flight. Kevin Patrick Chilton had been named to command the mission in November 1994, with the remainder of his crew announced several months later in April 1995. Today, Chilton holds the record for having achieved the highest rank of any military astronaut; a career Air Force officer until his retirement in February 2011, he eventually rose to become a four-star general, holding several directorial posts at the Pentagon and later commanding the Air Force Space Command and the US Strategic Command. Chilton came from Los Angeles, California, where he was born on 3 November 1954. After high school, he entered the Air Force Academy and earned a degree in engineering sciences in 1976, followed by a master's in mechanical engineering, on a Guggenheim Fellowship, from Columbia University in 1977. Chilton earned his wings at Williams Air Force Base in Arizona, qualified in the RF-4C – the reconnaissance version of the Phantom jet – and was deployed to Korea, Japan and the Philippines. He converted to the F-15 Eagle in 1981 and served as a squadron pilot, attending squadron officer school the following year at Maxwell Air Force Base in Alabama and emerging with the Secretary of the Air Force Leadership Award as the top graduate. Over the next several years, Chilton was a weapons officer, instructor pilot and flight commander, before entering test pilot school at Edwards Air Force Base in California in 1984. He graduated first in his class and in August 1987 was selected by NASA as an astronaut candidate. By this point in his career – aged only 32 – Chilton already held the rank of major and by the time of his first Shuttle mission on STS-49 in May 1992 he had reached lieutenant-colonel and was only months away from promotion to full colonel. He flew again on STS-59 in April 1994.

Seated to Chilton's right side on Atlantis' flight deck was Richard Alan Searfoss, making his second Shuttle mission. One of his claims to fame in the 1990 astronaut class was that he designed their official 'Hairballs' logo and patch. Searfoss was born on 5 June 1956 in Mount Clemens, Michigan. On graduating from high school in New Hampshire, he entered the Air Force Academy. Whilst there, he joined the Church of Jesus Christ of Latter-Day Saints and would thus become one of only a handful of Mormons ever to enter NASA's astronaut corps. Searfoss emerged from the Air Force Academy in 1978 with a degree in aeronautical engineering and gained a master's qualification in aeronautics from Caltech on a National Science Foundation fellowship the following year. His formal Air Force career commenced

2-18: Stunning view of Mir from STS-76.

shortly thereafter and he completed undergraduate pilot training at Williams Air Force Base in Arizona in 1980, then flew the F-111F Aardvark tactical strike aircraft at RAF Lakenheath in England, and at Mountain Home Air Force Base in Idaho. By 1987 – and having been selected as a finalist for Outstanding Young Men of America that year – he was serving as an Aardvark instructor pilot and weapons officer. His next assignment was Naval Test Pilot School at Patuxent River, Maryland, as an Air Force exchange student. Completion of the course in 1988 led to work as a flight instructor at the school and he was selected by NASA in January 1990.

Leading the cadre of three STS-76 mission specialists was Ronald Michael Sega, the payload commander, born in Cleveland, Ohio, on 4 December 1952. After graduation from high school, he entered the Air Force Academy and received a degree in mathematics and physics in 1974 and a master's credential in physics from Ohio State University the following year. Sega then entered active duty with the Air Force, learning to fly and serving as an instructor pilot at Williams Air Force Base in Arizona until 1979. He was then assigned to the physics faculty of the Air Force Academy, where he designed and constructed a laboratory facility to investigate microwave fields using infrared technologies. Concurrently with this research, Sega worked toward a doctorate in electrical engineering, which he received from the University of Colorado at Boulder in 1982. In the wake of his active military career, Sega remained an Air Force reservist. With his PhD in hand, he joined the University of Colorado at Colorado Springs on an assistant professorship in the Department of Electrical and Computer Engineering and by 1990, when he was selected as an astronaut candidate by NASA, he had secured a full professorial post. During this period, Sega also worked as Technical Director for the Lasers and Aerospace Mechanics Directorate of the Frank J. Seiler Research Laboratory at the Air Force and, in 1989-1990, he served as a research associate professor of physics at the University of Houston, affiliated with Alex Ingatiev, Director of the Space Vacuum Epitaxy Center (SVEC). Sega joined NASA's astronaut corps in January 1990 and flew his first Shuttle mission in February 1994. A few months later, in September, he was supporting the crew of STS-64, when he was asked to serve as the third director of operations in Russia. He arrived in Star City in November and remained there until his assignment to STS-76 in the spring of the following year.

It was already clear that the mission – "The Spirit of '76," as Kevin Chilton called it – would also involve an EVA, to be performed by mission specialists Linda Maxine Godwin and Michael Richard Uram Clifford. Both were embarking on their third Shuttle flights. Godwin came from Cape Girardeau, Missouri, born on 2 July 1952, but spent her formative years in Jackson and earned a degree in mathematics and physics from Southeast Missouri State University. She then completed a master's degree and a doctorate at the University of Missouri in 1976 and 1980, respectively, with a research focus on low-temperature solid state physics, including electron tunnelling and the vibrational modes of absorbed molecular species on metallic substrates at liquid helium temperatures. She worked for NASA long before her astronaut days, beginning her career in the Payload Operations Division of the Mission Operations Directorate to work on Shuttle payload integration. A keen saxophonist and clarinet player, Godwin entered the astronaut corps in June 1985. As with many astronauts, Godwin lavished glowing praise on her parents and teachers for their inspiration – in fact, one of her science teachers would attend *every one* of her four Shuttle launches – although she admitted in a pre-flight interview before her final mission, in 2001, that she was not particularly *drawn* to the astronaut business. "I grew up watching a lot of the coverage of the early US space programme," she said, "so that made me interested in NASA, but I never thought it was something I could do." That changed about two-thirds of the way through her PhD, in 1978, when the space agency hired its first group of female astronauts. Two attempts to enter the astronaut corps in 1980 and 1984 were unsuccessful, but Godwin found that her career took a third-time-lucky turn.

Clifford, nicknamed 'Rich', was born in San Bernadino, California, on 13 October 1952. He graduated from high school in Ogden, Utah, and entered the Military Academy at West Point. Clutching his science degree in 1974, Clifford was commissioned into the Army as a second lieutenant and served initially with the 10th Cavalry in Fort Carson, Colorado, before entering the Army Aviation School in 1976, from which he graduated as the top of his class. Designated as an Army Aviator, he was based in Nuremberg, West Germany, as a service platoon commander, and completed a master's degree in aerospace engineering from Georgia Institute of Technology in 1982. Clifford spent some time as a faculty member in West Point's Department of Mechanics and in December 1986 graduated from Naval Test Pilot School. Although he applied unsuccessfully for admission into the astronaut corps on two occasions, Clifford entered NASA's Johnson Space Center in July 1987 as a Shuttle Vehicle Integration engineer and worked extensively on design certification and integration of crew escape systems. He was also an executive board member of the Solid Rocket Booster Post-flight Assessment Team. Clifford was selected by NASA in January 1990.

The primary purpose of Godwin and Clifford's six-hour EVA on STS-76 was to install the MEEP hardware onto the exterior of the Docking Module. Its four experiments sought to investigate the frequency, sources, sizes and effects of both human-made and natural space debris striking Mir, capturing some particles for subsequent investigation, as well as evaluating materials for use aboard the ISS, including specimens of paint, glass coatings, multi-layered insulation and metals. The samples were stored in four Passive Experiment Carriers. MEEP was to remain attached to the DM for about 18 months and be recovered during another EVA on STS-86 in September 1997.

The launch of STS-76 was planned for early on 21 March 1996 from Pad 39B, but was postponed by 24 hours due to high winds and rough seas in the Cape Canaveral area, which threatened to interfere with a Return to Launch Site (RTLS) abort scenario, should one become necessary. Next morning, Atlantis suffered no difficulties and speared into orbit at the opening of the window at 3:13:04 am EST. The sole problem during the ascent was a minor hydraulic fluid leak and the STS-76 crew arrived in orbit, trailing Mir by 24,260 km and closing the separation distance by about 1,300 km with each circuit of Earth. Over the following day, Chilton and Searfoss oversaw several rendezvous and phasing manoeuvres to further gain on their quarry, whilst the remainder of the crew activated the Spacehab module and configured space suits and equipment for the EVA. Late on the evening of 23 March, Chilton executed the Terminal Initiation (TI) burn and after receiving authority to proceed from US and Russian flight controllers performed a perfect docking with the DM/Kristall module at 9:34 pm EST, a little over 42 hours into the STS-76 mission.

"Coming up to dock with Mir, approaching that tremendously large piece of machinery, it goes from being a very bright light to looking like a spider," Rich Clifford told a Smithsonian interviewer, years later. "Then, all of a sudden, the station's huge solar power array looks like it's ten *inches* from your window!" Clifford remembered thinking that it was a good thing that Atlantis was not slightly off-axis. Approaching slowly, he hardly felt a thing at the point of contact and capture. Chilton, working from the aft flight deck controls, was totally absorbed in the proximity operations and docking; so much so that when Clifford tapped him on the shoulder and offered congratulations, the commander jumped in surprise. "You can even see it on the video," said Clifford. "He was just so focused on the docking."

After capture, Clifford worked to pressurise the docking mechanism and hatches were opened to a flurry of handshakes, hugs and laughter as the two crews met in orbit for the first time. Traditional gifts of memorabilia, shirts and chocolate Easter bunnies were exchanged and Atlantis' crew gave Onufrienko and Usachev copies of Jim Lovell's book *Lost Moon*, upon which the 1995 movie *Apollo 13* was based. With five days of joint activities ahead of them, they got down to transferring equipment and supplies between the Spacehab module and Mir. The 10 m tunnel connecting the middeck to the Spacehab module provided the opportunity for a spot of impromptu space gymnastics, according to Rich Clifford. "We took a large rubber band, called a Dyna-Band," he recalled, "that we used for exercise and stretched it across the airlock hatch on the middeck. Then we'd shoot ourselves down the tunnel. We had a competition to see who could do it without touching the tunnel walls. Nobody ever made it."

Significantly, Shannon Lucid became an official member of the Mir-21 crew at 8:30 am EST on 24 March, when the custom-moulded seat liner that she would use in the event of a contingency landing was transferred to the Soyuz TM-23 descent module. To mark the event, STS-76 Flight Director Bill Reeves and Russian Flight Director Nikolai Nikoforov issued a joint "Go" statement. As transfer activities continued between the Shuttle and the station, the crew managed to spot Comet Hyakutake, discovered just a few weeks earlier, which would go on to make one of the closest cometary passes to Earth in 200 years. Its blue-green coma was at its most visible on 24 March and it made its closest approach to Earth about 24 hours later, temporarily eclipsing Hale-Bopp as the 'Great Comet of 1996'. At one stage, on the 26th, the STS-76 crew told journalists that Hyakutake was so brilliant that they could see it almost from horizon to horizon.

Later that day, preparations entered high gear for Godwin and Clifford's EVA and the hatches between Atlantis and Mir were closed, as was the hatch to the Spacehab module. The pressure in the Shuttle's cabin was then depressurised from 101.3 kPal to 70.3 kPal to prepare the spacewalkers for operating in the 29.6 kPal atmosphere of their suits, and they donned masks for a period of breathing pure oxygen to help clear nitrogen from their bloodstreams and avoid an attack of the bends. Godwin and Clifford departed the airlock at 1:36 am EST on 27 March and spent six hours and two minutes in the vacuum of space. It was the first American EVA outside a space station since the Skylab era, more than 20 years earlier, and the scale of Mir came as something as a surprise to Godwin. The Weightless Environment Training Facility (WETF) at JSC in Houston was not large enough to hold a full-size mockup of the DM and, in her own words, the STS-76 crew "had to pretend a lot in training". During underwater simulations, for example, she and Clifford left the airlock mockup and were then carried by divers to a DM mockup, lying on its side, although they were able to utilise virtual reality techniques to gain perspective for translating along its length. Spacewalking outside Mir was also novel in that the Shuttle's standard tethers were not large enough to attach onto the station's handrails, which required the spacewalkers to use new hooks and a foot restraint. Unlike the floodlit payload bay of the orbiter, Mir also provided little illumination, save for its flashing running lights, and it was often difficult to see in the gloom. In setting up their worksite, Godwin and Clifford first attached clamps and securely installed MEEP into place for its 18 months of operations. They also used cable cutters to remove the television camera from the DM, which was returned to Earth aboard Atlantis. Inside Mir itself, Onufrienko, Usachev and Lucid caught a few fleeting glimpses of the progress of the EVA, through the airlock hatch of the Kvant-2 module and a slightly better view through one of the cabin portholes in the base block.

Interestingly, as explained by Bryan Burrough in his book *Dragonfly*, Godwin and Clifford were explicitly forbidden to venture away from the DM and onto the structure of Mir itself, due to the "pin cushion" of experiments affixed to the exterior of Kristall and the sprouting, sharp-edged solar arrays and protruding solar sensors. "They said it wasn't safe," Clifford was quoted by Burrough. And it was true. "There are appendages all over Kristall. Some of them were visibly sharp. Snag points. Sharp edges. *Not* a clear translation path."

Spacewalking offered other challenges, too. "The hand work can be fatiguing on a spacewalk," Godwin pointed out, years later, and driven by adrenaline, it was not until she returned inside Atlantis that the realisation – *That was a long day* – finally dawned on her. "On some of these really long EVAs, you don't get anything to eat from the time you suit up until you take your helmet off, and that can be ten hours! You start thinking about it after a while. You look inside and see the crew eating. Of course, there's no reason for the people inside to starve out of sympathy. It's also tiring for the people inside. There's a mental focus that doesn't let up for *hours*, making sure they're on the procedures and that they're not leaving anything out or losing anything."

More than a decade after STS-76, Rich Clifford would reveal that he had been diagnosed with the early stages of Parkinson's disease in 1994 and that it presented him with great concern as he prepared for his EVA. He advised only Kevin Chilton of the situation and, upon his return from the mission, decided that he would leave NASA. "I didn't know how fast it was going to progress," Clifford acquiesced to an interviewer.

2-19: Atlantis during docked operations at Mir.

With the EVA behind them, the time came for what Chilton described as "a bittersweet moment"; the point to trade goodbyes with Onufrienko, Usachev and Lucid, close the hatches and prepare for undocking and departure. During their five days together, about 680 kg of water and 1,800 kg of scientific equipment, logistics and supplies had been transferred from Atlantis to Mir. "It will be sad to say farewell to such a good team," said Onufrienko at one stage. The hatches were closed at 8:15 am EST on 28 March and, following pressurisation and other checks, Atlantis and Mir parted company almost exactly 12 hours later, at 8:08 pm. After Chilton had manoeuvred the Shuttle to a safe distance, Rick Searfoss took control for the standard flyaround inspection and the "chameleon-like" appearance of Mir profoundly impressed him. "I was enthralled," he said later, "by the constantly changing and indescribably beautiful hues of white, tan, gold and blue." One hour after undocking, Chilton and Searfoss executed a separation burn to lower their orbit and steadily increase their distance from the station.

By this stage, the likelihood of rain and clouds over KSC on the morning of 31 March had obliged Mission Control to instruct the crew to prepare for a landing about 24 hours earlier than planned, at 7:57 am EST on the 30th. Predictions for a touchdown on the 30th included light winds and scattered clouds, but generally acceptable weather conditions. In anticipation for the early landing, Chilton, Searfoss and flight engineer Clifford checked out Atlantis' systems, but NASA revealed that they would use a circulation pump instead of the Auxiliary Power Units (APUs) to route hydraulic fluid to the ailerons, elevons, speed brake and rudder during re-entry. APUs No. 1 and 2 were activated as normal, but since APU No. 3 had been associated with the hydraulic fluid leakage during ascent, it was to be brought to life late in the re-entry and used only as a standby backup. In the meantime, Linda Godwin completed work in the Spacehab module and closed the hatch.

As circumstances transpired, both landing opportunities at KSC on 30 March – the first at 7:57 am, the second at 9:33 am – were called off when meteorologists described weather conditions as "too dynamic" to be acceptable. Following the decision to delay by 24 hours, the crew began reactivating systems and hit difficulties when reopening the payload bay doors. Microswitches showed that the positions of a set of centreline latches for the doors were not fully open, which prevented the opening process. Godwin compared the latch positions and described them as appearing to be fully open. Mission Control asked the astronauts to use manual deployment switches and the doors opened properly. For the remainder of 30 March, they conducted Earth observations, preparatory to a second attempt in the early hours of the 31st. With more stable weather expected at Edwards Air Force Base in California, it was decided to forego the two KSC landing opportunities on the 31st. Chilton and Searfoss pulsed the OMS thrusters for the irreversible burn at 7:25 am EST, committing to an hour-long hypersonic descent into the 'sensible' atmosphere. Atlantis touched down at Edwards smoothly at 5:29 am PST (8:29 am EST), just 11 minutes before sunrise, completing a nine-day flight and the start of an American human presence in space which would endure for the next 27 months.

NASA hoped that it might endure for even longer. Already, as highlighted by

Flight International in April, plans were advanced to expand the number of long-duration Mir astronauts to seven and continue Shuttle docking missions until the middle of 1998. With the first permanent ISS crew also expected to board the new space station at around the same time, it did not seem unreasonable to suppose that Shannon Lucid's mission might herald the start of a permanent US presence in space ... for good. As the situation evolved, this turned out not to be the case, and a 2.5-year gap opened up between the completion of the last long-duration Mir mission in June 1998 and the start of the first long-duration ISS mission in October 2000, but it indicated NASA's steady growth of confidence in the possibilities of spending extended periods off the planet.

Following Atlantis' landing, Shannon Lucid began her planned five-month mission aboard Mir with Yuri Onufrienko and Yuri Usachev. Before she joined them, Lucid described her colleagues as having absolutely no grasp of English, but that as her Russian skills improved, a measure of humour and banter developed between them. From time to time, whenever he was asked about their language 'barrier', Onufrienko would explain that they were "developing a *new* language ... a *cosmic* language". At one stage, in her first few weeks aboard Mir, the crew was asked by a journalist what language they spoke. Onufrienko replied that it was primarily Russian, but that the cosmonauts attempted to learn more English and use it from time to time, "so Shannon won't forget her English!"

Within days, she settled to focus on the Optizon Liquid Phase Sintering Experiment, which marked the first NASA investigation performed in a Russian microgravity furnace aboard Mir. Lucid also worked on protein crystal growth studies, took blood samples and acceleration, radiation and dosimetry measurements and observed the embryonic development of quails' eggs. However, much of her research equipment awaited the launch of Priroda, which had been delayed from March, due to its late delivery from the manufacturer and the priority granted to a commercial launch. The module finally roared into orbit from Tyuratam, atop a Proton-K booster, at 2:48 pm Moscow Time on 23 April. Over the next two days, the module's orbit was steadily raised to about 360×400 km and at 3:43 pm Moscow Time on the 26th Priroda performed a soft docking at the -X forward port of the multiple adaptor. Watching from inside Mir, Onufrienko described the final approach as having gone smoothly. Next day, via the Ljappa manipulator arm, Priroda was detached from the frontal port and, in an intricate, 50-minute operation, which began at 11:30 am Moscow Time, was manoeuvred to its permanent location on the starboard ($+Z$) radial port.

Following Priroda's arrival, Lucid spent a week conducting an inventory of all of the US research hardware aboard the entire station. One of the batteries used to power the Priroda module during its flight to Mir had fallen off-line during rendezvous and caused an internal fire. Due to concerns about the possibility of sulphur dioxide leaks, Priroda's atmosphere was tested before the crew opened the hatch to the new module. Onufrienko and Usachev then cleaned up the remaining contaminants, removed all 168 batteries, capped their connectors and bagged them for disposal. Lucid remembered times at which the three of them floated in Priroda, literally surrounded by bagged-up batteries, equipment and bits of scrap metal; in a few of her more pessimistic moments, she wondered if there was enough spare metal to build the ISS itself! Every so often, these fragments would clank and jangle against one another, producing what the crew described as "cosmic music". At length, they devised an impromptu assembly line to clear up the mess and Priroda was shipshape and ready for work in a sixth of the time anticipated by flight controllers.

From the windows of the new module, Lucid was able to photograph as part of her Earth observation work the wildfires which devastated parts of Mongolia during 1996-1997. The conflagrations, which arose from campfires set by inexperienced cattle herders, collectors of non-wood products and an ever-increasing number of road traffickers, affected 102,000 km^2 of forest and steppe. Lucid noted with concern that she had never seen fires of such ferocity on any of her previous Shuttle missions. She also photographed the devastation left by the volcanic eruption on the Caribbean island of Montserrat. Dormant for centuries, the Soufrière Hills volcano had erupted in July 1995, quickly destroying and burying the Georgian-era capital, Plymouth, and over the following years would lead to the creation of a complete exclusion zone in the southern half of the island.

In the aftermath of Norm Thagard's months of cultural isolation aboard Mir, Lucid was specifically asked by NASA managers what would keep her happy and fulfilled, aside from her work, whilst in orbit. For Lucid, having *worthwhile* work to do, rather than 'make-work', was of paramount importance. "If I didn't have science experiments to work, then I had plenty to do," she told the NASA interviewer. "I helped Yuri and Yuri all the time, because, obviously, I was living on Mir, just like they were, so it was important that I did my share of keeping Mir up, too." Books were a great companion and her daughter, an English major, selected a variety of titles for her to take to Mir. The books included Charles Dickens' *David Copperfield* and *Bleak House*. At one stage, Lucid mused at what the great English novelist might have thought about his works being read, more than a century after his death, aboard a Russian space station. "I thought about that a lot," she admitted, "about the power that authors have and his story was transcending the centuries." Her books, though, were no substitute for regular contact with family and friends in her native language. On 9 May, Lucid made a telephone call to her parents in Oklahoma, in honour of her mother, who was celebrating her 81st birthday. Next day, she also participated in a two-way video conference, in advance of Mother's Day, and had the opportunity to speak to her family in Houston.

A week later, the Progress M-31 supply craft arrived, bearing equipment and supplies for the crew, including 1,140 kg of propellant and 1,700 kg of cargo. It also brought a welcome supply of candy, into which the crew "made some good inroads" in very short order. The arrival of the cargo ship also assisted with the start of scientific research aboard Priroda, for Progress offered a place to temporarily store the trash from the module's packing materials. Onufrienko and Usachev worked to replace three nickel-cadmium batteries to bring its power system up to required levels. However, more trouble was afoot and another power controller aboard Priroda failed. Meanwhile, Lucid set up hardware for the Biotechnology System (BTS) and Canada's Microgravity Isolation Mount (MIM) and the QUELD liquid diffusion experiment. During the course of her mission, only one of her 28 planned experiments failed to achieve its results, due to an equipment malfunction, and in many cases they returned a huge volume of scientific data. She also tended to dwarf wheat seedlings in Mir's compact Svet greenhouse. This device offered a small, 930 cm^2 chamber and could accommodate plants up to 40.6 cm tall, as well providing them with nutrients, fluorescent lighting and injections of water into their growth material.

Also planned for 21 May was Onufrienko and Usachev's second EVA, which was tasked with transferring the Mir Co-operative Solar Array (MCSA) from the Docking Module to its permanent location on the Kvant-1 module. During their five hours and 20 minutes outside, the cosmonauts used the Strela boom which they had installed in March to move to the DM and gain access to the MCSA. They secured the array to Strela and moved it over to Kvant-1, at the opposite end of the station, and attached it to a mounting bracket. Onufrienko and Usachev then returned to the Kvant-2 airlock, via Strela, and assembled a 1.2 m Pepsi can replica from aluminium struts and nylon sheets, part of a commercial payload delivered aboard Progress M-31. The resulting video, which the cosmonauts taped of each other, was intended to appear in a rather tacky Pepsi marketing campaign. As circumstances transpired, it was never aired, reportedly because Pepsi changed the design of the can.

From inside Mir, Lucid videotaped the cosmonauts' activities, although the station's relative paucity of available viewing ports left her efforts somewhat frustrated at times. "I hear them exiting the airlock and leaving the station," she wrote in one of her letters home. "No sooner were they out the airlock than Yuri [Onufrienko] was yelling at me to look out the window and start taking pictures. I looked out and there was my commander, perched on the end of the [Strela], arcing over the blue and white Earth below." To Lucid, it reinforced in her mind that *The Future* was no longer some abstract concept, towards which she and NASA and the US-Russian partnership were working; *the future* was *now*. She had no need to imagine what the construction of the ISS would be like, for it was unfolding, right before her eyes. "This," she exulted, "is *real* space station work!"

Three days later, on 24 May, Onufrienko and Usachev were outside again for five hours and 34 minutes – slightly longer than anticipated – to fully deploy the MCSA, which they hand-cranked open to its full 18 m length. The array's 84 panel modules opened in a similar fashion to an accordion and the cosmonauts set to work attempting to link it to Mir's power supply. Unfortunately, the electrical cables permitted power to be supplied from only *half* of the array and the MCSA provided about 3 kilowatts, rather than six. A future Progress would deliver new cables to complete the work. On 30 May, the cosmonauts ventured outside for the fourth time in their mission to install the US-built Modular Optoelectronic Multispectral Scanner (MOMS) payload onto the exterior of Priroda. During their four hours and 20 minutes in vacuum, they secured it in place by means of female and male connectors which (in true Russian fashion) they described as "mammas and pappas". Ever the diplomat, Lucid described the terminology as making Mir "feel warm and homey". Onufrienko had also gone to the rather extraordinary length of placing red tape over controls that Lucid was not to touch during the EVA ...

Continuing their summer of spacewalks, on 6 June the cosmonauts spent three hours and 34 minutes working to replace sample cassettes on the KOMZA interstellar gas experiment and installed the Particle Impact Experiment (PIE) and Mir Sample Return Experiment (MSRE) onto the exterior of the Spektr module, as well as a construction materials sampling payload. Finally, on the 13th, Onufrienko and Usachev installed the four-segment, 5.9 m Ferma-3 experimental construction girder onto the lower surface of the Kvant-1 module and manually deployed the saddle-shaped Travers-III synthetic aperture radar instrument on Priroda. The antenna should have unfurled by commands from inside the module, but hadn't, hence the manual deployment. The two men completed their marathon of EVAs by filming the final segment of the Pepsi commercial.

Postponed several times, Progress M-32 finally arrived at the end of July, by which time Lucid's expected return to Earth at the beginning of August was becoming increasingly unlikely. Original plans called for Atlantis to launch on STS-79 on 31 July, bringing her replacement, John Blaha, to Mir for his four-month extended mission. However, following analysis of the STS-78 launch on 20 June, NASA discovered worrying damage to the field joints of the Solid Rocket Boosters (SRBs), caused by apparent leakage of hot gases. Ultimately, the discovery and the necessary repairs led to the postponement of STS-79 from late July until mid-September, thus extending Lucid's mission from about 140 days to over 185 days. On 15 July, she soundly surpassed Norm Thagard's 115-day US endurance record, although her delayed landing meant that she would miss both her son's 21st birthday in August and her daughter's 28th birthday in mid-September. As Lucid's mission stretched into September, she secured a new record on the 7th: exceeding the 169-day achievement of Yelena Kondakova, she broke the world record for the longest stay in space by a woman. Not until June 2007 and the flight of Suni Williams aboard the ISS would another female astronaut exceed Lucid's record. Touchingly, General Yuri Glazkov, a former cosmonaut and then serving as deputy commander of the Gagarin Cosmonaut Training Centre, praised her achievement: "I don't think you've taken the record from *us*," he told Lucid. "We have offered this record to *you*!"

In the meantime, throughout July, concern mounted over the impending launch of a new crew for Mir aboard Soyuz TM-24. Budget cuts had delayed the production of the man-rated Soyuz-U2 rocket, forcing Russia to assign the mission to a lower-performance Soyuz-U, which was normally utilised for unmanned flights. However, the Soyuz-U had failed in two successive launches in May and June 1996 and both of its payloads had been lost, which led to some anxiety over the wisdom of assigning one of these vehicles to a crewed flight. The situation was complicated by the weaker performance of the Soyuz-U. The Soyuz TM-24 spacecraft was 275 kg too heavy for the rocket, which required Mir's altitude to be lowered by about 11 km in order for the crew to reach it. They also had to shed 6 kg in harsh 'crash diets' and were no longer allowed to take 4 kg of personal effects with them. The spacecraft was also stripped of several backup systems and the amount of manoeuvring propellant was lower than normal.

The three-member crew initially consisted of cosmonauts Gennadi Manakov and Pavel Vinogradov, together with France's first female spacefarer, Claudie André-

2-20: Shannon Lucid, pictured aboard the Spektr module.

Deshays. Launch was planned for 14 August. Progress M-32 delivered much of the scientific gear for Andre-Deshays' $13 million mission, dubbed 'Cassiope', but trouble was in store for the two-man Russian crew. In early August, during routine medical examinations, a serious problem was discovered with Manakov's heart rhythm and he was hospitalised. In accordance with pre-flight procedures, both he and Vinogradov were grounded from the mission and replaced by their backups, Valeri Korzun and Alexander Kaleri. However, rather than replacing André-Deshays with her backup, Léopold Eyharts, it was decided to retain the Frenchwoman as part of the new crew. She had been intimately involved with the development of its scientific payload for more than 18 months. Should Kaleri or André-Deshays have also been grounded, closer to launch time, they would have been replaced by Vinogradov or Eyharts, although no 'second-backup' reserve commander existed for Korzun.

In his book *Dragonfly*, Bryan Burrough touched on the dynamic of this crew exchange upon John Blaha, who would be spending four months in orbit with different cosmonauts to those alongside whom he initially trained. He learned of the crew swap on 9 August, whilst at home in Houston, but upon his return to Star City a few days later Korzun and Kaleri took him out to dinner in Moscow and made their formal introductions. The commander, Valeri Grigoryevich Korzun, was making his first space mission. Born on 5 March 1953 in Krasny Sulin, within Russia's south-western Rostov Oblast, Korzun entered the Kachinsk higher military aviation pilot school and graduated in 1974. He then served for over a decade as an active-duty pilot in the Soviet Air Force, rising to become a flight section leader and squadron commander, flying the MiG-29 fighter. He also gained certification as a parachute instructor and accrued more than 370 jumps. In 1987, Korzun completed the Gagarin Military Aviation Academy and was selected for cosmonaut training in July of that year. Upon completion of his initial training in June 1989, he trained extensively on Mir and Soyuz-TM systems, prior to his assignment to the backup crew for Soyuz TM-23. It was a surprising reversal of fortune for Korzun, who initially assumed that he would be too tall to be selected for the cosmonaut corps.

Joining Korzun was Alexander Yuryevich Kaleri, who came from Yurmala, in today's Latvia, where he was born on 13 May 1956. He grew up with the Soviet space programme; he was five years old at the time of Yuri Gagarin's epic voyage and had decided that he wanted to be a cosmonaut as soon as he heard about Gherman Titov's day in orbit aboard Vostok-2 in August 1961. "Maybe I was *born* with that wish," he told a NASA interviewer, but his father's experience as a navigator, engineer and parachutist provided further stimuli. In 1979, Kaleri graduated from the Moscow Institute of Physics and Technology as a specialist in aircraft flight dynamics and control systems, then undertook postgraduate work on the mechanics of fluids and plasma. He was selected as a civilian cosmonaut candidate in early 1984, qualified in 1987 and was paired with Alexander Volkov on the backup crew for Soyuz TM-4, which should have led to a seat on the Soyuz TM-7 prime crew, but an injury temporarily grounded him. Again teamed with Volkov, he served on the Soyuz TM-12 backup crew with Britain's Tim Mace and eventually received assignment to the prime crew of Soyuz TM-13, only to be dropped in mid-1991 in favour of Toktar Aubakirov and moved to the next mission. He spent four months aboard Mir in 1992. Despite these disappointments, Kaleri later remarked that much of his life and career put him "at the right time in the right place". At the time of writing, he is one of only six cosmonauts to have flown as many as five times and, if one takes his ethnicity into account, he is the world's most experienced Latvian spacefarer.

Making her first flight on Soyuz TM-23 was Claudie André-Deshays, who became France's first (and so far only) female spacefarer. Born in Le Creusot, within France's eastern Saône-et-Loire department, on 13 May 1957, she studied medicine at the Faculté de Médecine (Paris-Cochin) and Faculté des Sciences (Paris-VII) and went on to obtain certificates in biology and sports medicine, aviation and space medicine and rheumatology. André-Deshays then spent eight years working in the Rheumatology Clinic and the Rehabilitation Department at Cochin Hospital in Paris and supported ongoing research into diagnostic and therapeutic techniques in rheumatology and sports traumatology. She also worked in the Neurosensory Physiology Laboratory at the *Centre National de la Recherche Scientifique* (CNRS, the French National Centre for Scientific Research) in Paris and was closely involved in the development of medical experiments for the Soviet-French Soyuz TM-7 'Aragatz' mission to Mir in November-December 1988, with a specific focus on the adaptation of human motor and cognitive systems in the microgravity environment. In September 1985, André-Deshays was selected by the *Centre National d'Études Spatiales* (CNES, the French National Space Agency) as an astronaut candidate. In conjunction with her CNES work, in 1986 she received a diploma in the biomechanics and physiology of movement and in 1992 a PhD in neuroscience. She served as the backup for the Russian-French Soyuz TM-17 'Altair' mission to Mir in July 1993, which was flown by her future husband, French astronaut Jean-Pierre Haigneré. In her later career, André-Deshays also flew to the ISS on Soyuz TM-33 in October 2001, becoming the first European woman to board that station.

With the removal of Manakov and Vinogradov and their replacement by their backups, the launch of Soyuz TM-24 from Tyuratam was postponed until 17 August and the three-member crew of Korzun, Kaleri and André-Deshays successfully rocketed into orbit at 4:18 pm Moscow Time. In anticipation of the arrival of the new crew, Progress M-32 was undocked from the Kvant-1 module on the 18th, but rather than being destructively deorbited, it was kept in independent flight, with the intention of redocking it at the station shortly after the departure of Soyuz TM-23 in early September. Two days after leaving Tyuratam, Korzun guided his ship to a smooth docking at Mir at 5:50 pm Moscow Time and the three new arrivals were welcomed with hugs and warm wishes by Onufrienko, Usachev and Lucid. A cheerful André-Deshays joked that she would use Priroda as her "guest-room", directly opposite Lucid's Spektr. During her two weeks aboard Mir, she focused on cardiovascular physiology, neurosensory response, embryonic development, fluid and structural dynamics and measurements of the effect of vibrations upon sensitive experiments. Early on 2 September, Soyuz TM-23 undocked from Mir, bearing Onufrienko, Usachev and André-Deshays and touched down at 10:41 am Moscow Time, about 107 km south-west of Akmola in central Kazakhstan. Onufrienko and Usachev completed almost 194 days in orbit and André-Deshays a little under 16 days.

With Mir's population thus returned to a long-duration crew of three (Korzun, Kaleri and Lucid), the Progress M-32 supply craft redocked at Kvant-1 on 3 September. With STS-79 expected to arrive later in the month, Lucid spent a considerable amount of her time making a thorough inventory of the Spektr and Priroda modules to prepare her successor, John Blaha, for his four-month mission. She packed experiment samples, data tapes and equipment, ready to be transferred over to the Shuttle for their return to Earth. By the end of their five days together, the STS-79 crew would have moved more than 1,800 kg of supplies over to Mir, including logistics, food and water generated by Atlantis' fuel cells. The astronauts would also transfer the Biotechnology System (BTS) for cartilage development studies, the Material in Devices as Superconductors (MIDAS) to measure the electrical properties of high-temperature materials and the Commercial Generic Bioprocessing Apparatus (CGBA), which included a self-contained aquatic system. Atlantis also removed over 900 kg of experiment samples and equipment from Mir, achieving a new record for the largest total logistical transfer between the Shuttle and the station, at more than 2,720 kg. The sheer volume of materials to be delivered to the space station and brought back home had already led NASA to manifest 'double' Spacehab logistics modules on at least the fourth, fifth and sixth Shuttle-Mir missions. STS-79 would mark the first outing of the enlarged facility, which, as has already been noted, was assembled from a single module and an Engineering

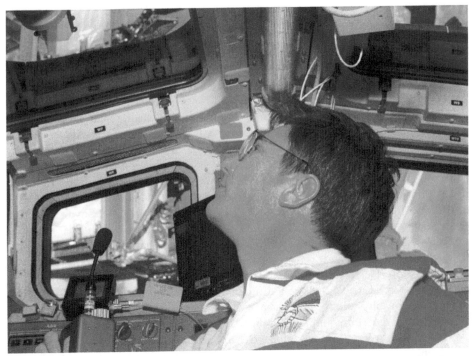

2-21: John Blaha peers through Atlantis' overhead windows as STS-79 draws towards a docking at Mir on 18 September 1996.

Test Article. The forward segment of the 4,775 kg module housed experiments to be performed during the Shuttle mission, whilst its aft area was occupied by logistics and supplies for Mir, including foodstuffs, clothing and spare equipment. Soft stowage bags were arrayed on the 'walls' and 'floor', with coloured labels to indicate their eventual destination: pink for items going *to* Mir, blue for items coming back *from* Mir and white for items which would remain aboard the Shuttle. In the topsy-turvy world of microgravity, the bags would remind the astronauts of a herd of sea-cows ...

Early in September, the STS-79 stack was rolled out to KSC's Pad 39A for the third time that summer. Figuratively and literally, it had been a long journey. Atlantis was originally transferred to the pad on 1 July for a launch later that same month, but with the discovery of STS-78's SRB leakage problems, NASA managers opted to roll the stack back to the Vehicle Assembly Building (VAB) for repairs. Adding to the urgency of the rollback was the anticipated landfall of Hurricane Bertha. By the middle of July, plans had crystallised to replace both STS-79 SRBs, but on the 25th the right-hand replacement booster failed a leak check at the junction between its aft-centre and forward-centre segments. Subsequent inspections of the booster's secondary O-ring revealed the presence of an applicator brush bristle, which was thought to be the root cause for the leakage. New O-rings were installed and the segment was restacked and leak-checked. In the meantime, Atlantis herself was removed from the stack and returned to the Orbiter Processing Facility (OPF). The External Tank was also de-mated in early August and mounted on a replacement set of SRBs; they were joined by Atlantis on the 13th and STS-79 was returned to the pad on the 20th. Launch was rescheduled for 14 September, but with the dissipation of Bertha the fury of Hurricane Fran threatened central Florida and on 4 September the stack was returned to the safety of the VAB yet again. On this occasion, though, it was a much shorter stay, and within 24 hours STS-79 was back on the pad, aiming for an opening launch attempt on the 16th. Early that morning, Atlantis' six-man crew, including John Blaha, was awakened, breakfasted, journeyed to Pad 39A and rocketed into the darkened Florida sky, precisely on time at 4:54:49 am EDT.

Not only was the STS-79 launch a spectacle for observers along much of the United States' eastern seaboard, but even Shannon Lucid was granted a ringside seat to witness it from aboard Mir. Atlantis' ascent was nominal, with the exception of a premature APU shutdown, which forced mission managers to briefly consider the option of docking with Mir a day earlier than planned. At length, it was decided to revert to the original timeline of docking on 18 September, but a close eye was maintained on the troublesome APU for the remainder of the mission.

Launching in the middle of the night meant that the STS-79 crew was awakened in the late evening and after their first sleep period they were roused to the sounds of Gene Chandler's 'Duke of Earl', chosen by mission specialist Carl Erwin Walz's wife in honour of their 20th wedding anniversary. Walz had already gained a measure of fame as an Elvis Presley impersonator and this had earned him admission into the astronaut corps' rock band, 'Max Q', as their lead singer. On a survival training expedition in Spokane, Washington, shortly after his selection into the corps in the spring of 1990, Walz and fellow astronaut candidate Terry Wilcutt were conversing in a bar. Walz happened to mention his previous singing experience in a Cleveland band, called 'The Fabulous Blue Moons'. Wilcutt challenged Walz to sing, right there and then. According to a 2009 article by Michael Cassutt in *Air & Space* magazine, Walz "talked with the band, agreed on a couple of Elvis tunes, then rocked out". The rest was history. Walz was born in Cleveland, Ohio, on 20 September 1955 and after high school entered Kent State University to study physics. With an upbringing in the 'Buckeye State', he grew up hero-worshipping fellow Ohioans John Glenn and Neil Armstrong. Walz got his degree in 1977, graduating *summa cum laude*, and followed up with a master's in solid state physics from John Carroll University in 1979. He was a Reserve Officers Training Corps student as an undergraduate, but took a two-year delay on 'inactive reserve' status to gain his master's degree before entering active service with the Air Force. After three years working on the analysis of radioactive samples from the Atomic Energy Detection System at the 1155th Technical Operations Squadron at McClellan Air Force Base in California, he trained as a flight test engineer at Edwards Air Force Base. Walz was drawn to the astronaut business, since it offered a combination of his interests in engineering and science, and also drove his desire to enter the Air Force. He knew of other flight test engineers – including Ellison Onizuka and Jerry Ross – who had become astronauts and so he sought to follow in their footsteps. After qualification as Distinguished Graduate of the 1983 class, Walz spent three and a half years at Edwards, working as a flight test engineer to the F-16 Combined Test Force, during which time he participated in airframe avionics and armament development efforts. In July 1987, he was appointed Flight Test Manager at Detachment 3 of the Air Force Flight Test Center. Less than three years later, Walz was accepted into NASA's astronaut corps. STS-79 was his third Shuttle mission.

It is interesting that Terry Wilcutt, who initially challenged Walz to sing Elvis, wound up as his future crewmate as the pilot of STS-79. For Terrence Wade Wilcutt, the urge to someday become a spacefarer was neither all-consuming, nor did it originate at an early age. "I didn't have a desire to be an astronaut until much later in my Marine Corps career, as I was finishing up my test pilot career," he later told a NASA interviewer, "but the desire to fly airplanes ... was always there." As a child, whilst on a Little League Baseball team, Wilcutt saw (and heard) jets breaking the sound barrier overhead and convinced himself that flying "as high and as fast as you possibly could" was the most exciting job in the world. Born in Russellville, Kentucky, on 31 October 1949, Wilcutt was raised and received his early schooling in Louisville, before entering Western Kentucky University to study mathematics. Whilst there, he was a member of the Lambda Chi Alpha fraternity, whose motto *Naught without Labour* guided his steps. Upon receipt of his degree in 1974, Wilcutt taught mathematics in high school for two years, then entered the Marine Corps. He earned his aviator's wings in 1978, flew the F-4 Phantom fighter, attended the Navy's Fighter Weapons School ('Top Gun') at Miramar, California, and was deployed overseas to Japan, Korea and the Philippines. His career took him through F/A-18 Hornet conversion training and in 1986 he graduated from the Naval Test Pilot School, after which he worked on numerous classified aircraft projects and was selected as a NASA astronaut candidate in January 1990. Thinking back over his life, Wilcutt was philosophical about the chances he had been given and how these ultimately put him in a position to apply for astronaut training. "Some of it's kind of happenstance," he said later, "or I just got lucky." By the time he flew STS-79, Wilcutt already had one prior Shuttle mission under his belt.

For much of the first two days of the mission, Wilcutt and STS-79 commander William Francis Readdy were preoccupied with the manoeuvring burns necessary to rendezvous with Mir. And for Readdy, it was an opportunity that he could hardly have anticipated. In his later career with NASA, he would rise meteorically through the ranks to become the agency's Associate Administrator for the Office of Space Flight – a position to which he had recently been appointed at the time of the Columbia accident in early 2003. Readdy came from Quonset Point, Rhode Island, where he was born on 24 January 1952, the son of a naval aviator who had flown in Korea. After high school, he entered the Naval Academy to study aerospace engineering and gained his degree in 1974. As a military aviator, Readdy flew the A-6 Intruder and served as an attack pilot aboard the USS *Forrestal* in the North Atlantic Ocean and Mediterranean Sea in 1976-1980. This experience was followed by test pilot school and a spell as an instructor at Pax River, and overseas deployments to the Caribbean and Mediterranean aboard the USS *Coral Sea*. His applications to become an astronaut was unsuccessful in May 1984 and June 1985 and he resigned from active naval duty and entered NASA as a research pilot in October 1986. Based at Ellington Field in Houston – home of the Texas Air National Guard's 147th Reconnaissance Wing – Readdy served as programme manager for the Boeing 747 Shuttle Carrier Aircraft and was selected as a Shuttle pilot in the June 1987 astronaut group. "A little footnote, there was an '85 group and then an '86 group" he recalled in his NASA oral history, "but after Challenger happened, that basically went away. There was no '86 selection. There were several of us that had interviewed for that and we got invited down to Houston to work in different support jobs until the next astronaut selection. The support jobs were all related to Return to Flight after Challenger." Readdy's support job on the Boeing 747 included the procurement and modification of the aircraft for Shuttle requirements. "Getting that all put together also gave us an opportunity to upgrade the existing Shuttle carrier airplane with better engines," he said. "It was a very involved structural modification with all the truss structure and everything else. You want to empty the airplane as best you can, so that you can support all that weight, but then you find out that because of *where* the orbiter is located on it – fairly far aft – that you need to keep the airplane in balance. We used to fly with baggage containers of pea gravel in all our ground support equipment in the forward cargo bay of the 747." Having twice flown on the Shuttle, STS-79 was Readdy's first command.

Readdy had succeeded Ken Cameron as NASA's second director of operations in Russia in the summer of 1994 and had been surprised when he walked into his Star City office one morning the following November to receive a fax from JSC in Houston. Simply *receiving* the fax was a miracle in itself. Yet this fax was different; it included his name in command of STS-79. "When the faxes were working over there," Readdy told a NASA interviewer, "when the *phones* were working over there, we were just ecstatic, and when the fax rolled out of the machine on that curly paper ... it was kind of a pleasant way to leave Star City with a mission assignment." Five months later, in April 1995, he received notification of his five crewmates on STS-79: Wilcutt as pilot and mission specialists Walz, Jay Apt and Tom Akers. At one stage in its development, STS-79 was also expected to feature a spacewalk by Apt and Walz. This was repeatedly mentioned by *Flight International*, as well as by NASA, throughout 1995, and when the crew produced its official mission patch they included, in pride of place, a pair of gloved hands shaking. However, the EVA was moved to STS-76 instead. "We didn't have an EVA," said Bill Readdy in his oral history. "As it turned out, that was one of the unpleasant surprises."

As the pilots focused on the Mir rendezvous, Apt and Walz were responsible for activating the bulk of the payloads in the first Spacehab double module. One of these was the Active Rack Isolation System (ARIS), which was flying as an ISS risk mitigation experiment to demonstrate a method for cushioning sensitive micro-gravity payloads from mechanical vibrations and other disturbances. Specifically, ARIS sought to expose the system to Shuttle- and Mir-induced low-frequency motions, including periodic thruster firings, of between 0.003 and 300 hertz. Despite the need to repair a bent push rod, the system responded well. Meanwhile, Carl Walz powered up the Mechanics of Granular Materials (MGM), whose objective was to better understand the behaviour of granular materials for construction purposes in dry and saturated states at very low confining pressures and effective stresses. Elsewhere, Apt also began work with McDonnell Douglas' Extreme Temperature Translation Furnace (ETTF), which could process materials at up to 1,600 degrees Celsius as part of investigations into how flaws form in cast and sintered metal compositions. He also presented a video tour of the Spacehab double module.

STS-79 was Apt's fourth Shuttle mission. Born Jerome Apt III, he entered the world in Springfield, Massachusetts, on 28 April 1949. He received his early education in Pittsburgh, Pennsylvania, and studied physics at Harvard University. After receiving his degree in 1971, Apt moved to the Massachusetts Institute of Technology and gained a PhD in laser spectroscopy in 1976. He then served as a staff member at the Center for Earth and Planetary Physics at Harvard, working on NASA's Pioneer Venus Orbiter mission, and was assistant director of Harvard's Division of Applied Sciences. He applied unsuccessfully for selection as an astronaut in 1978 and 1980. His NASA career began in 1980 as a planetary scientist at the Jet Propulsion Laboratory in Pasadena, California, and later as payload flight controller. It was third time lucky for Apt when NASA selected him into its astronaut corps in June 1985. In his memoir, *Skywalking*, astronaut Tom Jones – who later flew with Apt – described him as a consummate professional, highly experienced in a technical sense and a good friend and colleague, but admitted that he was "a man who had perfect confidence only in himself".

Readdy and Wilcutt accomplished a smooth R-Bar rendezvous and docking with Mir at 11:13 pm EDT on 18 September, a little more than two days into the mission, as the combined spacecraft flew over the Carpathian Mountains, to the west of Kiev. The hatches were opened about two hours later. A brief welcoming ceremony was quickly followed by John Blaha and Shannon Lucid assuming their respective roles as members of different crews: the former transferred his moulded seat liner over to the Soyuz TM-23 spacecraft, which he would use in the event of a contingency return to Earth, whilst the latter moved her equipment over to Atlantis and formally joined STS-79. For all six astronauts, it was the first time that they had seen Lucid in over six months and Terry Wilcutt, for one, was amazed at the extent to which she had adapted and become a creature of weightlessness. "All the long-duration crew members who spent time on Mir said that after about a month, they noticed a difference in themselves," he told a Smithsonian interviewer, several years later. "They'd truly adapted to zero-G. I could see it with Shannon Lucid. She was so graceful; there was never an unnecessary motion. She could hover in front of a display when anyone else would be constantly touching something to hold their position."

Almost immediately, the process of transferring the largest complement of equipment and supplies between the Shuttle and Mir got underway. It was a process overseen by STS-79 mission specialist Thomas Dale Akers, who employed a specialised inventory monitoring system and bar code recorder to effectively keep track of all of the items. "A lot bigger than my JSC office," was Akers' summary of the spaciousness of the double Spacehab. In his mind, the ability to handle *packages* in the soft stowage bags, rather than individual small items, made it easier to transfer materials to and from Mir. Akers compared the process to FedEx or UPS and felt that packaging items in groups made it far easier to inventory them. The sole difficulty was understanding the 'down-stowage' plan – what was coming back to Earth *from* Mir – in good time. "That was contingent on knowing how Shannon had packed all of her return items," Akers reflected. "We didn't get that information until a couple of days before launch, because of communications problems. Shannon had done a great job of packing, had everything ready, but the communication of finding out exactly where everything was, so we knew how heavy each package was, which determines how the loads are done, that's one thing we want to do differently."

Akers was making his fourth Shuttle mission on STS-79. Born in St Louis, Missouri, on 20 May 1951, he earned undergraduate and master's degrees in mathematics, both from the University of Missouri-Rolla. In 1973, the year after receiving his first degree, he found work as a National Park ranger at Alley Springs, and spent every summer there until 1976. By the time his ranger service ended, he had gained his master's degree and spent several years as a high school principal in his hometown of Eminence, before entering the Air Force in 1979. After initial officer training, Akers was assigned to Eglin Air Force Base in Florida as an air-to-air missile data analyst and during this period he also taught night classes in mathematics and physics for Troy State University. Selected in 1982 for the Air Force's Test Pilot School at Edwards Air Force Base in California, he graduated as a flight test engineer the following year and worked on a variety of advanced weapons programmes. Interestingly, his association with the University of Missouri-Rolla endured after both his Air Force and NASA careers and in 1999, after departing both services, he taught mathematics there.

During the first day of docked activities, Jay Apt, John Blaha and Alexander Kaleri moved the BTS and CGBA payloads over from the Spacehab module into Priroda and began setting them up and activating them. As part of the BTS experiment, Blaha was expected to take weekly samples of cartilage cells to examine their development in microgravity. By 22 September, the final 'powered' transfer item, the MIDAS semiconducting experiment, was installed aboard Priroda. Watching and participating in these full-on, 18-hour days, Blaha had nothing but praise for the Shuttle crew as they accomplished all of this work. "Each evening," he said, "the STS-79 crew and the Mir crew met for dinner either on Mir or Atlantis. These were unforgettable times. I will always remember how they all helped me move into my new home." A farewell meal in Mir's base block, early on 23 September, was followed by the movement of the two crews back to their respective spacecraft for undocking. At 8:00 am EDT, Valeri Korzun and John Blaha closed the Kristall hatch and the vestibule between the station and the Shuttle was depressurised.

After an abbreviated, six-hour sleep period, Readdy successfully undocked from Mir at 9:33 pm EDT and after departing to a safe separation distance Wilcutt took control of the orbiter for the customary flyaround inspection. Several hours later, the pilots executed a test of the Shuttle's thrusters to slightly lower their orbit, in order to mimic a manoeuvre planned for STS-82 in February 1997 to reboost the Hubble Space Telescope. With only two fully functional APUs, the weather criteria for a landing on 26 September were more restrictive than normal, with crosswinds required to be less than 18.5 km/h, cloud ceilings above 3,000 m and visibility above 11 km. Weather conditions at KSC appeared to be acceptable, with only scattered clouds and light winds in the vicinity of the SLF runway, whilst Edwards Air Force Base in California offered a backup capability if necessary. For Shannon Lucid, returning home after 188 days in space, much of her time had been spent exercising to condition herself for the punishing return to terrestrial gravity. At 7:06 am EDT on 26 September, Bill Readdy fired Atlantis' OMS engines to begin the irreversible deorbit burn. A little over an hour later, at 8:13, he crossed the eastern seaboard over the state of Georgia and guided his ship into Florida airspace and to a smooth touchdown on Runway 15.

"Depending on whether you want to look at the glass as half-full or half-empty, it meant that Shannon had to spend an awful lot longer on Mir than she had planned," said Readdy in praise of his crewmate's accomplishment. "I think Shannon's girlhood dream to run her own laboratory was such that ... it fit right in with her plan *and* it also allowed her to set the world record for time in space" for women.

Aboard Mir, John Blaha became the first long-duration US space station astronaut to directly follow a previous long-duration US space station astronaut, thus forging the first link in a 2.5-year 'chain' of Shuttle-Mir operations. Flying to a space station had been in Blaha's mind since October 1991, when he attended an Association of Space Explorers meeting in Berlin and had the opportunity to chat to a group of Russian cosmonauts and watch a video about Mir. Two years later, when Norm Thagard and Bonnie Dunbar began training for NASA's first long-duration mission, Blaha asked at one of the astronaut office meetings if pilots, as well as physicians, could be considered for places. Blaha had flown four Shuttle missions, twice in the commander's seat, and he knew that if he was ever going to have the opportunity to live and work aboard a space station before he retired, it would have to be Mir. For him, the ISS was too far into the future. Mir was the present.

Assigned as Shannon Lucid's backup, Blaha was initially pointed at the fourth long-duration mission, launching in December 1996 and returning to Earth in May 1997. In fact, the idea of a given astronaut backing up one mission, skipping the next to offer some rest and respite, and then serving as prime crew member on the third was in NASA's mind from the beginning. The removal of Scott Parazynski and Wendy Lawrence from training in the summer of 1995, combined with a distinct shortage of volunteers within the astronaut office, put paid to this idea and it was decided instead to fly astronauts consecutively, backing up one mission and flying the next. The Russians disliked 'consecutive' assignments, but NASA had no other choice. Also, the sequence of Blaha and Linenger's missions were switched at some point in 1995. Blaha was launched first to immediately replace Lucid. "John and I switched positions in the sequence," remembered Linenger in his memoir, *Off the Planet*, "in order for me to do a spacewalk."

Years later, Blaha was critical of the value of the 'handover' he received from Lucid. He felt that five days was woefully insufficient for the outgoing astronaut to provide the required depth of knowledge for the incoming astronaut. "I would use the word I'm *disappointed*," Blaha recalled in a NASA oral history. "I'll never quite understand why she did not give me all the information she had learned during her

2-22: Old friends John Blaha and Shannon Lucid are reunited in space.

long stay on Mir. She gave me a handover, but it was not a complete handover. It's not what I would do for my brother or sister if I were living in a cabin on a mountaintop. I would give them a thorough checkout of the cabin before I left. My goal would be to help them start with the knowledge that I had when I was finishing." In the case of subsequent Mir astronauts, Blaha stressed the need for much more time to be set aside during the docked phase of the Shuttle mission for the two long-duration crewmen to work on these issues. "The handover time is like gold," he said. "If you optimise the handover for the 'long' person who is going to be there, you could even save money in training on the ground." He felt that this understanding was not wholly grasped by Shuttle-Mir management and that it was a pity that it was not capitalised upon in readiness for the ISS.

Blaha's expedition did not begin well. Among his earliest work was the BTS, but he soon noticed that the vessel chamber was not rotating correctly. Since it housed living cells, it was imperative that it had be restored to fully functional status as quickly as possible. With the Shuttle crew still in residence, Blaha and Jay Apt took photographs of the hardware and downloaded them to Houston, for experts to analyse and suggest corrective actions. Initial repair efforts were unsuccessful, but eventually it was discovered that a control and data cable had become dislodged. Blaha was asked to power down the equipment to resolve the cable problem, but Houston had to pass its instructions through Russian mission control in Moscow, which took several days. Fortunately, during a brief communications pass, Mission Scientist John Uri was able to talk Blaha through the BTS repair, which proved successful. "It ain't Apollo 13," Uri quipped, "but from a scientist's perspective, we pulled it off and saved the experiment."

In spite of this, all was not well during Blaha's early time in orbit. His research encompassed various scientific disciplines, including fundamental biology, environmental radiation measurements, growing and monitoring wheat in Mir's Svet greenhouse and human life sciences and immune function experiments. In *Dragonfly*, Bryan Burrough noted that the astronaut was so overworked, even in these first few days, that he had hardly enough time to say goodbye to the STS-79 crew when the time came for them to depart. His daily routine was wholly governed by a minute-by-minute timeline, known as 'Form 24', whose schedule proved hopelessly at odds with the reality of living in the cluttered space station. In one of his early tasks at the greenhouse, Blaha was required to photograph the dwarf wheat seedlings, which he noticed were steadily maturing, but it took hours to even find the Nikon and Hasselblad cameras ... and when he found them, it turned out that they were not in flight-worthy shape. "The hardware was literally falling apart," he was quoted by Burrough as having said. "I probably spent about ten hours of time getting these two cameras in flight-worthy shape." From the perspective of his two Russian crewmates, Valeri Korzun and Alexander Kaleri, the sight of Blaha working feverishly as he would have done on a short-duration Shuttle flight was worrisome. Kaleri, who had a prior long-duration mission to his credit, told Blaha that he needed to rest and conserve his psychological energy. Blaha did not complain, but continued to work hard, until, by the middle of October, the exhaustion hit him. When the NASA ground team piled on more work, it was Korzun who eventually reacted with a few home truths. "I'm the commander of Mir," fumed Korzun, as paraphrased by Burrough, "and I can tell you what they are doing with John Blaha over the last ten days has really been *wrong*. The Americans really need to get better organised!" The comment appeared to have the necessary effect and within days Blaha's overworked schedule was cut down substantially, but all three crew members routinely found themselves working 14-hour days on experiments, maintenance and general housekeeping.

With the BTS now back up and running, Blaha set to work sampling mammalian cartilage cells for post-flight analysis and recorded the progress of the experiments on video. He completed work with a binary colloidal alloy test, which grew crystals of two separate materials together over time and regularly acquired microbial specimens from throughout Mir, including air, water, spacecraft surfaces and skin samples. The BTS woes then returned to the fore, when an air bubble developed in the liquid growth medium and problems arose with a computer-controlled pump, which required Blaha to replace the medium and reset the hardware. Each day, he recorded data and made manual changes to the water and lighting cycles for the dwarf wheat seedlings in the Svet greenhouse, as well as planting new crops.

Late in October, at a space-to-ground press conference, Blaha praised the performance of Korzun and Kaleri, who worked unbelievably hard, to such an extent that he rarely saw them in the warren of Mir's modules. "Every now and then, I would do something with one of the cosmonauts, but not often," he said. "Maybe there were 15-20 times in that four months. The reason was we all were too busy. We couldn't be together. All three of us had to be working on things to accomplish all the work." Watching movies helped Blaha to sleep at night, which aided his wellbeing, and he admitted missing his wife, Brenda. He also used ham radio extensively to talk to amateur operators around the world and, around Halloween, even engaged in a two-way videoconference with the next Shuttle-Mir crew, STS-81, who would relieve him in January 1997. These touches and reminders of home provided a critical psychological crutch. "All three of us were busy, from eight in the morning until ten in the evening," Blaha remembered. Korzun and Kaleri worked even later, often until midnight. "I quit at ten or ten-thirty, if I could," continued Blaha, "so that I could wind down and get a good night's sleep. I never just watched them do stuff, because I never had any time to do that. I was busy with all the things I was doing, which predominantly was the science experiments." Seven-day working weeks were the norm.

Problems continued aboard the aging Mir and in early November it became clear that the station's waste recycling system had malfunctioned and reserve containers were almost full. The next Progress cargo ship was several weeks away and financial problems had delayed the production of enough boosters for Soyuz-TM missions, meaning that the replacement crew for Korzun and Kaleri would not launch in December 1996, but in February 1997. The delays also impacted unmanned resupply flights, with Progress M-33 postponed from 15 October until at least the beginning, and later the middle, of November. In readiness for its arrival, Progress M-32 undocked on the 20th, that same day M-33 finally launched. The new craft docked on the 22nd, bringing supplies, fresh fruit, clean clothes, the rather euphemistically-named 'human waste containers' and early Christmas and New Year gifts for the crew. Blaha was in the Kvant-2 module, looking through one of its small windows, as Progress M-33 approached. "I finally saw the Progress at a distance of 30 km; a shining star, rising towards us at great speed from beneath the horizon," he recalled. He then moved to the opposite end of the station, to Kvant-1, right at the point where the docking would occur, and felt a firm thump – "five times stronger than I remembered the Shuttle docking with Mir" – as the cargo craft made its presence known. Blaha described the excitement of waiting up until well past midnight to open up the cargo ship. "Once we found our packages," he wrote, "it was like Christmas and your birthday, all rolled together, when you are five years old. We really had a lot of fun reading mail, laughing, opening presents, eating fresh tomatoes and cheese." They continued to work through the Thanksgiving holiday and the first week of December was punctuated by a pair of EVAs by Korzun and Kaleri.

The cosmonauts' task was to install extra cabling and contacts for the MCSA solar array on the exterior of Kvant-1, thereby bringing its electricity-generating capability up to the required six kilowatts. Korzun and Kaleri ventured outside Mir at 3:54 pm Moscow Time on 2 December and spent almost six hours in vacuum. The cable, which reached 23 m in length, was attached to the array and trailed to the socket for the dorsal array on Mir's base block, which was no longer in use. For Blaha, who co-ordinated their activities from inside the station, it was the first time he had beheld an EVA. "I will forever have images implanted in my brain of Valeri and Sasha ... preparing for the spacewalks, asking many questions to specialists on Earth and probing every possible scenario," Blaha reflected later. "I will forever remember the incredible views of these two cosmonauts floating in space, silhouetted against the black of space, with Planet Earth rotating by us below. I will forever remember the sounds of strain in their breathing when the workload was intense. And finally, I will never forget the incredible feeling of accomplishment after the job was complete and everyone was safely inside the Mir space station." A week later, on the 9th, the two men returned outside for another six hours and 36 minutes to fit a new omni-directional Kurs antenna onto Kristall's Docking Module. They also reattached a cable to an amateur radio antenna, which they had mistakenly knocked loose during their first EVA. As Mir passed over Chile and Brazil, Blaha was able to listen on the amateur frequency to confirm that the antenna was working, but due to a lack of traffic they had to wait until they came within range of European hams before confirmation could be gained. Later that evening, Korzun successfully made contact with a Portuguese amateur operator.

Two weeks later, Blaha became the first American to spend Christmas in space since the Skylab era, more than two decades earlier. In a news conference on the 20th, Korzun described Mir's Christmas menu as "outstanding", with traditional cakes and dishes, including lamb, pork and desserts, as well as Italian foods and cheeses. With STS-81 and his replacement astronaut, Jerry Linenger, scheduled to arrive in mid-January, Blaha set to work packing 15 bags of equipment to be returned to Earth. Trouble was not yet finished with his mission, for on 10 January he heard a loud clattering from the Spektr module; one of two cooling fans had broken in the freezer which housed all of his life sciences samples ... and no spares were aboard Mir. He removed the front door of the freezer and attached a temporary door to hold the temperature as stable as possible for as long as possible. He also removed a fan blade and reinstalled the primary door with only one fan in operation, which was adequate for about a week, barely sufficient for Atlantis to arrive. During a communications pass, he spoke with Flight Surgeon Pat McGinnis and engineer Matt Mueller, who located replacement fans and had them delivered to KSC and loaded aboard Atlantis' Spacehab double module on the launch pad.

If STS-79 had completed the largest-ever transfer of items between the Shuttle and Mir, it was not to be a long-held record, for STS-81 was tasked with moving approximately 2,800 kg of logistics and supplies to and from the station, mainly aboard the second Spacehab double module. This included 635 kg of water, 516 kg of US research equipment, 1,000 kg of Russian logistics and 120 kg of miscellaneous materials, as well as the return to Earth of almost 1,100 kg of unneeded equipment. Among the items to be returned home by Atlantis' crew were the wheat samples from the Svet greenhouse, which marked the first occasion on which plants had completed a full growth cycle in space, whilst STS-81 also delivered the Treadmill Vibration Isolation and Stabilisation System (TVIS) to be tested in readiness for its role aboard the ISS.

In command of STS-81 was veteran astronaut Michael Allen Baker, making his fourth Shuttle flight, and a former director of NASA operations in Russia. Baker came from Memphis, Tennessee, where he was born on 27 October 1953. He received much of his later schooling in California and earned his degree in aerospace engineering from the University of Texas in 1975. He then entered the Navy, completed initial flight training and gained his gold wings as an aviator at Naval Air Station Chase Field in Beeville, Texas. Baker served as an attack pilot, flying the A-7E Corsair from the USS *Midway*, homeported in Yokosuka, Japan, and later worked as an airwing signal officer. In 1981, he entered the Naval Test Pilot School and, after graduation, undertook carrier suitability structural tests and catapult and arresting gear trials for the Navy's fleet of A-7s. In the two years before his selection by NASA in June 1985, Baker worked as an instructor, both at the Naval Test Pilot School and as at the Empire Test Pilots School in Boscombe Down in England. After completing astronaut candidate training in July 1986, six months after the loss of Challenger, he was intimately involved in the critical redesign and modification of the Shuttle's landing gear, including its nosewheel steering mechanism, brakes, tyres and a new 'drag chute' to be introduced into the fleet in 1992.

Also announced in February 1996 to join Baker on STS-81 were pilot Brent Jett (a veteran of STS-72) and three mission specialists, all of whom had flown before.

Peter Jeffrey Kelsay Wisoff, nicknamed 'Jeff', was born in Norfolk, Virginia, on 16 August 1958, and recalled growing up as the first men set foot on the Moon. "I remember watching that live on TV and being very excited about it," he told the NASA oral historian, "and always thinking that would be a great thing to do. It wasn't really until the Shuttle programme came along that there were enough seats ... for someone with my background to participate, because I'm not a pilot." Wisoff's background was principally in laser physics. After high school in his hometown, he gained a degree in physics from the University of Virginia in 1980 and then a master's and doctorate in applied physics in 1982 and 1986, both from Stanford. His PhD encompassed laser physics and laser construction technologies. Wisoff next accepted a faculty position in the Department of Electrical and Computer Engineering at Rice University and worked on the development of new vacuum ultraviolet and high-intensity lasers, whose potential applications included the reconstruction of damaged nerves. STS-81 was Wisoff's third Shuttle mission.

Sitting behind and between Baker and Jett on Atlantis' flight deck as STS-81's flight engineer was John Mac Grunsfeld, who was making his second Shuttle mission and, in time, would become the only human being to visit the Hubble Space Telescope on as many as three occasions. Like fellow astronaut Scott Parazynski, he was an avid mountaineer, and as Parazynski became the first astronaut to summit Mount Everest, Grunsfeld became the first astronaut to summit Mount McKinley in Alaska in June 2004. Unlike many astronauts, whose pre-NASA profession became more 'generalised' after admission into the corps, Grunsfeld remained a devoted astronomer, to such an extent that he served as the agency's Chief Scientist in 2003-2004 and, at the time of writing, in 2014, he serves as Associate Administrator of the Science Mission Directorate at NASA Headquarters in Washington, DC. He was born in Chicago, Illinois, on 10 October 1958, the son of the distinguished architect Ernest 'Tony' Grunsfeld III and grandson of Ernest Grunsfeld Jr, who designed the Adler Planetarium, the first of its kind ever built in the western hemisphere. By his own admission, Grunsfeld yearned to be an astronaut from early childhood, although his interests later shifted towards physics and cosmology. "I had a vacuum cleaner with a hose that became my liquid-cooling unit," he told a NASA

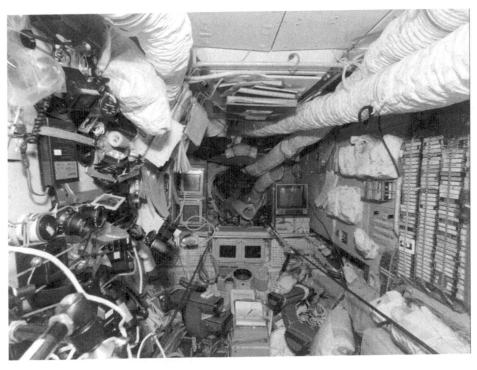

2-23: Interior of Mir's base block, seen by a member of the STS-81 crew.

interviewer, years later, "walking out to the space launch pad. I turned those big ice-cream tins, by cutting a hole and putting cellophane over it, into a space helmet. It seemed very natural at the time, because here was a nation starting the exploration of space." Exploring Highland Park, on the north side of Chicago, wading through knee-high snowdrifts and forests and into ravines, Grunsfeld pretended that he was an astronaut on another world. It was a love of nature and science which would remain with him.

After high school, he entered the Massachusetts Institute of Technology to study physics and earned his degree in 1980. Whilst there, he met a post-doctoral student named Jeff Hoffman, who was an astronomer. It was Hoffman's subsequent path in NASA which convinced Grunsfeld that it was accessible to him. To make ends meet as a student, he worked the graveyard shift in the control room for the Small Astronomy Satellite (SAS)-3, changing data tapes and monitoring strip-chart recorders. This led to a year-long post as a visiting scientist at the University of Tokyo, working with X-ray astronomer Minoru Oda. Grunsfeld returned to the United States and entered the University of Chicago, gaining master's and doctoral degrees in physics in 1984 and 1988, with a research focus upon energetic cosmic rays. After graduation, he worked as a senior research fellow at the California Institute of Technology, where he and his wife both learned to fly; in late 1991, when NASA invited Grunsfeld to attend an astronaut interview, he flew his own aircraft to

Houston. Until he received the call from the space agency, he thought his job at Caltech was the best in the world. "I was working with the Compton Gamma Ray Observatory," he said. "I had observatories, like Palomar, that I went to use big telescopes." It was electrifying to consider that *now* was the Golden Age of Astronomy and *he* was part of it, but that did not stop him from applying, unsuccessfully, to join NASA in January 1990. "I was studying neutron stars and black holes," he continued, "which are incredibly interesting and exotic objects. I had wonderful students. I thought life couldn't be better ... and then I got a call from Houston." He chatted it over with a colleague, Bruce Margon. The consensus was that whilst Grunsfeld could *always* be an astronomer, he might *not* always have the chance to become an astronaut. In March 1992, he was selected by NASA.

In his memoir, *Off the Planet*, Jerry Linenger reflected upon Grunsfeld's often wry humour. As an undergraduate, Grunsfeld had worked on his car in a self-help garage run by two MIT graduates, who later started a popular radio programme called 'Car Talk'. Prior to the STS-81 launch, he arranged with the producers to play a prank on the show's hosts. In orbit, Grunsfeld's telephone was patched through to the unsuspecting hosts. Without identifying himself, he complained about his government-issued vehicle, with its horrendous gas mileage, its habit of running extremely roughly for the first two minutes, with lots of shaking, after which it performed like a champ ... until eight minutes, when its engine died completely. The radio hosts were perplexed. Did the vehicle achieve good acceleration, they asked?

"Oh, yeah," replied Grunsfeld. "About 17,500 miles an hour!"

"Who *is* this?" the hosts asked.

Only then did Grunsfeld introduce himself as an astronaut aboard Space Shuttle Atlantis ...

The final STS-81 mission specialist, Marsha Sue Ivins, had already worked for NASA as an engineer for a decade before she was selected as an astronaut candidate in May 1984. Born in Baltimore, Maryland, on 15 April 1951, she grew up with dreams of one day flying into space. "When I was ten years old," she told an interviewer, shortly before her final Shuttle mission, "Alan Shepard made the first flight in the American space programme." Ivins remembered that the event captured her imagination and she refused to allow the fact that all of the astronauts were *male* and *military pilots* to get in her way. She realised that they were also engineers and, "for no other reason", went to the University of Colorado at Boulder to study aerospace engineering. Ivins received her degree in 1973 and began working at the Johnson Space Center in July of the following year. Her initial role was as an engineer for the displays and controls of the Shuttle, which was then in its earliest stages, and in 1980 she became a flight engineer in the Shuttle Training Aircraft. She applied for the astronaut programme on *three* occasions, was unsuccessful in 1978 and 1980, and made the cut four years later. STS-81 was Ivins' fourth Shuttle flight.

Originally scheduled for launch on 5 December 1996, the SRB problems throughout the summer months pushed each of the Shuttle-Mir flights back by approximately six weeks and STS-81 eventually settled on a target date of 12 January 1997. Atlantis' processing flow ran very smoothly, punctuated only by the need to replace a troublesome fuel cell, and the STS-81 crew and Mir crewman

Linenger arrived at KSC on 8 January, ahead of the formal start of the 43-hour countdown. Launch on the 12th took place at 4:27:23 am EST, at the opening of a short, ten-minute window, and by the time the Shuttle achieved orbit Mir was travelling over the Galapagos Islands, about 4,450 km to the south-west. About 25 minutes after liftoff, Korzun, Kaleri and Blaha were notified and were able to watch a videotaped uplink from Mission Control.

Embarking on a well-trodden, two-day rendezvous profile, Atlantis' crew set to work establishing laptops and laser ranging equipment and opening up the Spacehab double module in the payload bay. Wisoff, Grunsfeld and Linenger worked on the assembly of the TVIS and tested its restraint system, motor, running surface stability and effectiveness in reducing disturbances to the surrounding microgravity environment. Meanwhile, Ivins activated ESA's Biorack facility and Jett and Grunsfeld installed the centreline camera into the ODS, ahead of docking. On 14 January, Mike Baker greeted Mission Control's wake-up call with a chirpy "Good morning, Houston, 500 miles to go!" Following the R-Bar, directly 'up' from the centre of Earth towards Mir, Baker performed a successful docking at Kristall at 10:55 pm EST. The STS-81 crew had sighted Mir from a distance of about 64 km, but from the station John Blaha did not spot Atlantis until eight minutes or so before physical docking. Grunsfeld and Linenger then set to work on the process of equalising pressure between the two spacecraft, ahead of opening the hatches.

Through the porthole, inside Kristall, Linenger could see a jubilant Blaha and Korzun. As the hatches opened, in Linenger's recollection, the Shuttle crew was greeted by "an uninhibited laugh" from Blaha. "Then bedlam erupted," he continued, "as the six of us blundered our way through the hatch and bumped heads with the much more graceful (being fully adapted to space) threesome of Mir occupants. The scene was one of hugs, shouts, mixed language and laughter, feet dangling in all directions. Nine spacefarers embracing and floating in every which way. After the chaos calmed, we all migrated, single file, heads closely following feet, into Mir." In a similar manner to previous flights, after the welcoming ceremony, Blaha and Linenger swapped their moulded Soyuz-TM seat liners and joined their respective crews. The Shuttle crew gave Korzun and Kaleri new flashlights, because sections of Mir were being kept dark in order to conserve electricity. (In fact, a master alarm sounded with a low-electrical-power warning, midway through the welcoming ceremony, much to Korzun's embarrassment.)

"We're truly in the space station business," Blaha said at one point, as he entered the final days of his four-month mission. Keen to avoid the lack of preparation at the start of his own increment, he devoted a great deal of time to working with Linenger and readying him for his lengthy stay. In the weeks preceding Linenger's arrival, Blaha had spent his free time tapping out emails to NASA colleagues on Earth and updating the handover for his successor. With Linenger now aboard Mir, Blaha spent several hours with him each evening. Their discussions ranged from the technical to the mundane, but necessary, everyday tasks. "During my first few days on Mir," Linenger reflected, "John and I had many private conversations, tucked away in a corner of the station. We went through how the toilet works, how the treadmill works and he did his routine. He explained in detail how he cleaned himself after working out on the treadmill; there's a trick to it. Water is in short supply up there and therefore you need to use two or three thimblefuls, which you put on a little towel. He would cut the towel into about five or six sections to conserve towels, because there's no way to keep delivering new towels. He showed how you can make a couple of thimblefuls of water go a *long* way. Sanitation was not an easy task up there. During your treadmill sessions, you would definitely be sweating and our T-shirts would get soaked. Our supply of shirts and shorts was such that we could only change every two weeks!" In *Off the Planet*, he added that Blaha's "unalloyed frankness" and "uncensored remarks" were invaluable in allowing him to fine-tune his own mindset, ahead of four months aboard Mir.

In the meantime, transfers between the Shuttle and the station continued, in addition to ongoing research work inside the Spacehab double module. The 'loadmaster' was Marsha Ivins, whom Linenger described as having the daunting task of knowing where everything was stowed aboard the orbiter. "She ruled with an iron fist," wrote Linenger. "She demanded that every item be in its proper place." Yet Ivins was shocked by the clutter of Mir and at one stage, floating through the Kvant-1 module in search of Progress M-33 at its aft end, she became lost in a maze of garbage bags and broken equipment. One of the greatest ironies in her mind was that Kvant-1 was originally designed as an astrophysics module; now, after ten years in orbit, far from being filled with high-powered spectrometers and telescopes, it was a foul-smelling, engorged attic.

Blaha and Linenger stowed the crop of wheat from the Svet greenhouse aboard the orbiter to be returned to Earth. They also planted a fresh crop for Linenger's expedition. On 17 January, Baker and Jett pulsed Atlantis' thrusters to gather acceleration and vibration data for ISS operations and to support the Mir Structural Dynamics Experiment (MiSDE) and the Space Acceleration Measurement System (SAMS). They also began powering down several non-essential orbiter systems, in the hope of squeezing an additional day onto their ten-day mission. The additional day, if approved, would be added to the post-undocking phase of the flight and was requested following problems with the storage of TVIS data onto a faulty laptop computer. The Mission Management Team convened on the 18th to consider the extra day, but deferred their decision until the 20th, when it would be known whether the data collection could be completed within the framework of the existing flight plan.

Living and working aboard Mir was a joyful experience, aside from the hard work. John Grunsfeld likened it to "exploring a cave", and Jerry Linenger as not unlike an old, musty wine cellar, with its warren of cluttered passageways. For his part, Brent Jett remembered some advice from Alexander Kaleri: he should follow a line which ran through Kristall and offered a translation aid to 'swim' through bowels of the narrow module. "The Kristall is kind of like their attic," Jett said later. "They put a lot of extra equipment there, but then, once you got into the [multiple docking adaptor] and then into the base block, it's a lot more like what you would expect for a station." At times, the two crews would assemble for joint meals, as remembered by Jeff Wisoff. "It's customary when the Shuttle docks that the crews have a big meal together, and when our STS-81 crew arrived, the cosmonauts had recently had a supply ship come up. They had a big round ball of cheese and they had a fresh salami, or summer sausage. They had an axe and they were chopping this thing up and tossing out pieces to us. It was like being in a space deli." Jerry Linenger celebrated his 42nd birthday on 16 January and Marsha Ivins had brought up an inflatable cake, along with edible delicacies from Atlantis' pantry: dehydrated shrimp cocktails, red beans and rice. "That was a *very* fun occasion," said Wisoff.

After five days of joint activities, the hatches between Atlantis and Mir were closed at 7:46 am EST on 19 January. Following depressurisation and a good night's sleep, the Shuttle undocked from the station at 9:15 pm EST. Now in independent flight, the crew reassembled TVIS in the middeck and the treadmill was used by Baker, Jett and Blaha, acquiring the data to render it a success. The additional day was therefore declared unnecessary and Atlantis remained on track to land at KSC on the 22nd, after ten days in orbit. In their final days, the astronauts wrapped up the remainder of their scientific research aboard the Spacehab double module and tested a medical restraint system, which might someday be used aboard the ISS to move a sick or injured crew member from one module to another. An enormously successful mission ended with an enormously successful landing, albeit one orbit later than planned, due to dynamic weather activity, on KSC's Runway 33 at 9:23 am EST on 22 January. John Blaha had completed 128 days in space, eclipsing Norm Thagard's achievement and establishing himself as the most experienced US male spacefarer to date and the second most experienced US spacefarer of all time; that record being held by Shannon Lucid.

Mir was again returned to its normal, three-man strength, with Korzun, Kaleri and Linenger now settling into their own routine aboard the station. The Soyuz TM-25 crew of Vasili Tsibliyev and Alexander Lazutkin, together with German astronaut Reinhold Ewald, were scheduled to arrive in mid-February 1997, three weeks hence. For Linenger, the departure of the Shuttle carried with it a pang of loneliness, as he realised that his stay on Russia's space station had begun. "It was now just two Russian cosmonauts and myself," he wrote, "left to fend for ourselves, far removed from home and Earth."

It was the start of a troubled year and a trying time for the new partners.

3

Preparing for Space Station

VISITING A NATIONAL ICON

In April 1990, NASA launched what it hoped would be the jewel in the crown for astronomy: the $1.5 billion Hubble Space Telescope (HST). Today, of course, the telescope has earned itself a well-deserved reputation as one of the most successful space-based observatories ever launched. Across more than two decades of operations, its instruments have peered deeper into the cosmos than ever before. It has acquired images of distant galaxies, made breakthroughs in physics and cosmology by accurately determining the Universe's rate of expansion, detected planets around far-off stars, witnessed the impact of a comet into Jupiter, tracked cloud movements in the atmospheres of Uranus and Neptune and created the best currently achievable 'map' of the surface of Pluto.

With the advent of the Space Age, it was hardly surprising that plans for a space-based telescope would become an important step forward and an attractive option for the fields of astronomy and astrophysics. Yet the ideas long pre-dated even the launch of Sputnik. Shortly after the Second World War, physicist Lyman Spitzer of Yale University had argued that an orbiting telescope would offer enormous advantages over ground-based instruments, its abilities unimpaired by the distorting effect of Earth's atmosphere and its sensors able to detect high-energy emissions, including X-rays, from distant celestial sources. Following the creation of NASA, the first real efforts to develop a space telescope got underway and in 1975 the agency tried to sell the project to the politicians. Funding was initially denied by the House Appropriations Subcommittee, who reasoned that it was too ambitious, too expensive at around $400 million and lacked the required support from the National Academy of Sciences. This prompted large-scale lobbying from NASA and leading astronomers *and* a supportive report from the National Academy of Sciences. International co-operation was directed by Congress and the newly established European Space Agency (ESA) was invited to participate, with its role encompassing the creation of inexpensive solar panels for the telescope. The size of the mirror was reduced from 3 m to 2.4 m and together these measures halved the cost from $400 million to $200 million. There were other reasons for the reduction in mirror size. "The Shuttle could not lift a 3 m telescope to the required orbit," wrote Andrew Dunar and Stephen Waring in their book *Power to Explore*. "In addition, changing to a 2.4 m mirror would lessen fabrication costs by using manufacturing technologies developed for military spy satellites. The smaller mirror would also abbreviate polishing time from 3.5 years to 2.5 years."

In 1977 Congress granted approval for what was then known as the 'Large Space Telescope'. The primary candidates for the fabrication of the observatory's mirror were Perkin-Elmer Corporation, whose bid ran to $64.2 million, and Eastman Kodak, teamed with the defence contractor Itek, at almost $99.8 million. Despite being significantly higher, the Kodak-Itek joint bid included *two* independent tests of the grinding and polishing quality of the finished optics ... a 'double-checking' provision which Perkin-Elmer did not offer and which would not go unnoticed more than a decade later, when investigators dug into the cause of the telescope's unfortunate spherical aberration. Perkin-Elmer received approval from NASA to proceed with their bid in 1979. Meanwhile, Lockheed would build the spacecraft itself and the Europeans would make the solar arrays. In anticipation of the research bonanza, a Space Telescope Science Institute (STScI) was established at Johns Hopkins University in Baltimore, Maryland, in 1983 and the telescope itself was scheduled for launch by the Shuttle in 1985. By this time, it had been named in honour of the American astronomer Edwin P. Hubble, who, in the earlier part of the century, had not only conducted extensive research into the structure of stars and galaxies, but also made the surprising discovery that the Universe was expanding. The mirror was one of the most complex headaches of the project – both *before* and *after* launch. Optically, HST was a Cassegrain reflector and its two hyperbolic mirrors offered good imaging performance across what, for such a large telescope, was a wide field of view ... whilst also having shapes which were difficult to fabricate and test. Perkin-Elmer used custom polishing machines to precisely grind the mirror and, in case problems were encountered, NASA required the company to subcontract to Kodak to build a backup mirror using traditional polishing techniques. (The Kodak mirror is today on permanent display in the Smithsonian.)

In 1979, the construction of the Perkin-Elmer mirror began and was completed two years later, washed in hot, deionised water and coated with aluminium and protective magnesium fluoride. NASA remained sceptical about Perkin-Elmer's ability to competently fabricate the mirror and the delays ultimately pushed HST's launch back from April 1985 to first the summer and then the autumn of 1986. By this time, the total cost of the project had risen to a little more than $1 billion. At the time of its completion, HST housed five instruments: the Wide Field Planetary Camera (WFPC), the Goddard High Resolution Spectrograph (GHRS), the High Speed Photometer (HSP), the Faint Object Camera (FOC) and the Faint Object Spectrograph (FOS). These devices gave the telescope a range which encompassed not only the visible area of the electromagnetic spectrum, but also the ultraviolet. Physically, HST was a cylindrical spacecraft, measuring over 13 m in length and weighing 11,000 kg, which meant that it virtually filled the payload bay. It had been designed to be serviced by future Shuttle crews and, as such, was fitted with EVA-friendly hand holds, and would be deployed and retrieved using the Shuttle's Canadian-built Remote Manipulator System (RMS) mechanical arm.

By the time Challenger was lost in January 1986, further processing delays had slipped the launch of HST to October of that year. The problems faced by Perkin-Elmer have already been mentioned, but the manufacturer of the telescope's bodywork, Lockheed, had also suffered difficulties. By the end of 1985, HST was over-budget by 30 percent and three months behind schedule, bringing it dangerously close to breaking the 'ceiling' that Congress had imposed on its budget. If Challenger smashed the dreams of so many within America's space programme, it also provided breathing room for payload development. HST came through a major thermal vacuum test with flying colours in June 1986 and the enforced down time was used to add more powerful solar arrays, enhance redundancy capabilities, improve the software, install better connectors and replace the nickel-cadmium batteries that were prone to failure with nickel-hydrogen ones. As a result, by the start of 1990 the HST team felt supremely confident that their observatory heralded a new dawn in the study of astronomy. As long ago as 1983, NASA Administrator Jim Beggs had encouraged his subordinates to treat HST in terms of significance on a par with the Shuttle itself and had even labelled it "the eighth wonder of the world".

In the first few weeks after HST's April 1990 launch, the problems seemed reasonably benign: a few communications glitches, drifting star trackers and snagged coaxial cables were part and parcel in the process of wringing out a new spacecraft. More serious concerns arose when temperature changes bent materials in the solar arrays' booms, the effect of which was magnified by the orientation mechanism in such a way that it 'bounced' the whole telescope. The result was a 'jittering' in HST's images and, since the booms only stabilised in the final few minutes of orbital daylight, the pointing system was only able to meet its design specifications for a fraction of its orbit. Engineers at NASA's Marshall Space Flight Center (MSFC) in Huntsville, Alabama, worked with their counterparts at Lockheed to change the control program in the spacecraft's computer and successfully counteracted the vibrations. On 21 May, HST returned its first images of a double star in the constellation of Carina and these were lauded as being much clearer than were achievable with ground-based instruments.

Four weeks later, calamity befell the mission in a manner which could hardly have been anticipated. On 24 June, HST failed a focusing test. Its secondary mirror had been adjusted to focus the incoming light from a celestial source, but a fuzzy ring – like a halo – encircled even its best images, creating a blur. Additional tests revealed that the telescope was suffering from a 'spherical aberration' in its primary mirror; in essence, Perkin-Elmer had ground it to the *wrong* specification, removing too much glass and polishing it *too flat* ... by a mere *fiftieth* of the width of a human hair. The consequence was that HST was unable to acquire sharp images. With mounting horror, NASA realised that its attempts to sell its scientific showpiece on the basis of its ability to see further into the cosmos than ever before, with unprecedented clarity, now became very hollow indeed. The promised white knight of astronomy was now a white elephant. Even HST Program Scientist Ed Weiler of NASA Headquarters in

Washington, DC, admitted that it was comparable only to "a very good ground telescope on a very good night". MSFC staff were astounded and Senator Barbara Mikulski, a Democrat from Maryland, exploded that HST wasted taxpayers' money and was little more than "a techno-turkey". Meanwhile, Senator Al Gore – then a Democrat for Tennessee and later Vice President during the Clinton Administration – observed that, for the *second* time in less than *half* a decade, quality control shortcomings at NASA had been publicly exposed. The media had a field day. On 28 July 1990, the *New York Times* pointed out that, had Kodak-Itek's bid been accepted, then the mirror would have been subjected to *two* independent checks of its grinding and polishing accuracy, which certainly would have caught the error and enabled it to be rectified before launch. NASA responded by asserting that – with 20-20 hindsight – it would have cost in excess of $100 million to incorporate additional testing and independent checking of the telescope optics into Perkin-Elmer's contract, but the effect on the general public was the same. The once-proud agency was rendered a laughing-stock on late-night TV talk shows. David Letterman compiled a pejorative list of Top Ten Hubble Excuses, whilst others criticised MSFC for having been in charge of *both* the HST development *and* the Shuttle's Solid Rocket Boosters (SRBs) that caused the loss of Challenger. Several analysts noted that NASA's attitude had changed from the 1960s, in which problems were anticipated and incorporated into planning, into one where there was apparently little effort to prepare for unforeseen obstacles. In the words of space policy analyst John Logsdon of George Washington University, "the agency was not being honest with itself or with anyone else".

In orbit, the spherical aberration was particularly obvious in its effect on HST's WFPC and FOC instruments, both of which suffered in terms of spatial resolution and their ability to acquire images of individual celestial objects, including planets, star clusters and galaxies. Having said this, the aberration was well characterised and stable and, over time, astronomers were able to optimise the results obtained from HST by using sophisticated techniques, such as 'deconvolution', whereby software algorithms and microwave image processing methods were employed in an effort to remove many of the blurring effects of optical distortion. But the results were still less than ideal. Spectroscopy by the FOS and GHRS instruments was less severely affected, because the instruments required less focused light, and by increasing exposure times it became possible to gather valuable images. Nevertheless, by the end of 1991, HST had made almost 2,000 quality observations of hundreds of astronomical objects, including storms on Saturn and images of Pluto's moon, Charon.

At length, in December 1993, NASA triumphantly staged the first Shuttle servicing mission to HST, delivering the $50 million Corrective Optics Space Telescope Axial Replacement (COSTAR) device, whose ten, coin-sized mirrors corrected the spherical aberration and restored the potential of the affected instruments. STS-61 was expected to be the first of about five servicing missions, designed to extend HST's operational lifetime until at least 2005. (At the time of writing, in 2014, it is still fully operational.) In the aftermath of its successful repair, HST went on to observe some of the most significant events ever witnessed in the

3-01: The Hubble Space Telescope, pictured in Discovery's payload bay on STS-82.

history of astronomy and cosmology. Within months of STS-61, the rejuvenated images from the replacement Wide Field Planetary Camera (WFPC-2) revealed the first conclusive evidence of a Solar System-sized black hole, located 50 million light-years from Earth, in the M67 cluster. HST revealed a 'whirlpool' of hydrogen gas spinning around the black hole at an estimated 1.6 million km per hour. In June 1995, it was announced that the telescope had witnessed the final stages of stellar birth, yielding important insights into how proto-planetary disks and future planetary systems evolved around young stars. It also acquired the then-clearest views of Mars, as seen from Earth, across 103 million km of space, revealing wispy clouds and carbon dioxide 'frost' around its water-ice polar cap. Less than a year later, in March 1996, it was reported that HST had directly compiled the first map of the surface of Pluto, revealing significant differences in brightness (albedo), including a 'ragged' north polar cap, bisected by a dark strip, a bright, rotating spot, a cluster of dark spots and a bright linear formation. "It's fantastic," said Alan Stern of the Southwest Research Institute in San Antonio, Texas. "Hubble has brought Pluto from a fuzzy, distant dot of light to a world which we can begin to map and watch for surface changes." These observations alone served to cement the telescope's credentials as one of humanity's most important and culturally and intellectually significant creations.

"I would think the only other instrument that would rival it in historical value would be Galileo's original telescope," said astronaut Steve Hawley, who flew aboard both the initial Hubble deployment mission and the second servicing mission, "when he was able to look at Jupiter and detect the moons and fundamentally change the way we thought about the Universe at that time. I think it's not unfair to make that sort of comparison. Hubble is revolutionising how we think about the Universe we live in. I would say it's almost unrivalled in history."

At the time of STS-61, the second HST Servicing Mission (SM-2) was tentatively expected to occur on STS-88 in August 1997, featuring a seven-member crew to support four periods of Extravehicular Activity (EVA) on a nine-day flight. Launch of the $260 million mission was brought forward on the manifest, finally settling on STS-82 in February 1997, which marked the return to flight of Discovery following a lengthy period of modification and refurbishment in Palmdale, California. After arriving on the West Coast of the United States at the end of September 1995, Discovery underwent numerous upgrades, including the installation of the Orbiter Docking System (ODS) for missions to Mir and the International Space Station (ISS), which effectively replaced the middeck airlock. Consequently, she became the first orbiter to have the airlock removed from her middeck and the new combined ODS/airlock module was installed in the forward area of her payload bay. In addition, Discovery received improved payload bay floodlights, new shutters for her star trackers and general inspections, repairs of areas of structural corrosion and attention to her thermal-protection system. She returned to the Kennedy Space Center (KSC) in Florida in June 1996 to begin pre-flight processing for STS-82.

Like the first HST servicing mission, the four spacewalkers for STS-82 were announced some months before the assignment of the commander, pilot and flight engineer. In May 1995, astronaut Mark Charles Lee, a veteran spacewalker and then-chief of the EVA branch of the astronaut office, was named as the payload commander, with responsibility to oversee the development and planning of the four EVAs. His experience was immense. "Mark had worked Hubble tasks," recalled Steve Hawley, "back before we deployed Hubble in the first place." Lee came from Viroqua, Wisconsin, where he was born on 14 August 1952. After high school, he entered the Air Force Academy and graduated in civil engineering in 1974, then trained as a pilot at Laughlin Air Force Base in Texas. He flew the F-4 Phantom for several years at Kadena Air Base in Okinawa, Japan, as part of a tactical fighter squadron, and entered Massachusetts Institute of Technology in 1979 to earn a master's degree in mechanical engineering. His specialism was in graphite-epoxy advanced composites. Lee was then assigned to Hanscom Air Force Base in Massachusetts and resolved mechanical and material deficiencies that affected the combat readiness of Airborne Warning and Control System aircraft. In 1982, he upgraded to the new F-16 Fighting Falcon jet and served as an executive officer and flight commander. Two years later, in May 1984, he was selected as a member of NASA's tenth group of astronaut candidates. By the time of his assignment to the second HST servicing mission, Lee had flown three times. He performed the first test of the Simplified Aid for EVA Rescue (SAFER) backpack during his first spacewalk on STS-64 in September 1994.

Under Lee's direction, the four planned EVAs on STS-82 would install two key pieces of scientific hardware to further enhance HST's capabilities, both built by Ball Aerospace on behalf of NASA's Goddard Space Flight Center (GSFC) in Greenbelt, Maryland. The 318 kg Space Telescope Imaging Spectrograph (STIS) was designed to replace the GHRS, which would be removed from Hubble and returned to Earth. (The removal of the GHRS proved timely and fortuitous, for it shut itself down a few days before STS-82 launched, due to an internal electrical problem. A planned observation of Mars was cancelled, but managers decided it was not worth the effort to bring it back to life for just a handful of days.) With a spectral range which spanned the ultraviolet, visible and near-infrared wavelengths, the $120 million STIS could perform two-dimensional (rather than one-dimensional) spectroscopic 'mapping' across planets, stellar nebulae and entire galaxies and it had the potential to collect 30 times more spectral data and 500 times more spatial data than its predecessor. This high level of sensitivity was expected to resolve fine details in star formation regions of distant galaxies, identify supermassive black holes and investigate the distribution of matter in the Universe by studying quasar absorption lines. Also to be installed on STS-82 was the 347 kg Near-Infrared Camera and Multi-Object Spectrometer (NICMOS), which contained three separate cameras and acted as a spectrometer, a coronagraph and a polarimeter. Like STIS, it featured internal corrective optics to compensate for the effects of the telescope's spherical aberration. In orbit, the FOS would be removed from HST for subsequent return to Earth and the $100 million NICMOS installed in its place. The new instrument was designed to offer astronomers their first clear view of the Universe at near-infrared wavelengths (between 0.8 and 2.5 micrometres) using HST, thereby permitting studies of celestial objects previously too distant to be directly observed. Of particular interest to NICMOS investigators were brown dwarfs, which emit much of their light in the infrared. In order to attain the low operating temperatures of minus 215 degrees Celsius needed for its observations, the instrument featured a dewar of solid nitrogen and carbon dioxide, which would provide cooling for up to five years.

Although the perception of the public and some sectors of the media regarded SM-2 as 'routine', when compared to STS-61, it was actually one of the most complex Shuttle flights ever attempted. The crew would later stress that although the first servicing mission had removed several 'unknowns', there remained a tremendous amount of stress on their shoulders. "A number of the Orbital Replacement Units were made for [EVA] changeout when they were first designed," said Lee, "and they are fairly straightforward. Several of the others, though, require use of both hands and they get a little bit more difficult, because you don't know exactly what to expect. There are a lot of unknowns." In addition to the announcement of Lee as payload commander in May 1995, three other veteran astronauts were also assigned to join him in supporting the four spacewalks. Joining Lee on EVA-1 and EVA-3 was Steven Lee Smith, who was making his second Shuttle mission. Although his route to NASA followed a distinctly civilian path, Smith was intrigued by aviation from childhood. "When I was growing up in San Jose, California, my father and mother took me to the local airport to watch the airplanes take off and land," he recalled. "We'd sit at the end of the runway and watch those airplanes and *that* really first grabbed my aviation interest." Exploration was a powerful magnet for the young Smith and the images of America's first spacewalkers, floating in the void, prompted him to draw them as a boy. Remarkably, his parents kept all of his childhood drawings of rockets and astronauts. "That dream ... carried itself through high school and on to college," he told a NASA interviewer, years later. Little could he have foreseen that, by the end of his astronaut career, Smith would have established himself as the world's third most experienced spacewalker ... and although others have since surpassed him, at the time of writing in 2014, he remained within the worldwide 'Top Ten'.

Smith came from Phoenix, Arizona, born on 30 December 1958. His interest in science and mathematics started early and he would invite his former high school calculus teacher, Mr Lanborn, to the STS-68 launch. He "pushed us very hard," Smith recalled, reflecting on "being very inspired to understand math and technology, just by his incredible interest". Smith studied extensively at Stanford University, receiving his undergraduate and master's degrees in electrical engineering in 1981 and 1982, respectively, *and* a second master's credential in business administration in 1987. A keen sportsman, he was a seven-time high school and collegiate All-American in swimming and water polo and a two-time National Collegiate Athletic Association (NCAA) Champion at Stanford in water polo. He also captained the 1980 NCAA Championship team and learned to fly and scuba-dive. Betwixt his two advanced degrees, he worked for IBM as a technical group lead within the Large Scale Integration (semiconductor) Technology Group and, until 1989, served as a product manager with the Hardware and Systems Management Group. That same year, Smith joined NASA as a payload officer within the Mission Operations Directorate at the Johnson Space Center (JSC) in Houston, Texas, and was selected as an astronaut (on his *fifth* try) in March 1992. "Why do we fly in space?" he once rhetorically asked a NASA interviewer. "We fly in space to make people's lives better. *That's* the bottom line." On his first Shuttle mission, STS-68, Smith was part of a team which scoured Earth's surface with radar ... and potentially made millions of lives better.

With Lee and Smith assigned to perform the odd-numbered EVAs, their counterparts on EVA-2 and EVA-4 were civilians Gregory Jordan Harbaugh and Joseph Richard Tanner. The former, born in Cleveland, Ohio, on 15 April 1956, was a veteran spacewalker and had previously trained as a backup crewman for STS-61, the first Hubble servicing flight. After high school, Harbaugh gained a degree in aeronautical and astronautical engineering from Purdue University in 1978 and was immediately employed by NASA as an engineer at JSC. During the pre-Challenger era, Harbaugh supported Shuttle flight operations in Mission Control as a data processing systems officer and earned his master's degree in physical science from the University of Houston at Clear Lake in 1986. A year later, in June 1987, he was selected as an astronaut candidate. By the time of his assignment to STS-82, he had flown two Shuttle missions – including an EVA on STS-54 – and was only weeks away from his third flight, STS-71. After returning from his third mission, Harbaugh formally joined his STS-82 crewmates to begin preparation for one of the most challenging assignments of his career.

Harbaugh's EVA partner was Joe Tanner, who was born in Danville, Illinois, on 21 January 1950. Like Smith, he was a veteran of one previous Shuttle mission, but his knowledge of the reusable spacecraft's systems, even *before* joining the astronaut corps, was profound. Tanner had been an instructor in the Shuttle Training Aircraft (STA) and his whole aviation career seemed decidedly unusual. By his own admission, his family were "adventurers at heart". Growing up, Tanner drew excitement from the unfolding effort to plant American boots on the Moon in Project Apollo. A competitive swimmer at school and in college, Tanner earned a degree in mechanical engineering from the University of Illinois in 1973 and joined the Navy, won his wings and served as an A-7 Corsair attack pilot aboard the USS *Coral Sea*. He wrapped up his active military career as an advanced jet instructor in Pensacola, Florida, and entered JSC in 1984 as an aerospace engineer and research pilot. During the next several years, Tanner taught approach and landing techniques to astronaut candidates in the STA – a Grumman Gulfstream business jet, whose controls had been modified to closely mirror the Shuttle's handling characteristics during final approach – and instructed both pilots and mission specialists on the T-38 Talon. He also rose to become deputy chief of the Aircraft Operations Division. With almost 9,000 hours in military and NASA aircraft, Tanner had significantly more flight time than many Shuttle pilots and commanders. An unsuccessful application for admission into the astronaut corps in 1987 was followed by success in March 1992, but Tanner is distinct amongst NASA astronauts in that he had neither a background in military flight-test or an advanced degree. "My career path is a little bit different," he told a NASA interviewer, years later. "Academically, I'm not quite as qualified as most of the other people in the office. I guess I have to rely on my job experience!" He first flew on STS-66 in November 1994.

With four scheduled EVAs over four consecutive flight days, early planning foresaw the priority installation of the STIS and NICMOS instruments (each about the same size as a telephone booth) by Lee and Smith during EVA-1. Next day, Harbaugh and Tanner would venture outside to firstly install a pie-shaped Fine Guidance Sensor (FGS), replacing one of three existing devices, which was showing signs of mechanical wear. Positioned at 90-degree intervals around HST's circumference, these 227 kg sensors provided accurate pointing (to within 0.01 arc-seconds) towards desired astronomical targets, then held those targets in the field of view for observation. Additionally, the FGS were used for helping to determine the precise positions and motions of stars and other celestial objects, a field known as 'astrometry'. Following the FGS replacement, Harbaugh and Tanner would install a new Engineering/Science Tape Recorder (ESTR), replacing one of three units which

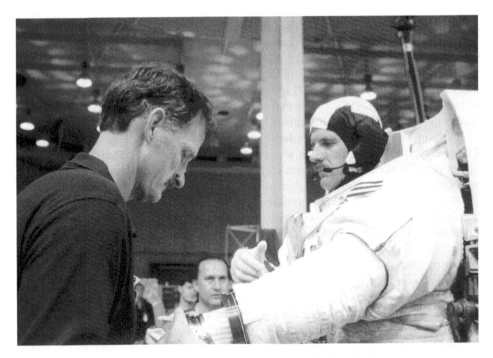

3-02: Astronaut Joe Tanner (left) assists crewmate Steve Smith with his suit, prior to an EVA training session.

stored data aboard the telescope whilst out of scheduled communication with the ground. Finally on EVA-2, they would fit an Optics Control Electronics Enhancement Kit (OCEK), which took the form of a cable to re-route signals to send commands through the Optical Control Electronics box to move the new adjustable mirror inside the FGS.

Lee and Smith would then venture outside for EVA-3 to replace another piece of HST's pointing system, known as a Reaction Wheel Assembly (RWA), together with one of four Data Interface Units (DIUs) that provided command and data interfaces between the data-management system and the telescope's subsystems. They would also remove a second ESTR and install a device known as a Solid State Recorder (SSR). Whereas HST previously stored its data on tape, the SSR had no reels or moving parts, which reduced the likelihood of wear, and its capacity was also ten times larger. It was intended that, after STS-82, the capacity and flexibility of the SSR would be utilised exclusively for scientific data, thus better accommodating the higher data rates of STIS and NICMOS. Finally, on EVA-4, Harbaugh and Tanner would replace one of two Solar Array Drive Electronics (SADE); built by ESA, this served to position HST's twin solar arrays. One SADE had already been removed and replaced by the STS-61 crew. That unit had been returned to Earth, refurbished to correct problems which had caused transistor failures and then launched aboard STS-82 to replace the second SADE. Finally, the spacewalkers would install Magnetic Sensing System (MSS) protective covers over key hardware to replace material which had degraded in the harsh atomic oxygen environment of low-Earth orbit. In the aftermath of the mission, a standard Servicing Mission Orbital Verification (SMOV) of the enhancements and new hardware – lasting between eight and ten weeks – was scheduled to take place, before HST returned to full science operations.

"Although the various servicing tasks are prioritised," NASA explained, "they have been designed to take into account the possibility that crew members may encounter unforeseen difficulties either in tasks or equipment that could change the pre-planned schedule of installation of various equipment components." It therefore made sense for Lee, Smith, Harbaugh and Tanner to 'cross-train' on all of their assigned tasks, and it also required no fewer than three space suit upper torsos and four lower torsos to be housed aboard Discovery for the mission. As the chief spacewalker ('EV1'), Lee would wear red stripes around the legs of his suit for identification, whilst Smith (EV2) was clad in a pure-white suit. The other two spacewalkers would carry slightly different patterns of red stripes: Harbaugh (EV3) carried a broken red stripe on the legs of his suit, whilst Tanner (EV4) had red-and-white diagonal hash lines. Smith and Tanner shared the upper torso of the third suit, but each had his own lower portion, complete with identification markings.

It came as something of a surprise to the spacewalkers from the first HST servicing mission that none of them would be carried over from SM-1 onto SM-2. "Working on STS-82 seemed like an obvious thing to do, to transfer knowledge to the next Hubble crew," recalled SM-1 veteran Jeff Hoffman in a NASA oral history. However, when his crew returned to Earth in December 1993, the EVA team was told, in no uncertain terms, that other astronauts would be given a chance on future HST missions. Nonetheless, Hoffman participated in several training exercises in the water tank with the STS-82 spacewalkers and later worked as the Capcom in Mission Control during their EVAs. "I think I was able to pass on a lot of information," he said. "I was basically the backup crew. In case one of them got sick, then I would have gone. Backup crews almost *never* fly, so that didn't happen, but it certainly made sense."

Joining Hoffman at the Capcom's console during STS-82 were Canadian astronauts Marc Garneau and Chris Hadfield. Since the Canadian-built RMS mechanical arm was a critical element of each of the Hubble servicing missions, Hoffman found 'interesting' notes from Garneau and Hadfield when he arrived on console at each shift changeover. "In each case, their notes were not all about the things that had been done by the EVA crew," Hoffman recalled, "but all the great things that the Canada arm had accomplished on the previous shift!" In fact, the importance of the RMS demanded that an experienced astronaut-operator was assigned to take control of it for the retrieval of HST, together with supporting the spacewalkers and deploying the telescope back into orbit.

By the beginning of 1996, with a little more than a year remaining before launch, the time came to assign the remainder of the STS-82 crew. Ken Bowersox, a veteran of three previous Shuttle flights, including piloting SM-1, was named as the mission commander, with fellow astronaut Scott 'Doc' Horowitz as pilot. The Mission Specialist Two seat would be taken by Steven Alan Hawley, who would serve as the

RMS operator and the flight engineer for ascent and re-entry. Nicknamed 'the Attack Astronomer', Hawley had been an astronaut since January 1978 and had flown three previous missions, including RMS operator for STS-31, the flight which deployed HST in April 1990. Hawley came from Ottawa, Kansas, where he was born on 12 December 1951. His love of learning traced its origins back at least two generations; *both* of his grandfathers were teachers – on the maternal side, in physics, and on the paternal side, in theology – whilst his father was a minister. As a child, Hawley found himself surrounded by physics books and his fascination with astronomy derived from the 'unconventional' nature of the subject. "All they were able to do," he told the NASA oral historian, "was look at whatever was out there and try to figure out what was going on by looking in clever ways or different ways". The fascination ran through at least one other family member, with his brother, John, becoming a theoretical astrophysicist. Hawley completed high school in 1969 and entered the University of Kansas to study physics and astronomy. Whilst there, he worked summers as a research assistant at the US Naval Observatory in Washington, DC, and at the National Radio Astronomy Observatory in West Virginia. Graduation in 1973 led him to the University of California at Santa Cruz for his doctorate, working at Lick Observatory on the spectrophotometry of gaseous nebulae and emission-line galaxies, with emphasis on chemical abundance determinations. At around the time that Hawley received his PhD in 1977, he spotted NASA's announcement for astronaut candidates and submitted an application.

He was interviewed at JSC and headed off for a two-year post-doctoral assignment at Cerro Tololo Inter-American Observatory in Chile. Several months into the post, in January 1978, he was at the observatory's headquarters, in the foothills of the Andes, when he received a call from George Abbey. He was in. Years later, fellow astronaut Mike Mullane paid homage to Hawley's immense brain-power: "Every checklist, including the two-volume malfunction massif," he wrote, "was in a virtual file drawer in his brain, ready for instant retrieval at the sound of a cockpit warning tone." He even did professional baseball umpiring during his summer vacations. After STS-31, his third Shuttle flight, Hawley left the astronaut corps to accept a position with NASA Ames Research Center in Moffett Field, California, but returned to JSC in August 1992 as deputy director of Flight Crew Operations. His immediate boss was former astronaut Dave Leestma. Early in 1996, Leestma asked Hawley to fly again on SM-2. "When I left to go to Ames, I expected that I would never fly again," Hawley admitted later to the NASA oral historian, "and when I got back from Ames, I expected I would never fly again, except knowing that ... there could be some chance that there'd be that opportunity if the right situation came along. But I remember thinking that it's not the sort of thing that it's appropriate for me in this job to lobby for." Leestma explained that the high-profile SM-2 flight required an experienced RMS operator, with advanced knowledge of HST. Hawley went home to talk to his wife and, with her blessing, returned to Leestma to accept the assignment. By early March, Bowersox, Horowitz and Hawley joined Lee, Smith, Harbaugh and Tanner to begin full-time training for the mission.

"For a robot arm operator, there's probably no greater task – no more rewarding task – than to do what we call a 'track-and-capture'," Hawley told the NASA oral historian. "When I deployed Hubble [on STS-31] . . . I picked it up out of the payload bay and let it go. It's a little different to go capture a *free-floating* object and to *berth* it and to move EVA guys around on the end of the arm. That's about as challenging as it gets in the robot arm world. Having been through dealing with Hubble once before, and knowing the dynamics of the big telescope on the end of the arm, knowing what to expect once you capture it, all that was helpful, so maybe I didn't feel as intimidated as I might have had that been my first experience with a large payload like that."

Training required the crew to spend a great deal of time in the water tanks of the 12 m-deep Neutral Buoyancy Simulator (NBS) at MSFC and the 7.6 m-deep Weightless Environment Training Facility (WET-F) at JSC. The construction of a large Neutral Buoyancy Laboratory (NBL) had begun in April 1995 at JSC, primarily to support EVA preparations for ISS assembly tasks, and according to Hawley STS-82 was the final Shuttle mission to use the NBS. Meanwhile, the NBL, later named in honour of the late astronaut Manley 'Sonny' Carter, was formally transferred to NASA ownership upon its completion in December 1996, just a few weeks before the launch of STS-82. Measuring 62 m long, 31 m wide and 12.3 m deep, and filled with 23.5 million litres of water, it represented the first NASA facility to be specifically engineered for the neutral buoyancy training of spacewalkers. "I didn't get much time in the NBL," Hawley said, "because we only had about a month before we launched." The training in the NBS, however, was highly realistic because the astronauts could operate in an 'integrated' manner with both the spacewalkers and the RMS. "You actually had the real space-suited crewmember on a real arm," Hawley recalled. "It's underwater, but you can actually manoeuvre him around. The experience was extremely valuable to learn the task that he had to do and the task that I had to do and learn how to communicate back and forth. In fact, we got to where you could run each EVA day pretty much end-to-end, as it would really work, with quite high fidelity." It contrasted sharply with Hawley's earlier STS-31 training, when the simulators depicted only computerised images of the telescope and the RMS and even HST itself was modelled with a large balloon, whose dynamic motions tended to lack realism.

To enable the multitude of servicing tasks, the STS-82 crew carried numerous tools and aids with them into orbit, which included handrails and handholds, transfer equipment, protective covers, tethers, grapple and stowage fixtures and foot restraints. A 43 cm titanium-aluminium Power Ratchet Tool (PRT), powered by a 28-volt battery, was designed for tasks requiring controlled torque, speed or turns and could be used in cases where right-angle access to HST components was required. The Multisetting Torque Limiter (MTL) was used in conjunction with the power tools and hand tools which interfaced with bolts or latches and was designed to prevent damage caused by the application of torque, whilst NASA's new Pistol Grip Tool (PGT) fulfilled a key recommendation from the SM-1 crew, who highlighted the need for a smaller, more efficient piece of equipment for very precise tasks. Computer-controlled, battery-powered and hand-held, the PGT could be

3-03: From the beginning of its development, Hubble was always intended to be refurbished by spacewalkers on Shuttle servicing missions.

programmed to control limits for torque, speed and number of turns. All told, and when including spares of the tools and various connectors, adjustable extensions and sockets, the STS-82 crew rose into orbit with more than 300 discrete pieces of equipment to service HST.

In spite of delays to several Shuttle flights in 1996, the mission held firm to its target launch date of February 1997. Originally scheduled to launch on the 13th, it actually flew two days earlier, on the 11th. At first glance, this seemed surprising, in view of numerous minor issues in the weeks preceding the mission. During Discovery's rollout to the pad in mid-January, the stack was halted whilst on the crawlerway when a 7.3 m, Y-shaped crack was identified on the deck plating of the Mobile Launch Platform (MLP). It ran from near one of the SRB flame openings to the corner of the platform, but despite its alarming appearance it was determined that the platform's integrity had not been compromised and the rollout was completed. Then, in the hours before liftoff, the loading of propellants into the ET met with delay, due to the need to assess the gaseous nitrogen purge system and monitor unusually high concentrations of trapped oxygen in the orbiter's midbody and payload bay. With the launch window due to open at 3:56 am EST on 11 February, at the opening of a 61-minute 'window', the precise time was adjusted slightly to 3:55:17 am at T-9 minutes, based on a final computation of HST's orbit. Exactly on time, to the very second, Discovery turned night into day across the Florida coast as she speared into one of the Shuttle's highest ever orbits, with an apogee of 574 km and a perigee of 475 km. A few hours later, they were trailing their quarry by 5,740 km, closing at a rate of about 560 km with each 90-minute orbit of the globe.

Following their first night's sleep, the astronauts began reducing the cabin pressure from its normal 101.3 kPal to around 70.3 kPal to prepare the spacewalkers for operating on the 29.6 kPal atmospheres of their space suits, as well as serving to clear nitrogen from their bloodstreams and thus avoiding a debilitating attack of the bends. (This protocol also eliminated the need for the EVA team to 'pre-breathe' pure oxygen for several hours.) Meanwhile, on Discovery's flight deck, Steve Hawley powered up the RMS arm and conducted a survey of the servicing equipment, housed on a Flight Support Structure (FSS) in the payload bay. Elsewhere, HST's controllers remotely closed the telescope's aperture door and secured its communications antennas.

Early on 13 February, about two hours before the scheduled retrieval, Discovery had reached a distance of about 14.8 km from the telescope, whose shiny surfaces made it literally glow and reflect the blues and whites of Earth. At about 1:00 am EST, Bowersox executed the Terminal Initiation manoeuvre, followed by a number of mid-course correction 'burns', to guide the orbiter toward its quarry from 'below', thereby minimising the risk of causing contamination from thruster firings. By the time he reached 730 m, at about 2:30 am, he assumed manual control and gradually brought the Shuttle to a position just 10 m away from the telescope. With the west coast of Mexico just coming into view, hundreds of kilometres below, Hawley then extended the RMS, grappled HST at 3:34 am and berthed it onto the FSS about 30 minutes later.

"You should have seen the expression on Dr Stevie's face," Bowersox told Mission Control. "It looked like he just shook hands with an old friend."

"We watched it from down here, Sox," replied Capcom Marc Garneau. "It was certainly an absolute thrill for us to see it on television and congratulations to all of you for an outstanding rendezvous and a great capture. We're looking forward to getting out there and starting work on that telescope."

Immediately after berthing on the lazy-susan-like FSS, a remote-controlled umbilical was mated to HST to provide temporary electrical power. With EVA-1 by Mark Lee and Steve Smith scheduled to begin late on 13 February, HST Program Scientist Ed Weiler was under no illusions as to the importance of the STIS and NICMOS installation. "All EVAs are important," he told journalists, "but [this] is really the Superbowl of EVAs. If that goes well, I think it will really put Hubble into a position of having world-class scientific capability well into the 21st century." Preparations for the first spacewalk proceeded briskly and it seemed likely that Lee and Smith would be outside around an hour earlier than planned. However, whilst in the external airlock, they were suddenly halted when one of the telescope's 12.2 m solar arrays *windmilled* through a quarter-turn, reorienting itself within about 60 seconds from a horizontal into a vertical configuration. It then stabilised.

"That was one of the more memorable things from the flight," Steve Hawley told the NASA oral historian, with a hint of understatement. "We coincidentally were trained to recognise an uncommanded slew of the solar arrays. If, for some reason, the solar array drive motor should fail in some manner, and they'll start to drive, you're trained to recognise that. You can send a command that will disable the motor so the solar arrays don't drive into something." Hawley and Joe Tanner were on Discovery's aft flight deck when the array moved. The two men exchanged glances. Both of them knew that the arrays were not supposed to drive *that* rapidly and it became clear that it was an uncommanded movement. With cameras focused on the airlock, few in Mission Control had any awareness of the motion and Tanner called the ground to advise them. "I was convinced that we probably wouldn't be going EVA today," admitted Hawley. "At the time, we didn't know what had caused it and they drove [the array] all the way to the stops."

It subsequently became clear that the new external airlock, which was flying aboard Discovery for the first time on STS-82, was part of the root cause. "There was actually an interior airlock that had been removed," said Hawley, "and they had replumbed the way the air is evacuated from the airlock volume. As luck would have it, the way the air exited was through a pipe that came out under the [telescope]. We didn't know it at the time, but what people on the ground figured was that air from the venting of the airlock impinged on the solar arrays and started them moving." The air had funnelled its way through thermal blankets in the payload bay, then vented directly onto the array, causing it to 'windmill' from horizontal to vertical. Fortunately, no damage was caused, but Hawley felt that if Mission Control *had* seen the event on television, they would have cancelled EVA-1. Fortunately, no damage was caused and about 75 minutes later than intended, at 11:34 pm EST, Lee and Smith floated out of the airlock to begin the first spacewalk of the mission.

Venturing into open space for the first time in his astronaut career, Smith was electrified. "Oh my *gosh* ... beautiful!" he radioed. "It was worth the wait!"

Quickly, the two men set to work, with Smith riding at the end of the RMS arm and Lee free-floating in the payload bay. They opened HST's aft shroud doors to remove the GHRS and FOS, both of which were stowed for return to Earth. Manipulating a suited crewman on the end of the mechanical arm was entirely new ground for Steve Hawley. "We had enough camera views that I could see what [he] was doing," Hawley remembered, years later. "I knew what his next step was going to be, so it was easy for me to put him where he needed to be." One of the biggest issues in training had been communication between Hawley and whoever was on the RMS, particularly if communication was *also* ongoing with the Capcom in Mission Control. "We developed some hand signals that we could use in the event that somebody else was talking on the radio," he said. "We didn't want to sit there and wait for that conversation to end before we could do the next task, because time is pretty critical." They also devised co-ordination systems, based on the orientations of the Shuttle and the EVA crewman, to ensure that movements were crisp and correct. "We spent a lot of time practicing being very disciplined in how we communicated," added Hawley. For example, in cases where the EVA crewman wanted to move to a different position within the payload bay, he might tell Hawley to move him port, starboard, forward or aft, relative to the orbiter. In other circumstances (for example, whilst working inside one of the telescope's bays), the EVA crewman might switch the point of reference to "body co-ordinates", with calls to Hawley such as "head-up", "feet-down", "left" or "right". Although complex at first, months of training and practice turned it into an elegant symphony. Hawley liked its lack of ambiguity, which eliminated errors. "My recollection," he said, "is that we never made a mistake the whole flight in terms of a bad command or going the wrong direction, because we thought it was very important and we practiced it a lot."

At about 2:00 am EST on the 14th, about two and a half hours into EVA-1, the STIS instrument had been successfully installed. Two hours later, NICMOS was also in place. Tolerances were incredibly tight, with no more than 1.2 cm clearance in some cases, requiring the free-floating Lee to verbally guide both Smith (who had a face-full of instrument) and Hawley (who was operating the RMS from the Shuttle's cabin). In fact, the question of *who* actually fitted the new devices proved a subject of some humour. Since Smith was physically *holding* the instrument, it might seem initially that he had installed it. Not so, joked Hawley, for it was *he* who was actually manoeuvring the RMS, with Smith *and* the instrument, into place. "All he can see is a face full of instrument," Hawley noted, "so my job was to manoeuvre him around. Steve Smith always said he inserted the instrument in the telescope, but I used to tell him, 'No, I really did. You were just *holding it*!'"

After installation, payload controllers verified that the health of the new instruments was good and after STS-82 they underwent several weeks of calibration. In the meantime, EVA-1 concluded at 6:16 am EST, after six hours and 42 minutes. The following night, it was the turn of Greg Harbaugh and Joe Tanner, who ventured into Discovery's payload bay almost an hour ahead of schedule at 10:25 pm

3-04: Greg Harbaugh and Joe Tanner work at the top of Hubble.

EST. Working quickly, the two men replaced the degraded FGS, which would be returned to Earth for refurbishment and re-installation by the SM-3 mission, then planned for November 1999. Next, they installed a replacement ESTR and OCEK. Late in the spacewalk, they noticed cracks and wear in the Teflon outer coat of the telescope's 17-layered thermal blanketing on the side facing towards the Sun and into the direction of travel. Some of the cracks were as long as 20 cm and were *not*, said Harbaugh, simply "tiny little spider cracks". Moreover, a small 'crater', caused by an orbital debris impact, was spotted in one of HST's antennas.

"In several places, it's cracked," said Tanner. "It's just gotten old, it looks like." Although there was no obvious evidence of crumbling, he recommended that care should be taken when touching the insulation. It was clear that although a more comprehensive fix would be necessary on SM-3, planning began to utilise some of the servicing time on STS-82 to effect repairs. In the meantime, at the end of EVA-2, Discovery's RCS thrusters were fired for 22 minutes to gently raise HST's altitude by about 3.3 km. The feasibility of executing such a manoeuvre had been demonstrated by the STS-78 and STS-79 crews and required Bowersox and Horowitz to fire the thrusters in a rocking, side-to-side motion to gradually ascend along the velocity vector. Discovery's tail-mounted RCS were fired continuously throughout the reboost, with those on either side of the nose firing sequentially at 60-second intervals. Two more reboosts were planned at the end of EVA-3 and EVA-4 to raise the telescope's orbit by a total of about 9.2 km. "If we were to have a camera outside actually looking at it, you would hardly notice this," said STS-82 Lead Flight

Director Jeff Bantle before launch. "Using the venier jets is going to be like normal attitude control. In fact, all of our reboost is planned while the [EVA] crew is outside doing the clean-up in the bay."

With the second spacewalk officially concluded at 5:52 am EST on 15 February, after seven hours and 27 minutes, the STS-82 crew fulfilled their minimum requirements for mission success. Lee and Smith were next, departing the airlock at 9:53 pm that same night to firstly change the DIU, which was never intended for orbital replacement. "The DIU is really a tough nut, because you have got a whole bunch of connectors you have to unfasten and reconnect and any one of them could be balky and create problems," said Harbaugh before launch. "It's is *not* a piece of cake." With Lee anchored to the RMS arm, and Smith free-floating in the payload bay, the DIU was replaced successfully. The spacewalkers then exchanged an ESTR for the new SSR and concluded EVA-3 by replacing one of the RWAs, which had failed a year earlier. Shortly after Lee and Smith returned inside Discovery at 5:04 am EST on 16 February, it was decided to insert an unscheduled fifth EVA to repair HST's damaged thermal insulation. A conversation between Bowersox and Jeff Bantle prompted the lead flight director to approve the 'short' EVA-5, lasting around four hours, to effect a fix.

In the meantime, Harbaugh and Tanner floated outside at 10:45 pm on the 16th to begin the fourth spacewalk, whose primary objectives included the replacement of the SADE for the solar arrays and the installation of covers over magnetometers at the 'top' of the telescope. This latter task required them to ascend about 18 m 'above' the payload bay and attach thermal blankets over two areas of degraded insulation around HST's light shield. From inside the crew cabin, Mark Lee compared Tanner's ascent to "riding your Harley", whilst the spacewalker admired the view and remarked that it was fortunate he did not suffer from a fear of heights. During the course of the EVA, Horowitz and Lee worked on Discovery's middeck to fabricate four new insulation 'patches' to be installed the following night. In total, 35 pages of instructions were transmitted up to the Shuttle, employing spare micrometeoroid insulation pieces, Kapton tape, parachute cord and alligator clips. By the time Harbaugh and Tanner returned inside the orbiter at 5:19 am EST on 17 February, after six hours and 34 minutes, the grand finale of EVA-5 had taken shape and was ready to go.

Following a good night's sleep, Lee and Smith left the airlock for what they expected to be the final time at 10:15 pm on the 17th. They attached thermal blankets onto three key equipment compartments at the top of the Support Systems Module, at HST's midpoint, where critical data-processing, electronics and instrument telemetry packages were housed. Specifically, Bay 7 carried mechanisms to control the solar arrays, Bay 8 held pointing electronics and a retrieval mode gyro assembly and Bay 10 accommodated the science instrument control and data-handling subsystems. All three required protection to prevent problems in the future. After Lee and Smith had cleaned up their work site and returned to the airlock, flight controllers noticed a potential problem with one of the four RWAs. Although the RWA fitted on EVA-3 was operating without problems, one of its older siblings had begun to exhibit discrepancies. "They would like to perform some testing on it that may take 15-30 minutes, just to assure themselves that nothing is wrong with it," Capcom Marc Garneau told the spacewalkers. "In the meantime, because we want to keep open the possibility of changing it out today, we'd like to hold off doing anything further." Lee and Smith entered the airlock and connected their suits' utilities to the orbiter's Servicing and Cooling Umbilical (SCU) to preserve their backpack resources.

Had the call come from the ground to replace the RWA, it would be have been necessary to repressurise the airlock, open the internal hatch and retrieve a spare unit from Discovery's middeck. Fortunately, engineers ran a series of tests and powered up the three 'old' RWAs to assess them for problems. "They go immediately into an idle mode, where if everything were perfect and there were no drifts or torques, they would just sit there at zero," explained Ed Weiler, "but all reaction wheels have a natural bias and two of them went up to about 20-40 rpm. This one sat about 0.3 rpm, then it went down to zero and then it slowly started to go up. We had to make a call." Engineers worked to modify software originally written to test the new RWA so that it could test the troublesome unit, which required a couple of hours. At length, when the software commanded the RWA to put high torque on the wheel, "the thing just took off," in Weiler's words, "and we knew it was just fine". A record-breaking sixth EVA on a single mission – never previously achieved by the Shuttle – evaded the STS-82 crew. Lee and Smith concluded EVA-5 after five hours and 17 minutes, bringing the SM-2 spacewalking total to 33 hours and 11 minutes. A final burn of the RCS thrusters increased their orbit by about 5.5 km to a final deployment altitude of 593×617 km. "We did 82 minutes of reboost," Steve Hawley recalled years later, "and got Hubble, I guess, as high as it had been." Ed Weiler was ecstatic, describing the observatory as no longer the 'original' Hubble Space Telescope ... but as a brand-new instrument: "You can call it Hubble-2".

After a job exceptionally well done, HST's solar arrays were oriented towards the Sun to provide electrical power and recharge its batteries. As Steve Hawley grappled the telescope with the mechanical arm, GSFC controllers commanded its aperture door to open and HST and Discovery parted company at 1:41 am EST on 19 February, a day later than planned, due to the extra spacewalk. Shortly afterwards, the telescope resumed standard operations and began processing commands to the ground through the Tracking and Data Relay Satellite (TDRS) system. Two days later, it was time for the final curtain to fall on STS-82. Originally, Discovery was scheduled to land at KSC at 1:50 am EST on 21 February, but Entry Flight Director Wayne Hale called off the first attempt owing to the presence of off-shore showers and low cloud cover over the Shuttle Landing Facility (SLF) runway. The next opportunity to land was at 3:32 am, requiring the irreversible de-orbit burn to occur at 2:21 am. At length, Bowersox was given the "Go" to begin the 3.5-minute burn, committing Discovery to a 71-minute-long descent. Re-entering the upper atmosphere over the Pacific Ocean, the Shuttle swept across the entire continental United States and appeared as a bright streak as it passed over JSC.

"I think we just flew over Houston," Bowersox radioed at one stage.

"You certainly did," replied Capcom Kevin Kregel, "and you lit up the entire sky with the orbiter and its trail. It was pretty impressive."

"It was a pretty good view from here, too," said Bowersox. "We almost saw the Astrodome."

The touchdown on Runway 33 at 3:32 am was aided for the first time by the presence of 52 halogen lights, positioned at 60 m intervals along the centreline.

Although overshadowed by the success of its predecessor, SM-1, the second HST servicing mission had proven equally as challenging for the crew and the thousands of ground personnel who put it together. Yet the work undertaken by Discovery's crew had turned the telescope from a 1970s spacecraft with 1980s optical technology into an observatory for the 21st century, with more modern instruments. "Three hundred years from now," explained project scientist David Leckrone, "none of us in all likelihood will be remembered as individuals, but certainly the Hubble Space Telescope will be remembered ... as a high point in human civilisation. That's an awe-inspiring thought and something that motivates us to do our very best for Hubble and for science."

The astronauts themselves were also pleased and relieved that their mission was complete. After almost two years of training, payload commander Mark Lee declared that he was ready to buy his crewmates a drink. "Up here," he said whilst in orbit, "we've got some orange mango and some lemonade, but that's as stiff as it gets. So I'm ready for a margarita!"

BROKEN ANKLE, BROKEN PLANS, BROKEN RECORD

Six weeks after STS-82, Columbia stood ready for a complex, 16-day research mission, one of whose astronauts had begun 1997 in the likelihood that he would not fly *at all* that year. By the end of July, however, Don Thomas and the entire STS-83 crew would have flown not once, but twice.

The payload for STS-83 – the Microgravity Science Laboratory (MSL)-1 – was expected to be not only the penultimate voyage of the Spacelab module, but also the last Shuttle flight to be fully devoted to materials processing, combustion science and fluid physics research. All future investigations in these disciplines, it was anticipated, would be conducted by long-duration crews aboard the ISS, whose inaugural construction flight was scheduled for mid-1998. In many ways, STS-83 was the last roll of the dice for many physical science investigators and its importance was illustrated by a highly experienced crew. In January 1996, NASA assigned payload commander Janice Voss, mission specialist Thomas and payload specialists Roger Crouch and Greg Linteris to support more than 25 experimental payloads inside the Spacelab module.

Their backgrounds highlighted the nature of the MSL-1 investigations. Engineer Janice Voss, making her third Shuttle flight, served as the payload commander. Also making his third mission was Don Thomas, who had served as a crewman aboard the second International Microgravity Laboratory (IML-2) Spacelab flight in July 1994 and, more recently, aboard STS-70. Three payload specialist candidates were also announced in January 1996 to support two positions on the crew. Assigned to fly on STS-83 were physicist Roger Crouch of NASA's Office of Life and

Microgravity Sciences and Applications at the agency's Washington, DC, headquarters, and mechanical and aerospace engineer Greg Linteris of the National Institute of Standards and Technology in Gaithersburg, Maryland. Backing both of them up was engineer Paul Ronney of the University of Southern California, who would support the MSL-1 mission as a crew interface co-ordinator at the Spacelab Mission Operations Control Center at MSFC.

Roger Keith Crouch was born in Jamestown, Tennessee, on 12 September 1940. ("One of the heaviest Southern accents I've ever had the opportunity to fly with," was STS-83 Commander Jim Halsell's light-hearted summary of Crouch.) He attended high school at Alvin C. York Institute – today the only comprehensive secondary school in the United States financed and operated by the state government – and completed his undergraduate studies in physics at Tennessee Polytechnic Institute. Receiving his degree in 1962, Crouch pursued an extensive research career with NASA's Langley Research Center in Hampton, Virginia, specialising in semiconductor crystal growth, electrical and optical properties of materials, electronic devices for remote-sensing and flat-panel displays and heat-shield protection systems for re-entering space vehicles. He went on to gain a master's credential and a PhD in physics, both from Virginia Polytechnic Institute, in 1968 and 1971, respectively. Crouch was a visiting scientist at Massachusetts Institute of Technology in 1979-1980. Several years later, in 1985, he was appointed chief scientist of NASA's Microgravity Space and Applications Division and served as Program Scientist for five discrete Spacelab missions. In January 1989 (the same year that NASA awarded him its Exceptional Performance Award), Crouch was selected as a candidate for the IML-1 mission, for which he served as one of the backup payload specialists. In his later career, he served as NASA's senior scientist for the ISS.

When the remainder of the seven-member STS-83 crew was announced in May 1996, it included STS-69 veteran Mike Gernhardt as the flight engineer and thus boasted no fewer than five PhDs. In command of the mission was James Donald Halsell Jr, who was embarking on his third Shuttle flight.

Seated to Halsell's right side on Columbia's flight deck for STS-83 was Susan Leigh Still, who became only the second woman ever to serve as a Shuttle pilot. The daughter of a prominent burns surgeon and one of ten children, Still was born in Augusta, Georgia, on 24 October 1961. After completing high school in Massachusetts, she attended Embry-Riddle Aeronautical University and gained a degree in aeronautical engineering in 1982, followed by a master's degree in aerospace engineering from Georgia Institute of Technology in 1985. Upon graduation, Still initially worked for Lockheed as a wind tunnel project officer, but later entered the Navy and was designated as a naval aviator in 1987. She served as a flight instructor in the TA-4J Skyhawk and later flew EA-6A Electric Intruders for Tactical Electronic Warfare Squadron 33 (VAQ-33) in Key West, Florida. Her next step was test pilot school at Patuxent River, Maryland (from which she emerged as a Distinguished Graduate), followed by training as an F-14 Tomcat fighter pilot. Selected as an astronaut candidate by NASA in December 1994, Still was the first member of her class to receive a flight assignment.

With all seven STS-83 crew members thus named, the first wave of ill-fortune hit the mission on 29 January 1997, when Don Thomas slipped down some stairs and broke his ankle, following an emergency egress training session at JSC. Unlike the Russians, NASA no longer assigned backup crews, but on this occasion (and barely nine weeks before launch) it was considered necessary to quickly train another astronaut to stand in for Thomas, if necessary. By mid-February, it was official: STS-73 veteran Catherine 'Cady' Coleman would serve as his backup. "We are hopeful that Don will be cleared for flight," said Director of Flight Crew Operations Dave Leestma. "He is an experienced astronaut with the majority of his required training for this flight already complete. The decision to assign Cady as backup was made to protect all available options." Thomas was also assigned, with Mike Gernhardt, as one of the two contingency EVA crew members; Coleman immediately began refresher classes and familiarisation sessions with more than 24 MSL-1 research facilities. "Cady's previous experience makes the amount of training required to bring her up to speed minimal," explained Leestma. Thomas, meanwhile, was determined to be ready in time for STS-83's scheduled 3 April launch date. "I'm in a period of pretty heavy physical therapy right now," he told journalists on 13 March, "spending about five or six hours a day walking in swimming pools, walking with the cast and without the cast, just getting my mobility strength back. We've got three weeks until launch and there's no doubt in my mind or the doctor's mind that I'll be ready in time."

By the beginning of April, the Mission Management Team opted to postpone the launch by 24 hours until the 4th, due to an improperly insulated water coolant line in Columbia's payload bay. There existed a possibility that it might freeze during the 16-day flight and technicians worked to install new thermal blankets. With liftoff now rescheduled for 2:00 pm EST, the only constraint seemed to be a slight chance of rain showers generated by sea breezes. Mission managers briefly considered, but later discarded, an option to launch STS-83 an hour earlier than planned at 1:07 pm to provide more daylight at the Transoceanic Abort Landing (TAL) site at Banjul in West Africa and alleviate concerns about delamination of a backup antenna there. Following a slight delay in evacuating the closeout crew from the launch pad's danger area and a problem with excessive concentrations of oxygen in the payload bay, Columbia thundered into orbit at 2:20 pm. Her ascent was the first to be monitored by a new Laser Imaging System, under development by Naval Research and Development. It was hoped that the new system, which would be provided to the Air Force's 45th Space Wing, would enhance the tracking of future launches. Prior to STS-83, range safety officers monitored Shuttle ascents by optical means, which were often impaired by engine plumes, low-level clouds and fog. By illuminating part of the vehicle with a non-invasive laser beam, it was hoped to acquire clear, defined imagery, even in low-visibility conditions. For the inaugural test of the hardware on STS-83, imaging equipment was located at three sites at KSC and Cape Canaveral Air Force Station, which illuminated Columbia's aft compartment and the SRB 'skirts' at specific stages during the countdown and ascent.

Within hours of achieving orbit, the 'blue' team of Gernhardt, Voss and Crouch set to work opening the hatch to the Spacelab module and activating its systems for a

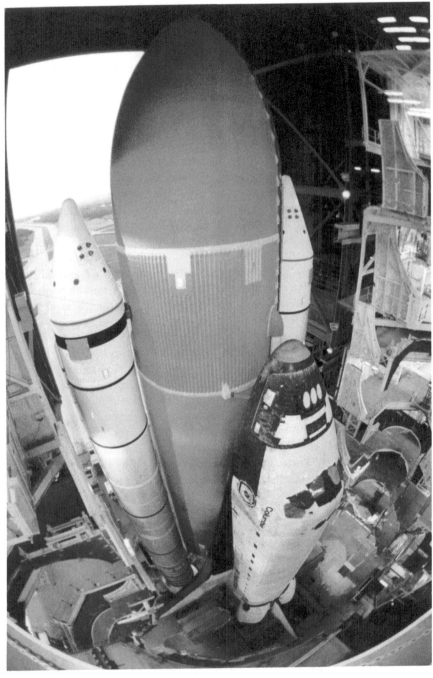

3-05: The STS-83 stack is prepared for rollout from the cavernous Vehicle Assembly Building (VAB), bound for the launch pad.

planned 16 days of around-the-clock, dual-shift research. The primary thrust of MSL-1 was threefold: crystal growth, combustion physics and the development of techniques to produce stronger and more resilient metals and alloys. Additionally, STS-83 would evaluate some of the hardware, facilities and procedures that NASA intended to employ aboard the ISS. It had already become clear that physical processes ordinarily masked by gravity on Earth were virtually eliminated in space, thus making it possible for scientists to conduct hitherto impossible experiments.

An important element of this work was the ability to grow larger and purer crystals of proteins ranging from insulin to HIV-Reverse-Transcriptase in order to determine their three-dimensional structural blueprints. By unlocking the crystals' structural details in this manner, biochemists hoped to better understand how they fitted into the overall biology of the human body. It was anticipated that MSL-1's growth facilities would be capable of processing upwards of 1,500 protein crystal samples, ultimately to address the 'social cost' of diseases such as cancer, diabetes, Alzheimer's and AIDS, which were estimated in 1997 to cost the United States around $900 billion per year.

Elsewhere, experiments focused on the behaviour of the combustion process in microgravity and its significance was highlighted by the presence of combustion expert Gregory Thomas Linteris as the mission's second payload specialist. Born in Englewood, New Jersey, on 4 October 1957 – the day that the Soviet Union launched Sputnik and inaugurated the 'Space Age' – Linteris was a specialist in combustion physics, chemical kinetics, spectroscopy and heat transfer. He earned his under-graduate degree in chemical engineering from Princeton in 1979, followed by a master's degree in mechanical engineering from Stanford in 1984, for which he gained fourth place in the James F. Lincoln National Design Competition. Following receipt of his master's qualification, Linteris' doctoral work focused on the high-temperature chemical kinetics of combustion reactions in a turbulent chemical kinetic flow reactor using laser-induced fluorescence and laser absorption. During this period, he received the Grumman Prize for Excellence in Research in 1988 and was awarded his PhD in mechanical and aerospace engineering two years later. He subsequently joined the University of California at San Diego, studying droplet dynamics and performing numerical and analytical modelling of the gas-phase interaction region of solid rocket propellants. At the time of his selection to join the MSL-1 crew, Linteris was working on advanced fire suppressants and the mechanisms of chemical inhibitors at the National Institute of Standards and Technology in Gaithersburg, Maryland.

It was anticipated that developing a better understanding of the peculiarities of different types of fuel, and the fires they produce, could lead to increased efficiency and reduced emissions within internal combustion engines. In the United States, in 1997, the annual expenditure on crude oil was estimated by the American Petroleum Institute at close to $200 billion. The process of combustion was not only a major player in converting the chemical energy in fuel into useful thermal and mechanical energy, but also a major contributor to air pollution through the emission of nitrogen oxides, carbon monoxide, unburned hydrocarbons and particulates. "Minimised emissions," said Fred Dryer of Princeton University, "and best miles-per-gallon require us to carefully control and tailor the combustion process." Prior to MSL-1, this control and tailoring could only be done with sophisticated computers, but experiments aboard STS-83 provided an opportunity to analyse theoretical predictions and develop new models. For example, theories held that small fuel droplets should go through three separate 'regimes' during combustion. One of these, known as a 'quasi-steady state', had been frequently studied on the ground: the square of the droplet/flame diameters decreased with time in a linear fashion and eventually extinguished itself when it became too small to support itself. The MSL-1 experiments not only provided additional data on that regime in far greater depth than was possible on Earth, but also investigated the other two regimes. On Earth, a mere one-percent increase in fuel efficiency translated into saving of 100 million barrels per year or $5.5 million per day.

One of the most important facilities aboard Columbia was the Combustion Module (CM)-1, developed by NASA's Lewis Research Center (today's Glenn Research Center) in Cleveland, Ohio, which supported the Laminar Soot Processes (LSP) and the Structure of Flame Balls at Low-Lewis Number (SOFBALL) experiments. Despite minor troubles with a cable configuration, LSP was activated by Linteris on 5 April. The experiment sought to gather data on flame shapes, together with the quantities, temperatures and types of soot produced under varying conditions. It supported ongoing efforts to develop methods of controlling fires on Earth and limit the number of deaths caused by carbon monoxide poisoning, associated with soot. Within a day of powering-up the experiment, Linteris was rewarded with his first glimpse of the concentration and structure of soot from a fire burning in the microgravity environment. Another major study in this area was the Droplet Combustion Apparatus (DCA), which occupied much of Janice Voss' time during one of her early MSL-1 shifts. It housed a variety of experiments to investigate burning drops of different fuels and monitor conditions at the instant of their extinction. Inside the DCA, which filled one rack in the Spacelab module, was the Droplet Combustion Experiment (DCE) to explore the fundamental combustion aspects of isolated fuel drops under varying pressures and oxygen concentrations. Each drop ranged between 2 and 5 mm in diameter.

In most practical combustion devices, liquid fuels are mixed with oxidisers and burned in the form of sprays. An essential prerequisite for an understanding of such 'spray combustion' and its application to the design of efficient and clean combustion systems is knowledge of the laws governing droplet combustion. In the absence of buoyancy-induced convection currents, a droplet ignited in microgravity burns with spherical symmetry and yields a simple, one-dimensional system capable of being very precisely modelled. Previous experiments on the ground had produced data for drops no larger than 3 mm in diameter, but the microgravity environment available in the MSL-1 module provided scientists with an opportunity to better investigate the complicated interactions of physical and chemical processes during droplet combustion.

"Everything's going great!" exulted MSL-1 Mission Manager Teresa Vanhooser as these and other experiments got underway. However, STS-83's fortunes were about to drastically alter for the worse. Shortly after reaching orbit, Halsell, Still,

Gernhardt and Mission Control had been monitoring erratic behaviour from one of Columbia's three fuel cells. Mounted beneath the payload bay floor, these cells used a reaction of oxygen and hydrogen to generate electricity in order to support the Shuttle's systems, run the MSL-1 experiments and, as a byproduct, yield drinking water for the crew. One cell was technically sufficient to support orbital and landing operations, but flight rules dictated that all three had to be fully functional for a mission to continue. Each cell had three 'stacks', made up of two banks of 16 cells apiece. In one of Fuel Cell No. 2's stacks, the difference in output voltage between the two banks had shown signs of increasing. It had been noted before launch, but had been cleared for flight. Late on 5 April, Halsell and Still adjusted Columbia's electrical system configuration to reduce demands on the ailing cell. This allowed mission controllers to stabilise it for ongoing analysis. Overnight, the rate of change in the cell slowed from five millivolts per hour to around two millivolts, but continued to exhibit a slight upward trend. "There's always a difference between the two halves of the stack," Mission Operations representative Jeff Bantle told journalists, "but we're noticing a changing difference. Actually, that changing difference has levelled off a lot, so the degradation was greater [in] the first 12 hours of the mission."

Bantle's main concern was that if the difference between the two stacks increased to 300 millivolts (and it was touching 250 millivolts by the evening of 5 April), it could force the crew to shut down Fuel Cell No. 2 entirely. "The concern is degradation in a single cell," he said. "If it degrades enough, rather than getting power out from the cell, you would have power output *into* that cell. You could actually have crossover and localised heating, exchange of hydrogen and oxygen within the cell, and could even have a localised fire. That's the very worst case. That's why we have flight rules that are very conservative to try to avoid and try to shut down and safe a fuel cell before you would ever get to that point." Early on 6 April, Halsell and Still performed a manual purge of the cell, lasting ten minutes, at Mission Control's request. However, as the situation showed no sign of improving, it was decided later that afternoon to declare a Minimum Duration Flight and bring Columbia back to Earth at the earliest possible opportunity. In spite of the disappointment, Mission Control retained its collective sense of humour, by faxing the astronauts a tongue-in-cheek list of the Top Ten 'real' reasons why they would be coming home early. Lead contenders on the list included running out of 'Columbian' coffee, forgetting to record the latest episode of *Friends*, forgetting to do their taxes before launch and – obviously – that the entire situation was an April Fool's prank.

Returning to the business at hand, Fuel Cell No. 2 – which had flown on three previous missions, without problems – was shut down on 6 April, followed by several other pieces of non-critical equipment to provide additional power to support the MSL-1 experiments for as long as possible. The lights in the Spacelab module were turned off and the astronauts found themselves running experiments by torchlight. According to one fluid systems engineer, Fuel Cell No. 2 had displayed a 500-millivolt discrepancy between its two stacks before it was even switched on, some 12 hours before STS-83 was launched. Similarly abnormal cell behaviour had been noticed during two missions by Atlantis, but on both occasions the discrepancy levelled off to well within safety guidelines shortly after it had been switched on and begun to take its full electrical load. With that prior experience in mind, engineers activated Columbia's fuel cells for STS-83 and, sure enough, No. 2 settled down to 'normal' levels. It performed normally throughout ascent, but the discrepancy reared its head again in orbit.

As a precaution, one of the fuel cells assigned to Atlantis' STS-84 mission, planned for May 1997, was removed for checks after displaying a similar behavioural signature. Aboard Columbia, Jim Halsell summed up the nature of the incident, describing the Shuttle as "one of the original electric fly-by-wire airplanes" and adding that the loss of a third of their electricity-generating potential had to be taken seriously. Nevertheless, the entire crew reacted with "shock and disbelief", according to Don Thomas, at the decision to cut short the flight. The fuel cell problem devastated the carefully choreographed 16 days of research, as scientists scrambled to reprioritise their schedules to make the most of the one or two more days available before coming back to Earth. Already, efforts were underway to lobby NASA to stage a rapid reflight of MSL-1, later in 1997.

Certainly, in spite of the troubles, the research work was proceeding superbly. Roger Crouch worked with the SOFBALL investigation, which was design to explore the conditions under which a 'stable' flame ball can exist and whether heat loss is responsible for its stability whilst burning. During the first run, a mixture of hydrogen, oxygen and carbon dioxide was burned within the facility for the entire 500-second limit. This was particularly significant, because they represented the lowest-temperature and weakest flames ever burned, with the most diluted mixtures, according to alternate payload specialist and principal investigator Paul Ronney. "These mixtures will not burn in Earth's gravity," he said. "We have known that burning weaker mixtures increases efficiency, but not much is known about the burning limits of these mixtures." As well as offering insights into the combustion process, it was anticipated that SOFBALL results would help to enhance theoretical models. By the time of MSL-1, anomalous 'flame balls' – essentially stable, stationary, spherically symmetric flames in combustible gas mixtures – were receiving significant attention. By examining their behaviour in microgravity, it was hoped that more efficient combustion engines, emitting fewer atmospheric pollutants, could be produced. During typical SOFBALL runs, a chamber was filled with a weakly combustible gas (hydrogen and oxygen, highly diluted with an inert gas) and ignited. The resultant flames and their motions were then imaged and recorded by video cameras, radiometers, thermocouples and pressure transducers. Unfortunately, the shortened nature of STS-83 meant that only two of 17 planned SOFBALL experiments could be completed.

Elsewhere in the Spacelab module, other experiments focused on materials processing, as part of efforts to develop techniques for manufacturing stronger and more resilient metals, alloys, ceramics and glasses. Two facilities used to support this work – the Japanese-built Large Isothermal Furnace (LIF) and Germany's Electromagnetic Containerless Processing Facility (TEMPUS) – had previously flown aboard the IML-2 mission in July 1994. The former heated and rapidly cooled materials, including ceramic-metallic composites and semiconducting alloys, in various temperature ranges to identify relationships between their structures and physical properties. Usually, the furnace followed a pre-programmed cycle, reaching a maximum temperature of about 1,600 degrees Celsius, and was activated by Don Thomas on 5 April. Among its early investigations were a study of the diffusion of impurities in molten salts and the Liquid Phase Sintering experiment, which tested theories of how liquefied materials form a mixture without reaching the melting point of the new alloy combination. Other LIF activities diffused molten semiconductors as part of efforts to explore how uniformly their constituents mixed during the cooling process. Diffusion studies have many terrestrial applications, from very small movements in plasma to massive depletions in Earth's ozone layer.

Meanwhile, the TEMPUS facility provided a levitation melting device for processing metals. It was activated by Roger Crouch, late on 4 April, a few hours after Columbia entered orbit. Containerless processing was also considered hugely advantageous, because Earth-based processes involving liquids can cause them to be affected by the properties of their holding vessel; in microgravity, on the other hand, positioning and control can be effected with greater precision. This, in turn, reduced motions within the sample liquid and was less intrusive upon the physical phenomena under study. Moreover, containerless processing was known to eliminate contamination of the sample caused by the material in the container's walls. TEMPUS was used to study the 'undercooling' and rapid solidification of metals and alloys, which typically occurs when a solid is melted into a liquid, then cooled below its normal freezing point without solidifying. The release of latent heat in the phase change is delayed and the material enters a 'metastable' state.

Following the decision to bring STS-83 home early, TEMPUS Assistant Investigator William Hoffmeister oversaw the effort to reprioritise several of its experiment runs to acquire as much data as possible. The crew was able to activate, observe and complete one run by melting a zirconium metal sample and levitating it, as part of efforts to examine the relationship between internal flows in liquids and the amount of undercooling that a sample can tolerate before it solidifies. TEMPUS provided the scope for physically manipulating samples, controlling rotations and oscillations and even 'squeezing' them by the application of an electric field. The experiments involving zirconium, in particular, were expected to determine the behaviour of this strong, ductile, refractory metal, which has found terrestrial applications in nuclear reactors and chemical processing equipment. With the impending landing of Columbia, the TEMPUS team had good knowledge of how to reprioritise their schedule to make the best of unexpected events. On the facility's IML-2 mission, a misaligned coil had forced investigators to shorten and replan several experiments.

Other areas of research aboard MSL-1 included plant growth employing the Plant Generic Bioprocessing Apparatus (PGBA) provided by BioServe, one of NASA's commercial space centres at the University of Colorado at Boulder. Previously flown aboard STS-77, the unit offered a highly controlled environment with lighting, temperature and gas-exchange functions, together with a 'nutrient pack' to supply water and other nutrients to nine different plant species. One of these, selected through a project with a Brazilian research group, was a member of the black pepper family. Also making its second Shuttle flight was the Middeck Glovebox, which supported experiments in fluid physics, materials processing and combustion science. In Columbia's payload bay, the Cryogenic Flexible Diode (CRYOFD) tested a pair of experimental heat pipes for future spacecraft and instrument-pointing applications.

In spite of the disappointment at the early return to Earth, informal plans were already afoot whilst Columbia was in orbit to refly the mission, later in 1997. "There were rumours already flying," noted Don Thomas on his website, OhioAstronaut.com, "that after fixing the problem NASA would be re-flying our crew in a few months to complete our Spacelab science mission. That definitely helped ease the sting of coming home early." His thoughts were echoed by Roger Crouch. "In some ways, this could make for a more meaningful flight in the long run," said Crouch, "but, certainly, *this* one was a bummer!" Ironically, delays to the International Space Station project had pushed the first Shuttle construction mission into the summer of 1998 and made the reflight possible.

So it was that when Jim Halsell departed Columbia after completing a picture-perfect 1:33 pm EST touchdown at the SLF's Runway 15 on 8 April, he was approached by KSC Director Roy Bridges with a handshake and the words "We're going to try to give you an oil change and send you back." With a duration of just three days, 23 hours, 13 minutes and 38 seconds, STS-83 established itself as the fourth-shortest successful Shuttle flight of all time. Not surprisingly, the seven astronauts were in favour of a reflight. Just three days later, on the 11th, Space Shuttle Program Manager Tommy Holloway released a statement, authorising planning for a reflight in early July. "The Shuttle manifest for the remainder of the year, while tight, appears able to accommodate a reflight of Columbia and its MSL payload," he explained, "with reasonable impact to Shuttle launch dates for the rest of calendar year 1997." Internally designated 'STS-83R' (for 'Reflight') in the early days, the mission was eventually assigned the number 'STS-94' and its target launch on the first day of July was formally announced by NASA on 25 April. "We're ready to go fly," said Halsell as these plans were being developed. "If it were up to me, I'd like to give the guys a week or two off to let them decompress from this flight and then we'll come back and start ramping up again for the next flight."

NASA normally spent around $500 million per mission, although a substantial proportion of that figure was devoted to hardware testing, processing, training, planning and simulations which did not require repetition. Holloway quoted about $60 million for STS-94 and stressed that flying Columbia within three months offered "a very good test of a capability we should have in place for the station, to bring an element of the station back, for whatever reason, and turn it around in as reasonable time as practical". True to the predictions, the MSL-1 reflight *was* cheaper: $55 million for the actual processing of Columbia, plus $8.6 million for expenses associated with the turnaround of the Spacelab payload itself. "Our approach," said Lead Flight Director Rob Kelso, "has been to treat this flight as a launch delay. The crew is exactly the same, the flight directors are all the same and the flight control team is almost identical. It's a mirror-image flight in many respects." Even the embroidered patch, worn by Halsell's crew, was exactly the same, albeit for a different-coloured border: red for STS-83, blue for STS-94.

Naturally, the MSL-1 investigators were ecstatic at the chance to fly again so soon. The Spacelab module remained in Columbia's payload bay at the Orbiter Processing Facility (OPF), although the tunnel adaptor was removed to allow technicians better access to its interior. Ordinarily, between flights, the modules were transferred to the Operations & Checkout Building, but during the short MSL-1 turnaround technicians successfully completed many critical tasks, including replenishing fluids for the myriad combustion science experiments. Normally, the Shuttle processing team supervised an orbiter for 85 days, but the reflight required a stay time of just 56 days in the OPF. To accommodate it, and ensure that necessary work – including the replacement of two Auxiliary Power Units and several Reaction Control System (RCS) thrusters in Columbia's nose – was completed, they deferred certain structural inspections until her next mission. Fuel Cells No. 1 and 2 were both removed and returned to their vendor, Connecticut-based International Fuel Cells, for analysis. Although the exact cause was not identified, it was believed to have been an isolated incident and engineers took steps to develop monitors to provide better performance data. Meanwhile, Columbia was rolled into the Vehicle Assembly Building for attachment to her External Tank (ET) and SRBs on 4 June and from there to Pad 39A on the 11th. Aiding the early July launch target, the orbiter was fitted with three main engines 'borrowed' from her sister ship Atlantis and two 'borrowed' SRBs, previously earmarked for Discovery's forthcoming STS-85 mission.

For Jim Halsell, flying twice in quick succession was "a marvellous, once-in-a-career opportunity". As forecasters continued to assess the Florida weather in the final days of June, the prospects of achieving an on-time liftoff seemed grim, with thunderstorms anticipated on 1 July and only marginal scope for improvement on the 2nd and 3rd. As part of efforts to avoid the thunderstorms, on 30 June NASA opted to bring the launch time *forward* by 47 minutes, opening the 'window' at 12:50 pm EDT, rather than 1:37 pm. This removed one end-of-mission daytime landing opportunity at Edwards Air Force Base in California, but also enabled two more opportunities at KSC. After a 12-minute delay, due to unacceptable conditions at the SLF, the crew was advised that the launch would go ahead and Columbia rose perfectly into the Florida sky at 1:02 pm.

"It looks like Columbia's performing like a champ," radioed Capcom Dom Gorie from Mission Control, shortly after orbital insertion.

"Roger, Houston, we copy," replied Susan Still from the pilot's seat, "and thanks to the whole ascent team for getting us to a safe orbit."

With four hours of launch, Janice Voss and Roger Crouch entered the Spacelab module to begin what they hoped would be 16 days of around-the-clock research. "It's good to be back," Voss reported to Mission Control, as Crouch cartwheeled and backflipped behind her in the voluminous lab. Among the most important unfinished business from the April flight was the completion of 144 scheduled experiments in the Combustion Module. In the event, more than 200 were completed during STS-94. Among the payloads getting a reflight was SOFBALL, whose off-the-shelf gas chromatograph performed flawlessly, successfully verifying the composition of a variety of pre-mixed gases prior to combustion and determining the remaining reaction and other combustion products. Unfortunately, in view of the constraints imposed upon the chromatograph by the Combustion Module, it was not possible to measure the gases to the required accuracy. Still, SOFBALL achieved the weakest flames ever burned, mixing a variety of gases which were too small to be flammable on Earth, but which burned for more than eight minutes in CM-1. During the inaugural SOFBALL run on 8 July, the mixture was so weak that it only produced what one researcher described as "flame kernels". A later test, using a richer mixture of hydrogen, oxygen and sulphur hexafluoride, proved more successful, burning for 500 seconds and allowing investigators to understand how different concentrations of fuel and oxidiser affected the flame balls' stability and existence. Many of the flames were so weak that they equalled barely a single watt of energy, compared to about 50 watts normally produced by a single candle on a birthday cake.

Working on SOFBALL, Don Thomas received the distinction of having a newly discovered combustion 'effect' named in his honour. One surprising finding from the experiment involved what happened when two small fuel droplets burn in close proximity: they initially moved away, then approached each other in a phenomenon dubbed the 'Thomas Twin Effect'. Elsewhere, the Laminar Soot Processes experiment used ethylene gas as part of research into more environmentally friendly fuel-burning engines. Already, STS-83 had offered an important insight into how soot particles formed in microgravity and the shortened nature of the mission enabled teams to enhance LSP for STS-94. One experiment run on 3 July involved a propane-fuelled study of soot, producing a "beautiful and steady flame", according to Greg Linteris. As well as using different fuels – propane, then ethylene – the experiment also burned them under differing atmospheric pressures to determine the amount of soot produced. Differing fuel types often made substantial differences. Natural gas tends to make little soot, whilst propane produces somewhat higher quantities and ethylene (which is typically used in diesel engines) generates even more. In total, on STS-94, no fewer than 17 separate experiment runs were conducted with LSP.

Related research focused on the Droplet Combustion Experiment, which involved several phases of observations of the burning characteristics of heptane fuel drops under a range of atmospheric pressures. These phases were 'normal' terrestrial sea-level pressure, together with half and one-quarter atmospheric pressure. The astronauts had a tougher time igniting the droplets at the lower pressures, but when they *did* ignite they burned stronger and more vigorously than anticipated in pre-flight models. Despite three brief malfunctions by DCE's computer, the experiment performed exceptionally well and by the time that it was shut down on the evening of 14 July heptane droplets ranging in diameter from 2 to 4 mm had been successfully burned.

Elsewhere, payloads were moved from Columbia's middeck to the Spacelab module and back, thanks to a new storage rack, dubbed 'EXPRESS', a rather meaty acronym for 'Expedite the Processing of Experiments for Space Station'. This was being tested on MSL-1 as a potential means of getting experiments and equipment quickly to the ISS. The idea behind the new facility was to enable researchers to essentially plug their hardware, electrical power and video and data connections into the EXPRESS rack, giving them an easier and more generic interface and the chance to get experiments into orbit just 10-12 months after conception, rather than 3.5 years ordinarily required for Spacelab missions. One possible advantage was that graduate students seeking to complete master's or PhD degrees in a few years could get their research projects into space in a reasonable time. For MSL-1, the rack supported the PGBA plant-growth facility and the Physics of Hard Spheres (PHaSE), both of which were transferred from middeck lockers to EXPRESS when Columbia entered orbit. Unfortunately, plans to observe plants growing on a live video feed to the University of Colorado hit a snag when the lens clouded up. Several plants, including a species of sage, native to south-east Asia, were flown as part of research into future anti-malaria and chemotherapy treatments. Meanwhile, PHaSE studied the transitions involved in the formation of colloidal crystals – collections of fine particles, suspended in liquids, such as milk or ink – in microgravity, as part of research into new 'designer' materials to produce future semiconductors, optics, ceramics or composites.

In materials processing, the Japanese-built LIF and Germany's TEMPUS facilities took centre-stage. Minor teething troubles hit LIF early in the mission, when it was found that the facility was using up helium from its cooling purge faster than predicted, although this did not impair its ability to collect data. The Liquid Phase Sintering study subjected tungsten, nickel, iron and copper to intense 1,500 degrees Celsius heating to create solid-liquid mixtures. On Earth, the process of 'sintering' is employed to form very hard, very dense solids, which can be used to make cutting tools, car transmission gears and radiation shields. With four times longer in orbit on STS-94 than on STS-83, the scientific teams associated with both LIF and TEMPUS were gathering so much data that they expected to spend at least a year analysing it all. During the mission, TEMPUS successfully yielded the first measurements of specific heat and thermal expansion of glass-forming metallic alloys and, in so doing, obtained the highest temperature (2,000 degrees Celsius) and the largest undercooling ever achieved in a space-based furnace. So revolutionary was the facility, which first flew on IML-2, that NASA managers guaranteed TEMPUS' German manufacturers a 'free' second flight on MSL-1 in exchange for US participation in using it. In total, 20 investigations were performed and almost a quarter of a million commands were transmitted during STS-94. The shortened flight in April had actually proven beneficial for TEMPUS investigators, since they were able to examine the characteristics of the facility and their samples and make adjustments and improvements where needed. On the whole, it performed superbly, with the exception of a problem with one of its video cameras, although a few experiments were terminated earlier than intended after their samples inadvertently came into contact with the side of the container.

The peculiar sense of *déjà vu* was not lost on Columbia's crew. "This is the first chance to refly a payload and a crew altogether in the same group so quickly," said Voss during a space-to-ground news conference. "It's been much easier to get back into the swing of things and all the experiments are going great and we all feel extremely comfortable and well-prepared because we've done this so recently." However, she added that the communications between the red and blue shifts were difficult at times. "There's a lot of issues, like where everything is stowed, and people always find their favourite places. On a single-shift flight, where we all sleep at the same time, it's a little bit easier to negotiate, because everyone's awake and you can ask them where they put something, but on a dual-shift flight you have to be very careful to work together as a team across those few hours when you're both awake."

In total, during the reflight, 206 fires were set, more than 700 protein crystals were grown and a variety of fluid physics and combustion science experiments were performed. A record 35,000 telecommands were relayed to the Spacelab research facilities and, although Columbia carried sufficient consumables to support a mission extension to 17 days, a request to lengthen the flight was not submitted. Early on 17 July, Jim Halsell and Susan Still performed the de-orbit burn to return to Earth. Trailing an orange streak across the clear Florida skies, Columbia settled onto Runway 33 at 5:46:36 am EDT, wrapping up a mission just eight hours shy of 16 full days. Summing up the mission, Halsell had nothing but praise for Columbia, which had performed flawlessly. "Days have gone by," he told journalists, "without having to do an 'error log reset', which is our way of saying there have just been no problems, whatsoever. Our flavour for this flight is that it has done what it set out to do."

TECHNOLOGY FOR SPACE STATION

A dark chord of tragedy lay at the heart of STS-85.

When its five-member 'core' crew of commander Curt Brown, pilot Jeff Ashby and mission specialists Jan Davis, Bob Curbeam and Steve Robinson were assigned in September 1996, they confidently anticipated an 11-day mission in July of the following year to deploy and retrieve the second ASTRO-SPAS variant of the Shuttle Pallet Satellite (SPAS), carrying a pair of instruments for infrared observations of the middle and upper atmosphere. The primary instrument was a German one, known as the Cryogenic Infrared Spectrometers and Telescopes for the Atmosphere (CRISTA), which was embarking on its second Shuttle flight to gather global data about medium- and small-scale disturbances in middle atmosphere trace gases at altitudes between 10 and 150 km. Provided by the University of Wuppertal, and first conceived in 1985, CRISTA was capable of scanning the horizon simultaneously in three directions, promising to provide new insights into disturbances induced by winds, wave interactions, turbulence and other physical processes.

"While the actual instrument was manufactured by industry," explained Professor Dirk Offermann, who co-ordinated CRISTA science from MSFC in Huntsville, Alabama, "students did the calculations, constructed the cryostat, designed the optics, then integrated the equipment with the help of university technicians." In fact, around 20 students from undergraduate to doctoral level, and representing 15 nations, had been involved in the project between 1985 and CRISTA's first Shuttle flight in November 1994. The instrument consisted of three telescopes with four spectrometers to measure emissions in the near-infrared and far-infrared portions of the electromagnetic spectrum. It had the capacity to acquire complete vertical profiles of trace gases within a minute as the lines-of-sight were scanned through the atmosphere. Its measurement speed and high sensitivity derived from a cryogenic liquid helium vacuum container, which cooled its optics and detectors. Abutting CRISTA was the Middle Atmosphere High Resolution Spectrograph Investigation (MAHRSI). Developed by the US Naval Research Laboratory, this instrument was designed to scan the horizon to observe ultraviolet emissions from nitric oxide and hydroxyl in the middle atmosphere and lower thermosphere, between the altitudes of 40 and 120 km; a region that is too high for high-altitude balloons to access and sounding rockets pass through it too quickly to gather useful data, making the capabilities of MAHRSI hugely important. The hydroxyl free-radical was known to contribute directly to the destruction of atmospheric ozone. "This information," explained NASA's press kit, "will be used to test many theories that have been based on assumed values and will provide the first global vertical measurements of hydroxyl in the stratosphere."

With CRISTA-SPAS-2 scheduled to be deployed by the RMS arm early in the STS-85 mission and remain in free flight for more than 200 hours, ahead of retrieval and return to Earth, numerous other payloads and experiments in Discovery's payload bay and middeck would occupy the crew's time. Of specific note was the Manipulator Flight Demonstration (MFD), provided by the National Space Development Agency of Japan (NASDA), which comprised three separate experiments aboard a support structure in the payload bay to evaluate components of a mechanical Small Fine Arm (SFA), which was at the time under consideration for inclusion aboard the Japanese Experiment Module (JEM) of the ISS. The 1.5 m arm included a shoulder roll and pitch joint, an elbow pitch joint and a wrist pitch and yaw joint and was supported by rotational and translational hand controllers and displays in Discovery's aft flight deck. "It is designed to do small, fine tasks," said Curbeam in an interview, "such as removing Orbital Replacement Units that will be on the outside of the Japanese Experiment Module and also opening and closing doors." During STS-85, it was intended that Jan Davis would manoeuvre the arm to grasp a simulated space station ORU box and to open a hinged experiment door.

Alongside CRISTA-SPAS-2 and MFD, a multitude of other payloads would accompany Discovery into orbit, promising a packed and busy flight plan. Then, abruptly, in mid-March 1997, NASA announced that one of the crew members had been withdrawn from the mission. Jeffrey Shears Ashby, STS-85's original pilot, was born in Dallas, Texas, on 16 June 1954, but his formative years were spent in Evergreen, Colorado, to the south-west of Denver, where he developed a keen love of the outdoors and for hiking, fishing, exploring and flying in particular. "Each part of my childhood," he once told a NASA interviewer, "has memories of spacecraft launches, so I'm sure that growing up with the space programme influenced me in some way to be where I am today. I believe that humans all have something inside

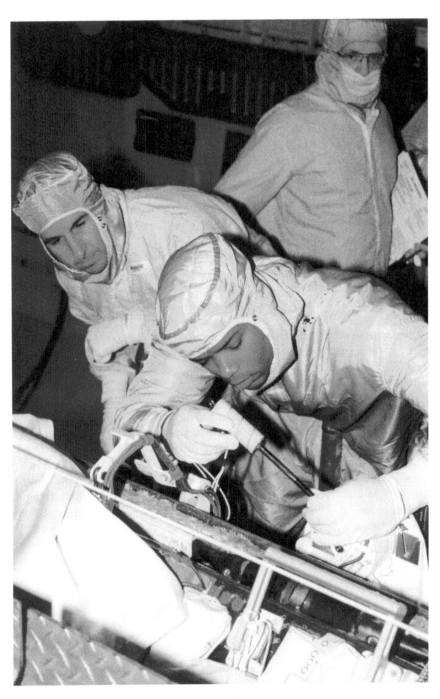

3-06: Steve Robinson (left) and Bob Curbeam were both embarking on their first space missions on STS-85.

that causes us to look upward at the horizon and wonder what is beyond. That feeling is very strong in me." When Neil Armstrong set foot on the Moon, it was already Ashby's desire to someday become an astronaut. After receiving a Navy scholarship out of high school, he entered the University of Idaho to study mechanical engineering and received his degree in 1976. Ashby then entered active service with the Navy, beginning a 25-year military career which would see him progress as a fighter pilot, test pilot and squadron commander, with more than 7,000 flight hours and over 1,000 carrier landings. He graduated from the Navy's Fighter Weapons School ('Top Gun') in 1986 and later attended test pilot school at Patuxent River, Maryland, then worked on the development of the F/A-18 Hornet aircraft, including smart weapons, night vision and electronic warfare systems. He went on to fly the aircraft in combat during Operations Desert Storm and Southern Watch over Iraq and during Operation Continue Hope in Somalia. Designated the Navy's Attack Aviator of the Year in 1991, Ashby subsequently commanded Strike Fighter Squadron 94 aboard the USS *Abraham Lincoln*. In 1994, his squadron was honoured with the coveted 'Battle E' award as the best F/A-18 unit in the Navy. By this stage, he had accrued a master's degree in aviation systems from the University of Tennessee. "At some point," he continued, "I realised that I was competitive for the [astronaut] programme and I began to apply and came down here for an interview."

Selected by NASA in December 1994, Ashby became the third pilot from his astronaut class to draw a flight assignment. However, a family tragedy forced him to resign from the STS-85 crew. He was replaced by veteran pilot Kent Rominger and NASA revealed only that Ashby had been reassigned to an administrative post at JSC, assisting the head of Flight Crew Operations. In reality, his young wife, Diana, was dying of cancer and Ashby requested removal from the mission in March 1997 to care for her in her final weeks. She passed away on 2 May. "Her one-liners and short quips began when she awoke each morning of her life and continued until her last breaths," Ashby later recalled on the webpage of the Melanoma Research Foundation, which his wife founded, "when Diana suggested to her sisters that they throw a big party each year in her honour and send the bill to me. Diana found humour to be a great medicine."

As circumstances transpired, the need to stage a reflight of the MSL-1 mission in July 1997 pushed STS-85 back until the first week of August, with the launch date finally settling on the 7th, which also happened to be Kent Rominger's 41st birthday. With two prior Shuttle missions under his belt, he joined Davis and three-flight veteran Brown as the experienced members of the crew. Nancy Jan Davis was born in Cocoa Beach, Florida, on 1 November 1953. Surnamed 'Smotherman' at birth, she later assumed her stepfather's patronymic and received much of her schooling in the Huntsville, Alabama, area. She earned two bachelor's degrees: one in applied biology from Georgia Institute of Technology in 1975 and a second in mechanical engineering from Auburn University in 1977. She then joined Texaco in Bellaire, Texas, to work as a petroleum engineer in tertiary oil recovery, before moving to NASA's Marshall Space Flight Center in Huntsville in 1979 as an aerospace engineer. Master's and doctoral credentials in mechanical engineering, both from the University Alabama in Huntsville, followed in 1983 and 1985, respectively. Her career with NASA continued to blossom and in 1986 she led a structural analysis team with a focus on two of the agency's Great Observatories: the Hubble Space Telescope and the Advanced X-ray Astrophysics Facility (later to be renamed the Chandra X-ray Observatory). Shortly before her admission into NASA's astronaut corps in June 1987, Davis worked as the lead engineer for the redesign of the attachment ring that linked the Shuttle's SRBs with the ET. Like Rominger, STS-85 was her third Shuttle mission.

Processing of Discovery for the flight ran very smoothly and the crew arrived in Florida on the evening of 5 August. No technical or weather issues offered any constraints and early on the 7th the astronauts departed the Operations & Checkout Building, bound for Pad 39A. With Curt Brown taking the commander's seat, and Kent Rominger occupying the pilot's position, the centre flight engineer's station was assumed by Robert Lee Curbeam. Nicknamed 'Beamer', he was only the eighth African-American astronaut and by the end of his NASA career he had also secured the record for the most experienced African-American spacewalker. At the time of writing, in 2014, it is a record which he continues to hold. Curbeam was born on 5 March 1962 in Baltimore, Maryland and earned a degree in aerospace engineering from the Naval Academy in 1984. "I was always interested in spacecraft and aircraft," he told a NASA interviewer. "I grew up being a big fan of Wernher von Braun and reading a lot of his work, but never really thought I'd be an operator. I thought I'd be an engineer, basically designing and building aircraft and spacecraft for other people to use." After leaving the Naval Academy, he entered active military duty, completed flight officer training and served as a radar intercept officer on several overseas deployments to the Mediterranean and Caribbean Seas and to the Arctic and Indian Oceans, aboard the USS *Forrestal*. He gained a master's degree in aeronautical engineering in 1990 and a second master's in astronautical engineering in 1991, both from the Naval Postgraduate School.

During this period, Curbeam attended the Navy Fighter Weapons School ('Top Gun') and was selected for Naval Test Pilot School at Patuxent River. It was at this stage in his professional life that he attended a lecture by astronaut Kathy Thornton and decided to pursue a career at NASA. "I had a chance to talk to her one-on-one afterwards," Curbeam told an interviewer, "and I just listened to what she had to say and the kind of things she did as an astronaut and I just decided that's what I wanted to do." Upon graduation in December 1991, having received the school's Best Development Thesis Award, he became project officer for the F-14A/B air-to-ground weapons separation programme. Curbeam was working as a Naval Academy instructor in the Weapons and Systems Engineering Department at the time of his selection by NASA as an astronaut candidate in December 1994. By his own admission, Curbeam did not know what *being an astronaut* entailed. "I thought it was more like a squadron atmosphere," he told the NASA interviewer. "I didn't realise how much of a part each individual astronaut had in the development of a mission, how it was planned, and that is what really interested me. It was almost like being a project officer back at Patuxent River, only now it was dealing with spacecraft."

A brief concern about the threat of ground fog threatening visibility at the SLF runway proved unfounded and STS-85 got underway with a spectacular liftoff at 10:41 am EDT, precisely on the opening of the 99-minute 'window'. For Bob Curbeam, launching on his first mission, the pre-flight training offered an approximate, but not exact, analogue of what he and his crewmates experienced. "There were a lot of surprises," he said later. "I was very surprised by how dynamic the whole thing was, because the simulator ... really doesn't do the trip justice. It's just extremely exciting. I can't describe how great it felt ... and how exhilarating the acceleration on my chest felt. I was just very excited during the whole thing." Within minutes, Discovery had been inserted successfully into a 57-degree-inclination orbit, ready and primed for 11 days of activity. For the first-time astronauts, though, it presented a brief opportunity for them to absorb the experience of where they were. "One thing that you do notice is that your problems, although they seem large to you, are very, very small in scope," said Curbeam, years later. "That is something that you realise, just how insignificant you are in the big scheme of things when you go up into space and look back on the Earth. At least, that was the feeling that I got."

There was precious little time for sightseeing, however, and Davis and Steve Robinson set to work activating the RMS mechanical arm for the scheduled deployment of CRISTA-SPAS-2 later on the afternoon of 7 August. The deployment was delayed by about 30 minutes owing to a minor communications glitch that prevented ground controllers from transmitting commands to the payload. STS-85 marked the first Shuttle mission to carry the OI-26 flight software, with the capability to support greater demands from multiple payloads. "We knew this flight was going to be command-intensive," said Lee Briscoe, the representative for the Mission Operations Directorate at JSC. "In the past, there's only been one path through our systems management computer software to get out to the payload. For this particular flight, we added a new pathway through that software specifically for payloads and we call it the payload throughput buffer. In the past, the payloads and the systems folks ... had to share a single command buffer, called our two-stage command buffer. It was a slow, but usable technique. As more and more people started communicating and wanting command capabilities and ground interface, we thought it would a good idea to put in that separate pathway, so we put in this new payload throughput buffer to allow the payloads to do all their commanding through that and not have to necessarily go through the kind of routine we were doing when we used the other buffer. The two-stage buffer would be free for orbiter use." The problem, Briscoe continued, was that as multiple commands started flooding the system, they generated several 'data-rejects' and commanding capability slowed down. Although a number of STS-85's experiments ran behind schedule in the early stages as a result of the commanding problem, Briscoe noted that it was quickly solved. Deployment of CRISTA-SPAS-2 eventually took place at 7:27 pm EDT, as Discovery orbited high above the north Pacific Ocean.

By the morning of the 8th, the satellite was trailing the orbiter by about 25 km, with the distance between the two spacecraft increasing by about 2.5 km per orbit. Over the following nine days, it remained at a distance of 80-110 km, and to complement its work a number of ground-launched rocket and balloon validation flights were conducted from Wallops Island, off the coast of Virginia. "Ours is really the first generation with the tools to understand the environment, with the information systems, with the technologies," said Mike Mann, NASA's deputy associate administrator for the Mission to Planet Earth programme, "and we have a responsibility not only to ourselves, but to our children, to apply those tools, including the Space Shuttle, including these sophisticated instruments, in every way we possibly can to understand the environment." CRISTA-SPAS-2 researcher Derel Offerman noted that the telescopes and spectrometers were quickly brought online and began returning valuable atmospheric data. Certainly, the MAHRSI instrument showed higher-than-expected levels of hydroxyl molecules in the upper atmosphere at high northern latitudes, indicative of high concentrations of water vapour, which University of Iowa space scientist Lou Frank speculated might be due to constant bombardment of the Earth by very small comets.

For nine days, Brown and Rominger executed periodic thruster firings to maintain their distance from the satellite, preparatory to the retrieval on 16 August. The final rendezvous began shortly after 5:00 am EDT on that date and was highlighted by the Terminal Initiation burn a few hours later, when Curt Brown assumed manual control of his ship. Unlike many previous Shuttle missions, which featured a so-called 'R-Bar' rendezvous, approaching the target from 'below' (along the Earth Radius Vector), STS-85 performed a Twice Orbital Rate Fly Around (TORFA) exercise. It started just like most previous rendezvous, approaching along the R-Bar, until Discovery reached a distance of 150 m from CRISTA-SPAS-2. Then, Brown inaugurated a fly-around of the target, to position the orbiter onto the 'V-Bar' (or Velocity Vector), a 90-degree angle change. "Once we get to the V-Bar, we'll stop there and fly in on the V-Bar from about [75 m] on into grapple range," explained Brown. "That is what we do on our normal space station rendezvous. We have corridor requirements and we have closure requirements that we must meet." In effect, added Rominger, the TORFA required Discovery to "actually hop over the top of CRISTA and do a full-circle loop, all the way back around it, stop and come up from below for a while, then continue that full-circle loop back around to being out in front of it again". By this point, the orbiter was about 100 m from the satellite and Brown guided his ship to about 10 m to practice station-keeping for the ISS, before commencing "a very specific closure rate ... on into simulated docking". By 10 m, the RMS was within range of CRISTA-SPAS-2 and at 11:13 am Davis grappled the satellite. Seventeen minutes later, she had berthed it back into the payload bay.

The rendezvous profile was designed to mimic the kind of manoeuvres that would be needed during Shuttle approaches to the ISS. In fact, Discovery's approach to CRISTA-SPAS-2 closely mirrored the manoeuvres which would be used on the first four Shuttle missions to install American-built elements of the station (Flights 2A, 3A, 4A and 5A). In essence, the satellite played the role of the station's first element, the Russian-built FGB module. "When we fabricate the space station, the rendezvous are going to vary quite a bit, depending on where we dock with it and what piece we're installing," continued Rominger. "Ideally, we'd like to approach it from 'below', because we have natural braking from Earth's gravity and we don't

3-07: The CRISTA-SPAS-2 is deployed from Discovery.

have to fire our jets toward the station." The data accrued from this so-called 'Velocity Vector' profile – approaching CRISTA-SPAS-2 from directly 'ahead' of the satellite – also demonstrated the fine controllability of the Shuttle. Additionally, a Canadian-built Space Vision System (SVS) was used to precisely position the RMS during retrieval activities and improve the astronauts' perception of large structures under unfavourable viewing conditions. "We'll also be characterising how we'll be able to [use] this vision system with different lighting conditions," added Davis. "We'll be using some sophisticated methods of looking at low-light conditions and how well the targets can be viewed under those conditions."

With CRISTA-SPAS-2 providing a constant presence throughout the mission, the crew tended to a multitude of other payloads. During their first day of orbital activities, they activated the other instruments in Discovery's payload bay. These included the Technology Applications and Science (TAS)-1, the largest and heaviest Hitchhiker payload ever carried aboard the Shuttle. Its eight discrete experiments focused on a wide range of scientific and technological disciplines, from studies of the solar constant to infrared observations of terrestrial cloud structures and from observations of the viscosity of xenon at its critical liquid/gas point to new thermal control and cooling systems, acceleration measurements and laser altimetry. The Belgian Solar Constant (SOLCON) investigation examined the effects of solar irradiance into Earth's atmosphere, whilst NASA's Mission to Planet Earth programme provided the Infrared Spectral Imaging Radiometer (ISIR) to evaluate the performance of an uncooled infrared detector array for cloud observations and the Shuttle Laser Altimeter (SLA), first flown aboard STS-72, as a pathfinder for future operational space-based laser remote sensing applications. On its maiden flight, SLA acquired the first measurements from space to be linked directly to the remote sensing of above-ground biomass and forest-canopy architecture. For its second mission on STS-85, the instrument was upgraded with a variable-gain amplifier to extend its range and its primary targets included the Amazon basin in Brazil, the Canadian boreal forest, the Kamchatka peninsula, the Patagonian icefields, West African tropical forests, high-latitude steppe deserts in China and large oceanic islands, including Tasmania, the Falklands and Madagascar. With both passive infrared and active laser sensing, ISIR and SLA also complemented each other to make STS-85 the first mission on which both passive and active remote sensing techniques were applied to cloud observations. The other five TAS-1 experiments observed the behaviour of xenon gas at its critical point, evaluated a new space-based liquid nitrogen cryocooler and a two-phase capillary pumped loop thermal-control system, monitored low-frequency accelerations in the Shuttle's structure during launch and landing and supported student investigations in the Space Experiment Module (SEM). The latter included specimens of coloured sand, yeast, seeds, paint, brine shrimp, software and X-ray films.

Also activated on the first day was the joint US/Italian International Extreme Ultraviolet Hitchhiker (IEH)-2, which was flying for the second time after having been carried aboard STS-69 in September 1995. In addition to its main objectives of studying the solar flux in the extreme ultraviolet region of the electromagnetic spectrum and the high-temperature plasma confined in a toroidal 'ring' tracing the orbit of Jupiter's moon, Io, the IEH-2 complement also included two Shuttle Glow experiments to measure emissions and aurorae in Earth's atmosphere and around the orbiter itself and a pair of technology investigations to study the Sun and provide investigators with interactive and intelligent control over their payloads.

On the morning of 8 August, Davis commenced work with Japan's $80 million MFD payload, during which she performed an inaugural checkout of the SFA. Although data conflicts between some of the MFD components prevented the ORU detachment and attachment test and the door opening and closure test, it was reported that Davis and Robinson's work provided good data on the system and its various control modes. By the fourth and fifth day of MFD operations, on 12-13 August, the crew handed over control to personnel in Mission Control, who successfully manipulated the robotic arm's motions, but did not attempt any ORU or door opening/closure tasks. "Our robot arm moves very smoothly, as expected," said Masanori Nagatomo, a NASDA project manager. "We haven't seen any malfunction in our payload system. We have more than 14 microswitches and every microswitch worked well." Veteran Shuttle astronaut Koichi Wakata, who flew aboard STS-72, served as NASDA assistant payload operations director for the MFD.

Other experiments in the middeck were tended by the crew members. Bob Curbeam focused much of his attention on the Bioreactor Demonstration System (BDS), which sought to grow colon cancer cells, larger than possible on Earth, as part of efforts to identify methodologies for halting their growth and killing them in the human body. Meanwhile, Steve Robinson worked on the Solid Surface Combustion Experiment (SSCE) and a unique device called the Southwest Ultraviolet Imaging System (SWUIS), the latter of which was provided by the Southwest Research Institute in San Antonio, Texas. It consisted of a wide-field ultraviolet imager to observe Comet Hale-Bopp, which passed through the inner Solar System in 1996-1997 and was visible to naked-eye astronomers for more than 18 months. Initially conceived in 1991, the SWUIS concept was approved by NASA and its hardware tested on a pair of high-altitude demonstration flights aboard an SR-71 Blackbird aircraft, which carried the instrument to an altitude of about 27 km and a speed in excess of 3,920 km/h. Following the discovery of Hale-Bopp in July 1995, and predictions that it might increase substantially in brightness during 1997, NASA solicited proposals from the astronomical community for secondary Shuttle payloads and SWUIS was manifested onto STS-85.

The crew member assigned to supervise SWUIS was Stephen Kern Robinson, who was making his first Shuttle flight. Born in Sacramento, California, on 26 October 1955, the son of Canadian parents, Robinson seemed from a young age to be destined for a career in the space programme, flying aircraft since the age of 13, when he fashioned his own hang glider from aluminium pipes and a broken sprinkler system. "I was fascinated with things that flew just about as soon as I could walk," he once told an interviewer. "I grew up just trying to make things fly and then trying to make myself fly." Fascinated by mathematics, he grew up with fleeting career aspirations – a biologist, a musician, an artist – but the one mainstay was that he wanted to fly and, preferably, into space. The young Robinson mowed his

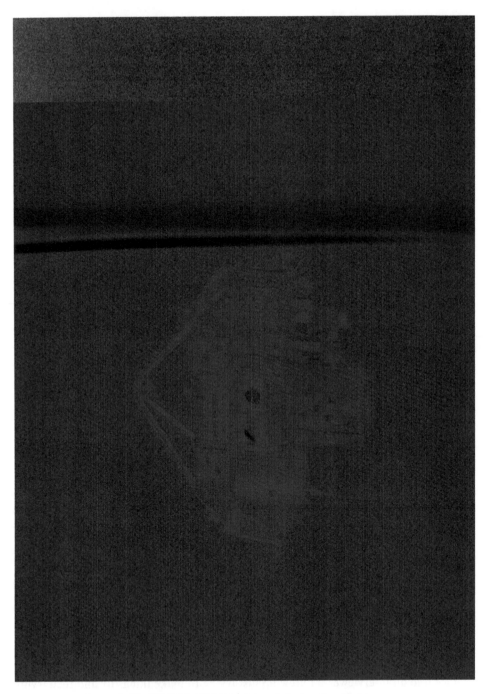

3-08: CRISTA-SPAS-2 pictured during free flight.

neighbours' lawns to earn enough money for flying lessons. "At the time, my father didn't know what I was up to," he admitted. "I rode my bike 60 miles just to have that chance again." Entering the University of California at Davis to study mechanical and aeronautical engineering, he interned as a student at NASA's Ames Research Center in Moffett Field, California, in 1975. During the course of his studies, he also worked as a graphic artist, surveyor, musician and radio disc jockey. (In his later NASA career, Robinson played in the all-astronaut rock band, 'Max Q'.) He earned his degree in 1978, then formally joined NASA a year later as a research scientist in fluid dynamics, aerodynamics, experimental instrumentation and computational scientific visualisation. "We were doing the final detailed design of the Space Shuttle at that point," he recalled. "It hadn't flown yet. We were using high-speed, heated, hypersonic wind tunnels to look at the re-entry characteristics of the Space Shuttle shape." In the early 1980s, he also founded and operated a computer graphics software firm in Silicon Valley. Robinson earned his master's degree in 1985 and a PhD in 1990, both in mechanical engineering, and both from Stanford University. Upon receipt of his doctorate, he headed the Experimental Flow Physics Branch at NASA's Langley Research Center in Hampton, Virginia, focusing on aerodynamics and fluid physics. In 1993, Robinson joined MIT as a visiting engineer in the Man-Vehicle Laboratory, conducting neurovestibular research on astronauts for the Spacelab Life Sciences (SLS)-2 mission. He also worked on EVA dynamics for space construction and conducted research on environmental modelling for flight simulation, cockpit human factors for GPS-guided instrument approach procedures and moving-map displays. Returning to Langley in September 1994, he was selected by NASA into the astronaut corps three months later.

After less than three years as an astronaut, Robinson found himself a crewman on his first Shuttle mission. For a mechanical and aeronautical engineer, the sheer velocity and the view were both phenomenal. "It's indescribable," he said. "Flying at five miles a second in space, the view changes rapidly. You see a wide spectrum of colour and light you don't see when down on Earth. It's a great place for an artist." In addition to his SSCE duties, Robinson was particularly excited by the SWUIS observations of Comet Hale-Bopp. Since the flight deck windows were all designed to filter out ultraviolet light, SWUIS was operated through the optically-pure side hatch window on Discovery's middeck. In readiness for his first observation run on 9 August, Jan Davis strategically positioned the RMS arm to provide shading from the effects of 'Earthshine' and Curt Brown manoeuvred the Shuttle to orient its side hatch toward the comet. "Back in March [1997], Hale-Bopp was nearer to the Earth and nearer to the Sun and so we were able to see it more brightly and it had a longer tail," said Robinson. "Now, it's about as far away from the Sun as it was then, so it's not as bright and the tail's not as big. Still, it's relatively close to us and we can see it much more clearly from space, because the atmosphere doesn't diffuse the light." All told, during STS-85, Robinson observed Hale-Bopp on nine separate orbits, one more than originally planned, and recorded more than 430,000 images in a variety of key ultraviolet emission bands. He also managed a few visual observations. "It was a very, very faint target to my eye," he said, "and I was not able to get the binoculars on it in time. There's a *lot* of Earthshine. Jan has done a really good job putting the [RMS] shadow right over the window, but there's a tremendous amount of Earthshine."

Overall, the scientific accomplishments of STS-85 were highly successful. On 11 August, Brown was awakened earlier than planned when a telemetry and command malfunction on the TAS-1 payload caused the SLA laser to become stuck in its 'on' position. He deactivated the avionics and SLA in an effort to regain command capability from the ground, which was restored within a matter of minutes. "We did all sorts of science," recalled Bob Curbeam. "Just about every kind of science that you can think of, we did some of it. Everything from oceanography to upper atmospheric science to meteorology, robotics, colon cancer research." Curt Brown agreed that STS-85 was "a perfect example of the versatility and the capabilities of the Space Shuttle programme", adding that "diversity" was his word for the flight. "We have over 40 different payload activities on board," he continued. "These payloads are from six different countries and are supported by many different colleges and universities all around the world."

One representative of those "different countries" who was physically aboard Discovery was a Canadian payload specialist, Bjarni Valdimar Tryggvason, who also happened to be the first person of Icelandic descent to journey into space. He was assigned to join the STS-85 crew in November 1996 to operate the Canadian-sponsored Microgravity Vibration Isolation Mount (MIM) experiment, a double-locker-size unit intended to isolate future ISS payloads from disturbances caused by thruster firings or crew activities. It was intended to operate for 30 hours and began operations on the middeck on 8 August. Tryggvason had waited 13 years for a flight assignment, having been selected as one of Canada's first group of six astronaut candidates in December 1983. Born in Reykjavik on 21 September 1945, he grew up in Vancouver, British Columbia, where he completed his high school education. Tryggvason earned a degree in engineering physics from the University of British Columbia in 1972 and undertook postgraduate work in engineering, with specialisation in applied mathematics and fluid dynamics at the University of Western Ontario. He initially worked as a meteorologist with the cloud physics group at the Atmospheric Environment Service in Toronto and from 1974-1979 as a research associate in industrial aerodynamics at the Boundary Layer Wind Tunnel Laboratory at the University of Western Ontario. Tryggvason subsequently travelled to Kyoto University in Japan and to James Cook University of North Queensland in Australia as a guest research associate and returned to his *alma mater*, Western Ontario, as a lecturer in applied mathematics in 1980-1982. His next career move was as a research officer at the Low Speed Aerodynamics Laboratory at the National Research Council of Canada (NRC) and in 1983 he was selected as one of six Canadian astronaut candidates by the NRC. Tryggvason trained as a backup payload specialist for the second set of Canadian Experiments (CANEX-2), which at the time of the Challenger accident was provisionally scheduled for launch in early 1987. Following the resumption of Shuttle missions, he served as backup to fellow Canadian astronaut Steve MacLean when CANEX-2 flew on STS-52 in October 1992. Tryggvason was also principal investigator for the MIM experiment, which was carried not only aboard the Shuttle, but also aboard NASA KC-135 and DC-9 aircraft and Russia's Mir space station. His Icelandic heritage led to the Icelandic President Olafur Ragnat Grimsson being at KSC on 7 August 1997 to witness Discovery's launch. "We are a culture of settlers created from the old days of the sagas," said Grimsson, "and we see Bjarni Tryggvason as a direct descendent of the great discoverers of the Viking period. Therefore, I think it's almost a divine indication that the name of the Space Shuttle should also be Discovery."

Originally planned as an 11-day mission, forecasters predicted acceptable weather conditions at KSC for an on-time landing at 6:14 am EDT on 18 August, although concerns remained about the possibility of ground fog developing in the vicinity of the SLF runway. After much consideration, Entry Flight Director Wayne Hale opted for the conservative approach and advised Brown and his crew that they would remain in orbit for another 24 hours, pending an improved situation early on the 19th. Precisely on time, at 6:08 am EDT, Brown fired Discovery's OMS engines for a little over two minutes for the irreversible de-orbit burn to commence their descent. The Shuttle entered the 'sensible' atmosphere high above the South Pacific Ocean and approached Florida from the south-west, alighting on KSC's Runway 33 at 7:07:59 am, after 11 days, 20 hours, 26 minutes and 59 seconds in flight. Declaring that it was good to be home, the six STS-85 astronauts could justifiably announce that their mission had achieved all of its scientific objectives, but for the first-timers on the crew, in particular, the view and the sensation of being in space far outpaced the science. "I think the biggest impression that I was left with from that flight," reflected Bob Curbeam, "although we did a lot of work in the sciences, was just the *view*. It's absolutely incredible!"

LAST CHANCE SPACEWALKS

When STS-87 commander Kevin Kregel arrived at KSC with his five crewmates in their T-38 jets on 15 November 1997, his words to the assembled media proved prophetic. "The crew is looking forward to launching on the fourth United States Microgravity Payload mission," he told then. "It's going to be a real exciting one. We're doing a lot of great science. Winston and Takao are going to do a spacewalk; Takao is going to be the first Japanese to do a spacewalk. Leonid will be the first Ukrainian to fly on the Shuttle. It's really going to be a super mission." Kregel's third flight into space, and his first in command, would indeed be exciting and would accomplish all of these tasks, but the STS-87 crew would return to Earth with mixed emotions: elation at having safely and successfully concluded 16 days of orbital research, tempered with bitter disappointment at the botched deployment of an important scientific satellite.

The crew, which included veteran astronauts Kregel and Winston Scott, together with four first-timer fliers, was international in nature, with representatives of four nations aboard Columbia. One of these 'rookies', payload specialist Leonid Konstantinovich Kadenyuk, had been announced in May 1997 as the first representative for independent Ukraine to fly into space. Interestingly, Kadenyuk had been a cosmonaut for much longer than any of his STS-87 crewmates. Born in Klishkivtsi, within Chernivtsi Oblast of the Ukrainian Soviet Socialist Republic, on 28 January 1951, he graduated from the Chemihiv Higher Aviation School in 1971 and completed the State Scientific Research Institute of the Russian Air Forces. This earned him the credentials of a test pilot, with proficiency in test aircraft flying, aerodynamics, aircraft construction and exploitation. As a test pilot, Kadenyuk flew 57 different types of aircraft and following his selection into the cosmonaut corps in August 1976 he trained to fly both the Soyuz spacecraft and the ill-fated Buran shuttle. In 1978, he graduated from the famed Yuri Gagarin Cosmonaut Training Centre and in his later career earned a master's degree from the Moscow Aviation Institute's department of aircraft construction.

In addition to his role as a cosmonaut, Kadenyuk performed numerous sky dives, several of which included live, in-flight reporting, and conducted Buran tests aboard the Su-27, Su-27UB and MiG-25 aircraft. He also chaired the State Committee on the Su-27M cockpit design. Hopes of flying aboard a Buran mission met with disappointment, when the project was cancelled in the harsh economic downturn surrounding the fall of the Soviet Union and in 1996 Kadenyuk moved to the Institute of Botany at the National Academy of Sciences of Ukraine in Kiev as a scientific investigator, developing joint US-Ukrainian space biology experiments. In March of that year, he was selected by the new National Space Agency of Ukraine (NSAU) for this Collaborative Ukrainian Experiment (CUE) and was joined as a payload specialist candidate by Yaroslav Pustovyi in September 1996 and several others. Pustovyi was a civilian physicist from the Institute of Magnetism at Ukraine's National Academy of Sciences. By December, Kadenyuk had been appointed as the prime payload specialist on STS-87, with Pustovyi as his backup. It had been a long road from his selection as a cosmonaut in the mid-1970s, through training to fly a Soviet shuttle in the mid-1980s and eventually flying aboard a US Shuttle in the late 1990s. "I believe that every person has his destiny," he wryly told journalists before STS-87, "and my destiny has been to wait for a long time!"

In fact, it was not until after Ukraine gained its independence from the former Soviet Union that the first hint of a chance to fly would arise. In July 1990, the fledgling Ukrainian parliament (the *Verkhovna Rada*) adopted a Declaration of State Sovereignty, which established the principles for self-determination, democracy, law and political and economic independence from the Soviet yoke. Following the unsuccessful attempt by Communist hardliners in Moscow to remove Soviet Premier Mikhail Gorbachev in the summer of 1991, the Ukrainian parliament adopted the Act of Independence and formally established the country as an independent democratic state. Presidential elections followed in December 1991 and the parliament's chairman, Leonid Kravchuk, was elected to lead the new government. The ensuing years, however, were difficult for Ukraine, with a sharp economic downturn and the loss of 60 percent of its gross domestic product between 1991-1999, coupled with five-digit inflation rates. Crime and endemic corruption also took their toll, but a new currency – the *hryvnia* – was introduced in 1996 and the economy steadily stabilised. Leonid Kuchma was elected as Ukraine's second president in July 1994 and during his 11 years in power, relations with Russia

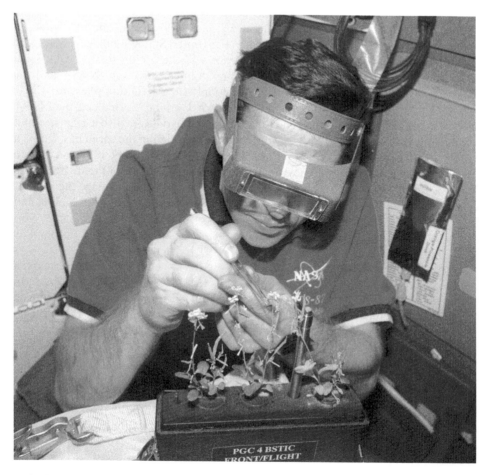

3-09: Leonid Kadenyuk works with the Collaborative Ukrainian Experiment during STS-87.

steadily improved, at the expense of allegations of corruption, electoral fraud, discouragement of free speech and concentration of too much power in his office.

During the course of 1994, NASA Administrator Dan Goldin and Ukrainian Deputy Prime Minister Valeri Shmarov met for talks on collaboration over remote-sensing and Earth sciences, space biology and welding, advanced concepts and technology and student and scientist exchanges between their two nations. Later that same year, several sites within Ukraine were observed during the Shuttle's second Space Radar Laboratory (SRL-2) mission. Of specific interest to NASA for future development was the Ukrainian Universal Hand Tool (UHT), which the space agency hoped to lease from its manufacturer, the Paton Welding Institute in Kiev. The institute was recognised as a world leader in welding technologies, which carried potential applications for the construction of future space structures, and it was announced in July 1994 that a political go-ahead before the end of the year could result in a Shuttle demonstration flight by late 1997. Four months later, in November, during a state visit by Kuchma to the United States, a bilateral civil space agreement was signed, one of whose provisions was the option to fly a Ukrainian cosmonaut aboard the Shuttle. Following this political direction, NASA and the Paton Institute concluded a $36,000 contract to initiate a definition phase for UHT. "While full details remain to be worked out with Ukraine," explained the NASA news release, "as currently envisioned the experiment would involve a US astronaut conducting sample welds on a variety of materials mounted in the cargo bay during a spacewalk. These sample welds would demonstrate the ability of the UHT to perform contingency or emergency repairs to simulated elements of the International Space Station." NASA's MSFC in Huntsville, Alabama, would manage the project. In 1995, the Collaborative Ukrainian Experiment was approved by Dan Goldin and his Ukrainian counterpart, Alexander Negoda of NSAU. Later that same year, President Bill Clinton formally announced that a cosmonaut would accompany CUE on a Shuttle flight.

Despite the historic significance of the venture for his country, Leonid Kadenyuk objected to the mistaken definition of himself as the 'first' Ukrainian cosmonaut. In the Soviet era, a number of ethnic Ukrainians had flown into space, the first being Pavel Popovich, aboard the Vostok-4 mission in August 1962. "I think that the first Ukrainian in space was our legendary Pavel Romanovich Popovich, who was Cosmonaut No. 4," he said, "but now, of course, since Ukraine has become independent, this will be the first flight of a Ukrainian. I believe the first flight of any cosmonaut, of any government, is a very important event in the life of that country. I am very proud that it has fallen to me to play this role and I will do everything I can to be worthy of this honour." He added that he did not see his flight aboard STS-87 as a one-off publicity stunt, but as the start of a vigorous programme of co-operation in space for Ukraine. Already, the country was playing a key role in the development of several important launch vehicles, most notably the Zenit.

Although he had blended seamlessly into Kregel's crew, the other STS-87 astronauts commented on a language barrier, early in their training. "Leonid took English lessons and improved, so it wasn't a problem," explained Indian-born mission specialist Kalpana Chawla. "Plus, we took 80-hour Russian courses, so we could speak a little bit of Russian. Not much, but enough that if a word was too big for him, you could open up your dictionary and show him what you were saying." Chawla originated from Karnal, India, with NASA revealing her date of birth as 1 July 1961. However, it later came to light that her actual date of birth was 17 March 1962 and her birth date had been adjusted in order to enroll her into school at a younger-than-normal age. The youngest of four children, she completed her schooling in Karnal and developed an early love for aircraft and aerospace engineering, opting to study physics, chemistry and mathematics. Occasionally, her father would take her for flights in a small Pashpuk aircraft, courtesy of a local flying club, and she grew up learning about the exploits of the French-born Indian aviator J.R.D. Tata (1904-1993), India's first licensed pilot, who completed some of the country's first mail flights. "The airplane that he flew now hangs in one of the aerodromes," she said, "and I had a chance to see. Seeing this airplane and just knowing what this person had done during those years was very intriguing. It definitely captured my imagination." Chawla received a degree in aeronautical engineering from Punjab Engineering College at Chandigarh, the capital of the Indian states of Punjab and Haryana, in 1982.

Upon receipt of her degree, Chawla moved to the United States and earned a master's credential in aerospace engineering from the University of Texas at Arlington in 1984, followed by a second master's in 1986 and a PhD in aerospace engineering in 1988. Her final two degrees came from the University of Colorado at Boulder. Following receipt of her doctorate, Chawla began working at NASA's Ames Research Center in Moffett Field, California, working on power-lift computational fluid dynamics. Her research concentrated on the simulation of complex air flows encountered around aircraft and she later supported research in mapping flow solvers to parallel computers and testing these solvers by carrying out powered-lift computations. In 1993, she joined Overset Methods, Inc., in Los Altos, California, as vice president and research scientist to form a team to simulate moving multiple-body problems. Chawla's responsibility entailed the development and implementation of efficient techniques to perform aerodynamic optimisation. She was selected by NASA as an astronaut candidate in December 1994. On STS-87, she became the second ethnic Indian to fly into space, after Rakesh Sharma, as well as India's first woman astronaut.

Alongside Chawla and Kadenyuk aboard Columbia were Kevin Kregel, STS-72 veteran Winston Scott and two more first-time astronauts, pilot Steven Wayne Lindsey and Japanese mission specialist Takao Doi, making this one of the most diverse Shuttle crews ever assembled. Lindsey, who later went on to serve as chief of NASA's astronaut corps and commanded the final flight of Discovery in March 2011, came from Arcadia, California, where he was born on 24 August 1960. He was raised in Temple City, a suburb of Los Angeles, and later praised the "small-town values and ethics and drive" as having driven him to excel. His father, a truck driver and farmer, who had grown up during the tough years of the Depression, had served in Korea and later worked doggedly to complete his degree in electrical engineering. This inspired the young Lindsey. "I wanted to be an engineer, like my father," he later told a NASA interviewer. "I was very influenced by him, but I also wanted to fly airplanes." After high school, he entered the Air Force Academy and gained a degree in engineering sciences in 1982. He was then commissioned into the Air Force, receiving his pilot's wings in 1983 at Reese Air Force Base, Texas, and qualifying in the RF-4C Phantom. For the next few years, Lindsey was a combat-ready instructor pilot and academic instructor at Bergstrom Air Force Base, Texas, and in 1987 he was selected to attend the Air Force Institute of Technology to pursue an advanced degree. He also attended Air Force Test Pilot School at Edwards Air Force Base in California, emerging as Distinguished Graduate and recipient of the coveted Liethen-Tittle Award of Class 89A and receiving his master's degree in aeronautical engineering in 1990. As a qualified test pilot, Lindsey conducted weapons and systems tests in the F-16 and F-4 aircraft and later served as deputy director of the Advanced Tactical Airborne Reconnaissance System Joint Test Force and as F-16 flight commander for the 3247th Test Squadron. In June 1994, he graduated from

Air Command and Staff College at Maxwell Air Force Base, Alabama, and was involved in the weapons certification for the F-16, F-111, A-10 and F-117 aircraft when he submitted his application to NASA as an astronaut candidate. "I hadn't really thought about being an astronaut for many years," Lindsey acquiesced, "and I found myself, as a test pilot, now qualified for the programme. I decided, *what the heck*, I'll give it a shot, and so I applied." His application succeeded and he was selected as an astronaut in December 1994.

On STS-87, in addition to a multitude of scientific and technological research payloads, Columbia would support her first spacewalk in 16 years of operations. On STS-5 in November 1982, an EVA was cancelled due to crew sickness and space suit problems, whilst on STS-80 in November 1996 a planned pair of excursions were called off due to a difficulty with the airlock hatch. On STS-87, Winston Scott and Takao Doi – who became Japan's first spacewalker – were to simulate a series of ISS construction tasks. Not surprisingly, Doi was excited by the opportunity, but despite months of training he was aware that the real thing might be different from the ground practice. "In order to move from one place to another," he told an interviewer, before launch, "you always have to grasp something to prevent yourself from floating away, but if your grip is too tight your body will move in the other direction. You must softly touch the handrails and move along. That's the basic point of moving in space. It's impossible to make a complete simulation of the space environment during training. Even the training in the [water tank at JSC] is different, because the water generates resistance and therefore body movements will be more stable and predictable than in space. The only way to master moving in space is to actually do this *in* space." Doi got his chance on 25 November 1997, a few days into STS-87, but his tasks that night proved somewhat different to what he had prepared for.

Takao Doi, Japan's third professional astronaut (after Mamoru Mohri and Chiaki Mukai) and the country's fourth spacefarer (when one includes journalist Toyohiro Akiyama, who flew to Mir), was born in Tokyo on 18 September 1954. He grew up with a fascination for astronomy. "I was watching stars when I was a kid," he later told a NASA interviewer, "and I still like watching stars right now." (In 2002 and 2007, he discovered two supernovae.) "While I was watching stars, I dreamed about travelling in space, visiting the centre of our galaxy, exploring a globular cluster, observing the birth of stars. I was just fascinated by the mysteries of space and what I could learn from space." During his high school years, he observed Mars with a borrowed telescope and was enamoured. "I was hooked even more so when I entered college," he said. "I wanted to make my own rocket, so I chose to study rocket propulsion." Doi earned his undergraduate engineering degree in aeronautics from the University of Tokyo in 1978, followed by a master's in 1980 and a PhD in aerospace engineering in 1983. Upon receipt of his doctorate, Doi spent two years at Japan's Institute of Space and Astronautical Science (ISAS), working as a researcher on space propulsion systems, before moving to the United States in 1985 to join NASA's Lewis Research Center (today's Glenn Research Center) in Cleveland, Ohio, as a National Research Council associate. In June of that year, Doi was selected by the National Space Development Agency of Japan (NASDA) as one of three payload specialist candidates for the Spacelab-J mission, at that time planned for launch in February 1988. The flight was extensively delayed in the aftermath of the Challenger disaster and Doi returned to academia, working on microgravity fluid dynamics at the Center for Atmospheric Theory and Analysis at the University of Colorado at Boulder. In 1990, alongside chemist Mamoru Mohri and physician Chiaki Mukai, he resumed training for Spacelab-J. Mohri ultimately flew the mission in September 1992 and Doi was selected by NASA as a Japanese mission specialist candidate in January 1995.

Less than three years later, aboard STS-87, one of Columbia's key payloads was the SPARTAN-201 research satellite, which was making its fourth flight. It would be utilised to investigate the mechanisms responsible for heating the Sun's corona and accelerating the anomalous 'solar wind', which originates in this region. It was also intended to conduct joint observations with another satellite, the Solar and Heliospheric Observatory (SOHO), which had been launched in December 1995 as part of efforts to expand physicists' understanding of conditions within the corona. The plan called for SPARTAN-201 to be deployed by Kalpana Chawla, using the RMS arm, on the second day of the STS-87 mission and be retrieved about 50 hours later. It was an important responsibility, but would quickly and unfairly turn into her nemesis.

Bad luck could not have seemed further away, however, when Columbia lifted off precisely on time at 2:46 pm EST on 19 November, beginning 1997's eighth and final Shuttle mission. The ascent, though perfect, was unusual, as it featured the first use of a novel 'heads-up' manoeuvre, about six minutes after launch. Intended for use on low-inclination missions, the GPCs commanded the main engines, over a 20-second period, to roll Columbia to the left by 180 degrees from a 'belly-up' to a 'belly-down' orientation, enabling the crew to communicate with Mission Control through the TDRS network, about 2.5 minutes sooner than previously possible. As well as eliminating the need for a ground station in Bermuda (the use of which had required the Shuttle to ascend towards orbit belly-up), the move saved NASA around $5 million per year. The computers pulled off the manoeuvre crisply and Columbia rolled at five degrees per second, resulting in a minimal 15-second communications loss with Mission Control. "In the roll, we transitioned from ground station telemetry to TDRS telemetry," explained Launch Integration Manager Don McMonagle. "That transition occurred in less than 15 seconds, which is precisely as we expected. We're very satisfied that the unique aspect of this ascent trajectory went as planned." The manoeuvre, which was to be used on future flights, allowed the Shuttle's cockpit antennas to lock onto the overhead TDRS.

"Nice roll," radioed Capcom Scott 'Doc' Horowitz from Mission Control.

"Copy and concur," replied Kevin Kregel from the commander's seat. The entire manoeuvre was controlled by the computers and the six STS-87 astronauts did not even know if they would roll to the left or the right until it actually happened; that decision was made in real time, based on incoming velocity and orientation data. Telemetry indicated that the manoeuvre, which took place 480 km downrange of KSC and at altitude of about 110 km, went smoothly, although it was not visible to long-range tracking cameras. Since it occurred outside of the sensible atmosphere, it was not considered risky and aerodynamic loads on the vehicle did not present a problem. Before launch, Kregel likened it to an E-ticket ride. "We had to do a fair amount of analysis to ensure that we weren't doing something dumb," said Flight Director Wayne Hale. "One thing we didn't want to perturb was the [Return to Launch Site] abort mode. That is a fairly intricate manoeuvre, it's been analysed to death and we spent a lot of money making sure it would work if we ever had to do that and it is based on a heads-down trajectory. We picked a time that was after Negative Return." The analyses concluded that, even with an electrical system failure or two main engine shutdowns, the vehicle would remain controllable. To ensure maximum safety, the manoeuvre was carried out before Columbia departed the range of the Air Force's Eastern Test Range radars, which allowed Mission Control to maintain communications with the crew until TDRS took over. For Steve Lindsey, sitting in the pilot's seat, it presented him with an impressive view of Earth. "We're nose-high, so all you can see is blue, blue, blue and then black," he said of the initial stages of ascent to orbit, "but about six minutes into the flight, we do a roll to heads-up. During that, the roll to heads-up is really the first time you get to see the Earth from space. I will never forget that first roll. It rolled my way, so I could see the Earth coming up and seeing the Earth from space for the first time was very memorable. Things like your home town are very memorable, but just seeing the beauty of the Earth and that it's actually round, like they say in the geography books, that's something I won't ever, ever forget."

Immediately after reaching orbit, Chawla checked out the RMS in preparation for the SPARTAN-201 deployment. It was at this stage that the first problems materialised. One of the satellite's tasks was to conduct simultaneous observations of aspects of the solar corona with SOHO, as part of efforts to recalibrate the latter's instruments and ensure that their sensitivity had not degraded after two years in space. However, at 9:38 am EST on 19 November, only hours before STS-87 launched, SOHO suffered a sharp reduction in power in its attitude-control system, which initiated a gradual drift in its pointing capability. This led to an Emergency Sun Reacquisition, in which it placed itself into a redundant 'safe mode'. The satellite resided in solar orbit, directly up-Sun of Earth and 1.6 million km closer than Earth, and it was hoped that the joint experiments with SPARTAN-201 could allow both spacecraft to gather data from their respective vantage points. Shortly before going to bed on the evening of 19 November, Kregel's crew were advised of the SOHO problem and told to postpone the SPARTAN-201 deployment until the 21st, by which time it was expected that the satellite would have been restored to full capability. Seven hours after the incident, a 'spacecraft emergency' was declared for SOHO, in order to gain additional coverage from the 34 m tracking antennas of NASA's Deep Space Network and normal control was finally restored at 7:11 pm EST on 20 November. However, Lee Briscoe of NASA's Mission Operations Directorate explained that even in the case of further problems with SOHO the SPARTAN-201 deployment would have still gone ahead. "There's both a science programme and a joint science programme," he said. "I can't exactly quantify for you what percentages are, but we would go ahead and complete the SPARTAN-specific science programme and we'd expect there would be no disruption of that."

Next day, Chawla grappled SPARTAN-201 to begin the deployment procedure and she released the small satellite at 4:04 pm EST, as Columbia flew above the Pacific Ocean. What happened next, however, carried the potential to damage her astronaut career. "We pick up the SPARTAN and take it to its deploy location," she told an interviewer before the flight. "It's critical that the location be right, because it's supposed to be looking at the Sun at that time. We let it go and we watch it for about two minutes when it does a pirouette. That tells us that all the automated tasks that are supposed to go on have started on time. If that does not happen, then we'll take it back and put it in the bay for another try on another day." After releasing the satellite, the crew was surprised when it failed to perform the 45-degree pirouette, implying an attitude-control system malfunction. Had Chawla missed a critical command during the pre-deployment checkout process? It was a question that would be revisited repeatedly in the coming weeks.

"Houston, Columbia," Chawla reported. "It's D1 + 3.5 minutes and we do not see SPARTAN doing a pirouette, so at D1 + 4 we plan to regrapple."

"Columbia, we concur," replied Capcom Marc Garneau from Mission Control.

"The command that we didn't receive on-board the spacecraft would have powered down some of the systems that were up in a standby health-check mode," said SPARTAN Mission Manager Craig Tooley later that day. "We basically power up a bunch of subsystems and check their status and then bring them back down into a quiescent state to await deployment, where the sequence is then picked up again and the flight programme kicks in. That step – that brought those down into a quiescent state and had them waiting for the deployment – wasn't received by the spacecraft. Therefore, the attitude-control system was still there, warming up, but not configured to deal with the error signals from the sensors or fire some pyrotechnic gas valves that would have turned on the thrusters, so it was basically 'idle' and it never got out of idle mode." Chawla duly moved in to regrapple SPARTAN-201, but did not receive a firm 'capture' indication on her panel on Columbia's aft flight deck. She backed the RMS away from the satellite, but in doing so she accidentally bumped SPARTAN-201 and imparted a spin on it. In her later recollections, Chawla explained that she thought the satellite was belatedly carrying out its pirouette manoeuvre, which should have occurred immediately after deployment. "It all seemed to go real fast," she said. "After I moved the arm back, I thought maybe SPARTAN was doing its manoeuvre. That was really my immediate reaction."

However, the satellite had entered a rotational spin of about 1.9 degrees per second, making it essential for the crew to match that rate with Columbia's thrusters in order to attempt a retrieval. "Houston, we have a tip-off rate. It's rolling around," Kregel reported. "We will go ahead and match it and regrapple." For much of the next hour, he used the RCS thrusters to manoeuvre the Shuttle in such a manner that its rotation matched that of the satellite, so that Chawla could make a second attempt to recapture it. The primary concern centred on timing. A second deployment opportunity had to occur within an hour of the first, in order to prevent a timing device from automatically closing the fuel valves to SPARTAN-201's pointing system. Had they managed to recapture the satellite within 60

3-10: The STS-87 crew inspect the SPARTAN-201 payload before launch.

minutes, the attitude-control system would have remained intact and permitted a redeployment shortly afterwards. However, Kregel's efforts to match the rotation rate exhausted the Shuttle's fuel budget and Flight Director Bill Reeves called off the attempt at 5:00 pm EST, instructing the crew to manoeuvre to a separation distance about 64 km 'behind' the satellite. By this stage, flight controllers had in the back of their minds a plan to re-rendezvous with SPARTAN-201 on 24 November and send Winston Scott and Takao Doi outside on a spacewalk to manually grab it. Although the rescue of the satellite was only expected to take a couple of hours, it was decided to run the EVA to the planned six hours and have Scott and Doi complete most of the other ISS construction tasks for which they had already trained.

Two key options were presented by fellow astronaut Jim Wetherbee, then serving as deputy director of JSC, in a conversation with Kregel. The first was that Scott and Doi would position themselves on opposite sides of the MPESS truss structure and grasp SPARTAN-201's instrument canisters when it came within reach, whilst the second called for one of the spacewalkers to ride at the end of the RMS. "In our final analysis," Wetherbee told Kregel, "the crew workload goes up significantly [with Option Two] because you have three people now in the loop: the crew members outside, the arm operator and you, flying the vehicle. The bottom line was that the crew co-ordination was just a little bit tougher and so our recommendation is going to be that we focus on Option One with the two crew members on the MPESS, grabbing onto the instrument canisters." In response, Kregel highlighted that the astronauts had already built a small model of the satellite in the middeck and rotated it in different axis to try to recreate the spin rate.

After checking equipment, Scott and Doi began donning their space suits on the afternoon of the 24th, as Kregel and Lindsey continued a series of RCS burns to close the gap between the Shuttle and the satellite. The most critical manoeuvre was the Terminal Initiation, which occurred at 5:51 pm EST and placed Columbia onto an intercepting path to arrive directly 'beneath' the slowly-rotating SPARTAN-201. Shortly after 7:02 pm, Scott and Doi became Columbia's first spacewalkers, as they pushed open the outer airlock hatch and floated into the payload bay. Commenting on a "nice view over the Earth", Doi and his partner quickly moved to their positions at opposite ends of the MPESS carrier structure as SPARTAN-201 tumbled, just 210 m away. When Kregel positioned the orbiter within physical reach, he gave the spacewalkers a "Go" to capture it.

"Winston," he told Scott, "if we're patient, it looks like the telescope is rotating. The top end of the longitudinal axis looks like it will come down to you and the bottom one right up to Takao. If we just wait, it'll come right to us."

"Okay, copy that, Kevin," replied Scott. "We'll just be patient and see what happens."

A few minutes later, the elation was evident in his voice as the satellite tumbled closer. According to Kregel, SPARTAN-201 appeared to be relatively stable throughout the rendezvous, which suggested that its backup attitude-control system had activated at some stage after Columbia's withdrawal on 21 November, damping out the rotation to leave a more tumbling motion. Right on cue, at 9:09 pm, the two spacewalkers reached out and grabbed the satellite. They encountered some difficulty when manually latching it back onto the MPESS, but Chawla, operating the RMS, succeeded in driving it onto its retention latches. The procedure was complete by 10:23 pm, with SPARTAN-201 secured in the payload bay. At this stage, Kregel was able to read telemetry to Mission Control to enable engineers to characterise the satellite's health. The response from Capcom Marc Garneau was favourable.

Yet the EVA had not even reached its midpoint, for Scott and Doi would remain outside for a total of seven hours and 43 minutes. Following the failure of Columbia's airlock hatch to open in November 1996, which prevented STS-80 spacewalkers Tammy Jernigan and Tom Jones from testing a series of prototype ISS tools, many of their tasks were offloaded onto the STS-87 EVA. Scott and Doi had trained to work with a small crane, which they would use to manoeuvre a dummy ISS battery, and a variety of different safety tethers. The crane's boom measured 1.2 m long when fully retracted, but telescoped outwards to 5.3 m when extended, and also had the capability to be pitched 'up' or 'down' and yawed side to side. It was hoped that the crane would prove useful in transporting ORUs with masses as high as 270 kg from translational carts on the space station's external truss structure to various work sites. In general, Scott praised the smoothness of the crane, although he noted that it did exhibit some unanticipated flexibility. After the crane had been installed into its socket on the port side of the payload bay, Doi evaluated its operating characteristics, whilst Scott removed a 225 kg dummy battery and its carrier from the starboard wall. These were attached to the end of the crane to assess its ability to move large masses. The crane's boom was extended by turning a ratchet fitting with a power tool, or, as a backup, by using a manually operated hand crank. The only difficulty was a jammed torque multiplier, used to tighten the battery onto its mounting location. However, Scott finally freed it by following instructions radioed to him from Mission Control. At one stage, the spacewalkers were advised that their proud wives were watching their exploits in real time, to which Scott passed on a message: "Tell Marilyn I just had to stop by and pick up a satellite, but I'll be home by suppertime!"

After stowing the crane and the mock battery, the spacewalkers cleared away their tools and headed back to the airlock. The EVA officially ended at 2:45 am on 25 November, when they repressurised the airlock and switched their suits off internal battery power. The astronauts later completed questionnaires to capture their initial thoughts, since STS-87 was the last Shuttle mission to involve a spacewalk in advance of the first ISS assembly mission in 1998. "This is our last chance to look at this before we employ it somewhere down the line on space station," said former astronaut Greg Harbaugh, then serving as head of the EVA Project Office at JSC. "We have to recognise that EVA is very much fundamental to the success of station. It is going to be an almost everyday occurrence for the next several years, so we'd better get used to it. If we have any misgivings about EVA, it's time to get those behind us. Starting with STS-88, we are into what we call the 'EVA Wall' and we had better be ready to step up to it. It's a gigantic undertaking, something like three times the amount of EVA work that we've ever done in the history of the American space programme."

Although their spacewalk had lasted far longer than the six hours originally allotted, and the astronauts had accomplished a significant amount of their objectives, a number of tasks could not be accomplished. One of them was an odd, beachball-like contraption known as the Autonomous EVA Robotic Camera/Sprint (AERCam-Sprint), which sought to demonstrate a free-flying camera to inspect the exterior of the ISS for signs of damage. The concept had attracted particular interest from NASA after the collision of an unmanned Progress freighter into the Mir space station in June 1997, which forced the evacuation and partial depressurisation of one of its research modules. Had a free-flying camera of AERCam-Sprint's capabilities been available, it might have been possible to investigate and acquire high-resolution photographs of the damage, without risking an EVA. Future applications could also include sophisticated sensors to sniff out coolant leaks or mechanical aids to hold spacewalkers' tools. STS-87 pilot Steve Lindsey, who oversaw AERCam-Sprint operations, praised its "limitless" potential. The focus of this attention looked innocuous enough: a 36 cm, 16 kg aluminium sphere, coated with soft Nomex felt. Its speed was constrained to no more than a few centimetres per second, allowing it to be used safely in close proximity to spacewalking astronauts. If its velocity somehow doubled above that, it would automatically shut itself down. AERCam-Sprint was designed to manoeuvre itself with a system of 12 cold-gas thrusters, operated by Lindsey, via a joystick, antenna and a pair of laptop computers on Columbia's aft flight deck. Housed in the airlock, the original plan called for Winston Scott to hand-release AERCam-Sprint for a half-hour journey. If time permitted, Lindsey hoped to fly it up to 50 m 'above' the Shuttle, taking a series of video images, and then guide it back into Scott's gloved hands. Two miniature television cameras, with 6 mm and 12 mm lenses, would provide stunning views for the crew and Mission Control.

"How high up above the bay we're going to go with it will depend on how much time we have," said Lindsey before the flight. "We'll be flying just above the payload bay most of the time, doing a lot of engineering evaluations, some flying qualities evaluations, as well as doing some simulated space station-type tasks where we'll be observing and evaluating an EVA crew member at work and trying to use the camera views to assist them in their work."

In comparison to the crane and battery tests, AERCam-Sprint was a low-priority task, but it was disappointing that the crew did not get the opportunity to test it. On 1 December, mission managers decided to add a second EVA to STS-87, although its primary focus was not AERCam-Sprint, but rather to repeat and gather additional data on the crane tests from the first spacewalk. According to Greg Harbaugh, planners had been unable to acquire all of the necessary data following the 24/25 November EVA. Specifically, Scott experienced difficulties connecting the ORU onto the crane, when its boom proved more flexible than expected. "Winston had some difficulty mating the large ORU, the big battery, to the end of the boom," explained Harbaugh. "That was not entirely unexpected, but now we have an opportunity to go back and refine our techniques, based on what we observed the first time."

Scott and Doi's first focus during the second EVA, which began at 4:09 am EST on 3 December, was working with a mock piece of ISS hardware, known as a 'cable caddy'. It was originally carried aboard STS-80 and provided a means of holding up to 20 m of replacement electrical cable. Scott began the task of mounting it onto the crane in the same manner as he had done with the battery, pushing it down onto the latches and then 'shoehorning' it into place. He also rotated the assembly by 90 degrees, using a tether to help anchor it in place, then forced the caddy directly 'down'. His efforts generated no obvious problems. Next, Scott tried locking the cable caddy onto the crane using a semi-rigid tether to keep the boom relatively stable. Two different safety tethers were also evaluated. The Body Restraint Tether (BRT) was designed to hold a spacewalker steady, whilst clamped to a handrail, and had the benefit of freeing the hands to conduct useful work. Scott had already tested it on his STS-72 EVA in January 1996 and Doi evaluated it on STS-87. The second tether was the Multi-Use Test (MUT), which, although similar to the BRT, had the capability of performing a wider range of tasks. For example, different end effectors could be attached to it in order to grasp ORUs, tools or handrails.

Near the end of the EVA, Scott and Steve Lindsey finally got the chance to test AERCam-Sprint. "What I'm going to do is retrieve the Sprint from the airlock and I'll make my way up to the foot restraint and mount myself in the foot restraint," Scott explained before launch. "Once I've done that, I'll power it on and look for a sequence of five flashing lights that tells me the power-up went well and everything's working properly. Then, when Steve gives me the go, I'll rotate it about one or more axes to perform another self-test. When we get the indication that the self-test is complete, I'll just stand by. Steve will go through a series of tests and checks from his console. When he gives me the word, I'll simply release it, sit back and watch and catch it once the flight is over. He will pilot it back to my location, I'll reach out and grab it, put a tether on it, power it down and that will be the end." The first flight of AERCam-Sprint ran like clockwork and after Lindsey had tested its nitrogen thrusters – which Scott described as a sensation of "very tiny thumps" against his space suit glove – the $3 million robot was released into the payload bay at 7:15 am EST. Spectacular images of Columbia and Earth were transmitted to Mission Control from its two tiny television cameras and the handling characteristics proved so good that Lindsey was granted additional time to test it. After 76 minutes of work, Lindsey gently manoeuvred AERCam-Sprint back towards Scott's gloved hands at 8:27 am. It used barely 65 percent of its nitrogen supply and a mere eight percent of the power from its lithium batteries, reaching a maximum distance of about 12 m from the orbiter. "Having this kind of floating eyeball in our hip pocket," said Greg Harbaugh, "available when we need it for space station is going to be worth its weight in gold over the next few years." Scott and Doi stowed AERCam-Sprint in the airlock and completed their second EVA at 11:08 am, after four hours and 59 minutes.

The two spacewalks and the SPARTAN-201 fiasco had thus far distracted attention from the research that was ongoing aboard the fourth United States Microgravity Payload (USMP-4). It had been activated by Chawla on 19 November. Three of its experiments – the AADSF and MEPHISTO solidification furnaces the IDGE dendrite-growth facility – were veterans of previous USMP missions, but it also featured the Combined Helium Experiment (CHeX), which sought to test theories of the influences of boundaries on matter by measuring the heat capacity of helium whilst confined to two dimensions. A clearer understanding of the effects of 'miniaturisation' on material properties was expected to lead to smaller and more efficient and cheaper electronic devices, including high-speed computers. CHeX was a refrigerated dewar coated with a magnetic shield. It held 392 crystal silicon disks that forced the liquid helium into very thin layers. Through these layers, precise temperature measurements were taken. "The temperature resolution," said USMP-4 Mission Scientist Peter Curreri, "is equivalent to the ability to measure the distance between Los Angeles and New York to the thickness of a fingernail. The information from this experiment can be applied to future generations of microprocessors."

Meanwhile, AADSF was used to process alloys of lead-tin telluride and mercury-cadmium telluride. Its operations ended on 30 November, when the science team noticed unexpected readings from several temperature sensors that were used to control the solidification process. This prevented investigators from completing the second growth of three lead-tin telluride samples. Elsewhere, the French-built MEPHISTO furnace processed three identical bismuth-tin alloy specimens, whilst sensors monitored temperature fluctuations and the position of the solid-liquid border, 'marking' them with electrical pulses. Its data was correlated with

3-11: Steve Lindsey poses with AERCam-Sprint in Columbia's middeck.

measurements from the SAMS accelerometer to determine the impact of Shuttle motions on the solid-liquid interface. The IDGE investigators had previously used ultra-pure organic crystal called succinonitrile, but on USMP-4 they conducted experiments with pivalic acid, whose 'face-centred-cubic' material solidified like many non-ferrous materials. Its transparent nature and low melting point made it an ideal sample for obtaining fundamental 'benchmark' observations. Other experiments in Columbia's cabin included the Middeck Glovebox, which Chawla used to explore ways of creating uniform mixtures of liquids to form specific metal alloys for potential use the bases of future semiconductors.

The success of USMP-4 had been tempered, though, by the SPARTAN-201 problems. However, it was hoped that all was not lost. On 29 November, Kregel performed a health check of the satellite to gather data in anticipation of attempting a redeployment, later in the mission. Six to 20 hours of solar observations might still be possible, as opposed to the 50 originally planned, with two options for deployment on 1 or 2 December, either of which would require STS-87 to be extended in length to 17 days. But the second deployment was called off, due to worries about insufficient propellant reserves aboard Columbia. Propellant for only one rendezvous (and not two) had been budgeted by mission planners and much of that had been used by Kregel during the manual retrieval on 24/25 November. The Mission Management Team concluded that the STS-87 crew would not be able to conserve enough fuel to ensure a successful retrieval. Furthermore, if a similar problem were to arise whilst attempting to grapple SPARTAN-201, or additional station-keeping time was required, then the propellant margins would run dangerously low. In all rendezvous missions, flight controllers ensured that propellant margins were sufficient to guarantee success, even if critical systems such as the Ku-band radar were to fail. Additionally, the high-priority USMP-4 experiments would be impaired by the rendezvous operation. "During the intensive part of the spacewalks, we had to pause some of our experiments," said USMP-4 Assistant Mission Manager Jimmie Johnson, "but we picked them right back up after the satellite was captured and secured." At the end of the day, SPARTAN-201 was secure and healthy and the scope existed to fly it again on a future mission. "Here's a case where we have the SPARTAN in the bay," said Briscoe. "We have it. It's a healthy spacecraft and we can bring it back. If you were to deploy it under those kinds of propellant margins, you could stand a 40-50 percent chance of not bringing it back if you had dispersions or failures as you tried to re-rendezvous with it. Based on that, the management team decided we would go ahead and forgo another deploy and retrieval."

Nor could SPARTAN-201 conduct its solar observations whilst attached to the end of the RMS arm. Columbia could be oriented with great precision, but not good enough to point the instruments with the required degree of accuracy. After landing, the satellite would be impounded as investigators dug into the cause of the problem. Already, some members of the media were pointing fingers at Kalpana Chawla and many were unafraid to question the astronauts during space-to-ground news conferences. On 2 December, Kregel told an *Orlando Sentinel* journalist that Chawla had taken the situation in her stride and he refused to discuss blame. "We'd be very foolish if we tried to second-guess or tried to figure out what the actual turn of events were without having all the information," he said. "We're six folks up here, we know what happened on our side, we'll get together with the folks on the ground and we'll put the whole story together and make sure it never happens again."

Whatever the exact cause, Flight Director John Shannon confirmed the formation of an investigative board on 4 December. Some KSC engineers felt that the issue was due to an oversight by Chawla, although Lee Briscoe came to her defence by speculating that a Payload General Support Computer (PGSC) might have malfunctioned and prevented her inputted commands from reaching the satellite correctly. "The guys have gone back and looked at the data, they've dumped the PGSC log files and when they look at those, they don't see some of the indications there that the command was issued to the spacecraft that puts it in the mode to be ready to deploy," said Briscoe. "Did the crew miss a step? That's a possibility. Was there something in the PGSC software that didn't issue it? That's a possibility."

As discussions and recriminations got underway, the STS-87 astronauts pressed on with the final days of their mission. In addition to the USMP-4 investigations and Scott and Doi's spacewalks, Leonid Kadenyuk worked on the Collaborative Ukrainian Experiment, which explored the impact of the space environment on the pollination and fertilisation of *Brassica rapa* (turnip) and soybean seedlings. Watching the experiment, and participating in identical 'control' studies on the ground, were around half a million students and teachers in the United States and 20,000 others in Ukraine. The underlying objective of CUE was to develop an understanding of how to grow food plants in microgravity, which will prove highly desirable when long-duration missions into deep space become a reality. Many of the plants returned to Earth were healthy, although some were smaller than expected, averaging 4-5 cm in length, rather than about 12 cm.

In spite of their heavy workload and unsympathetic questions from the media, the four rookie members of the crew enjoyed their first experience of space travel. "On the tenth or eleventh day, I wanted to do one full orbit and sit and watch the Earth," said Chawla. "Doing so was mind-boggling. It really instilled this huge sense of how small Earth is. An hour and a half and I could go round it. I could do all the math and logic for why that was, but in the big picture the thing that stayed with me is this place is *very small*! I feel every person needs to experience this, because maybe we would take better care of this place. The planet below you is our camp site and you know of no other camp ground around." It was a philosophical reflection and memory which would remain with her, long after Kregel brought Columbia to a smooth touchdown on KSC's Runway 33 at 7:20 am EST on 5 December, wrapping up a mission of just less than 16 full days.

The orbiter, however, had endured significant damage to her thermal protection system. Light spots were clearly visible on many tiles and her overall condition was described as "not normal". Ordinarily, around 40 tiles' worth of damage was anticipated, but on STS-87 over one hundred tiles were found to be irreparable. Such damage was mainly caused by fragments of ice or foam falling from the ET during ascent, but the impacts on STS-87 did not follow aerodynamic expectations and were far greater in number. No fewer than 308 'hits' were counted during post-flight inspections, over a hundred of which were larger than 2.5 cm in diameter and some penetrated 75 percent of the total depth of the protective tiles. The chief suspects were new, environmentally friendly products added to the ET, including an upgraded foam insulation, large quantities of which had impacted the Shuttle. It was determined through theoretical modelling and still photography that the foam fell from the ET and was carried by aerodynamic flows to 'pelt' the orbiter's nose and fuselage during her violent climb to orbit. Possible causes included a defective coating of primer and the new foam insulation on the tank. The worrisome STS-87 findings would take centre stage during the Columbia Accident Investigation Board, less than six years later.

Meanwhile, another investigation board – chaired by former astronaut Rick Hieb – had already gotten its teeth into the issue of what caused the SPARTAN-201 failure. "They will go through the normal process of isolating what the cause may be, as one normally does in an investigation," said Shuttle Program Manager Tommy Holloway. "We'll have the opportunity to start looking at the hardware next week and I would expect in fairly short order we'll be able to determine the hardware implications. Then we'll look at the computers the crew used and examine those and go on down the fault tree." Initially, the investigators blamed crew error and as late as April 1998 *Flight International* noted that "the whole crew . . . has been cited for failing to monitor and support their colleagues, because this had not been included in the training checklist". Specifically, "poor communications with ground controllers" and "insufficient software" were contributory factors. However, Chawla and her crewmates were eventually absolved from blame and NASA acknowledged that instructions transmitted to them may have been unclear. Lighting conditions had not been simulated and established rules for how to match rotation rates were not present. Kevin Kregel added that the failure gave the crew a chance to demonstrate their ability to develop alternative procedures in orbit, which would undoubtedly be needed to work around problems during ISS operations. "A big success was, in a short timeframe, all the folks got together and came up with a plan to retrieve a very valuable asset," he said. "If that doesn't show the ability of humans to adapt to changing situations in space, then I don't know what is."

SCIENCE FOR SPACE STATION

Before joining NASA, Scott Douglas Altman's claim to fame was that he provided a stand-in for Tom Cruise during the making of the movie *Top Gun*. In 1986, he was attached to Fighter Squadron 51 at Naval Air Station (NAS) Miramar in California, during which time he completed two deployments to the Western Pacific and Indian Oceans, flying the F-14A Tomcat. It was in this capacity that Altman performed several aerial manoeuvres for Cruise's character, Pete 'Maverick' Mitchell. In fact, Altman was the pilot responsible for having inverted his F-14 at the enemy 'MiG' . . . and for having been told to offer an obscene gesture to the enemy pilot. "*Top Gun* was a real thrill," he remembered, years later. "The word was going around town that Hollywood was coming to Miramar and they were going to do a movie and we were all excited. My squadron had just gotten back from a 7.5-month cruise, so our airplanes were at home, we were available, we weren't too highly tasked. It turned out they picked my squadron to fly the F-14s. Then the skipper got together and tried to pick four guys that he thought were mature enough to handle the capability that they were being given in working with the movie. The flying was incredible. Most Navy pilots don't get to buzz the tower, like in the movie. If you did, you could just peel your wings off and throw 'em at the door, because you probably wouldn't be flying anymore! But since it was Hollywood, they wanted the scene. I had to buzz the tower and, of course, they wanted *nine* different takes . . . so we did it *nine* times!"

Little did Altman know that a dozen years later, he would strap into a far bigger and more powerful vehicle, as pilot of Columbia on STS-90, for his first journey into space. In his later NASA career, Altman would go on to command the final Shuttle servicing mission to the Hubble Space Telescope. Born in Lincoln, Illinois, on 15 August 1959, he grew up with a love of flying. "I think I was watching the old *Sky King* TV show," he recalled. "It had a pilot as the star of the show and that's what I wanted to be." He graduated from high school in 1977 and entered the University of Illinois to study aeronautical and astronautical engineering. His original goal was to enter the Air Force – as an Illinois boy, he lived in the middle of the United States and there were "not many oceans around to *see* aircraft carriers", so the Navy did not appear to be an option at first – but it turned out that he was too tall to pass the initial screening. After this rejection by the Air Force Academy, in what Altman described as "a bump in the road", he discovered that the Navy had different height standards and he was accepted. Upon receipt of his degree in May 1981, he entered the Navy and received his wings of gold in February 1983. Altman was subsequently assigned to NAS Miramar, just north of San Diego, and in 1987, after *Top Gun*, he was selected to attend Naval Test Pilot School at Patuxent River, Maryland. Flying as a fighter pilot and later a test pilot was, in Altman's mind, the key which later unlocked an astronaut career for him. Upon graduation from test pilot school, and following receipt of a master's degree in aeronautical engineering from the Naval Postgraduate School in 1990, he flew the new F-14D as a maintenance officer, operations officer and strike leader over southern Iraq. Altman applied unsuccessfully to join the March 1992 astronaut class, but was ultimately selected by NASA in December 1994.

Standing just over 1.8 m in height, Altman formed part of the tallest crew ever launched into orbit on STS-90. Five of the seven astronauts exceeded 1.8 m, with a sixth coming close. With (almost) six strapping six-footers as her crewmates, the only STS-90 astronaut who missed the mark by a considerable margin was diminutive Kathryn Patricia Hire, the flight engineer. Born on 26 August 1959 in Mobile, Alabama, she watched the Apollo lunar landings as a child and observed the Moon through a telescope, "and just developed a love of aviation and aerospace in general". By her own admission, education was critical to unlocking future career opportunities. "When I went to high school, there were no women at the Naval Academy," she told a NASA interviewer. "That is something that became available before I graduated from high school, so I was able to apply for that position and was actually accepted." She earned a science degree in engineering from the Naval Academy in 1981 and entered active military duty. "Then, when I was at the Naval Academy, there were no female Naval Flight Officers, so that became available before I graduated from the Naval Academy, so I was able to take advantage of *that* opportunity." Hire earned her Naval Flight Officer Wings in 1982 and flew worldwide oceanographic research missions to 25 countries aboard the P-3 Orion aircraft. At this time, she was based at NAS Patuxent River, Maryland. Hire later taught airborne navigation at Mather Air Force Base in California and in January 1989 she left active duty – but remained a Navy reservist, based at NAS Jacksonville in Florida – to work for Lockheed as a Shuttle mechanical systems engineer at KSC. She later gained a master's degree in space technology from Florida Institute of Technology in 1991 and rose to become test project engineer and ultimately supervisor of Shuttle mechanical systems and launch pad access swing arms. She also

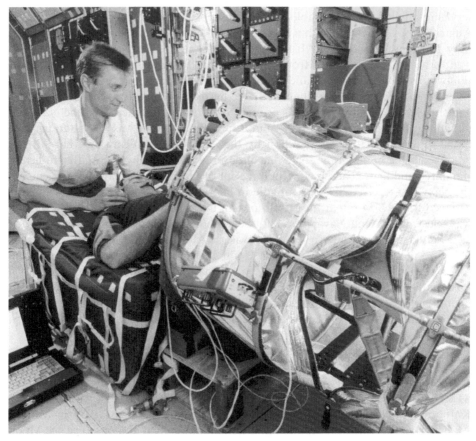

3-12: Jim Pawelczyk (left) and Dave Williams train for their research duties aboard a mockup of the Spacelab module for the Neurolab mission.

headed the checkout of Shuttle space suits and the Russian Orbiter Docking System. During her six years at KSC, Hire supported the processing of the orbiters from landing to ground preparations to launch on over 40 missions. Still a Navy reservist, she became the first woman in the US armed forces to serve aboard a combat aircrew in May 1993. This came in the aftermath of the Secretary of Defense's repeal of Title 10 United States Code combat aircraft restriction for female military aviators and Hire was assigned to Patrol Squadron 62, with which she served as a navigator/communicator aboard the P-3C Update III Orion maritime patrol aircraft in the Atlantic Ocean, Europe and the Caribbean Sea. She was selected as an astronaut candidate by NASA in December 1994 and on STS-90 became the first former KSC employee to fly into space.

Joining Altman and Hire were four science crew members – payload commander Rick Linnehan, mission specialist Dave Williams and payload specialists Jay Buckey and Jim Pawelczyk – together with veteran mission commander Rick Searfoss. As was normal protocol for Spacelab missions, the science crew was named in advance of the orbiter crew, with Linnehan, Williams, Buckey and Pawelczyk announced in August 1996 and Searfoss, Altman and Hire in April 1997. The credentials of the science crew (Linnehan was a veterinarian, Williams and Buckey were physicians and Pawelczyk a physiologist), together with the STS-90 payload, 'Neurolab', implied that the mission was a medical and biological research flight. In fact, it was also the final voyage of the Spacelab module and the last dedicated life sciences mission aboard the Shuttle before anticipated long-duration research commenced on the ISS. With this last throw of the dice for the life sciences community, Searfoss was keen for his crew to participate in as many experiments as possible. In the early days of the space programme, pilot astronauts typically avoided medical matters as if they were a plague of locusts. All were keenly aware that there were only two ways for a pilot to walk out of a physician's office: *fine* or *grounded*. Searfoss, however, had no concerns about involving himself in Neurolab's experiments. "This is research," he told an interviewer before launch. "No one's out there to ground you. No one's going to find something that's going to keep you on deck forever."

By the time of STS-90, he had flown two earlier Shuttle missions as pilot, including the second Spacelab Life Sciences (SLS-2) flight. "Test pilots comprise about a third of the astronaut corps," he said. "I think that diversity in every respect – both technical background and nationality – is very healthy and great for the programme. There is a saying that once you get into orbit, *everyone* is a mission specialist. Certainly, the test pilot skills are an absolute must for the launch and entry phases, but in orbit you're there to do a job."

To do that job, Searfoss praised the "four brilliant docs", Linnehan, Williams, Buckey and Pawelczyk, who undertook 16 days of Neurolab research. The science was among the most complex ever attempted, focusing on the nervous system and investigating the effects of microgravity on blood pressure, balance, movement and sleep regulation. Nor was the mission exclusively focused on its human subjects, for the crew also transported over 2,000 animals into orbit and, controversially, euthanised and dissected several of them to perform complex surgical procedures. However, Linnehan insisted that all were treated respectfully. "I can guarantee the animals are well-fed, well-housed and well cared for," he said. "It's my duty to check every day to make sure everything looks good as far as their food, water and general health. I have absolute authority, on-orbit, if I need to, to stop an experiment if an animal becomes sick."

The $99 million Neurolab came about following President George H.W. Bush's declaration of the 1990s as the 'Decade of the Brain', in recognition of advancements made in humanity's awareness of its basic structure and function. The 26 experiments inside the Spacelab module were broadly grouped into eight sets: four of which used Columbia's crew as test subjects, whilst the others focused on mice, snails, crickets and fish. The mission attracted participation from the United States' National Institutes of Health, NASA and several international sponsors, including ESA and the Canadian, German, French and Japanese space agencies. Key questions still required answers, in spite of 37 years of sending humans into space, including the mystery of how astronauts adapted so quickly to the weightless environment, even though all of our basic movements were learned in the presence of gravity. Moreover, physiologists were intrigued to learn how gravity-sensitive organs, such as the inner ear, cardiovascular system and muscles, coped in space and how astronauts' sleep patterns were affected as the Sun 'rose' and 'set' 16 times in each 24-hour period. It was as part of these studies that STS-90's animal passengers entered the equation. How would their inner-ear mechanisms – so vital for balance and motion – change in microgravity . . . and what of animals actually *born* in space? Would their gravity-sensitive organs develop differently to their Earth-born counterparts? The ultimate question was: *Must* gravity necessarily be present at the point in an animal's life when basic locomotion skills, such as walking or climbing, were being learned? The STS-90 crew would also participate in this research by way of studies before, during and after the flight.

Preparing Columbia for launch was as complicated an affair as the experiments themselves. After having been transferred to Pad 39B on 23 March 1998, tracking an opening launch attempt on 16 April, many of her mammalian and aquatic passengers had to wait until the final hours of the countdown before they could be loaded into their cages in the Spacelab module. This proved more difficult, since the Shuttle was oriented in a vertical position. Working from the middeck, two technicians were lowered, one at a time, in sling-like seats, down the 6 m tunnel into Neurolab in the payload bay. One technician waited in the tunnel, whilst the other entered the module itself to await the cages and aquariums, which arrived in separate slings. The delicate procedure was completed without incident, although telemetry from a couple of fish subjects – whose heads had been 'wired' with tiny transmitters – proved spotty. When one fish 'hid' in the corner of its tank, for example, the antenna used to gather telemetry data occasionally refused to function properly. Otherwise, preparing NASA's oldest Shuttle orbiter for her 25th space mission ran smoothly and on 13 April Searfoss and his six crewmates arrived at KSC, with an 80 percent likelihood of acceptable weather on launch day.

They were even treated to a call from President Bill Clinton, who was at JSC in Houston at the time. "I hope you find out a lot of things about the human neurological system to help me," Clinton told Searfoss with a grin, "because I'm moving into those years where I'm getting dizzy and I'm having all these problems and I expect you to come back with all the answers!"

"Well, thank you, Mr President," Searfoss replied. "We'll take that on board as one of the challenges we'll try to meet."

The answers Clinton requested would be another day coming, because Columbia's liftoff was postponed 24 hours to allow technicians to replace a faulty network signal processor, needed to relay voice and data between the astronauts and Mission Control. The device was over a decade old and a replacement was 'borrowed' from the orbiter Endeavour. By the evening of the 16th, the replacement unit had been fitted in an avionics bay in the middeck and tested. In the meantime, the animal lockers – carrying 18 pregnant mice and 1,514 crickets – were removed for maintenance and later replaced. Were another delay to be enforced on 17 April, STS-90 would stand down for four days, because many more of the animals would need to be replaced. Fortunately, launch on the 17th went without a hitch and Columbia speared into orbit at 2:19 pm EDT.

"Ascent was *incredible*," wrote Canadian astronaut Dave Williams in his diary. For launch, he sat on the flight deck, which offered him an all-round perspective of the dynamic climb to orbit. "The view was very impressive as we completed the roll program and progressed to SRB separation. Before SRB separation, there was a lot of vibration, but the ride quickly became very smooth as we climbed to our final orbital altitude." Making his first launch into space, Dafydd Rhys Williams came from Saskatoon, within Canada's western prairie province of Saskatchewan, where he was born on 16 May 1954. Although he knew that Canada was the third discrete nation to place its own satellite into orbit, the young Williams knew that there was little chance of becoming an astronaut, so he focused his attention on terrestrial exploration instead. He learned to scuba dive at the age of 13 and drew inspiration from the idea of underwater exploration, watching the exploits of Joe MacInnis, Jacques Cousteau and Phil Newton. He received much of his schooling in Quebec and graduated from McGill University in Montréal with a biology degree in 1976. After postgraduate training in advanced invertebrate physiology at the Friday Harbor Laboratories of the University of Washington, he switched to vertebrate neurophysiology for his master's research. "I was kind of blending my diving background with a research background," he admitted to a NASA interviewer, "and as I went on into graduate school and looked at the opportunities for the future, I thought I would like to take my knowledge of physiology and my desire to live and work in these unique environments and apply it clinically, so I applied to medical school." Williams earned his doctorate in medicine, together with master's degrees in physiology and surgery, from McGill's Faculty of Medicine in 1983 and completed a two-year residency in family practice at the University of Ottawa, then a residency in emergency medicine at the University of Toronto and a fellowship in emergency medicine from the Royal College of Physicians and Surgeons of Canada in 1988. He served as an emergency physician at Sunnybrook Health Science Centre, lectured in surgery at the University of Toronto and participated in the training of ambulance attendants, paramedics, nurses, residents and practicing physicians in cardiac and trauma resuscitation. Williams later directed the Advanced Cardiac Life Support (ACLS) programme at Sunnybrook and headed the centre's Department of Emergency Services, whilst also holding assistant professorships in surgery and medicine at the University of Toronto. He was selected by the Canadian Space Agency, alongside Chris Hadfield, Julie Payette and Mike McKay, in June 1992. "The selection process went over the course of six months," Williams recalled. "We started off with over 5,000 people applying and, as I recall, there were over 600 applications from kids, less than ten years of age, because they wanted to get their application in early! I think that's great." Eventually reaching the 20-strong group of finalists, he was summoned to a week-long intensive selection programme, featuring fitness and psychological assessments, aptitude tests and questions. "Looking at the other 19 people around me," he said, "any one of that group would have been an incredible astronaut. I still don't understand why they picked me." Within the CSA ranks, he participated as Medical Officer in the seven-day Canadian Astronaut Program Space Unit Life Simulation (CAPSULS) in February 1994. As part of CAPSULS, he was principal investigator for a study into the initial training and retention of resuscitation skills by non-medical astronauts. In January 1995, alongside Japan's Takao Doi, he was selected by NASA as an international mission specialist candidate.

Three years later, riding Columbia into orbit, a novel manoeuvre awaited Williams and his STS-90 crewmates. Ten seconds after SRB separation, and 2.5 minutes into the ascent, the Shuttle's OMS engines fired in a new engineering test. Normally only ignited in space, they were used for the first time in the 'sensible' atmosphere, in order to evaluate methods of allowing the vehicle to transport up to 230 kg of additional payload into orbit. "The goal," explained pilot Scott Altman of the so-called 'OMS Assist' manoeuvre, which lasted 102 seconds, "is to help to increase the Shuttle's performance margin to get more mass to orbit by using some of that performance of the OMS engines." In total, 1,810 kg of propellant was expended. Four minutes later, as STS-87 had trialled the previous November, Columbia's GPCs rolled the orbiter into a 'heads-up' orientation to improve communications through the TDRS network. All in all, it was an awesome ride for the crew.

The astronauts and 2,000 mammalian and aquatic passengers may have been joined by another in the final seconds before launch, as a wayward bat apparently attached itself to the ET. "We did take his body temperature with an infrared camera," said Launch Director Dave King. "He was 68 degrees and the tank surface

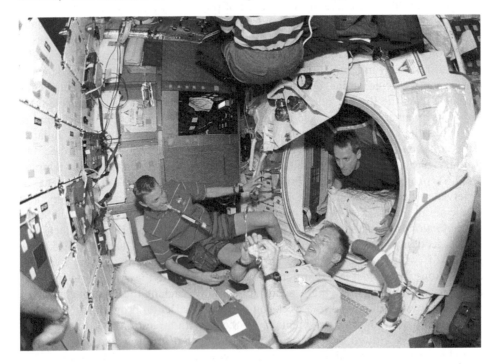

3-13: Unusual perspective of Columbia's middeck during STS-90. Jim Pawelczyk (left) and Rick Searfoss (lower) are joined by Rick Linnehan, who floats through the hatch from the Spacelab module. In the background are the astronauts' sleep stations.

was 62 degrees, so we've decided he was just trying to cool off. Some have said he may have heard the crickets in Neurolab, but it was his choice whether to hang around when we started the engines or not!" Hopefully, for the bat's sake, he took flight before Columbia's three main engines thundered to life, for its sensitive ears would not have survived the acoustic shock. It was the first time that crickets had flown into space. However, there was no possibility of the bat hearing chirping from Neurolab, for the crickets were not yet old enough have the wings needed to produce such sounds. As well as humans, rodents and crickets, STS-90 also carried several hundred snails and fish, whose gravity-sensing organs are either similar to our own or very simple and easy to analyse. It was hoped that the results of such experiments could lead to advances in the development of electrodes as connections to the nervous systems of people with deafness caused by hair-cell damage.

As with previous life sciences flights, Neurolab functioned on a single-shift system, with all seven astronauts waking up at the same time, to preserve their circadian rhythms and support the labour-intensive experiments. For five of them, it was their first experience of space travel and they were determined to enjoy it. "I feel a tremendous pride to be a Polish-American,". said payload specialist James Anthony Pawelczyk, who carried a small Polish flag into orbit with him, "and I wanted to provide some way to share that enthusiasm and pay a little bit of tribute back to the people of Poland." Born on 20 September 1960 in Buffalo, New York, Pawelczyk was the first US astronaut of full-blooded Polish descent to enter space. He completed high school in New York and earned two bachelor's degrees in biology and psychology from the University of Rochester in 1982, then a master's degree in physiology from Pennsylvania State University in 1985 and a PhD in biology and physiology from the University of North Texas in 1989. Following receipt of his doctorate, Pawelczyk completed a fellowship at the University of Texas Southwestern Medical Center in 1992, then accepted an assistant professorship in medicine until 1995. During this period, Pawelczyk also worked as a visiting scientist at the Department of Anaesthesia in Copenhagen, Denmark, and co-edited *Blood Loss and Shock*, which was published in 1994. He was then appointed as assistant professor of physiology and kinesiology at Penn State University in Pennsylvania in 1995, from which he took a leave of absence for the Neurolab preparations. In April 1996, he was selected, alongside veterinarian Alex Dunlap and physicians Chiaki Mukai and Jay Buckey, for payload specialist training.

Meanwhile, Dave Williams carried his wife's flying wings into space with him and Kay Hire, who had worked as a Shuttle engineer, reported that helping to prepare the orbiters on the ground had prepared her amply to actually participate in a real mission. Spotting her home town, and the entire Gulf of Mexico coastline, was particularly special for her. "The first time I had the opportunity to look back at Earth and saw what I thought was familiar, it seemed a little strange," she recalled, "because I thought the picture would be a little smaller, but I was able to see the entire Gulf Coast all in one view. That was surprising to me the first time I saw it, because usually when you see the pictures that are sent down from space, they're *smaller*, and we have a much *broader* panorama that we can see out the Space Shuttle windows. It was very easy to zero in on particular areas, especially Mobile Bay." For

Scott Altman, spotting his home town of Pekin, Illinois, was a particular highlight. "Every time we came over, there were clouds and I missed it," he told a NASA interviewer, years later. "Then, the last day before we deorbited, I looked ahead and there was a pass that was coming close. As I went up to the flight deck, I saw we were going over St Louis and I looked down and *there it was*. I could see the city and from St Louis you follow the Illinois River up and it took me right to Pekin, Illinois, and it was *there*. I couldn't believe my eyes, so I grabbed the camera as fast as I could. I was kind of shaking, taking pictures."

After a busy first day activating Neurolab, the crew was awakened on 18 April by Capcom Chris Hadfield, with the invitation to "get those neurons into action". Overall, for the first part of the mission, Columbia performed without incident, but on 24 April a valve malfunctioned in the Regenerative Carbon Dioxide Removal System (RCRS), which triggered an alarm and raised the very real possibility of a premature return to Earth. "It was a noise I did not recognise," admitted Kay Hire, upon hearing a strange sound at about the same time that the valve failed. Already running two hours behind schedule, the crew initially suspected one of two electronic controllers, but switching to a backup achieved little success. As a precautionary measure, Mission Control asked them to load two lithium hydroxide canisters, thereby providing ample time to evaluate and recover from the problem. Meanwhile, Searfoss and Altman opened the RCRS, removed a hose clamp and used aluminium tape to bypass a faulty check valve, which was apparently allowing cabin pressure to leak into the system and throw off its electronic controllers. The fix worked and Capcom Mike Gernhardt told the relieved crew that the risk of a shortened mission had been averted.

Elsewhere, the Neurolab research was progressing well, including the controversial dissection of several mice in the General Purpose Work Station (GPWS) by Williams and payload specialist Jay Clark Buckey on 18 April. Buckey's involvement with NASA predated most of the other STS-90 crew members, for he had been selected in December 1991 and served as a backup payload specialist for the SLS-2 mission. For his work in support of that mission, he received a meritorious service award from the University of Texas. He was born in New York City on 6 June 1956 and earned a degree in electrical engineering in 1977 and a medical doctorate in 1981, both from Cornell University. Buckey interned at New York Hospital-Cornell Medical Center and completed his residency at Dartmouth-Hitchcock Medical Center and later joined the Air Force Reserve, qualifying as a Distinguished Graduate from the School of Aerospace Medicine and training as a flight surgeon. From 1987 until 1995, he was based at NAS Joint Reserve Base in Fort Worth, Texas. The early part of his medical career was spent at the University of Texas Southwestern Medical Center, as a NASA space biology fellow, research instructor and assistant professor of medicine. In 1996, Buckey was appointed associate professor of medicine at Dartmouth Medical School and took leave of absence to train for STS-90.

In orbit, Buckey and Williams oversaw the decapitation of four adult mice to recover tissue samples from their brains and inner ears, which were needed by members of Neurolab's neutral plasticity team to investigate how nerve cells 'rewire' themselves under the stress of a new and unfamiliar environment. It was hoped that such work could lead to new insights into neurological disorder, such as stroke,

Parkinson's disease and balance problems. Jim Pawelczyk added that the surgical work had to be performed quickly to avoid the onset of degradation in certain nerve fibres. Although it was not the first time that animal dissections had been performed aboard the Shuttle – the SLS-2 crew had done similar experiments in October 1993 – many scientists considered the previous work to have been messy and time-consuming. Neurolab demonstrated the ability of trained medical specialists to conduct intricate surgery in a weightless environment. Later, embryos were taken from euthanised pregnant mice as part of studies into the role of gravity in their early development. The results of such studies were expected to prove crucial in determining whether or not humans or animals could someday be born in space, thus allowing colonies to be established and long-duration voyages to the stars, which would require several generations of space travellers. Before dissecting them, the astronauts injected the pregnant mice with cell 'markers' to label the brain cells in their embryos, track the development and migration of the cells and compare the results with data gathered from Earth-born mice.

Typically, the pea-sized foetuses ranged between 10 and 14 days old and, after removal from their mothers, were preserved whole until the end of the mission. Other mice, with 'hyperdrive' units fitted to their heads, connected to the hippocampus area of their brains, were placed into the GPWS on two maze-like 'racetracks' as part of the Escher Staircase Behaviour Testing experiment. The hippocampus helps to develop spatial maps to help the mice to navigate from place to place and the investigation sought to explain the disorientation frequently experienced by Alzheimer's sufferers. The hyperdrive units did not create unpleasant side effects for the mice, because their brains had no pain receptors, and they had been taught to use the racetracks before launch. They received rewards of sweetened condensed milk after completing a certain number of 'correct' turns. In space, however, the rules changed and investigators monitored their neural activity as the mice struggled to recognise 'home base' after making fewer turns than they had been taught. Several of the mice exercised their 'space legs' for the first time on 22 April, using their forelimbs to scoot around a small jungle gym, but hardly using their hindlimbs. As they moved, the astronauts videotaped the mice and marked their joints with ink to allow each motion to be meticulously analysed after landing.

"It's amazing to watch these animals behave in orbit," said Linnehan on 25 April. "They act just like we do. They learn very fast how to get around their cages and get to food and water. Just last night, we were checking on some of the younger rats, watching them eat. It was kind of akin to the way we eat. We float around and hold our food to feed ourselves. One of the rats was holding onto a piece of food with his front paws, munching on it leisurely, letting it go and going over to drink some water, coming back and grabbing the food again." Later in the mission, the first survivable surgical procedure was conducted on six of the mice. Buckey and Williams anesthetised and injected them with dye markers in a muscle development study. Although the procedure was relatively straightforward, it paved the way for more complex work aboard the ISS.

However, problems were brewing. Electrical problems were noted with transmitters implanted in four oyster toadfish and on 27 April NASA revealed that dozens of the neonatal mice had died unexpectedly. Others were in such bad shape that they had been euthanised by the astronauts, whilst a few were resuscitated and nursed back to health with bottle-feeding of nutritional and other supplements; at one stage, Linnehan thanked his crewmates for helping to hand-feed and care for the ailing mice and checking on them each evening, before bedtime. At first, Linnehan suspected that the deaths may have been caused by the design of the cages; in the free-floating environment, perhaps the rodents found it harder to move around and get to their mothers, eventually succumbing to dehydration and depression. Back on Earth, the immense death toll – more than 50 percent of the neonatals – had drawn condemnation from animal rights groups, including People for the Ethical Treatment of Animals, which charged NASA with abuse and pressed Congress to ban future experiments aboard the Shuttle.

The disaster could not have occurred at a worse time for the neonatals, who were particularly vulnerable and just learning to move and search for solid food in their new environment. The poor state of some of them obliged Linnehan to cancel several experiments, telling Mission Control pointedly that "there is no meaningful data to be gained with these animals at this point". In total, more than 55 out of a total complement of 96 mice were lost. "It was an unforeseeable event," said Linnehan, "and it's regrettable that it happened. However, we still got back most of our primary science." It later became apparent that the ill-fated neonatals were victims of a higher-than-normal rate of maternal neglect, from mothers who had themselves become dehydrated and unable to lactate adequately. As the youngest mice became anaemic, the mothers stopped feeding and grooming them. Linnehan drew praise from animal rights campaigners on 1 May, after defying an instruction to destroy a rodent when ultra-fine electrodes implanted in its brain broke loose. "Based on his expertise and professional opinion," said Neurolab Mission Scientist Jerry Homick, "he determined the animal was not in any danger and determine it was appropriate to return that animal to its housing."

By now, the intense pace of the flight and the additional free time given up to care for the sick rodents meant that most of the crew averaged three or four hours of sleep each night. Linnehan started calling it 'Blurrolab', rather than 'Neurolab', "because I just could never remember what day it was or where we were, we just kept on going". Years later, he described Neurolab as one of the hardest parts of his professional life, not least because, as payload commander, the primary responsibility for the entire science mission fell on his shoulders. "I don't think I'm ever going to work as hard in terms of the time I put in and the midnight oil burned," he said. However, he had nothing but praise for Williams, Buckey and Pawelczyk – "Smart, smart people," he recalled, "*Much* smarter than me" – but highlighted that the crew frequently worked in excess of 16 hours, every day. Nevertheless, despite the neonatal fatalities amongst the mice, the mission was shaping up to be a huge success. "We went into this mission facing a number of challenges," said Homick. "We knew we had a difficult timeline to work with, we knew that we had a number of complex hardware systems ... to acquire the data and we knew we were going to be implementing a number of very difficult experimental procedures, using cutting-edge technologies. With all of that in mind, we did expect to achieve a great deal of success with this mission and I'm pleased to report that I think we've exceeded our expectations."

Although the astronauts had averaged just a handful of hours of sleep each night during the rodent crisis, Dave Williams was able to rest fitfully, which aided another of Neurolab's investigations: a series of measurements of brain waves, eye movements, respiration, heart rate, internal body temperature and snoring. As well as improving the performance of astronauts in space, such experiments were expected to support the sleep-wake cycles of workers on Earth. Not only did the hectic workload of the astronauts carry the potential to impair their performance, but so did the experience of sunrises and sunsets on no fewer than 16 occasions in each 24-hour period. It was recognised that around 20 percent of astronauts on single-shift missions needed to take sleeping pills, which represented between three and eight times as high a number as the general population on Earth. Linnehan, Williams, Buckey and Pawelczyk also donned body suits and sensor-laden skullcaps before going to sleep in order to monitor their rest patterns. Although Williams admitted that his sleep was reduced, he added that he was still able to 'turn over' in his sleep, despite having no pressure points on his body; it was a purely psychological function.

Two of the medical experiments were provided by Canada. The Visuo-Motor Co-ordination Facility (VCF) studied changes in movement during weightlessness which affected the astronauts' pointing and grasping capabilities, whilst the Role of Visual Cues in Spatial Orientation created 'fake' gravity by applying pressure to the soles of the feet to find out if it somehow 'overrode' visual cues and enabled them to readapt to a terrestrial-type environment. Developed by CSA, the VCF projected visual targets onto a screen and the crewman grasped and pointed at them and tracked them as they moved, using an instrumented glove. The motor skills thus demonstrated were recorded at various stages of the 16-day mission to evaluate changes in the nervous system as it adapted to microgravity conditions. A rotating chair, the Visual and Vestibular Integration System, mounted along the centre aisle of the Spacelab module, was designed to stimulate the astronauts' vestibular systems with spinning and tilting sensations. Performed by all four of the science crew, the 45 rpm chair was used six times during the mission, with the astronauts' eyes shielded from external stimuli, giving their nervous systems no visual cues, whilst a video camera, trained on their faces captured their reactions to the sudden motions.

As well as proving somewhat dizzying in nature, a few investigations generated a sting for the crew. One experiment demonstrated an innovative technique known as 'microneurography', whereby a fine needle – about the same diameter as an acupuncture needle – was inserted into a nerve, just below the knee. This allowed nerve signals from the brain to the blood vessels to be measured directly, whilst the cardiovascular system was monitored with the Lower Body Negative Pressure apparatus. Results were expected to aid sufferers of autonomic blood pressure disorders, including 'orthostatic intolerance', an inability to maintain proper blood pressure whilst standing for long periods. To prepare for the microneurography procedure, Linnehan, Williams, Buckey and Pawelczyk spent two months at Vanderbilt University in Nashville, Tennessee, training with the hardware. (In fact, as early as September 1996, astronaut Rhea Seddon – a veteran of SLS-2 – had been

3-14: Beautiful orbital sunrise, featuring the Spacelab module, as viewed from Columbia's aft flight deck windows.

assigned to Vanderbilt's Center for Space Physiology and Medicine as a NASA liaison to evaluate flight equipment and operating procedures.) The insertion of the needle, which all four astronauts typically achieved within about 40 minutes during several runs in orbit, was delicate and difficult. "Finding a nerve," said Principal Investigator David Robertson, "is a lot more difficult than finding a vein. This is a very difficult thing to do on Earth and the idea that it can be done in space is a little bit astounding to many people."

The orbiter crew of Searfoss, Altman and flight engineer Kay Hire participated in several Neurolab experiments. Hire worked with the Bioreactor Demonstration System on the middeck, growing cultures of renal tissue and bone marrow, for possible applications in the treatment of kidney disease, AIDS and other immune-system ailments, as well as for the chemotherapy of cancer patients. The orbiter crew also tended to the very few malfunctions with the Shuttle itself, including a blockage in a waste water dump line.

As STS-90 entered its final stages, it was decided not to extend the mission to 17 days, after reports from the Neurolab team that they had accrued sufficient data and the additional time would be unnecessary. This was a pity, for the conservation efforts of the seven astronauts had added sufficient consumables for a full 36 hours of extra research. Another deciding factor was the weather at KSC in anticipation of landing opportunities on 3 and 4 May. Although it was expected to be reasonable, forecasts on the 4th were not as good as for the 3rd. "There is a front that we expect to start to affect the Cape on Monday [4th]," said Entry Flight Director John Shannon. "It will bring rain showers and low cloud ceilings." The option of landing at Edwards Air Force Base in California was undesirable, due to the need to remove the animals and perform data collection at the KSC facilities immediately after Columbia's touchdown. As circumstances transpired, Searfoss guided his ship to a smooth touchdown on Runway 33 at 12:09 pm EDT on the 3rd, completing a mission of almost 16 days in orbit.

Even before STS-90 launched, there existed a possibility that Neurolab might be reflown, with the same crew and aboard the same orbiter, in August 1998. At a press conference on 15 April, Shuttle Program Manager Tommy Holloway noted that engineers were exploring the feasibility of flying the payload again in the late summer. This was because further delays to the ISS construction effort had opened up a three-month 'gap' in the Shuttle manifest. But with the first station flight scheduled for launch on 4 September it was eventually decided to protect that option and the Neurolab reflight was shelved. (Ironically, ISS construction met with further delay and did not ultimately commence until November-December.) When STS-90's inaugural results were published in April 1999, they offered promising insights into the neurological mechanisms responsible for Alzheimer's disease and epilepsy and the astronauts' cognitive work was expected to lead to diagnoses and rehabilitative measures for brain-injury sufferers. It was also anticipated that the sleep studies would find terrestrial applications for shift workers, the elderly, jetlag sufferers and insomniacs.

Summing up the mission, the Neurolab crew were philosophical and jocular about their accomplishment. "If we don't figure out how to stem nerve-muscle degeneration," said Rick Linnehan, "we're not going to be able to travel to other planets or live in space stations. We would face the risk of fractures when returning to Earth." On the other hand, explained Jay Buckey, who later returned to Dartmouth Medical School, at least in space "you can put your pants on both legs at the same time!"

END OF THE OLD ERA

Three weeks before the New Era began, the Old Era ended.

Today, 1998 is remembered as the year in which construction of the ISS got underway, with the launch of Russia's Zarya control module in November and the first Shuttle assembly mission, STS-88, with the Unity connecting node, in early December. However, another highlight of the year was the return to space of one of America's 'Original Seven' Mercury astronauts, John Glenn, on a flight which secured him the title of the oldest human ever to be rocketed into space. It is a title which Glenn retains at the time of writing in 2014 and it was an event which served to unite the old era with that of the new. Yet, aside from the historic nature of his mission, Glenn's return to space and the motivations of NASA itself were greeted with praise and criticism in equal measure.

Late in the summer of 1997, the rumour mill became rife with speculation that Glenn – more than three decades after becoming America's first man to orbit Earth – might fly a Shuttle mission as part of efforts to understand the process of aging in the microgravity environment. Several months later, in January 1998, amid much public fanfare, Glenn was formally assigned as a payload specialist on STS-95, a scientific research mission, scheduled for launch in late October. In flying STS-95, the 77-year-old Glenn would become history's first septuagenarian astronaut and would soundly surpass the achievement of 61-year-old Story Musgrave, the previous record-holder. However, some saw the flight as little more than a publicity stunt, with questionable scientific value, and others argued that risking the life of Glenn, a US national hero, was inexcusably rash.

Having been selected into NASA's first group of astronauts, and having flown the Friendship 7 mission in February 1962, John Herschel Glenn Jr later moved into politics. He served with distinction for a quarter of a century as a senator for Ohio and ran unsuccessfully for the presidency in the 1984 election. To the media, Glenn was a hero from the very day that he and the other members of the Mercury Seven were introduced in April 1959. Freckle-faced, witty, articulate and charismatic, he was described by some journalists as epitomising 'all-American' qualities and many were surprised when he did not secure the first American suborbital flight. Glenn was born in Cambridge, Ohio, on 18 July 1921, although he grew up and received his education in the town of New Concord, studying engineering. Already, as a youth, he had undergone flight training and took the Army Air Corps' physical examination and passed. However, when no orders materialised, he took the Navy's physical, which he also passed and was sworn into the Naval Aviation Cadet Programme. Initial training at the University of Iowa was followed by preparation at Olathe, Kansas, and finally at Corpus Christi, Texas. It was whilst stationed at the latter base that Glenn learned of his eligibility to apply for the Marine Corps, which he did, winning his wings and lieutenant's bars in 1943. That same April, he married Annie Castor.

After a year of training, Glenn joined Marine Fighter Squadron 155, flying F-4U combat missions in the Marshall Islands of the Central Pacific during the Second World War. He returned to the United States shortly before the end of the conflict to begin test pilot work at Patuxent River, Maryland, evaluating new aircraft. Subsequently, he served as an instructor in advanced flight training at Corpus Christi from 1948-1950, completed marine amphibious warfare training and flew 63 combat missions during the Korean conflict. It was whilst in south-east Asia that Glenn shot down three MiGs along the Yalu River, earning himself the nickname 'MiG Mad Marine'. (He was also tagged 'Magnet Ass' for his ability to attract flak.) Overall, in the Second World War and Korea, he flew 149 combat missions and his chestful of medals proved it: six Distinguished Flying Crosses and an Air Medal with 18 clusters. After Korea, he joined the Navy's Test Pilot School at Pax River, later serving as project officer for a number of advanced fighters. Whilst serving in this capacity for the F-8U Crusader, he set the transcontinental speed record by flying non-stop from Los Alamitos Naval Air Station in California to Floyd Bennett Field on Long Island. The attempt arose from Glenn's desire to kill two birds with one stone: running the Crusader's engines in afterburner at full combat power whilst at high altitude and seizing the Air Force-held transcontinental speed record, which then stood at three hours and 45 minutes. "We could do the test," he wrote in his memoir, "and also call attention to the fine plane the Navy had purchased as its frontline fighter." Under the name 'Project Bullet' – so-called because the Crusader flew faster than the muzzle velocity of a bullet from a .45-calibre pistol – Glenn volunteered himself as pilot for the attempt. "The plane flew beautifully," he wrote of the epic flight on 16 July 1957. Glenn, who beat the previous record by 21 minutes, was awarded another Distinguished Flying Cross, his fifth in total.

In his 1999 autobiography, *John Glenn: A Memoir*, he described the process by which he came to be considered for the STS-95 mission. "In early 1995," he wrote, "I prepared for debate on [NASA's] budget by reviewing the latest NASA materials, including *Space Physiology and Medicine*, a book written by three NASA doctors. As I read, a chart jumped out at me." It described the physical effects upon astronauts in space, including aspects of muscular change, osteoporosis, disturbed sleep patterns, balance disorders, a less responsive immune system, cardiovascular changes, loss of co-ordination, a decline in drug and nutrient absorption and differences in blood distribution patterns. After several years working on the Special Committee on Aging, Glenn pondered the possibility of an older person venturing into space to evaluate these effects. In his mind, such research carried potentially great implications for future ISS crews, who would spend many months in orbit. Late in 1995, he approached NASA Administrator Dan Goldin for the first time with the possibility and was received with warm enthusiasm, but it would appear that Glenn's support for President Bill Clinton in his effort to secure re-election in 1996 immeasurably aided his campaign. After discussing the idea with the president, Glenn described a flight with Clinton in Ohio, whilst on the campaign trail. "We

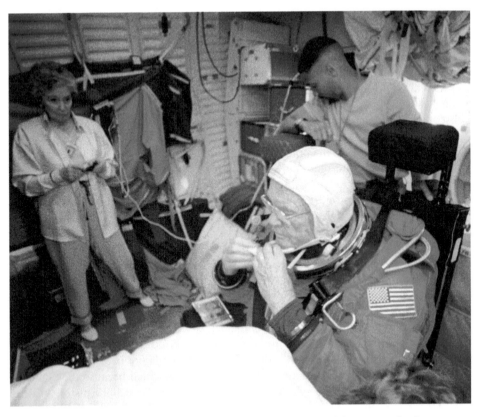

3-15: With his wife, Annie, looking on, John Glenn adjusts his communications cap ahead of an STS-95 training session.

were returning to the airport," Glenn wrote, "when he [Clinton] brought the subject up again. He grinned, leaned over and slapped me on the knee. 'I hope that flight works out for you,' he said." Yet, according to space policy analyst John Logsdon, Goldin could not have been pressurised by the Clinton administration. "There was no intervention that caused it to happen," insisted Logsdon. "More to the point, the one thing the White House could have done was veto it and they obviously did not do that."

Of course, it was not enough to have the backing of the president. To convince the nation and the world that this was not simply an old senator and US hero getting a joy ride into space required Glenn to ensure that his mission pulled its own weight in terms of scientific gain. "My campaign for a shuttle flight had taken me to Dr John Eisold, the US Navy admiral who was the attending physician for Congress," wrote Glenn, "and had an office in the Capitol." At the Bethesda Naval Hospital in Bethesda, Maryland, he was put through heart exams, liver, kidney and pancreatic scans, a whole-body MRI scan, and a head scan. At the end of the gruelling tests, Eisold told Glenn that he saw no physical problem which would prevent the former astronaut from flying again. At a follow-up meeting with Dan Goldin – and carrying his medical results in hand – Glenn reiterated his interest. This time, Goldin took serious notice of his words and gave Glenn two conditions: first, that his mission should be scientifically valuable, and second, that he should be able to pass all of the exams required by active-duty astronauts, many of whom were half his age.

NASA had been roundly criticised in the aftermath of the 1986 Challenger disaster for its practice of flying 'passengers' aboard the Shuttle, the most tragically famous of whom was schoolteacher Christa McAuliffe. In announcing his decision, Goldin stressed that Glenn's assignment was based upon a desire to "expand the reach" of the astronaut corps, with McAuliffe's backup, schoolteacher Barbara Morgan, already selected as a member of the 1998 class of Shuttle pilots and mission specialists. "We are *not* reviving the civilian in space programme," said Goldin of Morgan's selection. "We are asking people to go through the formal process. As we're trying to get biologists and geologists because of the tremendous findings we're having in planetary science, education is a prime mission of NASA. We want scientifically developed educators to participate. Any future astronauts will go through the full training programme at NASA." This declaration made it easier for Goldin to silence the naysayers who felt that Glenn's mission and the selection of Morgan was a stunt; for Glenn was, after all, undertaking the full Shuttle payload specialist syllabus and would be qualified by the end of the process. Moreover, Morgan would undergo the full mission specialist training syllabus and fly as a fully-qualified NASA astronaut.

In *John Glenn: A Memoir*, Glenn stressed that his wife, Annie, and grown children were unhappy with the notion of him venturing into the cosmos again. (Annie's initial response was "Over my dead body!") However, it was turning inexorably from a dream into a reality. "Back in Washington, on 15 January 1998," Glenn wrote, "an aide interrupted a meeting with the word that I had a phone call." It was Dan Goldin. "You're the most persistent man I've ever met," said the NASA Administrator. "You've passed all your physicals, the science is good and we've called a news conference tomorrow to announce that John Glenn's going back into space!" Next day, the assignment of Glenn to STS-95 was revealed to the world's media. "What a great day for America," exulted Goldin, "because the man who almost 36 years ago climbed into Friendship 7 and showed the boundless promise for a new generation is now poised to show the world that senior citizens have the right stuff."

If there was an important, global message to emerge from Glenn's mission, it was the fact that 'elderly' people no longer had to be barred from life experiences on the basis of their age. In his memoir, Glenn remembered an elderly couple stopping him at an airport whilst training and telling him that he had changed their lives for the better. For years, the couple told Glenn, they had been putting off a planned trip of a lifetime to climb Mount Kilimanjaro, as life, work and children entered their lives. Eventually, they considered themselves too old ... until NASA announced that it was flying Glenn on the Shuttle. The husband then turned to his wife and told her that if John Glenn could return to space at the age of 77, then *they* could certainly climb the fabled African peak. They duly booked tickets to do it. "More than anything," wrote Glenn, "I think the excitement surrounding my return to space was due to that redefinition of what people could expect of the elderly and what the elderly could expect of themselves."

By the time the remainder of the crew was named, STS-95 was already being dubbed 'The John Glenn Flight' by sections of the media. It was a nomenclature to which Glenn himself was vehemently opposed. "I'm here to be a member of this crew and work with everybody else," he told an interviewer. "I'll be doing some of the experiments myself, I'll be backing up some of the other people. I'm here as a working crew member and that's *it*. I hope everybody concentrates on the science of this thing." However, *the science* remained a key stumbling block. Although STS-95's medical experiments carried a clear slant towards the study of aging in space, some journalists doubted that having a single data point – Glenn – was in itself sufficient to begin a long-term programme. Unsurprisingly, Glenn disagreed. "You've got to start *somewhere*," he told one probing enquiry. "I hope eventually we have a very large database out of all this. If I'm fortunate to be the one to start, I'm honoured to be the first data point on the new chart." He compared it to his orbital flight in 1962 – just "one flight," he admitted, then rhetorically asked "Was it statistically significant or not?" – and expressed a fervent wish that more 77-year-olds could be sent into orbit. Former astronauts added to the torrent of doubt. Mike Mullane wondered in *Aviation Week & Space Technology* why, if research into aging was necessary, older candidates had not been selected in recent astronaut classes. Story Musgrave, who flew aged 61 in November 1996, added that NASA needed to be honest about why it was flying Glenn. "We are flying a legislative passenger, as we have in the past," he said. "It's John Glenn. Marvellous. But it *is* a legislative passenger." Musgrave also pointed out that Glenn did not devote himself, full-time, to the mission, an assertion which the senator admitted. "We in the Senate normally go for about three to five weeks, then we have a week or two where people are back in their states," he told a NASA interviewer in August 1998, "and I've used those periods to be in Houston or over at the Cape, training with the crew. And all this month of August, the Senate is out of session, so I'll be here. I'm here all during this month and then we will go out of session in very early October and I'll be here training continually from that time on."

Three weeks after Glenn was announced to join the STS-95 mission, in February 1998 the remainder of the crew was announced. Commanded by veteran astronaut Curt Brown, it comprised pilot Steve Lindsey, payload commander Steve Robinson, mission specialists Scott Parazynski and Spain's first spacefarer, Pedro Duque, and Japanese payload specialist Chiaki Mukai. These 'professional' crew members had nothing but praise for their septuagenarian comrade. Brown was nearly six years old when Glenn flew Friendship 7 and recalled his pioneering mission as "a big, significant event in my life". Lindsey added that he was surprised to even *meet* one of the Original Seven Mercury astronauts, much less *fly* into space with one of them. Robinson agreed, pointing out that if he had been told that he would be flying with the first American to orbit Earth, "I would have said that is just impossible, *squared*". In the later stages of his STS-95 training, Robinson sketched a design for the crew patch that included a silhouette of the Shuttle and Friendship 7. "I remember drawing pictures of his rocket and entering them in art contests,"

Robinson said of his childhood exploits, "developing my skills as an artist." Thirty-six years later, as he drew the design for the STS-95 patch, his crewmates kidded him that he was *still* drawing pictures of John Glenn's rocket as an *adult*! For Parazynski, flying into space with Glenn was like climbing Mount Everest with Sir Edmund Hillary or playing baseball with Babe Ruth or football with Pele. "This, for an astronaut," Parazynski said, "is about as exciting as it gets!"

The seven astronauts of STS-95 would support more than 80 experiments in a pressurised Spacehab module and in the orbiter Discovery's middeck and payload bay. They would also refly the SPARTAN-201 satellite, whose deployment during STS-87 had been hampered by technical difficulties. Brown noted that STS-95's large number of simultaneous experiments offered a close analogy to forthcoming ISS research. "We have an international team," he said. "We have a lot of experiments, ranging from protein crystal to life sciences to hard-core science, so we have basically everything that you'll see on the International Space Station, other than we won't be staying up there for four, or five, or six months."

With the SPARTAN-201 reflight planned, it seemed hardly surprising that Lindsey – a member of the STS-87 crew – had been selected to join the mission. "After STS-87, we did a thorough investigation into all the things that happened to cause this, to try to prevent it when we fly again," he told a NASA interviewer. He noted that SPARTAN's software on STS-87 did not provide the Shuttle crew with any feedback (or 'talkback') of commands having been received. "Since then, the software's been modified, not in how it operates with the satellite, but the feedback it gives to us. Now, when we send a command, it tells us that we *did* send that command and we can actually go back to the screen later and verify this." Similarly, if SPARTAN-201 again failed to perform its pirouette manoeuvre, the release position of the RMS was adjusted to prevent the arm from bumping the satellite. "Whereas before it was kind of an off-angle, such that when you go in to regrapple it you get a lot of arm dynamics," said Lindsey, "and the arm bounces it around quite a bit, now it's in more of an orthogonal or a square position, so that we go straight in." To Curt Brown, it illustrated an important aspect about the space programme: that each mission posed its own array of questions and that the astronauts and planners "never quit learning". He added that STS-95 boasted a new RMS end effector camera, to avoid the trouble experienced on STS-87, when the view 'bloomed' and did not return consistently good images to the astronauts. In turn, this made manipulating the mechanical arm and grappling SPARTAN-201 difficult. Brown hoped that the enhancements and changes would be of benefit on STS-95. "I don't think there's anything you could point to and say this is a major shift in policy or philosophy," said STS-95 Lead Flight Director Phil Engelauf at a press briefing on 15 October, "just a lot of little improvements over the whole operation."

Launching only a few weeks before the first elements of the ISS, all seven STS-95 crew members were keenly aware that theirs was a precursor for space station research, albeit condensed from a period of several months into a duration of just a handful of days. "STS-95 has the look and feel of a space station," said Steve Robinson, whose role as the mission's payload commander encompassed responsibility for its scientific success. "We have an international crew, we have an international complement of payloads, sponsors and participants from all over the world who are involved in our mission. And the reason we're going up and flying is to learn things we didn't know before, the same reason we're going to fly International Space Station." The research extended from life sciences to protein crystal growth to materials processing to solar physics, astronomy and space technology.

Of the crew, only Duque was making his first flight. Born Pedro Francisco Duque in Madrid on 14 March 1963, he became the first Spaniard ever to journey into space; a fact which led his STS-95 crewmates to nickname him *Juan Glenn*. As a young child, Duque saw the first humans walk on the Moon and was enthralled by it. "I remember where that TV was located," he said, even though he was only six years old in July 1969, "and where I was that day." He graduated from the Universidad Politécnica de Madrid (UPM) in 1986 with a degree in aeronautical engineering and initially worked for Spain's Tres Cantos-based Grupo Mecánica del Vuelo (GMV) aerospace and defence company. Duque later joined ESA's European Space Operations Centre (ESOC) in Darmstadt, Germany, supporting the Precise Orbit Determination Group, and was a member of the flight control team for the Earth Remote Sensing Satellite (ERS)-1 mission and the European Retrievable Carrier (EURECA). Selected by ESA as an astronaut candidate in May 1992, he later trained as backup crewman to Germany's Ulf Merbold for the EuroMir '94 mission to the Mir space station. In May 1995, he was named as a backup payload specialist for STS-78 and in August 1996 joined NASA's astronaut corps as a European mission specialist candidate. He was formally certified as a qualified mission specialist in April 1998, two months after his assignment to STS-95. "I always tell everybody that I have won the lottery many times in a row for this," he once told a NASA interviewer, "but only I get to fly in space."

As the first full-blooded Spaniard ever to venture into orbit, Duque represented a nation whose star had diminished in international brightness from its glory days of empire. "It's exactly 100 years since we lost that war around Cuba with the United States," he explained in the weeks before the STS-95 launch, "and Spain really decidedly started to go down in the consideration of the world and everybody started to think of Spain as a very small and not very important country. We're celebrating in some way this hundred years, taking into account that now Spain is back: a country that joins the front and is no longer in the tail of the group." Duque highlighted Spain's involvement in the ISS and in the Ariane programme and during the course of STS-95 he spoke to Education Minister Esperanza Aguirre and took questions from schoolchildren representing the country's 17 regions.

The crew arrived at KSC several days before their scheduled 29 October liftoff, coincidentally on Robinson's 43rd birthday, which prompted Brown to lead an impromptu chorus of greetings. "I hope these guys fly better than they sing," quipped Robinson in response. "Thank you very much." Weather remained questionable, particularly in view of the presence of Hurricane Mitch to the south of the launch site and a strong high-pressure system to the north, which produced a 40 percent likelihood of unacceptable high winds. With two opportunities to fly on the 29th and 30th, any delay beyond that would necessitate a 96-hour stand-down to allow the ground crews to rest, before picking up the campaign for further attempts on 4 and 5 November. And after *that*, the Leonids meteor shower precluded any further chance of launching until the 17th.

So it was that around 250,000 wellwishers packed the roads and beaches of KSC in anticipation of good weather and a good launch on 29 October. President Bill Clinton and more than 3,500 journalists – one of the highest numbers ever present for a human launch – were on hand. Glenn's former Mercury crewmate Scott Carpenter, who famously wished him well before Friendship 7, was also on hand to extend his own greeting, although it was on another NASA television circuit during the T-9 minute hold in the countdown. In the days and weeks before launch, it surprised Glenn that so much attention was being given to the mission. With 1998 having already generated immense political scandal and lasting damage to President Bill Clinton's reputation, thanks to the Monica Lewinsky affair, Annie Glenn told her husband that people in America needed someone in a position of authority to respect and look up to. Moreover, elderly citizens were inspired to no longer regard themselves as couch potatoes, idly waiting out their remaining years. "The idea of an ancient guy like me going into space," wrote Glenn in his memoir, "was exhilarating."

Although the launch was delayed slightly as flight controllers evaluated an alarm during cabin pressure checks and shooed away an aircraft that had strayed into the danger area, the seven astronauts rocketed smoothly into orbit at 2:19 pm EDT and got straight to work activating the Spacehab module. Only one issue of serious note occurred during ascent. A small aluminium panel, located underneath the Shuttle's vertical stabiliser, and meant to cover the drag chute compartment during flight, had somehow become detached and fallen away, seconds after liftoff. The panel weighed 4.9 kg and measured 46×56 cm. As it fell, it was observed on launch pad camera imagery hitting the uppermost main engine, then vanishing from view. The crew was advised of the incident by Capcom Susan Still, who stressed that Mission Control did not expect it to impact the remainder of STS-95.

However, no one knew if the drag chute itself was still intact, whether it remained stowed or partially melted or, indeed, whether it had fallen out and been destroyed during ascent. It was not expected that the chute would inadvertently deploy during flight, but a range of contingency plans were developed to cater for any eventuality. The pilots were instructed not to deploy the chute during landing operations, but Lindsey received specific details for how to jettison it in the event of a problem. At speeds of less than Mach 2.8, it was pointed out, the chute would probably not inflate at all, even if it somehow managed to pop out of its container. At lower speeds, it would most likely inflate or tear away entirely and Brown was advised that should it deploy at an altitude of less than 15 km he might notice Discovery's nose pitch upwards. If that occurred, he would take his hands off the stick and Lindsey would hit the ARM, DEPLOY and JETTISON buttons to get rid of the chute. Lower still, should the chute deploy at an altitude of less than 50 m, it would require a particularly quick response from Lindsey. If that happened, and the crew did nothing, the chute would inflate and pull the nose upwards, increasing the sink rate to some 5 m/sec and Discovery would hit the runway hard enough to cause severe damage and perhaps crew injury. However, in the morning mail, Lindsey was told to keep his hands on the control panel, with the covers of the ARM, DEPLOY and JETTISON switches up, and he should be able to flip all three before the canopy could even unreef and inflate.

In the meantime, three hours and ten minutes into the flight, John Glenn sent his first message to Mission Control, as Discovery flew above Hawaii. "Hello, Houston, this is PS2 and they got me sprung out of the middeck for a little while," he said. "We are just going by Hawaii and that is absolutely gorgeous." In Mission Control, Capcom Bob Curbeam replied that he was glad that Glenn was enjoying the show. "Enjoying the show is right," continued the world's first septuagenarian astronaut. Then, quoting a famous line from his Friendship 7 mission, more than 36 years earlier, he said: "The best part is ... a trite old statement, *Zero-G and I feel fine!*" Less than two hours later, as Glenn exceeded the four hours and 55 minutes he had spent in flight aboard Friendship 7, Curt Brown noted that STS-95 had helped the senator to double his spaceflight endurance log time.

"Let the record show," Brown reported, "that John has a smile on his face that goes from one ear to the other and we haven't been able to remove it yet!"

Shortly after reaching orbit, the activation of the Spacehab module in Discovery's payload bay got underway. Its 30 experiments were sponsored by US, Japanese and European researchers. Having flown four research missions aboard the Shuttle between June 1993 and May 1996, Spacehab, Inc., had contracted with NASA to utilise its modules for logistics to Russia's Mir space station in 1996-1998 and invested $15 million of its own capital into the development of a 'double' logistics module to be used "as a laboratory and a cargo carrier". Late in 1997, the company was awarded a $42 million contract by NASA to provide support for three Shuttle missions, including a single module on STS-95 and double modules aboard two ISS assembly flights.

Although he had experienced weightlessness during his Mercury mission, Glenn was unused to floating in an area as spacious as the Shuttle's middeck or the Spacehab module. "Floating around took a little getting used to," he wrote in his memoir. "When I moved across the middeck or through the 25-foot tunnel leading to Spacehab back in the payload bay, just a tiny amount of pressure was enough to start the process. Pushing off without the right alignment could send me spinning. The tunnel to Spacehab was only three feet wide and I learned to adjust my course as I floated through it. Reaching for items that were hovering nearby, sometimes I bumped them and then had to chase them down. I learned right away not to push too hard off the wall or to reach for things too fast."

Aboard STS-95's Spacehab were a multitude of experiment facilities, devoted to a broad range of scientific disciplines. The Advanced Gradient Heating Furnace (AGHF) was designed for directional solidification and crystal growth, whilst John Glenn tended to the Advanced Organic Separations (ADSEP) payload, which provided the capability to separate and purify biological materials in the microgravity environment with minimal interaction from the astronauts. It supported three commercial investigations: the Haemoglobin Separation experiment, the Microencapsulation experiment and the Phase Partitioning experiment.

3-16: The STS-95 crew is pictured in the tunnel leading to the Spacehab module.

"Starting ADSEP meant moving its various modules from storage into active bays," wrote Glenn, "and setting switches and turning dials according to detailed instructions in our flight data files." The Advanced Protein Crystallisation Facility (APCF) supported the crystallisation of several protein solution samples. In other research, low-density 'aerogel' (nicknamed 'frozen smoke') was produced as part of an experiment into the capacity of a substance to transmit light and insulate against sound and electricity. The Astroculture facility supported several plant investigations, focusing on how the microgravity environment affected the composition of volatile oils important to the flavour and fragrance of plants and how genes could be transferred from a soybean seedling more efficiently in the microgravity environment. Biobox housed human bone and cell growth samples, whilst the Biological Research in Canisters (BRIC) studied the effects of microgravity on embryo formation in orchard grass and cultivated seedlings of rice and arabidosis to understand the physical properties of the cell walls. The Commercial BioDynamics payload featured a bioreactor, with a rotating culture vessel, to support six simultaneous cell tissue samples. The Commercial Generic Bioprocessing Apparatus (CGBA) investigated protein crystal quality in support of future drug design processes, sought new insights into the cause of enhanced antibiotic production, tested magnetic force as a method of separating and sorting cells within the human immune system and explored the transfer of plant genes from one plant type to another as part of understanding the growth process. Elsewhere, commercial providers supported a variety of crystal growth investigations and the Facility for Adsorption and Surface Tension (FAST) observed surface phenomena at liquid-to-liquid and liquid-to-gas interfaces in order to understand convection processes that are driven by the influences of buoyancy and gravity.

Glenn focused on a multitude of experiments during his nine days in orbit. His aging-related activities on STS-95 had drawn much publicity and focused upon understanding how and why bedrest and elderly patients on Earth experience steady losses of bone mass. Similar symptoms had been noticed in the bodies of astronauts in the 1960s and during the following two decades NASA's database on human adaptation to microgravity increased during the early Shuttle era. Similarly, Russian data from long-duration expeditions provided greater support for the existence of similar symptoms between the effects of space flight and those which accompany the aging process. In 1989, NASA and the National Institutes of Health (NIH) sponsored a joint Conference on the Correlations of Aging and Space Effect on Biosystems, whose participants included leading experts in the fields of gerontology and space physiology. Their findings identified a number of physiological and psychological areas in which space flight and aging shared significant characteristic symptoms and the conference addressed the question of whether microgravity exposure, high-energy radiation and stress levels could provide an accelerated, acute environment for research on aging. In the aftermath of the conference, NASA and the NIH continued joint research, one of whose consequences was the production of *Space Physiology and Medicine*, the tome which drew Glenn's attention in 1995. Two years later, in February 1997, the two agencies sponsored an 'Aging and Space Flight: Expanding the Science Base' workshop and signed an agreement the following September to commit to collaborative ground-based and space-based research activities. The announcement of Glenn as a payload specialist on STS-95 was made several months later, in January 1998.

Together with Chiaki Mukai, Glenn participated in the Sleep-2 investigation to evaluate sleeping patterns, as part of a wider effort to identify the factors which may contribute to treatments for sleep disturbances on Earth. "The study Senator Glenn will participate fully in will require him wearing 28 different electrodes and sensors on his body, for four nights during the mission," explained Steve Robinson, who carried primary responsibility for the scientific goals of STS-95. "Our other payload specialist, Dr Mukai, will also participate that way." Glenn later remarked that sleeping with the elaborate headgear and instrumented vest was actually easier in orbit than it had been during ground trials. "Imagine sleeping with a dozen buttons over half an inch thick stuck on your head that you feel every time you roll over," he wrote in *John Glenn: A Memoir*. "Weightlessness removed the irritating pressure." In highlighting the overall importance of Sleep-2, Glenn stressed that in 1998 America had 34 million citizens beyond the age of 65, a statistic which was predicted to triple within half a century. "Half of those people over 65 have some sleep problems," he said, "to the point that it interferes with their day activity. Sometimes their lives are actually shortened because of this." As part of Sleep-2, Glenn and Mukai's brain waves, heart rates, circulatory and respiratory systems and body core temperatures were measured by a multitude of leads, sensors and an instrumented skull cap.

Mukai was making her second Shuttle flight, having become Japan's first female astronaut during the STS-65 mission in July 1994. "To go into space became my dream when I was 32 years old," Mukai told a NASA interviewer, "because when I was a child Japan didn't have a space programme, so my first dream was to be a medical doctor." She was born in Yayebayashi, in the far-western part of Gunma Province, on the central Japanese island of Honshu, on 6 May 1952. After completing senior high school in Tokyo, she entered medical school at Keio University – Japan's oldest extant university – and received her doctorate in 1977. This was followed, in 1988, by a PhD in physiology, also from Keio. Between the receipt of her two doctorates, she was board-certified in medicine, served a residency in General Surgery and worked variously as a general surgeon at Keio University Hospital and as an emergency surgeon at Saiseikai Kanagawa Hospital. By the beginning of 1985, Mukai was an assistant professor in Keio University's Department of Cardiovascular Surgery. She then spotted an advertisement placed by the National Space Development Agency of Japan (NASDA) that called for payload specialist candidates for a joint US-Japanese Spacelab-J mission on the Shuttle. "I thought the space programme is such a wonderful area to use my medical expertise," she told the NASA interviewer, years later. "I applied and was lucky enough to be selected." On her admission into the first group of NASDA astronauts in June 1985, she was joined by chemist Mamoru Mohri (later to become the first 'professional' Japanese astronaut) and physicist Takao Doi (later to become the first Japanese spacewalker), and the trio began work on Spacelab-J. Mukai continued her work as a cardiovascular physiologist and as a visiting surgeon at Keio University and in 1989 she was board-certified as a cardiovascular surgeon. Although she did not fly on Spacelab-J, Mukai was assigned in October 1992 as the prime payload specialist for the IML-2 mission and also served in a backup capacity on the STS-90 Neurolab flight.

Working with Mukai, John Glenn also participated in the Microgravity Encapsulation Process (MEPS) experiment to study the formation of capsules to contain two types of anti-tumour drugs which could be delivered directly to solid tumours as part of chemotherapy treatments. Early on 31 October, he began a regime of providing ten blood samples and 16 urine samples to investigate the influence of the microgravity environment on his body, as part of efforts to better understand how the absence of terrestrial gravity affected his sense of balance and perception, the response of his immune system, the density of his bones and muscles, his metabolism and blood flow and his sleep patterns. This was part of the Protein Turnover Experiment, in which crew physicians Parazynski and Mukai oversaw blood sampling of Glenn and Duque to study the effects of space flight upon whole-body and skeletal muscle protein metabolism. "Scott became my Count Dracula," Glenn recalled, "after he floated in my direction wearing a set of plastic Halloween fangs. By a few days into the mission, he started grinning whenever he came my way with the syringe – or maybe it was just my imagination that he got to look more maniacal than ever."

In addition to studying the adaptation of humans to weightlessness, STS-95's research also focused on a wide range of other living creatures, including a pair of toadfish. The Vestibular Function Experiment Unit (VFEU) electronically monitored these electrode-implanted marine fish to determine the influence of gravitational changes upon their inner-ear (or 'otolith') systems. The toadfish had been loaded aboard Discovery about 35 hours before launch and were continuously monitored throughout the mission, until their removal from the orbiter about five hours after touchdown on 7 November. Supporting the VFEU and BRIC payloads was the Oceaneering Spacehab Refrigerator Freezer (OSRF), which maintained experimental samples at temperatures of 4 degrees Celsius and whose status was checked by the crew on a daily basis.

Early on the second day of the mission, Brown, Lindsey and mission specialist Scott Parazynski oversaw the deployment of the Petite Amateur Naval Satellite (PANSAT) that had been developed by the Naval Postgraduate School in Monterey, California. Designed to enhance the education of the school's military officer candidates through space operations, the non-recoverable PANSAT was intended to capture and transmit radio signals which would normally be lost because the original signals were too weak or were subjected to too much interference. Shortly after being released, PANSAT departed the vicinity of the orbiter at a rate of about 14 km with each 90-minute circuit of Earth.

Meanwhile, within hours of reaching orbit, preparations for the SPARTAN-201 deployment process got underway, with Brown, Lindsey, Robinson and Parazynski based on the flight deck. "The four of us in the cockpit are very, very involved in this deploy sequence," explained Robinson in a pre-flight interview. "It's rather delicate, it's carefully choreographed. Curt will be in charge of flying Discovery, moving it around relative to the satellite. Steve Lindsey will be backing him up, watching Shuttle systems and making sure that everything we need both for the arm, the satellite and the Shuttle are in good shape. Scott and I will be in charge of actually deploying the satellite. Scott's primarily responsible for the health of the satellite. I'm primarily responsible for the operations of the arm itself and we back each other up. We think that's a good teaming strategy." On 30 October, Robinson checked out Discovery's RMS arm by using it to survey a small piece of loose insulation on one of the OMS pods. Using two hand controllers, he grappled the satellite on 1 November and drove latches to unlock it from its carrier structure, then drew it 'upwards' into the deployment position. After Parazynski had powered up SPARTAN-201 and verified its systems, Robinson deployed it at 1:59 pm EDT. Shortly thereafter, it successfully performed the 45-degree, 90-second pirouette which had proven elusive on STS-87, prompting Brown to radio triumphantly: "Houston, Discovery, SPARTAN is in the manoeuvre."

Brown and Lindsey fired Discovery's thrusters to withdraw from the satellite, maintaining a distance of 10-16 km for about nine hours in order to test an experimental communications system aboard the SPARTAN. Known as the Technology Experiment Augmenting SPARTAN, or TEXAS, the system employed the orbiter as a communications relay and, despite a few initial problems, it worked exceptionally well. Later that day, Brown manoeuvred the Shuttle to a distance of 48 km from the satellite, whereupon SPARTAN-201 commenced two days of solar physics research with its White Light Coronagraph (WCS) and Ultraviolet Coronal

Spectrometer (UVCS) instruments. At 3:45 pm on 3 November, Robinson grappled the satellite and berthed it into the payload bay. During the rendezvous, the crew tested the Video Guidance Sensor (VGS), which was a component of an automated rendezvous and capture system for the ISS. Developed by MSFC, it employed on-board video cameras and dual-frequency lasers, together with computers and navigational inputs, to provide ranging and attitude controllability. Sensors and optical targets mounted on SPARTAN-201 and Discovery allowed for precise determinations of their respective positions and distances from each other. Also affixed to the satellite was the SPARTAN Auxiliary Mounting Plate (SPAM), which housed a small accelerometer.

Other payloads were mounted onto SPARTAN's support structure in the bay. Among them was the Space Experiment Module (SEM), flying for the fifth time and supporting eight student-sponsored investigations. Blue Mountain School of Floyd, Virginia, provided samples of film, soap, motor oil, bone, nails, Coca Cola and popcorn for young students to examine in weightlessness, whilst Dowell Elementary School of Marietta, Georgia, sought to compare the weight, mass and other physical properties of objects such as chewing gum, popcorn, bread, stickers, bubble wrap, chalk, paper clips, erasers and crayons. High schools were also involved in the SEM research, with Don Bosco Technical Institute of Rosemead, California, supplying an experiment to study the solidification of low-melting-point solders and solder wetting on plate gold, together with the structural integrity and tensile strength of adhesives in microgravity. Still other educational institutions focused on the influence of cosmic radiation upon Wisconsin fast plants, wheat seeds and brine shrimp, and studied the growth of bread mould and the germination and growth of perennial rye, Kentucky bluegrass, corn, oats, barley, lentils and sunflower. Nor were the involved schools all based in the United States. One experiment to study lettuce and cicoria seeds to compare their germination across two or three generations featured student investigators from Instituto Technico Commerciale Riccati in Treviso, Italy, whilst Colegio Santa Hilda in Buenos Aires, Argentina, contributed to a study of seed growth and survival.

In the meantime, Parazynski worked with the Orbiter Space Vision System (OSVS), which was embarking on its final test ahead of use on the first ISS assembly mission in December 1998. Elsewhere, Duque and Chiaki Mukai worked with the Middeck Glovebox, which on this mission supported three fluid physics investigations to observe the internal flows within free-floating liquid drops and test fundamental theories of atomic behaviour within mixtures of fine particles suspended in fluid (known as 'colloids'). It was hoped that this colloid research might lead to enhancements in the processing of materials on Earth to manufacture new materials and the STS-95 research focused on binary alloy colloids with different-sized particles, colloid polymers with chain-like molecules and fractal colloid aggregates with repeating structural patterns or networks. Duque began shutting down the glovebox and its experiments on 5 November.

Parazynski also worked with the Hubble Space Telescope Orbital Systems Test (HOST) payload, which provided a testbed for hardware destined to be used aboard the third HST servicing mission, then planned for June 2000. Four spacewalkers –

Steve Smith, John Grunsfeld, Mike Foale and Claude Nicollier – had already been assigned to the flight and HOST was designed to demonstrate that actual electronic and thermodynamic equipment would perform acceptably in the radiation and microgravity environment of low-Earth orbit. Key elements of the payload included a test of the 486 computer memory components for HST's new data processor, which it was hoped would overcome the difficulties experienced by earlier software, which had proved susceptible to heavy-ion radiation levels or single-event upsets. The new Solid State Recorder (SSR), planned for installation by the SM-3 crew to replace the earlier engineering science tape recorder, was also tested to understand its efficiency and performance. Although an SSR had been installed by the STS-82 crew, it was experiencing non-serious problems, most likely triggered by single-event upsets or high-energy protons in the South Atlantic Anomaly. Although STS-95's orbit was somewhat lower and more benign than HST's 590 km orbit, it was high enough – averaging about 550 km – to provide the best conditions to compare the two radiation environments.

Thirty-six years after his last mission, John Glenn found the view of Earth exhilarating and terrifying in equal measure. Shortly before STS-95 launched, Hurricane Mitch made landfall in Honduras and Nicaragua, causing landslides, sweeping away entire villages and killing thousands of people. "One of the laptops on the flight deck was set up to track Discovery on its orbits around the world," Glenn later wrote. "By following the track on the screen, you could anticipate when you were approaching an area that needed to be photographed. You couldn't wait until you *recognised* Honduras, for instance, because at 17,500 miles an hour, the photo angles you wanted would have slid by already." The high orbit presented the crew with a spectacular vantage point, offering Glenn glimpses of the Florida Keys and, looking north, as far as Lake Erie and into Canada. One night, he sat up and watched as thunder crashed over South Africa, producing lightning flashes hundreds of kilometres long. He described the flashes as "looking like bubbles of light breaking by the hundreds on the surface of a boiling pot".

For Curt Brown, whose four previous space missions had all flown into high-inclination orbits of 39 degrees or more, STS-95 operated at a relatively restricted 28.5 degrees, but he expected to see more of Earth from the higher altitude. "It should be more round, more of a globe," he told a NASA interviewer. "The horizon, the limb, should be more arched, so that will be quite a treat."

Other HOST experiments included analyses of fibre-optic performance and the test of a new cooling system for the Near-Infrared Camera and Multi-Object Spectrometer (NICMOS) which was to be installed by the SM-3 crew. Supporting HOST was the Space Acceleration Measurement System, a frequent Shuttle flier, which tracked tiny vibrations in the NICMOS cryocooler test hardware. After STS-95 ended, researchers analysed and correlated its data with other ancillary data from the mission as part of efforts to assess the cryocooler's usefulness for HST.

Also mounted in Discovery's payload bay was the International Extreme Ultraviolet Hitchhiker (IEH), flying for the third time to investigate the magnitude of the absolute solar extreme ultraviolet flux and emissions from the torus of plasma that occupies the orbit of Jupiter's moon, Io. It was also employed to observe Earth's thermosphere, ionosphere and mesosphere, utilising the University of Southern California's Solar Extreme Ultraviolet Hitchhiker (SEH) and the twin detectors of the joint US-Italian Ultraviolet Spectrograph Telescope for Astronomical Research (UVSTAR). Co-mounted on the same cross-bay Hitchhiker bridge as IEH-3 were five other payloads, one of which, the Naval Postgraduate School's PANSAT, was successfully deployed early in the mission. The others remained fixed in place. The Solar Constant (SOLCON) experiment was used to calibrate satellite-mounted solar irradiance instruments to provide a measure of 'quality control' capability, whilst the Spectrograph Telescope for Astronomical Research (STAR-LITE) studied astronomical targets in ultraviolet wavelengths. Its key targets included diffuse background emissions, scattered dust and recombination emission lines from the interstellar medium, supernova remnants, planetary and reflecting nebulae, star-forming regions in external galaxies and the Jupiter-Io plasma torus. The two other experiments were based upon Getaway Special hardware: the Cosmic Dust Aggregation (CODAG) was to simulate the aggregation of dust particles in an attempt to understand the dynamics of dust clouds which occurred in the early stages of the formation of the Solar System, and the Roach Experiment was to measure the effects of microgravity on the entire life cycle of the American cockroach. Managed by students from DuVal High School in Lanham, Maryland, the roach investigation included a specially designed habitat module for adults, nymphs and eggs that supplied air, heat, water and food provisions.

Although the mission specialists and payload specialists oversaw the bulk of the scientific research, Curt Brown and Steve Lindsey participated in a number of experiments. One of Brown's responsibilities was the 'Electronic Nose', a miniature

3-17: Discovery touches down at the Kennedy Space Center on 7 November 1998.

environmental monitoring instrument to identify a wide range of organic and inorganic molecules down to the parts-per-million level. It was employed to monitor the atmosphere of Discovery's middeck, targeting ten toxic compounds and including a package of alcohol wipes, a display and control palmtop for crew interface and a valve assembly for capturing samples. It was activated within hours of reaching orbit on 29 October.

In spite of the cutting-edge research, the links between the Old Era and the New Era nevertheless pervaded STS-95. On 4 November, Glenn spoke to fellow Mercury astronaut Scott Carpenter, telling him jokingly that the Shuttle would make a good retirement home. "If you spill food, it doesn't go on your necktie, it just floats out away from you," he told Carpenter. "In fact, I got some oatmeal on my glasses the other day. You don't need a walker up here, you don't need to worry about osteoporosis or canes or anything like that, because you just *float* across the room. There's no such thing as broken hips or anything like that. If you have trouble sleeping at night, it's no problem because you have *another* night coming up in not more than 45 minutes."

Other light moments included an interview involving Glenn, Brown and Lindsey with comedian Jay Leno.

"Does Senator Glenn keep telling you how tough it was in the old days, how cramped it was, how small it was, how lucky you young punks are?" asked Leno.

"Well, Jay, actually, no," replied Brown. "He doesn't always do that. Only when he's *awake!*"

Nine days later, a normal entry, approach and landing were expected, but precautionary plans were in place to accommodate an inadvertent deployment of Discovery's drag chute during re-entry or on the runway. Brown and Lindsey fired the OMS engines for the irreversible de-orbit burn at 10:53 am EDT on 7 November. From the higher-than-normal orbit, the burn lasted almost five minutes and slowed the Shuttle sufficiently to drop it into the 'sensible' atmosphere and on track for a Florida homecoming. The Shuttle Training Aircraft, flown by chief astronaut Charlie Precourt, had already performed routine weather observations and during the final approach would visually monitor Discovery's drag chute. Touchdown on the SLF Runway 33 occurred flawlessly at 12:03 pm EDT and the missing chute door posed Brown and Lindsey with no controllability problems.

"Houston, Discovery, wheels stopped, KSC," reported Brown.

"Welcome home, Discovery, and a crew of seven heroes from a mission dedicated to improving life on Earth. Beautiful landing, Curt," replied Capcom Susan Still.

"One-G and I feel fine," said Glenn after wheelstop, borrowing from and slightly changing another of his Friendship 7 quotes. In fact, as he admitted in his memoir, Glenn's stomach was actually in revolt against a salt-loaded lemon-lime drink which the crew had been obliged to drink prior to re-entry as a means of acclimatising themselves to terrestrial gravity. He felt dizzy and shaky, but was determined to walk out of the orbiter with the rest of the crew, "if it killed me". He wrote that he had to keep his feet wide apart for balance and that his crewmates, and particularly Brown, stayed close to him for support. "It was that same mutual concern and camaraderie," wrote Glenn, "that make NASA and the space programme so special."

Although a little unsteady on his feet after emerging from Discovery (a not-uncommon condition known as 'ataxia'), Glenn had experienced virtually no negative effects as a result of nine days of weightlessness in orbit. During an in-flight news conference he had pointed out, "I guess I came up expecting to be a little nauseous. I think there's something like 65-70 percent of the people who come up have some sort of problems with stomach awareness. I haven't had any of that. It's been great, I've been quite comfortable. I've been sleeping pretty well and everything's been going along fine." He took the anti-nausea drug Phenergan, as did most astronauts, and it seemed to have done the trick. Upon his return to Earth, Glenn joined his crewmates for the traditional walk-around of the orbiter on the runway, then faced four hours of medical checks. During the course of STS-95, he had endured 11 blood draws, given up 17 blood samples and 48 urine specimens, received two blood infusions and ingested tracer chemicals to monitor his muscle changes in space. He also underwent several days of body core temperature measurements, four nights of sleep monitoring and four days of cognitive testing. He also wore a portable electrocardiogram and a blood pressure sensor and recorder through re-entry.

In hindsight, more than a decade after STS-95, John Glenn remains the single data point on the chart for septuagenarian astronauts, as he remains the oldest person – by far – ever to have journeyed into orbit. Second place is still held by Story Musgrave, who flew aged 61 in November 1996. Although it seems likely that septuagenarians will fly in the future, for now Glenn's flight can be regarded for what it was: a remarkable scientific episode, with important medical research objectives, but at the end of the day Glenn was, as Musgrave remarked, a "legislative passenger". However, it is important to not become cynical. Space exploration remains a subject outside of the life experiences of most people on Earth and opportunities to enhance its stature in the eyes of the public should and must be utilised.

"Back when we were trying to get the space programme off the ground, Senator Glenn did something that got us going," said Curt Brown in a pre-flight interview. "It got us on the track to go to the Moon. We were looking for successes at that time and he was the first American to orbit Earth and that was what we needed and away we went. We've been flying Shuttles for a number of years now and International Space Station's coming up and in a few months we'll be launching our first Shuttle to go up and start assembling the space station. It's really nice to have this kind of attention on the space programme and Senator Glenn's helping us take that next step. If we can get the enthusiasm about our International Space Station, like we had during the early years of NASA, it's going to really help us and so I'm hoping everyone will get excited about that."

POWERFUL X-RAY EYES

"It's great to be back in zero-g again," said STS-93 commander Eileen Collins, early on 23 July 1999, as she and her four crewmates set about preparing Columbia for five days of orbital activities, but added darkly that "a few things to work on ascent kept it interesting". Those *things*, within seconds of liftoff, almost forced Collins – the first woman to command a space mission – and pilot Jeff Ashby to perform a hair-raising abort landing back at KSC and the incident effectively grounded the Shuttle fleet for almost six months.

Columbia's 26th flight was a long time coming. More than a year had elapsed since her Neurolab mission, but the cause of the delay had nothing to do with NASA's venerable old workhorse herself. Originally scheduled for launch in August 1998, it was postponed until December, then January 1999, then April, and ultimately midsummer, by a chain of technical problems with STS-93's primary payload, the $1.5 billion Chandra X-ray Observatory (CXO) and the Boeing-built Inertial Upper Stage (IUS) which would propel it into an unusual orbital location. Given the delays, it is a pity that Shuttle managers did not opt to refly Neurolab in the summer of 1998, because with only five missions achieved that year – as opposed to the normal seven or eight – the manifest was remarkably light. That was not to imply that no missions awaited the fleet: more than 40 flights to assemble the ISS were waiting in the wings, as was much of the already-built hardware, but repeated delays to crucial Russian components had temporarily stalled the construction effort. By the time the first Shuttle assembly mission, STS-88, lifted off in December 1998, the project was already running billions of dollars over budget and years behind schedule. More trouble was afoot. The service module, again Russian-built, would require a further 18 months before it finally reached orbit, delaying the arrival of the first long-duration crew until October 2000.

Sadly, Columbia was excluded from the ISS construction effort. As NASA's oldest Shuttle, and despite numerous weight-saving measures during her numerous overhauls since 1984, she remained considerably heavier than Discovery, Atlantis and Endeavour and thus could not comfortably transport large ISS components into orbit. (It was more saddening and bitterly ironic that this decision was later partially reversed and had Columbia not been lost during re-entry on 1 February 2003, her very next flight, the following November, would have been an ISS construction mission.) In the meantime, she was restricted to non-ISS flights, such as the deployment of Chandra, the repair and servicing of HST and a series of 'research missions' utilising a pressurised Spacehab module. Clearly, she remained a highly prized asset to NASA and the Chandra mission, STS-93, demonstrated this, for she was the only member of the Shuttle fleet capable of transporting the gigantic observatory and its IUS booster into orbit.

The reason for this was Chandra's sheet size. When affixed to the IUS and mounted on a supporting 'tilt table' in the payload bay, the observatory consumed 17.4 m of Columbia's 18.3 m payload bay and weighed more than 22,600 kg. In anticipation of their back-to-back ISS flights, the other three orbiters had already had their internal airlocks removed from the middeck and replaced inside the docking mechanism in the forward quarter of their payload bays. The result was that there was not enough room in the bays of Discovery, Atlantis or Endeavour to house Chandra. Yet even Columbia herself had to lose 3,200 kg of additional mass before she could transport the new observatory. To achieve some of these savings, engineers used older, lighter main engines, which lacked the newer and more rugged high-speed fuel pumps and combustion chambers. "We put Columbia on a strict diet to get to this mission," said processing manager Grant Cates. "That work actually began [in 1996] with the identification of this mission and the weight reduction that would be required." Nonetheless, the orbiter's weight would still creep above NASA's normal safety limit if an emergency landing was required, shortly after launch.

In such an eventuality, Columbia would tip the scales at 113,000 kg, some 590 kg heavier than safety rules mandated as the maximum allowable landing weight. In the case of STS-93, a one-time-only waiver was granted to this rule, based on a detailed analysis of the payload, the Shuttle's centre-of-gravity constraints and a host of interrelated factors. For Eileen Collins, the challenge of possibly having to perform a heavier-than-normal emergency landing, known as a Return to Launch Site (RTLS), did not faze her. "We would land at 205 knots, which is very close to the maximum certification around 214," she explained in a pre-launch interview. "There are some challenges there, but I feel very confident we've looked at the abort landings and they're well within the safe limits of landing the Shuttle."

Collins' confidence in her abilities and those of her crew would come close to being tested. Joining her and rookie pilot Jeff Ashby were mission specialists Catherine 'Cady' Coleman, Steve Hawley and Frenchman Michel Tognini. Although the latter was a veteran of one previous space mission, STS-93 marked his first Shuttle flight. Born in Vincennes, within the eastern suburbs of Paris, on 30 September 1949, Michel Ange-Charles Tognini had become France's third man in space when he flew a two-week mission to the Russian Mir space station in July 1992. Like so many spacefarers, he had grown up with aviation in his blood. "I wanted to be a pilot," he recalled in a NASA interview, "because my family was in the Air France company and they were dealing with planes and aircraft." His initial aspiration was to fly military transport aircraft and he earned a mathematics degree from Military School Grenoble in 1970, but after graduating in engineering from the *Ecole de l'Air* (French Air Force Academy) in 1973 Tognini broadened his scope into fighter and test piloting. He was posted for advanced fighter training at the Normandie-Niemen squadron and from 1974-1981 served in the French Air Force as an operational fighter pilot, flying Mirage F-1 aircraft out of Cambrai Air Base. By 1979, he had risen to become a flight commander. Three years later, Tognini was selected for the Empire Test Pilot School at Boscombe Down in Wiltshire, England. At the end of his studies in 1982, Tognini was awarded the Patuxent Shield Trophy – an award won by two other future French astronauts, including Patrick Baudry (in 1978) and Jean-Pierre Haigneré (in 1981). After graduation, Tognini returned to France as a chief test pilot at the Cazaux Flight Test Centre. His work included weapons system testing for the Mirage and Jaguar aircraft, as well as responsibilities associated with the safety of pilots, experimenters and flight engineers. He was selected by France's *Centre National d'Études Spatiales* (CNES, the French National Centre for Space Studies) in September 1985 and trained as a backup crewman for the Soyuz TM-7 mission in late 1988, before serving as prime crewman on Soyuz TM-15 in July 1992.

As an ESA astronaut, in March 1995 Tognini and fellow Frenchman Jean-Loup

Chretien joined NASA's 15th group of astronaut candidates, "to exchange information and expertise they have obtained while training for … separate missions aboard the Russian Mir space station". It was added that they would support NASA astronauts in their preparations for Mir flights, participating in "selected aspects" of the training of the new class, for Shuttle mission specialist certification. In November 1997, Tognini became the first member of the STS-93 crew to be announced by NASA and was joined by Collins, Ashby, Coleman and Hawley in March 1998. "I was very happy because I had been waiting for such a long time to get this assignment," Tognini told a NASA interviewer. "At one point, I was expected to have a flight on Shuttle-Mir, because I spoke Russian and I flew previously on Mir and Soyuz. Instead of flying to Mir, they asked me not only to fly on STS-93, but to also be in charge of the deployment of Chandra with Cady Coleman. I was very surprised and proud to have such a challenge and such responsibility."

Making his fifth Shuttle, Steve Hawley had been newly appointed as deputy director of JSC's Flight Crew Operations Directorate when he was approached to join the STS-93 crew. "I think they were looking for an experienced person for that mission," he said. "It was a relatively junior crew. I think they were looking for somebody experienced to add to the mix of relatively less experienced people." When he was first asked, Hawley declined the invitation to join the crew. "I thought I really hadn't been back in my real job that long yet," he told the NASA oral historian, "and I didn't think that it was appropriate for me to step aside and fly again that soon, but they pestered me and so I did it, but I honestly told them that I didn't think it was the right thing to do."

In many ways, STS-93 harked back to the early days of the Shuttle programme. Not only would it last just five days, making it NASA's shortest planned mission for almost eight years, but it would also feature the deployment of the orbiter's final IUS payload. Built by Boeing and provided by the US Air Force, the IUS had originally been developed as a short-term stand-in for a reusable 'space tug' and was initially dubbed the 'Interim Upper Stage', before changing to 'Inertial' in recognition of its sophisticated internal guidance system. Losing this 'interim' status also reflected a growing awareness, when the space tug was cancelled in late 1977, that the IUS' services would be needed throughout the 1980s. In fact, not until the early years of the present century did it fly for the final time as a 'standalone' booster. Boeing began developing the two-stage vehicle in August 1976 and supported its first launch aboard a Titan 34D rocket six years later. Measuring 5 m long and a little under 3 m in diameter and weighing some 14,740 kg, the cylindrical booster – made from Kevlar-wound aluminium – was capable of delivering 2,270 kg payloads to geostationary altitudes. Its first stage carried 9,700 kg of solid propellant and a large motor that was capable of firing for up to 145 seconds, making it the longest burning solid-fuelled engine ever used in space applications. Meanwhile, the second stage carried 2,720 kg of propellant. Both the first and second stage nozzles, commanded by redundant electromechanical actuators, could steer the former by up to four degrees and the latter up to seven degrees. Although solid rockets were known to generate a harsh impulse, the separation mechanism between the first and

3-18: The STS-93 crew was the first in history to feature a female commander, Eileen Collins (at left).

second stages employed a low-shock ordnance device in order to avoid damaging its payload. Moreover, solid propellant was chosen over a liquid-fuelled booster because of its simplicity, safety, high reliability and low cost. Hydrazine-fed reaction control thrusters provided the IUS with additional stability during the 'coasting' phase between the first and second stage firings, as well as ensuring accurate roll control during the engine firings and assisting with the satellite's insertion into geostationary orbit on satellite-deployment missions. Situated between the two stages was an equipment section with avionics systems to provide guidance, navigation, control, telemetry and data management services to Chandra on STS-93. Importantly, most critical components, except the bellows for the gimbal actuator, were fully redundant to provide a reliability of more than 98 percent.

The booster's first Shuttle use on STS-6 in April 1983 put a key NASA communications satellite into a lower-than-planned orbit. This failure led to several missions being postponed, but the IUS eventually established itself as an exceptionally reliable booster. However, it was still prone to difficulties. On 9 April 1999, a Titan IV rocket was launched, carrying a missile early-warning satellite, but both the first and second stages of the IUS failed to separate properly and the $250 million payload was lost. Ordinarily, NASA would have watched the resulting investigation with interest and incorporated its findings into its own plans to ready

Chandra's IUS for launch. This was complicated, however, by the fact that the investigation itself was classified. Moreover, said Scott Higginbotham, who oversaw Chandra's pre-flight processing at KSC, the IUS assigned to STS-93 was impounded as part of the investigation. Original plans had called for it to be attached to Chandra on 23 April, but this was delayed as investigators set to work. Columbia's launch, then scheduled for 9 July, came under review and the impounded IUS had a domino-like effect on the training of the astronauts and their ground teams.

"We were planning a two-day-long sim, starting 14 April, that would involve all the different control centres, a joint integrated simulation, with everybody," explained Lead Flight Director Bryan Austin. "That was going to be a big deal. That has been postponed because the Air Force folks and Boeing IUS people were going to be taken away initially to be part of the investigation. That exercise has been put on hold. That kind of put a wrench in things in terms of our sim schedule." The launch of Chandra was also critical because it was the Shuttle's first IUS launch since STS-70 in July 1995 and, according to Austin, "there has been a lot of change in the expertise level, collectively. For the most part, IUS deployment procedures are the same, but something that, to me, has been a struggle for this flight with some of our IUS friends is to get them to realise that the payload on the other end of the IUS is not the typical thing that the Shuttle has been doing." His point was that, as soon as the Chandra/IUS transitioned to internal battery power, just minutes before deployment, the observatory would be on its own. On previous missions, if something went awry at the last moment, the IUS and its payload could be retracted for another attempt or returned to Earth. However, mainly due to power and temperature constraints, mission managers had just a single shot at a deployment. "It's either going to become orbiting space trash," said Austin, "or it's going to go out and do its mission."

In fact, Chandra was the third in a quartet of 'Great Observatories' which NASA had been planning for more than two decades to explore the Universe with sensors that jointly covered virtually the entire electromagnetic spectrum. The first two observatories – HST and the Compton Gamma Ray Observatory – had been launched by two Shuttle crews in the early 1990s and focused on visible and ultraviolet studies, as well as measurements of high-energy gamma rays. Two others would then cover X-rays (Chandra) and infrared (the 2003-launched Spitzer Space Telescope). "Hubble revealed the visible side of the Universe," said theorist Michael Turner of the University of Chicago, "but most of the Universe does not emit visible light. It's only visible by other means, in particular the X-rays. Chandra will give us the same clarity of vision as Hubble does, but for the 'dark' side of the Universe we know the least about." Comparing astronomers' capabilities before and after Chandra, astronaut Steve Hawley likened it to the difference between the small reflecting telescope he had used as a child and the enormous observatory on Mount Palomar in California.

"We can make Superman jealous with *our* X-ray vision!" claimed Ken Ledbetter, NASA's head of mission development for Chandra. The observatory, which received Congressional approval in 1987, used the most precisely figured X-ray mirrors ever built. It was hoped to equal, or even surpass, HST by studying some of the most exotic phenomena in the known Universe, including quasars, black holes and white dwarfs. The spacecraft was a tapering cylinder, 13.8 m in length, with two solar arrays at its base to provide electrical power. At the opposite end, mounted in the telescope's primary focus, were its two scientific instruments: the High Resolution Camera and the CCD Imaging Spectrometer. Chandra's concentric nest of cylindrical mirrors were coated with reflective iridium and gave it ten times the resolution of existing X-ray detectors and 50 times the sensitivity.

Of the many astronomical targets for Chandra, black holes were sure to seize the public's interest, even though so little was known about them. By definition, they are 'unobservable', but the capability existed to indirectly study them by analysing radiation emitted by material being sucked into them. As interstellar gas and dust is accelerated, for example, it collides with increased energy levels and emits X-rays before vanishing across the 'event horizon'. By examining such emissions in unprecedented detail with Chandra, astrophysicists hoped to identify black hole signatures. Additionally, supernova explosions thought to lead to black holes were placed under scrutiny. One of the observatory's first celestial targets was the remnant of a massive star in the Large Magellanic Cloud that was seen to explode in 1989. On a far larger scale, it was anticipated that Chandra would focus on the amount of dark matter present in the Universe, by carefully examining galactic clusters. Such galaxies which make up these clusters are deeply embedded in huge clouds of hot, X-ray-emitting gas and are held in place by gravity generated by all the components of the cluster. Its observations were thus expected to enable astrophysicists to refine their numbers of how much 'normal' matter is present in a given cluster and thus how much 'dark' matter must also be present in order to generate gravity needed to hold it together.

Chandra was originally conceived in the 1970s, when NASA envisaged a Large Orbiting X-ray Telescope. That was later descoped and became the HEAO-2 Einstein Observatory, whose success prompted the space agency to include an X-ray telescope on its wish list for a four-spacecraft flotilla of Great Observatories. Early plans called for it to be launched into low-Earth orbit and periodically serviced by Shuttle spacewalkers for a projected 15-year lifespan, but in 1991 escalating costs and technical problems forced a rethink of the mission. It was effectively split into two halves. One half would transport a high-resolution camera and imaging spectrometer into a highly elliptical orbit, which, although beyond the reach of Shuttle repair missions, would enable it to gather data from 55 hours of each 64-hour orbit. Meanwhile, the second half would be equipped with a super-cooled X-ray spectrometer and would be launched into a lower orbit, but was cancelled by NASA in 1993 in another round of budget cuts. Even with the cancellation of the second mission, Chandra – known at the time as the Advanced X-ray Astrophysics Facility (AXAF) – was still expected to cost in the region of $3 billion during its first eight years of operations, including the spacecraft itself, the Shuttle and IUS launch costs, annual mission control and data analysis fees, use of the TDRS communications system and a one-time charge to test the mirrors at NASA's X-ray Calibration Facility (XRCF) at MSFC in Huntsville, Alabama. Nevertheless, even this cost was far cheaper than the $7 billion envisaged for an all-in-one AXAF in low-Earth orbit.

The observatory's two largest mirrors (each 122 cm in diameter) for its High Resolution Mirror Assembly were completed in June 1991 and tested at the XRCF in September of the same year. By January 1995, all eight nested mirrors had been completed, polished and measured. "The first mirror took nine lengthy polishing cycles to complete," said AXAF Telescope Project Manager John Humphreys. "We then applied a process of continual improvements to get the job done much faster and were able to complete the final mirror in only three polishing cycles." The reflective chromium-iridium coating was applied in May 1996.

The decision to launch the observatory into a location far from Shuttle repair crews was a tough sell, particularly in the wake of HST's spherical aberration problems in the early 1990s. Still, the highly elliptical orbit ranging between 10,000 and 140,000 km from Earth offered several advantages. It was thermally benign, eliminating the problematic cycle of light-to-dark and warm-to-cold, every 90 minutes, and thus removed the problem of temperature cycling, which tended to wear out electrical and mechanical subsystems. Additionally, the observatory would spend 85 percent of its time above Earth's radiation belts, allowing it a large, uninterrupted portion of each orbit for celestial observations and avoiding interference from energetic particles which might otherwise overwhelm its sensitive instruments.

As with the other Great Observatories, it was always intended that Chandra would be named after an eminent astrophysicist whose own research had helped to pave the way for the work it would perform. Chandra honoured the Indian-born scientist Subramanyan Chandrasekhar (1910-1995), affectively nicknamed 'Chandra', who has been widely labelled as the father of modern astrophysics. A Nobel Prize winner for his contributions to astronomy, he also conducted valuable theoretical work on stellar evolution that established a basis for the existence of neutron stars and black holes – the very objects that his mechanised namesake would observe from its high orbit. "Chandra thought black holes were the most beautiful things in the Universe," said Lalitha Chandrasekhar, his 88-year-old widow, who attended the observatory's launch on STS-93. "I hear a lot of people say they are bizarre, they are exotic, but to Chandra they were just beautiful."

It was not, however, only the redesign of the observatory and the troublesome IUS which kept the mechanised Chandra on the ground, but also technical glitches with the spacecraft itself. Problems completing its construction at prime contractor TRW's Redondo Beach facility in California had pushed the launch date from August to December 1998, then January 1999, and ultimately April owing to a number of computer software errors. On 20 January, only days before the spacecraft was due to be shipped from California to Florida, NASA announced yet another delay, caused by the need for TRW to evaluate and correct a potential problem with several printed circuit boards in the command and data-management system. A number of other, TRW-built satellites had turned up faulty copper circuitry and, fully aware of Chandra's high-priority status and the fact that its orbit would render it irreparable, the decision was taken to remove and replace the boards. Although the replacement process delayed the spacecraft's delivery to Florida by only a matter of days, it pushed STS-93's launch back by five weeks to late May 1999, due to the requirement to conduct lengthy tests at KSC. This target, however, conflicted with Shuttle mission STS-96, the first ISS docking flight, and the decision was taken to postpone STS-93 until early July. Then, with only two weeks to go, Chandra engineers were alerted to yet another problem, this time with 20 electrical capacitors. Fortunately, these were cleared for flight, but it offered another heart-stopping moment in the observatory's tumultuous development.

This would not be the final show-stopper, for Columbia's launch was twice postponed. On 20 July, the countdown was halted, just seven seconds ahead of liftoff, when high concentrations of hydrogen gas were detected in the Shuttle's aft compartment. It was a particularly hazardous moment, coming shortly before the ignition of her three main engines. If the halt had been called *after* ignition, the result would have been a pad abort and probably a month-long delay in readying the vehicle for another attempt. The cause of the problem seemed to be a hydrogen 'spike', which a sharp-eyed launch controller spotted briefly peaking at 640 parts per million, or double the maximum allowable 'safe' limit. During the momentary crisis, the mood in KSC's Launch Control Center was tense, as indicated by voices on the communications loop. Sixteen seconds before launch, it seemed, one of two gas detection systems indicated the 640 ppm hydrogen concentration and even though the second device showed a more normal level of 110-115 ppm, launch controller Ozzie Fish radioed his colleague, Barbara Kennedy, at the Ground Launch Sequencer (GLS) console to manually stop the countdown. To the assembled spectators at KSC, listening to spokesman Bruce Buckingham's commentary, all seemed normal at first.

"T minus 15 seconds," announced Buckingham, then "T minus 12 ... ten ... nine ... "

Inside the Launch Control Center, Fish urgently radioed: "GLS, give cutoff."

" ... eight, seven ... " continued Buckingham.

"Cutoff. Give cutoff!" interjected NASA Test Director Doug Lyons.

"Cutoff is given," replied Kennedy at the GLS console.

"We have hydrogen in the aft [compartment]," Fish reported, "at 640 ppm."

By now past what would have been a 'normal' ignition of the main engines, Buckingham announced the disappointing news to the public. Back in the Launch Control Center, with the hydrogen concentration decreasing back toward normal levels, Lyons polled his team, asking them if any emergency safing procedures were needed, such as evacuating the crew from the Shuttle, and was told that this was unnecessary. Within ten seconds of the call for cutoff, the indication of high hydrogen levels had dropped to 115 ppm. Engineers would later blame the problem on faulty instrumentation and flawed telemetry. Although disappointing, the abort had, at least indirectly, shown that NASA was not making special provisions to get STS-93 away on time for the sake of several high-level spectators in the audience. Eileen Collins' presence on the crew, as the first woman ever to command a space mission, had dominated the news. Sitting in the VIP area at KSC was none other than First Lady Hillary Clinton, her daughter Chelsea and representatives of the United States' women's football team. Clinton had formally announced Collins' assignment to command STS-93 in March 1998 at a press conference in the

3-19: Chandra is prepared for deployment on STS-93.

Roosevelt Room at the White House. For Collins herself, the assignment was representative of having worked her way through the ranks, just like the male Shuttle pilots, but for Clinton it was a public relations boon. In fact, some observers remarked that the naming of a Shuttle commander *from* the White House was rare, if not unprecedented.

"Eileen's just trying to do her job," said STS-93 mission specialist Cady Coleman. "At the same time, I'm actually very excited about the historical significance – not for Eileen, not for me, but for the little girls out there. It's really bringing home the fact to them that Eileen can do what she has set out to do. If all of us can become astronauts, it will help them to realise that the world can be theirs." Steve Hawley, who sat on the astronaut selection board in late 1989, which ultimately picked Collins for training, remembered that she would possibly become the first female Shuttle commander. "All of us that were part of that decision," he said, "take pleasure in seeing it happen. As Eileen said herself, I think another opportunity is clearly available to young girls growing up." Added pilot Jeff Ashby: "Eileen has made me feel very comfortable and treated me not like a rookie, but as somebody who has flown before and I respect her for that." For Michel Tognini, the wind of change carried even greater significance. When he began his French Air Force career, his superiors doubted that women could even fly aircraft. "Just recently," he told a NASA interviewer in mid-1999, "I saw in the French Air Force newspaper that we have the first female French fighter pilot, even though we said 30 years ago that it would never happen. Never say never."

Collins likened her leadership style to the flavour of a family. "It was important for me as a commander to learn what their talents and interests were," she said of her crew. "With that, we were able to decide who would do what duties on the flight, keeping in mind that we would be able to change that later, if we found that we needed to spread the workload around a little bit better. As we started working eight to 16-hour days together, we really started to become a family. We got to know each other so well that we became like brothers and sisters. That's one of the strengths I see in my crew. We listen to each other, we get along well, we really understand and can focus on the mission, and when we make decisions for the mission, we do what would make it most successful. One thing that comes to mind is how we work together when we do simulations and we're given malfunctions. You really get to see how people work under stress and you really get to know each other that way. That's why it's so important to train together. You need to know each other really well when you go up on a mission like this one."

Yet the issue of being the first woman to command a space mission was not lost on Collins. Nor did she forget the other female giants, upon whose shoulders she stood. "I wouldn't be sitting here today if it weren't for all the people who've gone before me and set the stage to bring women into aviation," she explained before the launch. "In the beginning of the century, it took a lot of courage to fly as a woman, when that really wasn't a woman's place. During World War II, there were the Women Air Force Service Pilots and the women who ferried aircraft. In the late 1950s and early 1960s, women competed to be astronauts. In the later 1960s, we started getting more women in the military. In the 1970s, women were offered the opportunity to fly in the military, active duty. That's when I first became interested in flying. We had our first women selected as astronauts in 1978. Since then, we've had more and more women become astronauts. There are three women Shuttle pilots now, including me."

The 20 July launch scrub demonstrated that NASA was unwilling to compromise safety to get Columbia off the ground, even with the First Lady in attendance. Fortunately, the fact that the main engines had not ignited meant that another attempt could be made on the 22nd. A third opportunity was also available on the 23rd, but after that a three-week delay until mid-August would be unavoidable, because the Air Force had scheduled a major upgrade of the tracking radars on its Eastern Test Range. In eager anticipation of another attempt on 22 July, eight middeck payloads were removed and serviced and hydrogen sensors in the aft compartment were recalibrated. Additionally, the hydrogen igniters – the system that cleared unburned hydrogen from beneath the engines, ahead of ignition – was replaced. The second effort to get STS-93 airborne also seemed afflicted by misfortune when lightning strikes were recorded, just 5 km from the launch pad. According to flight rules, no lightning was permitted within a 12 km radius. The countdown was held at T-5 minutes, in the hope that conditions might improve, but when they failed to do so, the attempt was scrubbed.

"Eileen, we gave it our best shot with this storm today," said Doug Lyons, "but it didn't agree with us, so our best bet is to give it another try another day."

"Okay, CDR copies," replied Collins, "and we though you guys did a great job tonight. We're proud of the work and the crew will be ready to go at the next opportunity."

NASA managed to convince Boeing to postpone a scheduled Delta II launch from nearby Cape Canaveral Air Force Station in order to give STS-93 another opportunity on 23 July. A safe launch was paramount and former astronaut Don McMonagle, then serving as head of the Mission Management Team, noted that STS-93 would slip until 18 August if this third attempt was scrubbed. In the late evening of 22 July, Collins and her crewmates clambered back aboard Columbia. Right on time, and true to form, Columbia sprang from Pad 39B at 12:31 am EDT on the 23rd, turning night into day across the marshy Florida landscape.

However, the commentator's excitement-tinged announcement – "We have ignition and *liftoff* of Columbia, reaching new heights for women and X-ray astronomy" – masked a serious problem brewing in the Shuttle's main engines. It came to the attention of Collins and Ashby five seconds after leaving the pad, when they noted a voltage drop on one of the electrical buses. This caused one of two backup controllers on two of the three engines to abruptly shut down. The third engine was unaffected by the problem and, luckily, all three performed nominally, propelling Columbia into a 246 km orbit, inclined 28.45 degrees to the equator. Nonetheless, the scare was significant. On no other mission had a Shuttle crew come so close to having to perform an RTLS abort landing. Had the primary controllers, which immediately assumed command, also failed, an engine failure was likely and *that* would have required Collins to wait for SRB separation, flip Columbia over, fly 'backwards' at ten times the speed of sound in order to bleed off speed and head back west, then jettison the ET and guide the orbiter down to the SLF runway ... in darkness.

"We were prepared for that," she said later. "We were listening for the engine performance data calls [from Mission Control] on ascent. This crew would have been ready to do whatever was needed." Fortunately, Columbia made it safely into space, but was travelling at 4.5 m/sec slower than expected. Although this discrepancy was tiny in view of her 28,000 km/h orbital velocity, it was enough for puzzled engineers to question whether there might have been a 1,800 kg shortfall in the liquid oxygen pumped into the ET before liftoff. NASA confirmed on 24 July that the loading of propellants had been done correctly, although it would become part of the investigation into the electrical short. Analysis of still video imagery during the ascent also revealed another problem: a leak of hydrogen gas from one of the main engines. The images, particularly those from cameras mounted on Pad 39B, revealed a narrow, bright area inside the nozzle of the right-hand engine, possibly indicative of a weld-seam breach in one of more than 1,000 stainless steel hydrogen recirculation tubes. Although Wayne Hale, then serving as Columbia's mission operations director, stressed that few conclusions could be made until the engine was back on Earth, he speculated that the leak might explain the 'missing' liquid oxygen. As hydrogen was lost at a rate of about 1 kg every second, the main engine controllers compensated by gobbling oxygen at a higher rate.

It had been one of the most potentially hazardous ascents in Shuttle history and would lead directly to the grounding of the rest of the fleet for the next six months. For Eileen Collins' crew, savouring their inaugural moments of weightlessness, their business was getting the $1.5 billion Chandra X-ray Observatory primed for deployment. The first few hours were spent checking the health of both Chandra and the IUS, primarily under the direction of Coleman and Michel Tognini, before the stack was tilted up to its deployment angle of 58 degrees. "Michel and I work as a team," Coleman explained before launch. "I put my finger on a switch, he verifies it's the right switch and that is very, very helpful to me. We also have a third person in the background – Steve Hawley – whose job is the big picture of the deploy. It's very human to make a mistake and we cannot afford that, so we're doing everything we can to prevent that." Interestingly, assignment to STS-93 was not Coleman's first involvement with Chandra. Earlier, she had met the team which ground the mirrors for the telescope and presented them with a NASA award. "I also visited Kodak, where they assembled the entire telescope," she said. "This was new for me, to learn about this amazing telescope that was going to be launched. Suddenly, I was assigned to the mission and I thought, 'You know, I'm *supposed* to do this!'"

If the crew had missed their first deployment 'window', matters would have become complicated. "If we have to keep [the payload] in the bay overnight, it really constrains things," said Bryan Austin. "If they lose any power to the heaters that keep the [propellant] lines from freezing, it really gets dicey in terms of being able to still possibly support a mission, because we cannot put them in a warm-enough attitude to keep everything warm without hurting [Chandra]." Thankfully, all went well on the first attempt. After a critical, Go-No Go decision by flight controllers at JSC in Houston, Texas, and at the Chandra Operations Control Center in Cambridge, Massachusetts, the IUS was transferred to its own batteries for power and cables routing electricity to the spacecraft were severed. At 7:47 am EDT on 23 July, seven hours and 16 minutes into the mission, as Columbia flew high above Indonesia, Coleman commanded the Chandra-IUS stack to be spring-ejected from its cradle in the payload bay. The deployment occurred precisely on time, at the opening of a 'window' that ran for eight minutes and 45 seconds.

"Houston, we have a good deploy," reported Collins. "Chandra is ready to open the eyes of X-ray astronomy to the world."

After the mission, Coleman told an interviewer that she was so taken aback by the beauty of the observatory disappearing into the inky blackness that she was rendered "almost too excited to video". Shortly after deployment, Collins and Ashby manoeuvred Columbia into a 'window-protection' orientation, with the orbiter's belly pointed towards the IUS nozzle. An hour later, at 8:47 am, with the Shuttle about 50 km 'behind' the stack, the first-stage engine of the IUS ignited for just over two minutes. Approximately 60 seconds after the completion of its 'burn', the first stage separated and the second stage took control and performed its own burn for another two minutes. The booster's next task was to keep Chandra properly oriented as its twin solar arrays unfurled. Shortly before the separation of the second stage, and after insertion into a preliminary elliptical orbit, at 9:22 am, Chandra's solar arrays were deployed with perfection. The separation of the second stage occurred without incident at 9:49 am. The IUS team was delighted with the performance of their booster.

By this point, the observatory was in an orbit with an apogee of almost 74,000 km and a perigee of 325 km. This was adjusted by Chandra's own thrusters over the following three weeks to achieve a final elliptical orbit with an apogee of 140,000 km and a perigee of 10,000 km.

Another development was the discovery of what Wayne Hale called "our smoking gun" for the mishap during ascent: Collins found a tripped circuit breaker in the cockpit for the centre main engine controller. Its discovery persuaded mission managers that the controller of the engine, or at least its wiring, had been responsible for the electrical short.

With the deployment of Chandra behind them, the five astronauts focused on a variety of experiments in the middeck. One of these was the SWUIS telescope, previously carried on STS-85. Steve Hawley employed it to conduct ultraviolet observations of Mercury, Venus, Jupiter and the Moon. In particular, he was able to examine faint emissions from Jupiter's upper atmosphere, supporting simultaneous observations from the Galileo spacecraft, map features on the surface of the Moon and image the clouds of Venus. Hawley operated SWUIS through the middeck access hatch window, in support of which Collins and Ashby executed frequent attitude manoeuvres. Meanwhile, Coleman monitored a number of protein crystal growth studies and Tognini tended a biological cell culture experiment. Following five days in orbit, Collins fired the OMS engines as planned at 10:19 pm EDT on 27 July to commit Columbia to a descent back to Earth. Passing over Baja California and north-western Mexico, the orbiter bisected Texas from west to east, crossed southern Louisiana and alighted onto the SLF Runway 33 at KSC in darkness at 11:20 pm EDT. In the aftermath of the mission, the STS-93 crew had nothing but praise for their commander. "I get asked a lot how it is to fly with Eileen," said Jeff Ashby. "I think the women's soccer team captain summed it up best: *Eileen Rocks!*"

REVISITING A NATIONAL ICON

Despite having staged just two Shuttle flights by the summer of 1999, NASA anticipated a return to regular operations in the autumn and winter of that year, with Endeavour targeted to fly the STS-99 Shuttle Radar Topography Mission (SRTM) in September, Discovery to fly the STS-103 HST servicing mission in October and Atlantis to resume ISS construction work on STS-101 in December. These would form the prelude to a ramping-up of Shuttle flights in 2000, with major elements of the new space station destined to be launched, preparatory to the arrival of the first US-Russian long-duration crew to begin its permanent occupation. However, former astronaut Brewster Shaw – then serving as Boeing's head of the ISS programme – expressed doubt that high flight rates were achievable. In March 1999, he told *Flight International* that although United Space Alliance (USA) was "doing quite well" with Shuttle operations, extending the capability beyond seven missions per annum would "tax the system" and carried implications for flight safety.

By the end of July 1999, that flight safety question mark hung firmly over wiring troubles experienced during STS-93's problematic ascent, when an electrical short, four seconds after liftoff, knocked out two of Columbia's critical main engine controllers and left them running on their backups. This precipitated a lengthy series of inspections of the orbiter's 375 km of wiring. Moreover, Columbia had also suffered a small leak of liquid hydrogen from her right-hand main engine. Although the leak was relatively minor in nature, it was serious enough for the engines to consume liquid oxygen at a higher-than-normal rate to maintain their thrust during the remainder of the ascent to orbit, causing them to shut down several seconds earlier than intended when the oxygen level finally ran dry. This, in effect, left the STS-93 crew in an orbit about 11 km lower than expected. Fortunately, the rest of the mission proceeded smoothly.

In the immediate aftermath of STS-93, an already planned Space Shuttle Development Conference was held at NASA's Ames Research Center in Moffett Field, California, during 28-30 July. It was the first such conference organised by the space agency and sought to map out plans for the evolution of the reusable fleet of orbiters, including proposals for liquid-fuelled SRBs, equipped with oblique wings, capable of gliding back to KSC after separation for refurbishment and reuse. Other upgrades included a transition of the orbiters' main engine propellant valves from pneumatic to electromechanical actuation, the development of electric APUs to replace the maintenance-intensive hydrazine-powered units and the installation of more powerful proton-exchange membrane fuel cells. Further enhancements centred on a Main Engine Advanced Health Management System to increase safety and reduce turnaround costs, more durable thermal protection system tiles for the Shuttle's belly, better brakes and tyres and an uprated avionics suite on the flight deck. NASA Associate Administrator for Space Flight Joseph Rothenberg noted during the conference that the agency expected to make a decision about "looking at alternatives to the Shuttle" by 2005 and that within three or four years after that date, a firm conclusion would be made about "what will be possible to replace the Shuttle or whether to keep it going". Henry McDonald, the head of NASA-Ames, added that the lack of a realistic replacement for the Shuttle in the late 1990s meant that the agency was looking at keeping the fleet of orbiters in operational service for at least ten or even 20 more years. Little could NASA have foreseen that tragedy would engulf the fleet, changing the 'game plan' considerably and bringing the ultimate retirement of the Shuttle fleet much closer.

Within weeks of Columbia's return from STS-93, in mid-August 1999, NASA revealed that *three* liquid hydrogen coolant tubes (rather than the one originally suspected) had been punctured by a small plug which worked its way loose during the ignition sequence of the main engines. The plug was designed to seal a liquid oxygen injector tube. In total, the punctured tubes leaked more than 1,500 kg of liquid hydrogen, causing the affected main engine to overheat and use more than 1,800 kg more liquid oxygen than necessary and producing a slightly lower than expected orbit. It was highlighted that a larger leak from maybe 20-40 of more than 1,000 coolant tubes would have raised the likelihood of an engine shutdown, seconds after liftoff, which in turn might have enforced a hairy Return to Launch Site (RTLS) abort.

The short circuit, meanwhile, was traced to a wiring defect within Columbia's fuselage, as electricity arced between exposed wire and a metal screw which had worn through insulation during 25 previous missions. The wire extended from the orbiter's forward avionics bay to the aft compartment. Suspicion centred on the wiring having been inadvertently damaged by technicians standing on payload bay access platforms during pre-flight activities, therefore procedures were modified to prevent future problems. To avoid subsequent damage, flexible plastic tubing was installed over some wiring, rough edges were coated to prevent chafing and protective shielding was fitted. Given this discovery, NASA prudently opted to inspect Atlantis, returned to KSC late the previous year, following several months of modifications, but initially did not focus on Endeavour, which was heavily into her processing flow for the STS-99 launch on 16 September 1999. The large SRTM radar instrument had already been installed into her payload bay and it was felt that, since Endeavour had only flown 13 times, there was less likelihood of such wear and tear. However, as the inspections gathered pace, Endeavour quickly entered the equation and by the end of August numerous wiring 'nicks' had been identified and the STS-99 launch was postponed until no earlier than 10 October and STS-103 until early November as the issue steadily engulfed the entire Shuttle fleet. In addition to the wiring issue, a technician reported a crimped freon coolant line for the SRTM payload, but concern that it would need to be replaced proved unfounded and it was instead secured with a brace. In order to allow engineers to reach Endeavour's fuselage wiring, the SRTM payload was removed. Also coming under the microscope was Discovery, whose planned 14 October launch on the STS-103 HST servicing mission received a minimum two-week delay. Yet it remained unclear whether STS-103 or STS-99 would fly first. HST teams stressed that having only half of their gyroscope control capability, placing them one failure away from a total shutdown, meant that a mission to the telescope carried priority. Supporters of SRTM, on the other hand, argued that valuable radar mapping data would be degraded by winter snow cover in the Northern Hemisphere and funding from the Department of Defense for the payload was rapidly running out.

In the middle of September, it was decided to switch the launch order of the two missions, with STS-103 rescheduled for 28 October and STS-99 for no earlier than 19 November. This displaced the STS-101 mission to January 2000. It was stressed that even these revised dates were dependent upon the completion of the wiring inspections. Also contributing to the delay were preparations at KSC for the impending onslaught of Hurricane Floyd, together with the need to attend to a corroded OMS propellant valve aboard Discovery. NASA confirmed the decision to fly STS-103 first, but targeted its launch date no sooner than 19 November. Meanwhile, NASA appointed an independent industry team, led by Henry McDonald, to review Shuttle safety and maintenance practices. In response to the wiring delays, USA agreed to pay "several million dollars" in penalties for the disruption of the Shuttle flight schedule. "United Space Alliance was aware of the financial impact when we made the decision to undertake a more thorough review of the Shuttles and thus delay the schedule," admitted former astronaut Mike McCulley, then serving as USA's vice president and deputy programme manager. By the end of October, the situation had brightened somewhat, with STS-103 firmly

3-20: Four members of the STS-103 crew discuss procedures before a training session. Seated is Steve Smith, with (from left) Jean-François Clervoy, John Grunsfeld and Claude Nicollier standing.

rescheduled for no earlier than 2 December and STS-99 postponed until mid-January 2000.

Technical troubles had not yet finished with STS-103. In early November, NASA managers ordered the replacement of one of Discovery's three main engines, following concerns about a 1.3 mm drill bit fragment lodged inside it. This pushed the launch until no earlier than 6 December. Attached to her ET and SRBs, the Shuttle stack was transferred to Pad 39B on 13 November, following other work, but suspect wiring between the tank and boosters prompted another delay as corrective actions were implemented. Then, the newly replaced main engine failed a series of leak checks, pushing launch back to no earlier than 9 December and forcing NASA to the conclusion that if STS-103 was not off the ground by the 14th (a so-called "drop-dead date"), it would most likely be postponed until mid-January 2000, owing to concerns about the much-hyped Year 2000 (Y2K) computer bug. Discovery needed to be on the ground, and powered-down, before the end of the year. A leak in a hydraulic system quick disconnect fitting on one of the Shuttle's APUs required replacement and inspections of potentially cracked wiring harnesses in the main engine compartment produced a further delay, first to 11 December and then to the 12th.

Frayed wiring and a crushed hydrogen conditioning line on one of Discovery's main engines produced another headache, by which time NASA managers had extended the drop-dead timeframe available to launch STS-103 until no later than 18 December. However, it was stressed that waiting so close to the end of the year would require one of the four planned EVAs to be deleted and the mission duration shortened. In the meantime, USA determined that they could replace the crushed conditioning line at the pad and in time for a 16 December launch. With the successful replacement completed by the 13th, managers pressed ahead with a revised launch attempt on the 16th. Unfortunately, despite a 90-percent likelihood of acceptable weather conditions, this ultimately came to nothing, when a potential problem arose with the ET. Paperwork and X-rays indicated that oxygen line welds on the tank might have come from a 'bad' batch which did not meet NASA's stringent specifications. Although it was concluded that all was fine, the incident was followed by concern about the quality of propellant lines within Discovery's main engines, which conspired to effect a 24-hour postponement until 17 December. Bad weather put paid to *that* launch attempt, with low clouds, rain and gusty winds at KSC, and any chance of acceptable weather on the 18th also proved hopeless. By this stage, Rothenberg authorised another attempt on 19 December, but launching so late would require a shorter flight of eight days (rather than ten), the deletion of one of STS-103's four spacewalks and shortening the length of time required to drain toxic propellants and power down Discovery after her mission. Under its worst-case deservicing plans on the runway at Edwards Air Force Base in California, USA decided to reduce the post-landing safing time by simply powering the orbiter down. Instead of draining residual liquid oxygen and hydrogen supplies for the fuel cells, these cryogens would be allowed to vent overboard.

Launching on the 19th would produce a rendezvous and retrieval of HST on 21 December, followed by three EVAs on the 22nd, 23rd and 24th, a deployment of the iconic telescope on Christmas Day and the return to Earth of STS-103 on the 27th. Former astronaut Don McMonagle, then serving as chair of the Mission Management Team, pointed out that a fourth EVA could not be attempted on the shortened flight. "We do not intend to extend this mission beyond the 27th, unless there are mitigating circumstances that require us to do that," he told journalists. "Mission success can be achieved in three EVAs. That is the plan for this mission." Having said this, HST Project Scientist David Leckrone stressed that lists of get-ahead tasks had been formulated for the three EVAs, such that the crew could perform other work, such as replacing insulation blankets on the telescope.

By now, there remained no contingency time or opportunity to shorten the mission any further; if Discovery was not off the ground on the 19th STS-103 would be postponed to no sooner than 13 January 2000. As circumstances transpired, weather and technical conditions came together on 19 December and Discovery and her seven-man crew rocketed into orbit at 7:50 pm EST, right on the opening of that night's 42-minute 'window'. The successful launch came as a relief, after nine frustrating delays and nine agonising weeks later than originally planned. It had been a peculiar launch campaign, with camps arguing on the one hand that a flight in December was not mandatory, whilst on the other that every possible attempt should be made to fly before the end of 1999. Although Shuttle Program Manager Ron Dittemore had repeatedly stressed that there was "no pressure" to fly STS-103 in December, it was also true that postponing the mission until January 2000 would not have been a casual decision. A great deal of processing work on the Shuttle had already been completed, some of which would need to be repeated in the event that the mission was stood down for another three weeks. "We would actually have to back out of certain activities we have already performed on the vehicle," he said, "and then we also believe our teams would suffer some because they would have a three-week layoff and we'd have to bring them back up to their peak training. There *is* some loss you suffer by going into January." Dittemore also reflected on the words of a friend, who told him that "the road to success is always under construction". The success that STS-103 was to become, during its eight days in space, had certainly lived through its own fair share of construction. Describing the successful launch as "a very nice Christmas present indeed", David Leckrone was "thrilled to pieces" as the poor weather conditions finally cleared and Discovery speared into orbit.

STS-103 could not reach HST soon enough. When the four-man EVA team (known as the 'payload crew') was announced in July 1998, it was tracking an ambitious mission in June 2000, which encompassed a record-breaking six spacewalks to install new hardware and imaging equipment aboard HST. All four men were veteran astronauts: payload commander Steve Smith had previously served aboard the SM-2 mission, whilst Mike Foale had made EVAs from the Shuttle and Russia's Mir space station. It was their responsibility to ensure that all 300 tools were in the proper places and configurations for the required tasks for their respective EVAs. "We put those together," said Foale, "we check them out, make sure the batteries for the power tools are charged up and gather those together in the airlock." The other two astronauts, John Grunsfeld and Switzerland's Claude Nicollier, had no spacewalks to their credit at that time, but both brought immense expertise to SM-3, because Nicollier had served as RMS operator on the SM-1 mission in December 1993 and Grunsfeld was a professional astronomer.

"The ambitious nature of this mission," said Dave Leestma, then-director of Flight Crew Operations, "made it important for the payload crew to begin its training as early as possible." Original plans called for the six EVAs to replace HST's Faint Object Camera (FOC) with the Advanced Camera for Surveys (ACS), replace Fine Guidance Sensor (FGS)-2, replace both solar arrays with rigid, high-efficiency ones and replace an Engineering/Science Tape Recorder (ESTR) with a new Solid State Recorder (SSR). Ancillary tasks for SM-3 included fitting a cooling system to the telescope's aft shroud in order to upgrade the thermal protection capacity of its systems, installing a new-technology cryocooler for the NICMOS instrument and attaching six voltage/temperature improvement kits to enhance HST's battery-charge capability. Repairs to the telescope's aging thermal insulation on its Sun-facing side were also planned, following the discovery by the SM-2 crew of areas of peeling. However, six back-to-back EVAs was more than had ever been attempted in the Shuttle's history and, as David Leckrone noted, it was "shaping up to be quite a lengthy and complicated mission" that pressed the envelope in terms of capability. With SM-3 assigned to Columbia, all eyes were on getting STS-93 off the ground in mid-1999, such that NASA's oldest orbiter could complete several months modification and refurbishment in Palmdale, California, to meet a mid-2000 launch date with HST. However, the lengthy delays to STS-93 and the need to tend to extensive wiring inspections meant that Columbia would be out of service for far longer than intended.

"I'm not an astronomer," admitted Steve Smith, "but from a layman's perspective, the Hubble Space Telescope has been like a time machine. It's given us views of things that have happened years ago in remote places, and in that respect it is kind of the thing of fiction. It *is* a time machine. It has given us an insight into things that have already happened before we were even alive and some of the views have been absolutely incredible. The Hubble Deep Field is my favourite image. Hubble was pointed towards what was considered a vacant piece of the sky for ten days, which is very unusual. When they got the image back, in this very small piece of the sky, about the size of the sky that you would see looking through a straw, they found 1,500 galaxies, just like the Milky Way. If you expand that finding over the *entire* sky around us, the implications are incredible."

Equally incredible for Smith was the prospect of visiting HST for a second time, having previously served aboard the STS-82 SM-2 mission in February 1997. "I'd always heard that once you see Hubble, you'll never forget it," he said, "that it really is this magical-looking space ship. And it really is a spectacular sight to see ... and *huge*, much larger than I had ever imagined. Most of us who have gone to the Hubble Space Telescope have never seen it before and that was my challenge before STS-82. Even though we have mockups of it, it just doesn't look the same in the water."

The situation with the Shuttle's wiring problems in 1999 was compounded further by technical problems with HST itself. The telescope carried six gyroscopes to support its precise pointing requirements, but one had failed in 1997, another in 1998 and a third in early March 1999. Since HST needed a minimum of three gyroscopes to remain operational, this third failure placed it just one malfunction away from having to place itself into a protective 'safe mode' and suspend astronomical research. Under HST flight rules, NASA was required to consider a 'call-up' Shuttle servicing mission before a fourth gyroscope could fail. "Hubble is a priceless, irreplaceable resource and it's got a finite life," said David Leckrone, "so every month lost is a month lost forever to science." For Steve Smith, who was responsible for orchestrating the EVA timeline for SM-3, the news that their mission could no longer wait until June 2000 was met with a combination of dismay and euphoria. "In the second half of 1998, we spent a lot of time in the water tank, working towards this six-EVA Hubble flight," he said. "Most of the tasks we worked on were very difficult, very new tasks. When we finally got the news, we were in a meeting called the Cargo Integration Review and, just on the voice loops from NASA-Goddard, we heard that Hubble had suffered another gyroscope failure. It was at that moment that I thought to myself, 'Gosh, I don't think they can wait for us to go back up in the year 2000'. Although Hubble was still working perfectly well, it was a greater risk now, and we met the news with disappointment for the Hubble community, because we knew that their gem was now at a lower level of redundancy." This was juxtaposed with another realisation in Smith's mind: "My gosh, we're gonna be part of this historic, rushed, launch-on-need effort to go restore Hubble's redundancy."

In spite of the gyroscope situation, others countered that HST was in a secure condition and could wait until after STS-99. "We can go get it and service it even if there are *no* gyros remaining," said Shuttle Program Manager and former astronaut Bill Readdy. "It's not a question of the safety of the spacecraft. It's a question of losing science. The probability is that by the summer of 2000, one or more of the gyros would have failed. That's what's driving us." Consequently, it was decided to split the SM-3 mission into two halves, with the first half scheduled for October 1999 and the second in September 2000. The so-called SM-3A mission would install, in order of priority, six new gyroscopes and six new voltage regulators to prevent battery over-charging, a new computer, a new FGS, a new S-band radio transmitter and a new SSR. A year later, SM-3B would install the new solar arrays and new cameras and cryocooler. With SM-3A just seven months away, the three final members of the STS-103 crew were hurriedly announced by NASA. Commanding the mission would be Curt Brown, joined by pilot Scott Kelly and French RMS operator Jean-François Clervoy. Brown would be embarking on his sixth Shuttle flight, and Clervoy his third, whilst Kelly – one member of the only set of identical twins even selected by NASA for astronaut training – was making his first mission.

Scott Joseph Kelly was born on 21 February 1964 in Orange, New Jersey, together with his twin brother, Mark, who was six minutes older. Their parents were both police officers in a town which already counted Thomas Alva Edison, General George McClellan and Whoopi Goldberg as past residents. "When I was a kid, I always liked to do the things that were difficult," he said, "and as I got older, those desires really peaked in me." Having said this, becoming an astronaut was not an all-consuming life goal for Scott Kelly. As he grew up, he aspired to play professional baseball, then drive race cars, with flying gradually taking centre stage. After completing his schooling in West Orange, Kelly attended the University of Maryland for a year, with initial plans to enter the medical profession, but decided that naval aviation was his calling in life and he moved to the State University of New York Maritime College to study electrical engineering. "Our grandfather was an officer in the US Merchant Marine and a fire boat captain in New York City," he reflected, "so that played a small part in the interest and having a Third Mate's licence, which I actually still have." Upon graduation in 1987, he received his commission through the Navy's Reserve Officer Training Corps. "When I was younger, I had a desire to fly in the military," he told a NASA interviewer, years later. "I chose the Navy because I thought landing on aircraft carriers would be kind of the most challenging type of aviation in high-performance jets, which I wanted to fly. And I was right."

A year later than his brother, Scott Kelly was designated a naval aviator in 1989 at Naval Air Station Chase Field in Beeville, Texas, then reported to Fighter Squadron 101 for initial F-14 Tomcat training. Following the completion of his early training, he served deployments aboard the USS *Dwight D. Eisenhower* in the North Atlantic Ocean, Mediterranean, Red Sea and Persian Gulf with Fighter Squadron 143 ("the world-famous *Pukin' Dogs*," according to Kelly). In January 1993, he was selected to attend Naval Test Pilot School in Patuxent River, Maryland (by coincidence, in the same class as his brother), and upon graduation he worked within the Strike Aircraft Test Squadron at Pax River's Naval Air Warfare Center. During the next two years, Kelly flew variants of the F-14 Tomcat, F/A-18 Hornet and KC-130 Hercules tanker; he was the first pilot to fly the F-14 with an experimental digital flight control system and performed high-angle-of-attack and departure evaluations. He received a master's degree in aviation systems from the University of Tennessee at Knoxville in 1996. "Being an astronaut became tangible to me when I was working as a test pilot in Pax River," he said. "The commanders and pilots of the Shuttle are typically military or former military test pilots, so having had that background made it much more likely that I would be considered for the job."

Together with Mark, he was selected by NASA as a Shuttle pilot candidate in April 1996 and upon his assignment to STS-103 in March 1999 became the first of the siblings to draw a mission assignment. However, despite their intense competitiveness in life goals, neither of the Kelly brothers appeared to have been competitive among themselves. "We certainly encourage each other to do our best," said Scott Kelly, "but if one person is better at certain things than others, it's really not a big deal to us." Although other 1996 astronaut candidates had been assigned before him, the contorted launch schedule meant that Kelly became the first US member of the class to actually reach space. For his brother Mark, far from being disappointed that he came second, the idea that his twin brother was the first American from the 1996 group to enter space was intensely satisfying.

Selection to join STS-103 was also a particularly complex first assignment for Scott Kelly, requiring him to participate actively in the rendezvous with HST. "Once we get very close to the telescope, the commander will start flying the vehicle manually," he told a NASA interviewer in the weeks before launch. "Prior to that, we've traded off on who performs some of the burns, but as we get closer I jump in the commander's seat. I'll perform the last few rendezvous burns. Curt will get in the back of the vehicle, preparing to fly the orbiter manually. We'll come up from 'below' Hubble and slowly decrease our closure rate until we're travelling at a very slow closure and to close proximity. It's very critical that we conduct these burns very precisely, because, if we didn't, we wouldn't be able to catch up. We could overshoot the vehicle or undershoot it. Whether we would be able to recover from that or not is hard to say, so it's a very precise task that we do."

With three gyroscopes having failed aboard HST, it was recognised that a further failure would force mission managers to place the telescope into protective safe mode until the Shuttle's arrival. As it cost NASA about \$21 million per month to keep HST operational, it was not a minor problem. On the morning of 13 November 1999, a Saturday, these dire fears were realised when a fourth gyroscope failed. "The failure mode here is the breakage of flex wires, the little hair-sized wires that carry current into the motors," explained David Leckrone of the cause of the gyroscope problems. "We think the breakage comes about because those things get embrittled because of oxygen that's trapped in the fluid in which they're suspended, and that oxygen has been there since the things were built." Since the gyroscopes were designed to perform with very low levels of vibration, their operating fluid was extremely thick (about equivalent to 10W-30 motor oil) and tended to hold onto its oxygen. "It's that oxygen that's causing the embrittlement," said Leckrone. "That says the failure is not run-time related. It's age-related."

3-21: Discovery thunders into space on 19 December 1999, kicking of the third servicing mission to the Hubble Space Telescope.

By the time that HST's fourth gyroscope failed, the STS-103 stack was on the pad, but liftoff remained at least three weeks away, forcing controllers to place HST into 'Zero-Gyro Sun-Point Safe Mode'. Its aperture door was closed and the twin solar arrays were oriented parallel to the body of the spacecraft. HST was then positioned such that the arrays were perpendicular to the Sun. Until the arrival of Discovery, HST would return only engineering data, slowly spinning at a single revolution per hour to maintain stability. "We don't like having a spacecraft sitting there not working," said HST Program Manager John Campbell, "but hopefully it'll only be another three weeks to the servicing mission and we'll be back online again shortly thereafter."

It was actually closer to six weeks by the time Discovery reached orbit on 19 December 1999 to begin a slightly truncated eight days of operations, and three EVAs, to bring the telescope back to operational service. On the second day of their mission, they reduced the Shuttle's cabin pressure from its normal 101.3 kPal to around 70.3 kPal to prepare the four spacewalkers for the 29.6 kPal pure oxygen of their space suits. With Steve Smith and John Grunsfeld scheduled to perform EVA-1 and EVA-3, it was Mike Foale and Claude Nicollier who had drawn the short straw by having just one spacewalk to perform. They were tasked with EVA-2. Their EVA-4 had been scrubbed owing to the repeated delays to Discovery's launch. (They had already lost EVA-6 from the six-spacewalk flight, when SM-3 was broken into two missions in March 1999.) In a space-to-ground interview on 20 December, Foale took the news in his stride. "Well, they come, they go, and you have to kind of roll with it," he said. "I try to look at the big picture. Down the road a bit, there'll be other flights, other chances. Certainly, I would regret not getting as many EVAs as I could, being greedy as I am, but this is fine and I understand why the programme made that decision."

Two days into the mission, at 7:34 pm EST on 21 December, Curt Brown manoeuvred Discovery to within range of HST and Jean-François Clervoy grappled the telescope with the RMS mechanical arm. "Houston, Discovery, we have a good capture," Clervoy reported. "We have Hubble grappled!" Based upon pre-flight estimates, the retrieval occurred seven minutes ahead of schedule. At the time of capture, the two spacecraft were flying over the Gulf of Mexico and Houston, prompting Capcom Steve Robinson to tell the crew that observers had seen them in the sky, a little dimmer than Jupiter and a little brighter than Saturn.

Kicking off an ambitious trio of EVAs, nearly an hour ahead of schedule, Smith and Grunsfeld switched their space suits onto internal battery power and floated into Discovery's floodlit payload bay at 1:55 pm EST on 22 December. Their primary task was the replacement of all six gyroscopes, housed in pairs within three Rate Sensing Units (RSUs), in order to restore the telescope's stability for science operations. Described by HST Program Manager John Campbell as "Job One", the task was completed with crispness and perfection, as Smith worked inside HST to electrically disconnect the RSUs and Grunsfeld – anchored on the end of the RMS – drove out the bolts with a power tool. Smith then installed battery voltage regulator kits to keep the telescope's batteries properly conditioned. Following this activity, mission controllers conducted 'aliveness' tests to verify that all of the RSU electrical connectors and pins were in place. Whilst these tests were underway, Smith and Grunsfeld moved to open a pair of valves to allow the residual nitrogen to vent from the NICMOS instrument. This had been installed by the STS-82 crew in February 1997, but suffered a thermal short a few weeks later and had been out of action since January 1999 due to its dwindling coolant supply.

It was anticipated that during the second half of the mission, SM-3B, tentatively scheduled for launch in the spring of 2001, a gaseous neon cooling system would be installed to enable NICMOS to continue scientific research. Unfortunately, when the spacewalkers tried to remove the valve covers, they refused to budge. Mission Control advised them to photograph the covers and press on with the voltage regulator work, but were quickly called back to make another attempt with a high-torque wrench. This time, Smith succeeded in removing the covers and loosening the valves. Although it was not considered a critical task, it was part of the much-desired plan to get NICMOS operational. The valve problem and difficulties closing and latching the aft shroud doors left them behind schedule and the intended 6.5-hour EVA had run to eight hours and 15 minutes by the time Smith and Grunsfeld repressurised Discovery's airlock at 10:10 pm EST. This made it the second-longest spacewalk in Shuttle history at that time, eclipsed only by the eight hours and 29 minutes which astronauts Pierre Thuot, Rick Hieb and Tom Akers spent outside Endeavour during STS-49 in May 1992, the programme's only spacewalk involving three astronauts. Congratulated on a job well done, EVA-1 had completed virtually all of its tasks. The only item undone was the installation of handrails, ahead of the FGS-2 replacement by Foale and Nicollier during EVA-2. However, the replacement of the six gyroscopes had already secured completion of 80 percent of STS-103's critical tasks. "The gyros are in!" exulted HST Program Manager John Campbell. "We watched as each of them got its power on, cheered each one of the six and were really pleased to see all six come up. We know from the aliveness tests that the central computer can use all six gyros when [it] needs them."

Next day, at 2:06 pm EST on 23 December, Foale and Nicollier entered the payload bay. Their first task was to install the $7 million DF-224 computer, which comprised a radiation-hardened Intel 486 processor to replace the 1970s-era device aboard HST. In making his first spacewalk, Nicollier became the first ESA astronaut selected by NASA to perform an EVA, an accolade to which Jean-François Clervoy made reference in a congratulatory message. Built around the Intel 80486 DX2 processor, and running at 25 MHz, the new, chair-sized computer was made up of three identical boards for redundancy and was 20 times faster and carried six times as much memory as its predecessor. Additionally, it could operate on just 30 watts of electrical power, as opposed to 100 watts for the earlier computer. NASA engineers joked that it represented the most expensive 386-to-486 upgrade in history. "A 486 by most people's standards is out of date, but this is a pretty special 486," Foale explained before the mission. "It's able to withstand all the radiation and there's very strong radiation up at the high altitude the Hubble flies without causing the program to crash."

It was also a difficult upgrade, since it was hard for Foale to see critical connectors on the left side of the box. He had to take exceptional care not to impact connectors on a nearby data management unit. "All the connectors are on the side of the box where you can't see them," said John Grunsfeld before the mission, "so Mike has to do that basically without the aid of stereoscopic vision. He'll have one eye as he's reaching around to do that and on the left of him, on the [aft shroud] doors is the data management unit and huge bundles of delicate cables. The challenge there is how do you jam yourself as close as you can to that without touching it, so you can see the connectors you have to disconnect, without damaging the cables." Providing clearance cues, Nicollier positioned himself close to the aft shroud doors and by 4:30 pm EST Foale had successfully hooked up the new computer. Shortly afterwards, aliveness tests confirmed that it was functioning as expected. "Most excellent!" Grunsfeld told the spacewalkers from his station on Discovery's aft flight deck. "The brains of Hubble have been replaced." Foale and Nicollier's next task was the installation of FGS-2, which had been removed during SM-2 in February 1997 and returned to Earth for refurbishment. It took two attempts to properly seat the sensor in its alignment rails, but once it was installed Steve Smith asked Mission Control if the spacewalkers could press on with replacing the Optical Control Electronics (OCE) package, a task scheduled for EVA-3. Steve Robinson replied from Houston that because Foale and Nicollier's spacewalk was already heading towards the seven-hour point, they should call it a day and begin cleaning up their work site. By the time the spacewalkers returned inside the airlock at 10:16 pm EST, they had been outside for eight hours and ten minutes, concluding the third-longest EVA in Shuttle history.

The final spacewalk got underway at 2:17 pm on 24 December and was the first US EVA performed on Christmas Eve. Smith and Grunsfeld replaced one of the Engineering/Science Tape Recorders with a new Solid State Recorder and installed equipment to aid with fine-tuning the performance of the new FGS-2. They also installed a new S-Band Single Access Transmitter (SSAT), to replace one of two identical units which had failed in 1998. The transmitter, which funnelled data from HST through NASA's Tracking and Data Relay Satellite (TDRS) network to ground stations, was not designed for orbital replacement by spacewalkers, and including numerous fiddly screws and cable connectors, which quickly placed the astronauts behind schedule. By five hours into EVA-3, they were 45 minutes behind the timeline. "We have to work with tiny little bolts," said Grunsfeld. "It's kind of like if you had to repair a watch wearing winter gloves. I'm trying to deal with little screws that are non-captive and if you drop one, obviously it'll float away." The protracted procedure to replace the transmitter obliged several other tasks, including the installation of stainless steel foil insulation panels over six of HST's equipment bays, to be deferred to the next servicing mission. As circumstances transpired, they managed to install covers over two bays, before they were called to begin cleaning up their work site and closing out the spacewalk. By the time Smith and Grunsfeld returned to Discovery's airlock at 10:25 pm EST, they had secured the third-longest EVA in Shuttle history, at eight hours and eight minutes. This placed STS-103 in the record books for having the second-longest, third-longest and fourth-longest EVAs in the Shuttle programme and the mission boasted more than 24 hours and 33 minutes of spacewalking time in total.

Although much attention had focused upon the spacewalkers, the task of Jean-François Clervoy as RMS operator was a monumental one. His task was incredibly delicate, requiring him to manoeuvre the EVA crewmen and their tools with great precision and stop and start their motions at specific times. However, he could not apply the arm's brakes. "The arm will be in running mode at all times, so I will have to protect the [hand-controller] sticks to ensure that nobody bumps into them," he said. "I will have to stay concentrated for several hours and Claude Nicollier will be my backup for that job when Steve and John are outside, and John Grunsfeld will be my backup when Claude and Mike are outside." By his own admission, Clervoy believed that Nicollier or Grunsfeld would probably take over about 20 percent of the whole EVA time flying the RMS to allow him to take a breather. As EVA-3 came to a spectacular conclusion, Clervoy could celebrate a job well done.

Pausing before he returned inside Discovery, Smith took a moment to congratulate flight controllers, engineers and scientists who had made the repair work possible. "In the last three days," he said, "we put 13 new [components] in the Hubble Space Telescope with 100-percent success and we appreciate greatly the great folks who are working on this holiday, Christmas Eve 1999." Grunsfeld added his own words of praise for their space suits, which he described as "incredible machines" which enabled the HST repair work. This proved a slightly premature statement, for whilst in the airlock, shortly before repressurisation, his suit failed to transition from internal battery power to Discovery's power supply. At this point, Grunsfeld had only about 30 minutes of battery power remaining in his suit. Ultimately, the suit was left on battery power throughout the repressurisation process and Claude Nicollier later inspected it and noticed a bent pin in the multi-pin electrical connector.

Only hours after the completion of EVA-3, it was Christmas Day and STS-103 became the first (and only) Shuttle mission to remain in orbit over the festive period. In fact, counting Apollo 8 in 1968 and the third Skylab crew in 1973, this was only the third occasion that American astronauts aboard an American spacecraft had spent Christmas off the planet. The day began with season's greetings for Curt Brown as the STS-103 crew was awakened to the sound of Bing Crosby's 'I'll Be Home for Christmas'.

Christmas in space was a fact that Scott Kelly had taken time and pains to discuss before launch with his five-year-old daughter. "I told her that we were going to point the telescope at the North Pole and get a picture of Santa," he said. "She was all excited and really didn't mind too much her dad being away for Christmas." The crew received messages from their families and shared a meal of duck liver and other delicacies from south-western France, supplied by Jean-François Clervoy. Speaking to Mission Control, Brown conveyed his own take on the Christmas message. "The familiar Christmas story reminds us that, for millennia, people of many faiths and cultures have looked to the sky and studied the stars and planets in their search for a deeper understanding of life and for greater wisdom," he said. "We, the Discovery crew, in this mission to the Hubble Space Telescope, are very proud to be part of this ongoing search beyond ourselves. We hope and trust that the lessons the Universe has to teach us will speak to the yearning that we know is in human hearts everywhere; the yearning for peace on Earth and goodwill among all the human family." John Grunsfeld added that he wished for peace on Earth in the new millennium.

Later in the day, NASA Administrator Dan Goldin sent his congratulations to the crew. "This proves once again you need *people* to do things," he told them. "You just can't send a robot up there to go fix that telescope. It was like watching a ballet. It was just fantastic what you folks have done. I'm very, very pleased."

"Well, sir, thank you for the words, but we must put that credit on the folks who trained us and prepared us for this," replied Curt Brown. "They did a great job and I think it shows. During the EVAs, we did exactly as planned and had a number of surprises that Hubble threw at us with bolts that were a little tighter and things that were a little bit different than planned. As you mentioned, humans are able to accomplish the changes without any trouble and robots, obviously, wouldn't be able to do that."

On Christmas afternoon, the final steps ahead of redeploying HST into space got underway. By 4:19 pm EST, the latches holding the telescope to its support structure in the payload bay were released and at 6:03 pm Jean-François Clervoy successfully released HST back into free flight to continue its voyage of scientific exploration, with the first data from the rejuvenated observatory expected about two weeks later. As it drifted away into the inky blackness, ground controllers reported that it was operating normally on its own, holding steady with its new gyroscopes, which prompted a unanimous cheer from the STS-103 crew.

3-22: As two spacewalkers work outside Discovery, this view back towards the orbiter's cabin and external airlock shows their crewmates watching through the rear and overhead windows.

Owing to the year-end rollover problem associated with the much-hyped Y2K computer bug, it was essential that Discovery was back on the ground, and powered down, well ahead of New Year's Eve. Having said this, NASA engineers were confident that the date rollover from 1999 into 2000 would not pose any problems for the Shuttle's computers. "We have done everything that we could do to make sure we are ready for the year-end rollover," said Entry Flight Director Wayne Hale. "We have tested, we have done everything we know to do to make it as seamless as possible. As far as we know, everything would work just fine if we worked right through 12 midnight on 31 December." Two landing opportunities existed at KSC at 5:18 pm and 7:01 pm EST on 27 December, with further options in both Florida and at Edwards Air Force Base on the following two days. Although NASA had announced before launch that KSC and Edwards would both be kept on standby as 'joint primary' landing sites, Hale stressed that Florida carried the edge, because of the $1 million cost of ferrying the orbiter across country from the West Coast to the East. Either way, it was mandatory that Discovery made landfall before 29 December.

Early on the 27th, concern about high crosswinds at KSC forced mission managers to wave off the first landing attempt. At length, Brown and Kelly were instructed to fire Discovery's OMS engines at 5:48 pm EST to begin the 73-minute descent to Earth on the second landing opportunity of the day. The burn lasted almost five minutes and produced a spectacular fireshow for ground-based observers, as the Shuttle's glowing plasma trail enthralled skywatchers from southern Texas to New Orleans and the coast of the Gulf of Mexico. Discovery's ghostly touchdown on Runway 33 at 7:01 pm ended a mission less than one hour shy of eight full days in orbit. It was a fitting and spectacular end to the millennium, and closed out almost the first four decades of human space exploration. It would have been difficult to countenance on 1 January 1900, that, by the end of the century, humans would have the ability to fly space vehicles and deliver and repair space telescopes, hundreds of kilometres above Earth, and the crew of STS-103 offered their own moment of reflection on the KSC runway. "We shouldn't forget this is almost the end of the century, this is the end of a millennium," said Mike Foale. "We started this century dreaming about leaving the planet Earth and flying. And human beings have achieved that."

It had been a hard, tortuous road. The opening years of the new century would be harder still.

4

Steps to the future

FIRE!

In his memoir, *Off the Planet*, Jerry Linenger expressed no concerns whatsoever with Valeri Korzun and Alexander Kaleri, his cosmonaut comrades aboard Mir. He had trained and played badminton in Star City with Korzun, whilst Kaleri "was still on an even keel, psychologically, and continued to work efficiently", even more than five months into a six-month space mission. The men shared photographs and thoughts of their wives and families and Linenger modelled his working day on that of Kaleri: a follower of strict routine, but by no means a blind slave of the minute-by-minute Form 24 schedule. If repairs became necessary, Kaleri would remain on-task until it was complete, rather than rushing away to tend to another chore. Exercise was important, twice daily, as was a regular bedtime, to hedge against the difficulty of ascertaining 'day' and 'night' cycles when the station was subjected to 16 orbital sunrises and sunsets in each 24-hour period.

Three weeks after the departure of STS-81 on 19 January 1997, Soyuz TM-25 was primed to roar into orbit from Tyuratam. In anticipation of the new launch, Progress M-33 had been undocked from Kvant-1 on 6 February and was destined to complete an autonomous flight until early March, whereupon it would return to Mir. Twenty-four hours after the Progress departure, Korzun, Kaleri and Linenger boarded Soyuz TM-24 and undocked it from the forward port of the multiple adaptor, flew around the station and redocked at the aft port of Kvant-1. This freed up the station's forward port for Soyuz TM-25. In making the flyaround, Linenger became the first American to board a Soyuz spacecraft during independent flight and he described the event in his memoir. The three men powered down all non-essential equipment aboard Mir, then donned their custom-made Sokol launch and entry suits and boarded the spacecraft. In a process typical of spacefarers, Linenger's spine had stretched by about 5 cm, making him taller than when he was originally fitted for the suit. "Only by jamming my chin into my chest and having Valeri and Sasha cram me into the suit could I pull the hood over my head," he wrote. "I was fortunate not to sprain my neck." At 9:28 pm Moscow Time, Soyuz TM-24 undocked from the 'front' of Mir, performed the flyaround and redocked at the Kvant-1 port at 9:51 pm. "Contact was firm, but not frightfully so," wrote Linenger. "After powering down the spacecraft, we got out of our space suits and opened the hatches." Particularly noticeable was the *smell* of space in the vestibule between the two spacecraft: "a distinct, burnt-dry smell," remembered Linenger, "that I can't describe any other way". Three weeks into his stay aboard Mir, the experience transformed the station from a mechanical object into a real home for Linenger. "It was the only time," he wrote, "during my stay in space that Mir looked warm, inviting and spacious. It reminded me of opening the door to a summer cottage that had been boarded up for the winter, looking inside and seeing familiar surroundings." In the next few months, however, Mir would turn into something distinctly less homely.

The launch of Soyuz TM-25, which took place at 5:09 pm Moscow Time on 10 February 1997, was already two months late, due to financial difficulties involving the procurement of new boosters and equipment. Such problems had dogged the Russian space programme since the early 1990s, with an 80-percent cut in government funding, but had become particularly acute in the middle part of the decade, even prompting *Flight International* to speculate in January 1997 that the once-proud country "may have to abandon its manned space programme" for that year, due to severe shortages. Around half of Russia's orbiting communications satellites were at risk of failure and the launch of Russia's service module for the ISS had been delayed by more than eight months, pushing the construction sequence for the multi-national space station into disarray. The service module would enable the first long-duration crews to live aboard the new station, but had been delayed from April 1998 until the end of that year, and perhaps later still. NASA imposed a deadline of 28 February 1997 on the Russians to produce the funds to finish the service module and even threatened to otherwise relegate the proud spacefaring nation to the rank of an ISS subcontractor. To hedge against the spectre of Russia's failure, NASA was in the process of building and testing its own $100 million Interim Control Module (ICM) to sustain the ISS until the arrival of the service module. Ironically, at about the same time, NASA paid Russia an additional $20 million in what Administrator Dan Goldin described as "seed money" to support factories to provide ISS modules and to enable two more US astronauts to occupy Mir until mid-1998.

Launching towards Mir in this new era of desperate uncertainty were two Russian cosmonauts and a German researcher, the latter destined to spend 18 days aboard the aging station under a $60 million contract with the *Deutsches Zentrum für Luft- und Raumfahrt* (DLR, the German Aerospace Centre). In command of Soyuz TM-25 was Vasili Vasilyevich Tsibliyev, making his second space mission. He was born into a poor collective farming family in Orehovka, in the Crimea, on 20 February 1954. Excited by the mission of Yuri Gagarin, he finished school and entered the army in 1971, later training as a Soviet Air Force fighter pilot. Tsibliyev graduated from the Kharkov Military School of Aviation in 1975 and later flew MiG jets along the Inner German border. His first attempt to enter the cosmonaut corps in 1976 was rejected, but after a deployment to an Odessa squadron on the Black Sea and eventual (after five applications) admission into test-pilot school he was successful in March 1987. He commanded a six-month mission to Mir in 1993-1994. His flight engineer aboard Soyuz TM-25 was 'rookie' cosmonaut Alexander Ivanovich Lazutkin, who came from Moscow, born on 30 October 1957. In addition to a background as world-class gymnast, he studied mechanical engineering at the Moscow Aviation Institute, graduated in 1981 and later joined the Energia spacecraft design bureau. Lazutkin was selected for cosmonaut training in March 1992.

Germany's ninth spacefarer was physicist Reinhold Ewald, who came from Mönchengladbach, in the far western part of the country, between Düsseldorf and the border with the Netherlands, on 18 December 1956. A karate, theatre and football enthusiast, Ewald received a degree in physics from the University of Cologne in 1977 and a master's credential in experimental physics in 1983, followed by a doctorate in 1986, with a minor degree in human physiology. He worked as a research scientist at the University of Cologne, specifically focusing on the construction and utilisation of a radio telescope at the Gornergrat Observatory at an altitude of 3,100 m, near Zermatt in Switzerland, in which he studied the structure and dynamics of interstellar molecular clouds, the birthplace of new stars. He also oversaw projects in stratospheric research and was selected into the astronaut team of the newly reunified Germany in October 1990, to serve as backup for his countryman Klaus-Dietrich Flade on the Soyuz TM-14 expedition to Mir in March 1992. In recognition of his work in support of the mission, Ewald received Russia's Order of Friendship medal. In 1995, he began training for a second German Mir flight that was planned to begin at the end of the following year, but budget problems pushed the flight into February 1997. His backup was fellow German astronaut Hans Schlegel.

Trouble seemed to be brewing for the mission almost as soon as Soyuz TM-25 reached orbit. Conditions inside the cramped vehicle were unbearably cold and, Ewald later told Jerry Linenger, it made no difference how much clothing they applied. The spacecraft commenced its final rendezvous with Mir two days after launch, on 12 February. During the latter stages the Kurs automated docking system failed. Inside Kvant-2, Linenger had been watching the approaching craft, but was perplexed when it appeared to be retreating away from Mir. "I quickly assessed the condition of the craft," he wrote. "I could detect no stuck-on thruster or other problem that might cause the spaceship to lose control and then watched as the Soyuz stopped moving, stabilised and then began to approach the Mir space station once again." At one point, Valeri Korzun floated into Kvant-2, yelling to Linenger, asking what he could see. "His face was red with rage," wrote Linenger. "He looked helpless, almost tearful." Linenger quickly understood; Korzun had been aboard Mir for almost six months and the arrival of Tsibliyev and Lazutkin marked the beginning of his journey back to Earth and a reunion with his family. For Linenger, on the other hand, this mission was *not* his ticket home; he would be aboard Mir until the next Shuttle flight in May 1997, so he was far calmer. He told Korzun that he had seen no instability and – just at that moment – the two men's talk was arrested by the successful soft-docking at 6:51 pm Moscow Time. After checking that all was well, Korzun now "looked as crazily happy as he had only moments before looked forlorn". Alexander Kaleri was also keenly excited. "It was obvious," wrote Linenger, "that both of them had been repressing their heartfelt *I-want-to-go-home* emotion until this moment."

With six men now aboard Mir, the two crews celebrated with the traditional Russian welcome of bread and salt and Tsibliyev, Lazutkin and Ewald brought gifts of fresh bananas, apples, lemons and oranges. For Linenger, the aroma of the citrus reminded him of home. For the new arrivals, after two days inside the Soyuz, having the full volume of Mir to float around in quickly brought sensations of nausea and space sickness. Tsibliyev performed a weightless barrel-roll for the television audience, but when the cameras were switched off all three newcomers sought to keep themselves as still as possible, to acclimatise to their new environment. In conversation with Tsibliyev, Linenger learned that all had gone well with the automated rendezvous until Soyuz TM-25 was about 2.5 m from Mir. "Then [Tsibliyev] noticed that the alignment was out of limits," Linenger wrote. "His spacecraft's probe was about to miss hitting the awaiting drogue on Mir, so he took over manually and backed out. He flew out the error and then steered the craft back in for the docking."

However, even with Mir at its maximum size, with the base block and all of its additional modules – Kvant-1, Kvant-2, Kristall, Spektr and Priroda – in place, the space station was too crowded to effectively support a crew of six for more than a few weeks. Its ailing life-support system was pushed to the limit and Linenger was not alone in looking forward to the end of the 18-day handover period and the return of Korzun, Kaleri and Ewald to Earth aboard the old Soyuz TM-24. With Linenger based primarily in Spektr, and Ewald conducting the bulk of his research mission in Priroda, the four Russians worked throughout the complex. In general, Ewald's Mir '97 mission proceeded very smoothly, with a few minor hiccups; as part of one medical experiment, for example, he had to drink 200 ml of tomato juice with salt, but this was ruined by the "bad quality" of the drink itself. Other investigations into materials processing and the behaviour of liquids in the microgravity environment went well. Described by Tsibliyev as "busy as a bee", Ewald also filmed a self-guided television tour of Mir for German audiences and pinned up drawings from German schoolchildren all over the station. On 20 February, Tsibliyev celebrated his 43rd birthday, which also happened to be the 11th anniversary of the launch of Mir's base block. In the final days of the joint mission, the crew fixed a malfunctioning gyrodyne in Kvant-2. Then, less than a week before Korzun, Kaleri and Ewald were due to return to Earth, something happened aboard Mir which made headlines around the world.

Fire.

On 23 February, Russia celebrated 'Army Day' (properly 'Defender of the Fatherland Day') in honour of its armed forces and the six men aboard Mir were encouraged to spend more time at rest. That evening, Korzun, Kaleri, Linenger, Tsibliyev, Lazutkin and Ewald gathered around the dinner table in the base block for a rare whole-crew meal, eating cheese, sausages, jellied pike-perch, borscht and red caviar and listening to traditional Russian folk music. The expensive caviar was not a normal part of their rations, but, according to Lazutkin, had been purchased

4-01: Mir's cluster of four radial modules in their final configuration, with Kristall and the Docking Module at the 9 o'clock position, Kvant-2 at 12 o'clock, Priroda at 3 o'clock and Spektr at 6 o'clock.

by the Russians as a treat. After eating, Linenger headed back to Spektr to begin donning electrodes for a sleep study, whilst the others lingered around the table. Then Lazutkin made his way into Kvant-1 for a mundane, but essential, routine. Under normal conditions, Mir's two Elektron units generated sufficient oxygen for a three-member crew, but during a six-man changeover period they required augmentation. Three times daily, a cosmonaut had to install a cylindrical cassette, nicknamed a 'candle', into the station's solid-fuel oxygen generator. The candle measured about 30 cm in length and employed lithium perchlorate, which, when heated, generated supplemental oxygen for the crew. Lazutkin installed the candle, checked it and was about to return to the base block, when, all at once, he was startled as *sparks* began to emerge from its top. It appeared as if an entire box of firework sparklers had gone off, all at once.

The time at which the outbreak occurred was later recorded as about 10:35 pm Moscow Time. Within seconds, the sparks had been replaced by a small column of orange-pink flame, which shot upwards from the candle, to a height of between 23-30 cm. In *Dragonfly*, Bryan Burrough made reference to Lazutkin's initial thought that the fiery candle resembled a "baby volcano". Reinhold Ewald quickly saw the flame, spitting into Lazutkin's hand, and mouthed *"Pozhar"*, the Russian word for 'fire'. At once, Korzun flew into Kvant-1, "like a giant hawk", according to Lazutkin, but their combined efforts to switch off the malfunctioning candle had no effect. Attempts to smother it with a towel were also fruitless. "The oxygen from the canister is obviously fuelling the fire," wrote Burrough, "creating the blowtorch effect. The flame is now shooting up into the open air in the centre of the module, flashes of sharp red and pink, at a 45-degree angle in front of him." It had now reached perhaps 60 cm in length. Korzun dismissed everyone from Kvant-1 and told Tsibliyev, Lazutkin and Ewald to prepare Soyuz TM-25 for departure. He also called for more fire extinguishers and oxygen masks. Tsibliyev and Kaleri passed two extinguishers into Kvant-1, which Korzun would use to dispense foam over the burning candle. Unfortunately, because Soyuz TM-24 was docked at the aft end of Kvant-1, it was unreachable until the fire was put out.

At first, the extinguisher failed to function properly. By now, thick black smoke was beginning to pour through the hatch from Kvant-1 into Mir's base block. Next, the piercing buzz of a fire alarm in the multiple docking adaptor echoed through the station and triggered an automatic shutdown of the ventilation system to prevent smoke from being blown into the neighbouring modules. By now, Linenger had floated out of Spektr and the crew began to don oxygen masks. However, his attempt with Tsibliyev to remove fire extinguishers from the walls in Priroda had failed, because they were still secured by their launch straps. It was a dangerous situation which would not go unnoticed in US Congressional hearings later in 1997 to determine Mir's suitability for a continued presence by American astronauts. The two men darted quickly out of Priroda, through the multiple docking adaptor and into Kvant-2 to grab another pair of extinguishers and head into the base block. During this time, Korzun had a few fleeting seconds to reflect on the eerie silence of Mir. "With all the ventilators shut off, the station has gone quiet," Burrough wrote. "The only sound, other than the occasional muffled comments of Kaleri or one of the others behind him in the base block, is the sound of the fire", which hissed like eggs frying in a pan. Korzun's efforts with the foam seemed to have little effect upon the ferocity of the fire; he turned to water instead and after expending his third such extinguisher the flame appeared to have diminished to a small ember.

At the next available communications pass with Russian Mission Control, Korzun reported their status. "We have had a fire aboard," he began. "We managed to extinguish the fire. The oxygen canister began burning. We extinguished it after using the third fire extinguisher. The crew is wearing oxygen masks. The pressure is O_2 equal to 155. PCO_2 is equal to 5.5. There are nine masks left. After we take off the used masks, if we feel worse we will put a new set of masks on and go to the [Soyuz] capsules. We will be situated in the capsules. The smokiness of the atmosphere is below average, but we don't know the level of toxic gases ... " Ground controllers approved Korzun's plan and after several questions, assured him that Mir's atmosphere remained acceptable. Gradually, the smoke cleared and the crew assembled in Kvant-2, where the air was cleanest, and it was here that Jerry Linenger set up a makeshift first-aid station and distributed rubber filter masks. "I treated the skin burns, most minor, but some second-degree," he wrote in *Off the Planet*, "by cleaning them as best I could, applying ointment and then covering them with a gauze dressing." Despite some bronchial irritation, dry throats and a small burn on Korzun's hand, none of the crew was seriously injured, but the fire had left a film of grime across many surfaces in Kvant-1 and the base block and they spent several hours wiping and cleaning to bring the station back to life. In total, they remained on oxygen masks for about 2.5 hours. "Although we could talk to each other, our words were muffled because of the masks," Linenger wrote in *Off the Planet*. "Hand signals proved universally understandable and eliminated the need to mentally translate from Russian to German or English."

Later, Korzun told ground controllers that the situation had stabilised, although the scent of burning remained all around them. "We will observe the security measures" when turning on the next oxygen candle, he explained, stressing that the temperature was so high during the fire that even the body of the canister was burned and its metallic closure device had partially melted. For Reinhold Ewald, midway through his mission, it came as a relief that he was not forced to evacuate Mir and return to Earth. "You trained with the Soyuz to believe you can escape in your Soyuz in all circumstances," he was quoted by Burrough. "Return to Earth is assured in your mind. Even in the worst circumstances, in face masks, you think like that. You *can* come down. What I thought, after some seconds, after some action, I thought this would be the end of my two-week science mission. I wouldn't get *any* results from my science. I would get a hug, a big clap on the shoulder, but the results would have been zero for my flight. This was the end of my mission." Thankfully, that eventuality had been narrowly averted.

Much to NASA's dismay, it was another 12 hours before the agency learned any details of the fire. In the days and weeks which followed, estimates of the duration of the fire ranged from as little as 90 seconds to as long as 14 minutes. Tsibliyev, while admitting that "nobody was watching the clocks", guessed it to be about nine minutes. In its initial press release, NASA described "a problem" which "set off fire alarms and caused minor damage to some hardware on the station" and cited 90 seconds, but added that "the crew was exposed to heavy smoke for five to seven minutes and donned masks in response". It became clear that a crack in the oxygen candle's shell had allowed its contents to leak into the outer hardware and the fire required three extinguishers – around six litres – of water to put out. However, this water caused an increase in atmospheric humidity and temperature and the crew were told to take vitamin tablets and milk products. Linenger performed his own toxicological tests of the state of Mir's atmosphere and declared the air to be good. Yet the question of 'blame' followed hard on the heels of resolving the cause and in mid-March NASA was informed that "operator error" had caused the problem. Lazutkin had opened the seal for the oxygen candle and let it sit for hours in the dampness of Kvant-1, which allowed water to seep into its chemical mixture. A damp plug of chemical then somehow triggered the fire. Linenger poured scorn on this idea in his memoir, but the notion of blaming the crew whenever a situation went awry would arise on several other occasions as the unlucky mission of Tsibliyev and Lazutkin continued. Two months later, in April 1997, the US House Science Committee adopted an amendment to the NASA Authorisation Bill, sponsored by Republican chairman James Sensenbrenner of Wisconsin, which stated that the agency must certify that Mir met or exceeded its own safety standards. Sensenbrenner also later asked NASA's Inspector General to conduct a thorough review of the safety of the Shuttle-Mir programme. In the weeks which followed, the Russians would retort angrily that the fire had come about only after thousands of candles had been used without problems for many years.

On 2 March, the time came for Korzun, Kaleri and Ewald to board Soyuz TM-24 at the Kvant-1 aft port and begin preparations for their return to Earth. The three men undocked at 6:24 am Moscow Time and their descent module touched down safely in Kazakhstan, not far from Arkalyk, about three hours later, at 9:44 am. In completing their mission, Korzun and Kaleri had spent more than 196 days in orbit and Ewald about 19 days and 16 hours from his launch aboard Soyuz TM-25 until his landing aboard Soyuz TM-24. "Six months in space is a long, long time," Jerry Linenger wrote to his infant son, John, in one of his *Letters from Mir*. "Sasha [Kaleri] has a little boy about your age: what a great surprise he's in for. His baby boy was about six months old when he left, and is now just over a year. I saw how you changed during that time of your life, up to you eating your first chocolate cake on your first birthday, and I could hardly keep up on your progress, observing you day-to-day. I would love to see Sasha's face when he first sets eyes on his little guy!" Linenger's words highlighted the difficulty imposed upon long-duration crews for Mir, and for today's ISS, and underlined the reality that, despite email and family video conferences, spacefarers remain physically and psychologically isolated from their loved ones and that was bound to take its toll. Looking at photographs of John brought such feelings to a head. "There is nothing like our own flesh and blood, feeling joyous one moment and longingly sad the next," Linenger wrote in *Off the Planet*. "A pang of loneliness and guilt, for not being there for him, shot through me."

Two days after the departure of the old crew, early on 4 March, Progress M-33 returned to Mir, with the intention that Tsibliyev would use the Russian-built TORU remote-controlled system to guide it to a docking at the Kvant-1 aft port. A camera aboard the Progress would provide live television imagery, whilst Tsibliyev used two hand controllers from inside the station to remotely 'fly' the cargo ship. The left-side controller provided translational motions, whilst its right-side sibling supported acceleration and deceleration. "In order to do this successfully, the cosmonaut must have visual information," wrote Linenger. "In addition to this video image, the cosmonaut stands in front of a portable Progress control panel, [which] reflects the status of the on-board systems, such as fuel tank pressures and thruster status." By operating the two hand controllers, Tsibliyev generated commands which were transmitted via radio from Mir to Progress M-33's computers. However, TORU was classed as a backup docking system, secondary to the Ukrainian-built Kurs. Since the dissolution of the Soviet Union, Ukraine was no longer an integral part of Russia, but rather a competitor, and its refusal to accept late payments (or none at all) from the increasingly cash-strapped Russian government had led it to demand hard currency for each new Kurs system it produced. As a result, Russia wanted to phase out the use of Kurs to save money. Requiring Tsibliyev to use the TORU backup system as his primary method of docking the Progress was, in Linenger's words, "a significant departure from standard procedure".

With Lazutkin and Linenger providing extra pairs of eyes to look out for the incoming Progress, Tsibliyev was surprised when his TORU display showed nothing but static. By the time they achieved their first visual sighting, it appeared to be approaching them far more rapidly than it should. Tsibliyev, lacking a constant image on his display, was able to issue some commands to the spacecraft, although they were performed 'in the blind'. Every few seconds, he shouted questions to Lazutkin to ask if his actions helped matters, if the Progress had slowed or if it was beginning to veer away from them. Realising that there was a strong likelihood that Progress M-33 would hit Mir, Tsibliyev yelled to Linenger to get to Soyuz TM-25 for an evacuation. "I flew to a window that faced the same general direction as the window Sasha and Vasili were using and did so just in time to see the Progress go screaming by us," Linenger wrote later. "Fearing the very real possibility of collision, instinct told me to brace for impact. I gritted my teeth, held my breath, and hoped for a miss. Although the Progress had disappeared from view under the edge of the window, I quickly calculated that, having felt nothing, the Progress must have missed hitting the base block." It turned out that the spacecraft had passed Mir at a distance of 220-230 m and it was eventually decided to send it to a destructive re-entry in the atmosphere, rather than risk a second attempt to dock. In his later conversations with Linenger, Vasili Tsibliyev told him that the TORU had never once displayed a reliable image from the incoming spacecraft.

With a fire and a near-miss already behind them, the mission of Tsibliyev and Lazutkin seemed snakebitten by bad luck. Their luck offered no promise of improvement, as Mir was caught up in several weeks of equipment failures throughout March and into April. Firstly, on 5 March, one of the Elektron oxygen generators experienced an air bubble blockage in its electrolysis canal and was shut down. "Ordinarily, a shutdown of the cranky Elektron is no cause for alarm," explained Bryan Burrough, "but this time is different. The station's only other Elektron, in Kvant-1, has been out of commission for weeks and needs replacement parts that must be sent from Earth." As a result, although Tsibliyev and Lazutkin eventually managed to bypass the clogged filter with replacement hoses to restore the Elektron to life, they were forced to rely on the replaceable oxygen candles. Coming so soon after the fire, Tsibliyev offered no objection to this plan, but ensured that fire extinguishers and blankets were at the ready and the escape route through Mir to Soyuz TM-25 was always clear in case an evacuation became necessary.

In addition to the intensive work on the Elektron system, Tsibliyev and Lazutkin were faced with a significant coolant loop failure, which allowed about 1.6 litres of alcohol and ethylene glycol to leak from cracked pipework into the station's atmosphere, reaching what Russian officials described as "dangerous levels". With the failure of the coolant loop, temperatures inside Kvant-2 began to rise. Ground controllers reoriented Mir, which took Kvant-2 out of direct sunlight, but in doing so also placed the base block *into* direct sunlight. Temperatures inside the station over the following days regularly exceeded 30 degrees Celsius. Tsibliyev and Lazutkin redirected ducting to channel cooler air from other parts of the station, but a failure of the Vozdukh carbon dioxide removal system added to their woes and prevented them from regularly committing to their exercise sessions on the treadmill. When the Omega orientation sensor in the Spektr module failed, attitude controllability was temporarily lost. Power was also lost at times, reducing the modules to darkness. "I've been in dark places before," wrote Linenger, "but this was *unearthly* dark. Darker than any dark I've ever seen. Dark is not even the proper word for it." NASA was so concerned by the deteriorating situation aboard the space station, that it planned to equip the next Shuttle-Mir mission, STS-84, with an extra (eighth) seat, in case it was decided that Linenger's replacement astronaut, Mike Foale, could not remain aboard. In early April, Progress M-34 was launched to carry much-needed supplies, including air filters, replacement candles, lithium hydroxide canisters in case of further Vozdukh problems and repair equipment. It automatically docked at Kvant-1's aft port two days after launch. During the first part of its stay, the new vehicle was used to boost Mir's orbit by about 5 km. Finally, on 11 April, the Vozdukh system was brought back on line.

The second significant event in April was the first joint US-Russian EVA by Tsibliyev and Linenger. It was originally planned for the 17th, but was slipped to the 29th by the ongoing problems with Mir's angular rate sensors and the need to carry out intensive repairs on the Elektron and Vozdukh systems. With the coolant issue still only partially resolved, at 8:10 am Moscow Time on 29 April, Tsibliyev and Linenger ventured outside Mir, clad in upgraded Orlan-M space suits, for a historic EVA, one of whose primary aims was to install the 225 kg, dresser-sized Optical Properties Monitor (OPM) on the exterior of the Kristall docking compartment. It was a significant event, marking the first occasion that an American had spacewalked outside a non-American spacecraft, wearing a non-American space suit. As Linenger emerged from the Kvant-2 airlock, he was greeted by the intense rays of orbital sunrise. "Even with my gold visor down, it was just blinding," he recalled later. "I was basically unable to see for the first three or four minutes."

At length, the two men set to work on their respective first tasks. With Linenger positioned at the end of the Strela crane and Tsibliyev at its controls, they positioned themselves for the OPM installation procedure. "The OPM and myself were then swung away on the tip of the pole," Linenger wrote in one of his letters to his son. "Upon arrival [at Kristall], I attached the pole to its new location, we rejoined at my end and installed the OPM." Inside Mir, Alexander Lazutkin confirmed that the cable connections were satisfactory. Tsibliyev and Linenger then set to work retrieving a pair of NASA sampling experiments, the latter of which would be returned to Earth aboard Atlantis in May, and installing the Benton Radiation Dosimeter onto the exterior of Kvant-2. At times, they worked with "tranquillising music" in their earphones. The two men were expected to remain outside for about 5.5 hours, but they accomplished all of their tasks in exceptional time and the spacewalk ended after four hours and 57 minutes. The new suits, which had been delivered by Progress M-34, boasted a modest upgrade over their predecessors, the Orlan-DMA, and included a second visor atop their helmets. Their 40.7 kPal operating pressures were described as not causing a hindrance or fatigue and Tsibliyev, who had worn the older suit during five EVAs in late 1993, added that the gloves were easier and more comfortable. Having said this, the assembly of the suit, prior to the spacewalk, was very labour-intensive ("like rebuilding an engine") and Tsibliyev even pulled a muscle whilst doffing his ensemble after the EVA.

"A spacewalk on the surface of a sprawling space station has a different flavour than one conducted inside the cargo bay of the Space Shuttle," wrote Linenger to his son, John. He felt like a scuba-diver, but could not dispel the overwhelming sense that he was *falling*, and the sensation of tremendous orbital velocity was more apparent than it had been from inside Mir. "Crawling, slithering, gripping, reaching. You are not falling from the cliff; the *whole cliff* is falling and *you* are on it. You convince yourself that it is okay for the cliff and yourself on the cliff to be falling, because when you look out you see no bottom. You just fall and fall and fall. The Sun sets swiftly. Blackness. Not merely *dark*, but absolutely *black*. You see nothing." Adjusting his eyes to this orbital darkness, Linenger was able to make out Tsibliyev's silhouette. The cosmonaut was making his sixth career EVA and the uninhibited view of Earth never failed to move him. "Inside the station, you cannot see it, only *parts* of it," Tsibliyev was quoted by Bryan Burrough in *Dragonfly*. "When you get out, you can see the *whole thing*. It's so unusual, so dramatic, so emotional."

For Linenger, positioned at the end of the Strela boom, he felt that he was atop a fishing pole, which got longer and thinner and that *he* was the fish at the flimsy end. "In the midst of all this, you carry out your work calmly, methodically," he concluded. "You snap a picture or two and, below, notice the Strait of Gibraltar narrowly opening to the Mediterranean." The views from within Mir were also captivating. During his four months in orbit, Linenger beheld huge dust storms in the Sahara Desert, the drying-up of Lake Chad further to the south and the obvious triangular shape of the Sinai Peninsula and the long, snaking finger of the River Nile, cutting south from the Mediterranean coast into the heart of Africa. He also worked

4-02: Jerry Linenger was the fourth US long-duration occupant of Mir.

extensively on his research experiments, taking blood samples and studying flame-flow behaviour in microgravity, although for the bulk of March and April he joined Tsibliyev and Lazutkin on near-continuous repair tasks.

By the end of April, with US lawmakers steadily losing patience with the Russians over both Mir and their ISS commitments, there seemed a very strong possibility that NASA's next long-duration resident, Mike Foale, might not remain for his own four-month mission, but would launch and return home on STS-84 in May 1997. Meanwhile, Tsibliyev and Lazutkin were informed that their planned mission would be extended by almost a month and that their Soyuz TM-26 replacements, Anatoli Solovyov and Pavel Vinogradov, would launch no sooner than 6 August, due to the ongoing problem with the supply of boosters. In late April, *Flight International* explained that all future Russian members of Mir crews would remain aloft for durations in the region of 200 days, the maximum guaranteed service lifespan of the Soyuz-TM spacecraft.

The pitiful state of the space station at this stage made the arrival of the sixth Shuttle-Mir docking mission, STS-84, even more urgent. Aboard the Spacehab double module in Atlantis' bay would be 3,300 kg of water and logistics, including science equipment and replacement parts for Mir. Original plans had called for Scott Parazynski to be aboard as the fifth long-duration NASA resident after Norm Thagard, Shannon Lucid, John Blaha and Jerry Linenger. However, as described in Chapter 2, Parazynski was dropped from training in the summer of 1995, when it became clear that he was too tall to safely fit aboard the Soyuz-TM spacecraft. Later that same year, veteran astronauts Mike Foale and Jim Voss travelled to Star City in Moscow, with the former pointed towards Parazynski's former seat and the latter assigned to serve as his backup. In *Dragonfly*, Bryan Burrough explained that after STS-63 Foale worked on the development of a pair of joint US-Russian EVAs, scheduled for STS-86, the seventh Shuttle-Mir mission, planned for September 1997. "Foale had his sights focused squarely on STS-86," Burrough wrote, adding that he "expected to be named to its crew." That changed with the removal of Parazynski and his replacement by Foale. At the time of the assignments, it was anticipated that Foale would be the final US long-term resident of Mir, returning to Earth aboard STS-86 in September 1997, whereupon construction of the ISS would begin in December 1997 and achieve a permanently manned capability by the middle of the following year. However, delays to the ISS construction schedule led NASA to expand the number of Shuttle-Mir docking missions from seven to nine and add two more long-duration crew members. Wendy Lawrence, after several changes of fortune, rejoined the training flow and was expected to follow Foale in September 1997 and would herself be succeeded by Dave Wolf in January 1998 for the final expedition. "We had to use Jim Voss as a backup and then ... not fly, which is where he was going to end up," said Dave Leestma, head of Flight Crew Operations at the time. "That was hard. He stayed there *twice* as backup and then he came back." To soften the blow, Leestma assigned Voss in late 1997 as a crewman aboard the second ISS long-duration mission. As circumstances transpired, the crewing plan after Foale's mission would change considerably.

A TROUBLED SUMMER

In the meantime, the assembly of the sixth Shuttle-Mir crew got underway in February 1996, when STS-71 veteran Charlie Precourt was assigned to command the flight. Several months later, in July, he was joined by pilot Eileen Collins, who had earlier served on the STS-63 'Near-Mir' flight, and mission specialists Jean-François Clervoy of France and NASA astronauts Carlos Noriega and Ed Lu. Shortly afterwards, in September 1996, it was also announced that another mission specialist, Soyuz TM-20 veteran and former Mir resident Yelena Kondakova, would also join the crew. This formed part of a changing direction within NASA of attaching more cosmonauts to Shuttle crews, partly to offer their expertise during Mir operations and also in preparation for the ISS era. In April 1997, just weeks before the STS-84 launch, it was further announced by NASA that two more cosmonauts would be assigned to the last two Shuttle-Mir missions in January and May 1998.

Jean-François André Clervoy's assignment to STS-84 as the first ESA astronaut to fly aboard a Shuttle-Mir mission was unsurprising, in view of the large number of European payloads assigned to the flight. These included the European Proximity Operations Sensor, a package of global-positioning system receivers to provide long-range tracking of Mir during the rendezvous and serve as a pathfinder for ESA's Automated Transfer Vehicle (ATV), destined to resupply the ISS. Also aboard STS-84, carried inside the Spacehab double module, Europe's multi-purpose Biorack facility provided a temperature-controlled environment and a centrifuge for manipulating biological specimens, on this occasion housing a total of ten discrete experiments. These included observations of lentil roots, single-celled organisms, white blood cells and bone cells, studies of tadpole and fish embryonic development in microgravity and measurements of the effect of high-energy radiation on yeast samples. In addition to Biorack, ESA's Morphological Transition and Model Substances (MOMO) payload explored fundamental solidification processes in the transparent liquid alloy succinonitril-acetone, whose behaviour was similar to that of metals. It was expected that around 1,000 images of the solidification process would be acquired during STS-84. Clervoy was assigned as the payload commander on the mission and, together with 'loadmaster' Ed Lu, had responsibility for the transfer of equipment and supplies in both directions between the Shuttle and Mir and for overseeing the complex research objectives.

Born on 19 November 1958 in Longeville-lès-Metz, north-eastern France, Clervoy grew up with a desire and a willingness to someday fly into space. "When I was in second grade," he explained, "my teacher used to tell us that when we would be grown up we would be able to fly in space, the same way he was able to buy a ticket to go to the United States." As his education advanced, Clervoy realised it would not be so straightforward, but with a fighter-pilot father and a family yearning for the romance of adventure, the urge to explore space remained alive. He read and was inspired by the adventures of the French volcanologist Haroun Tazieff and the Atlantic explorer Alain Bombard. Clervoy gained his baccalauréat from the Collège Militaire de Saint Cyr l'École in 1976 and passed Math. Sup. and Math. Spé. M' at

Prytanée Militaire, La Flèche, in 1978. He graduated from École Polytechnique, Paris, in 1981, École Nationale Supérieure de l'Aéronautique et de l'Espace, Toulouse, in 1983, and qualified as a flight test engineer from École du Personnel Navigant d'Essais et de Réception, Istres, in 1987. By this stage in his career, Clervoy had already begun to gravitate towards becoming an astronaut. He was selected with the second group of French astronauts – or *spationautes* – in September 1985 and entered ESA in May 1992. The eventual paths of this seven-strong group are fascinating in their breadth: two never made it into space, whilst Jean-Jacques Favier flew aboard the Shuttle, Clervoy and Michel Tognini flew aboard both the Shuttle and Mir and Claudie André-Deshays and Jean-Pierre Haigneré (who subsequently married) participated in lengthy Mir missions. Clervoy's first Shuttle flight was aboard STS-66 in November 1994.

By contrast, both Carlos Ismael Noriega and Edward Tsang Lu were making their first space missions on STS-84. Both had been selected into NASA's astronaut corps in December 1994. Noriega originally came from Peru, born in Lima on 8 October 1959, although he moved with his family to Santa Clara in California when he was just five years old. Neil Armstrong's landing on the Moon in 1969 attracted the interest of the young Noriega. "I remember, as a small kid in elementary school, seeing this man dressed in white, jumping down a ladder, making all this dust, walking around on the Moon for the first time," he told a NASA interviewer, "but I was a young kid, just arrived in this country, barely spoke English." It was just a dream; far from something Noriega could actually achieve. However, his parents instilled a strong work ethic in him, as well as the importance of education, and "unwittingly" he prepared himself on the road to become an astronaut. After high school in Santa Clara, he entered the University of Southern California and gained a degree in computer science in 1981, then joined the Marine Corps as a Naval Reserve Officer Training Corps student. He completed flight school and flew CH-46 Sea Knight helicopters from 1983-1985 at Marine Corps Air Station Kaneohe Bay in Hawaii and supported two ship-based deployments to the Western Pacific and Indian Oceans to support the multi-national peacekeeping force in Beirut, Lebanon. In 1986, Noriega was transferred to Marine Corps Air Station Tustin in California as aviation safety officer and an instructor pilot at Marine Helicopter Training Squadron-301. He completed two master's degrees in computer science and space systems operations from the Naval Postgraduate School in 1990 and was then assigned to US Space Command in Colorado Springs, Colorado, as a Space Surveillance Center commander with responsibility for software development and integration of major space and missile-warning computer system upgrades for Cheyenne Mountain Air Force Base. At the time of his selection by NASA in December 1994, Noriega was serving on the staff of the First Marine Aircraft Wing in Okinawa, Japan.

Joining Noriega on STS-84, Ed Lu's background was quite different. Born to Chinese-American immigrants in Springfield, Massachusetts, on 1 July 1963, he attended high school in New York and earned his degree in electrical engineering from Cornell University in 1984. "I'd always been interested in space, even as a little kid," he told an interviewer, "and flying things – although it wasn't something that I

4-03: Atlantis docked to Mir on STS-84.

thought that I was ever going to do – it kind of happened by chance." Five years later, in 1989, he secured a doctorate in applied physics from Stanford and began working as a research physicist in solar physics and astrophysics. Lu was a visiting scientist at the High Altitude Observatory in Boulder, Colorado, from 1989-1992 and later held a joint appointment with the Joint Institute for Laboratory Astrophysics at the University of Colorado. For the next three years, he was a postdoctoral fellow at the Institute for Astronomy in Honolulu, Hawaii, and Lu's research has contributed significantly to the understanding of the physics of solar flares.

The seven-strong crew of STS-84, including Mike Foale as NASA's fifth long-duration astronaut for Mir, made their way to Pad 39A in the early hours of 15 May 1997. Their launch window opened at 4:07 am EDT. In order to accommodate the possibility of technical problems during the seven-minute window, the pre-programmed hold at T-9 minutes was lengthened from ten minutes to 40 minutes. It proved a beautiful night for flying and, with no constraints reported by any of the launch team, Atlantis rocketed spectacularly into orbit at 4:07:48 am. At the time of liftoff, Mir itself was far to the west of Australia, although it was not for almost an hour that Tsibliyev, Lazutkin and Linenger were advised that Atlantis was on her way. "That's great!" came the excited response from Linenger.

During the opening stages of their mission, Precourt and Collins oversaw the first of several phasing and orbit-adjustment burns to align the Shuttle and Mir for a docking on 17 May. Rendezvous tools, including laser range-finders, were set up and checked out and Ed Lu installed the centreline camera in the ODS, which Precourt would use during the final proximity operations. In the meantime, Clervoy took charge of the Spacehab, with particular emphasis upon bringing Biorack to life for several days of scientific research in orbit.

Precourt became the first US pilot-astronaut to fly two Shuttle-Mir dockings at 10:33 pm EST on the 17th, when he gently and perfectly brought Atlantis to a firm metallic embrace with the Russian station. At the time of contact, the two spacecraft were flying high above the Adriatic Sea. Following the standard pressurisation and leak checks, the hatches were opened and gifts were exchanged. The STS-84 crew offered baseball caps to Tsibliyev and Lazutkin, whilst ESA sent up boxes of food, including apples, oranges, chocolates, ice cream and *foie gras*. Yelena Kondakova also offered bread, salt, pretzels and fresh tea to the long-duration crew, who were clearly jubilant to welcome their first human visitors in more than ten weeks. Tsibliyev took Precourt to see the interior of Kvant-1, where the fire started, and gave the STS-84 commander one of the pins from one of the extinguishers, as well as one of his EVA gloves, as souvenirs. Soyuz-TM seat liners were swapped by Linenger and Foale and the five-day process of moving 3,320 kg of logistics and supplies between Atlantis and Mir got underway. Arguably the most important transfers were a new gyrodyne and a new Elektron oxygen generator from the Spacehab over to the station. An added element of urgency came on 17 May, when a pump in support of the sole operating Elektron aboard Mir temporarily shut down and knocked the device out of action. Although the Shuttle provided sufficient oxygen for both spacecraft during the docked phase, this made it increasingly important to get the new Elektron transferred to Mir, a day sooner than planned, on the 18th.

The new unit, which weighed 136 kg, was to be installed in Kvant-2, whilst the previously repaired Elektron would be moved to Kvant-1 and stored as a backup. Meanwhile, the broken Elektron from Kvant-1 was transferred to Atlantis for return to Earth. Original plans called for the installation and testing of the new Elektron to occur during the STS-84 docked phase, which would have necessitated a ten-day mission, but with the ongoing issue associated with the Kvant-1 coolant leak it was decided to install and test it at a later date, when the problems had been fully resolved. Consequently, STS-84 remained in its original baseline configuration as a nine-day flight.

As transfer activities entered high gear, NASA received a request from the Russians, to use Atlantis' waste water dump nozzles to remove about 270 kg of condensate from Mir in order to free up volume. "Normally, on board the Mir, they recycle the water," explained Jim van Laak, NASA's deputy manager of the Shuttle-Mir effort. "The condensate gets cleaned up and used for drinking water, the urine is reclaimed and use in the Elektron system. Because of the leaks they had earlier this year in the cooling systems, there was some ethylene glycol got into the condensate and they have not been drinking the water." However, it was decided that procedures were insufficiently defined to attempt the task and NASA did not want to add more work to the STS-84 crew as their docked mission drew to a close. "Any time you've got large volumes of water in a spacecraft, leakage is a concern, if it were uncontained, such as a split in a bag," said Shuttle-Mir Program Manager Frank Culbertson. "We also didn't know exactly what is in the water and, although not a major concern, it's still an issue we have to make sure we've addressed before we introduce anything into the Shuttle." Having said this, Culbertson added that the contaminated water would be sampled and analysed to provide the Russians with "a better evaluation" of whether or not it could be reused in the future or would indeed have to be discarded.

In space-to-ground news conferences, the astronauts and cosmonauts reiterated that Mir was a perfectly safe place for Linenger and Foale to live and work, despite its recent problems. Atmospheric samples at various points throughout the station were taken with the Cosmic Radiation Effects and Activation Monitor (CREAM) and the mechanical response of the complex was further tested through the Mir Structural Dynamics Experiment. The international nature of the mission was underlined by various national anthems played to the STS-84 crew as wake-up music on consecutive days, including the French *La Marseillaise* for Clervoy, the Peruvian *Himno Nacional del Perú* for Noriega and the *Patrioticheskaya Pesnya* (Russia's 'Patriotic Song') for Kondakova. Later in the mission, Clervoy talked to students at Arianespace's South American launch site in Kourou, French Guiana, whilst Noriega had the opportunity to speak to Peruvian President Albert Fujimori and US Ambassador to Peru Dennis Jett.

For the long-duration crew members, specific time was set aside for Linenger to hand over his tasks and offer advice to Foale, including, grimly, how to use the respirators, how to find his way to the Soyuz spacecraft and how to find the fire

4-04: Charlie Precourt (left) shakes hands with Vasili Tsibliyev, shortly after Atlantis' arrival on 17 May 1997.

extinguishers. "Don't be fooled by the illusion that this is all okay while the Shuttle is here," Linenger warned. "It'll change." Following his 'Near-Mir' rendezvous mission in February 1995, Foale described his first view of the station as "like seeing the Great Wall of China from a distance; you don't relate to it", very much like a tourist on a bus. Two years later, as he boarded Mir, he recognised that it was in far better condition than he had anticipated. It was quite different from the "dull, cellar-like" contraption that he had expected; rather, it was "a warm, welcoming, cosy place" and "looked like a home". In spite of Mir's troubles, Foale had no qualms about spending several months aboard the old space station. "What I'm preparing myself for is being away from my family," he told a journalist before launch. "The difficulties you may have heard about on the Mir, these are typical in a programme that's very mature. The station has been up there a long time and there's a lot of work to do to keep repairing it, maintaining it." He intended to fully integrate himself into the crew with Tsibliyev and Lazutkin and actively participate in the repairs and maintenance; he enjoyed working with them and he enjoyed speaking Russian. A mechanically-minded man, he added that he looked forward to off-nominal situations, which he said made "life more interesting". It was an assertion which would be put to the test later that summer. "Mike was the perfect diplomat," wrote Linenger in *Off the Planet*. "He must have gotten a better briefing than I did to the effect that the programme was not primarily concerned with doing good science or advancing our expertise in space operations, but rather was conceived and thrust down NASA's throat by the Clinton Administration as a form of foreign aid to Russia."

From the perspective of Charlie Precourt, who would go on to become the only American to fly as many as three times to Mir, the attitude of the media, the public and the politicians on the ground to Mir was acutely unfair. "There *were* a lot of problems," he told the NASA oral historian later, "but these are the kind of problems that we all experience in our homes. I'm right now in the process of changing the coil to the air-conditioner on my second floor. My family's without air conditioning. These are normal, everyday life events." Precourt stressed that, although parts of Mir were old and in urgent need of repair, most of the elements of the complex, including Spektr, Priroda, the DM, several of the solar arrays, the regularly exchanged Soyuz-TM and Progress vehicles, were essentially brand-new. To Precourt, having the necessity and ability to repair things, rather than bringing them home and replacing them, was the most important lesson from Shuttle-Mir. If the United States was to operate the ISS on a continuous, long-term basis, failures *would* happen and ongoing repairs of equipment were going to become a necessity.

At times during STS-84's docked phase, the crews had fleeting opportunities to relax. Before the mission, Precourt had worked hard to set aside at least two hours, on one evening, for the two crews to share a joint meal, including foods from their respective homelands: Russia, the United States, Peru, China and Britain. "It was one of the things I insisted on," he explained, "that we find the time in the flight plan for both the Russian flight controllers and the American flight controllers ... to leave us totally alone. With this one meal, I made sure they gave us this total freedom to be together for about ... two hours and it was just a really, really memorable experience." For his part, Precourt brought along Texan barbecue items for the multi-national, multi-cultural meal aboard the first 'international' space station.

At other quiet times, Clervoy and Linenger wadded up a lump of duct tape into a ball and played catch with it in Atlantis' middeck, whilst Eileen Collins remembered the homely nature of Mir itself, with traditional Russian music always playing. At length, on 21 May, the time came for the two crews to depart. At 8:43 am EST, the hatches between the DM and the Shuttle were closed and the connecting tunnel depressurised. Following an abbreviated sleep period, Atlantis undocked at 9:04 pm, whilst above the Ukrainian capital of Kiev, and began to move away from the station, her thrusters operating in Low-Z mode to avoid inducing contamination or disturbances. Although it was the sixth time that an orbiter had separated from Mir, the undocking of STS-84 carried several significant differences. Collins guided Atlantis 'below' Mir and halted for periods of about five minutes apiece at distances of 30 m, 90 m and 450 m, in order to test ESA's Proximity Operations Sensor in readiness for ATV missions to the ISS. In order to support the test, there was no fly-around inspection of Mir and the laws of orbital mechanics alone allowed the Shuttle to separate from the station naturally and without additional thruster firings.

Two days later, on the evening of the 23rd, in keeping with the STS-84 tradition of national anthems, the seven-strong crew was awakened by the sounds of 'The Star-Spangled Banner' to begin final preparations for re-entry and landing at KSC. Early the following morning, Entry Flight Director Wayne Hale polled his control team and received a unanimous "Go" to proceed with the irreversible de-orbit burn which would slow Atlantis sufficiently for her to plunge into the 'sensible' atmosphere, on an hour-long hypersonic descent, bound for a Florida touchdown. However, the first opportunity was skipped, due to the presence of low clouds over the SLF runway. Finally, above the Indian Ocean, Precourt fired the OMS engines at 8:19 am EST for a little over three minutes to bring his ship back to Earth. Travelling from the north-west towards the south-east, Atlantis crossed the Pacific and the continental United States and finally entered Florida airspace, touching down on SLF Runway 33 at 9:27 am. The landing was successful, in spite of a stiff tailwind and crosswinds from the right side that required Precourt to steer back towards the runway centreline.

Earlier in the flight, Linenger had been questioned by journalists about his plans after returning to Earth. "Going fishing" had been top of the list, but he had also received a touch of motivational advice from Carlos Noriega on maintaining his physical and psychological fitness. "I've been able to exercise very strongly," Linenger said. "I've got Carlos over here, who's a Marine, and I'm going to have him yell in my ear to get up and *walk*!" Thus, *walking* away from the orbiter brought STS-84 to an end for Jerry Linenger, concluding a 132-day mission which made him the most experienced male US spacefarer to date, slightly eclipsing the accomplishment of his predecessor, John Blaha. It was an achievement which would not last long, for Mike Foale's difficult five months aboard Mir would secure their own records ... and their own troubled place in the history books.

Foale's mission began relatively calmly. During his training, he had established a reputation for himself as the most 'Russianised' of the NASA astronauts, both in terms of language and his ability to ingratiate himself into a Mir crew, and ensured that he ate dinner each evening with Tsibliyev and Lazutkin and supported them with the station's chores. Each day, he also shared a ten-minute tea break with them. He tied his sleeping bag to the wall inside Spektr, although the noise of the two drive motors that slowly rotated the module's solar arrays frequently kept him awake at night. To initiate his experiments, Foale set up a containment unit for 64 black-bodied beetles under special lighting conditions and loaded the Svet greenhouse with samples for a rapeseed investigation. However, the problems aboard Mir continued, with the crew routinely working late into the evening to rectify the cooling system leaks.

Late in June, the first murmur of trouble turned into a storm which would engulf and come close to destroying the Shuttle-Mir effort. At 1:22 am Moscow Time on the 25th, the Progress M-34 cargo craft was undocked from the aft port of Kvant-1, preparatory to an autonomous flight and re-rendezvous and docking to test the TORU remote control apparatus. During the previous attempt, on 4 March, the system provided Tsibliyev with no visual information and the attempt failed. After separating from Mir, as planned, the cargo ship flew about 45 km 'above' and 'behind' Mir to a distance of about 7 km, whereupon it was expected to be guided back to a TORU-commanded docking by Tsibliyev at approximately 12:20 pm Moscow Time.

Initial tests of the equipment went well, despite a few minor teething troubles, and the TORU displayed a blurred, flickering television picture of what appeared to be

4-05: Jerry Linenger (right) assists Mike Foale with his Soyuz suit and seat liner during transfer activities on STS-84.

Mir, captured by a camera aboard Progress M-34. Unfortunately, since the station was backdropped by clouds, it was intensely difficult for Tsibliyev to make it out, much less make out how rapidly or slowly it was moving. "What Vasili was seeing on his screen was an image that didn't change in size very fast," Foale explained later. "That's the nature of using a screen to judge your speed and your distance. He couldn't determine accurately from the image that the speed was too high." Neither Foale or Lazutkin, glued to Mir's windows, could even *see* Progress M-34. Foale was tasked with using a laser ranging device, but could not do so without a visual 'lock'. Tsibliyev estimated the distance between the two spacecraft at about 2.9 km, but Lazutkin's efforts to see the Progress through windows in Kvant-2 and Kristall proved fruitless. The cosmonauts' concern intensified. By this stage, Mir should have filled an entire square on the checkerboard-like overlay on the TORU screen, but it actually occupied less than half a square, meaning Progress M-34 was actually approaching the station too slowly. As Tsibliyev and Lazutkin watched, the station grew steadily larger, eventually filling about three-quarters of one of the overlay squares.

Shortly afterwards, at 12:04 pm Moscow Time, the commander began to issue braking commands to the Progress, via the TORU, in order to slow the cargo ship by about a metre per second. A second period of braking followed and by 12:07 pm the crew expected Progress M-34 to be less than a kilometre 'above' them, closing towards the Kvant-1 aft port. At 400 m, Tsibliyev would apply further braking, move it slowly to 50 m, then prepare for a soft docking. He was unable to do so, because none of the three men could even *see* Progress M-34 from any of their windows.

At this point, with shocking suddenness, Lazutkin saw the cargo ship ... albeit *much* closer than it should have been and its speed was excessive for its close range. It emerged from behind a solar array which, until that point, had obscured his view, and it appeared to be heading directly towards the base block. "The distance is 150 m!" shouted Lazutkin. "It's moving closer! It shouldn't be coming in so fast!" The two men were powerless to do anything as Progress M-34 passed 'over' the Kvant-1 port and along the length of the base block, heading towards the multiple docking adaptor and the Spektr module. It was now becoming obvious that a collision was likely and Tsibliyev instructed Foale to make his way to the Soyuz TM-25 spacecraft, which was docked at Mir's forward port, and prepare for an evacuation. As he moved, according to Bryan Burrough, one of Foale's feet accidentally whacked Tsibliyev's arm, prompting some suggestion afterwards that it may have been a factor (for better or worse) in the disaster which would shortly engulf them all.

At 12:18 pm Moscow Time, as the two spacecraft flew high above Niger, Progress M-34 struck the Spektr module, sending a shudder through the entire space station. Lazutkin described the last-second sight of the cargo ship as "full of menace, like a shark". The master alarm instantly began booming. A popping in the men's ears confirmed in their minds that the station's hull *had* been punctured and that there *had* been a depressurisation, although probably not a catastrophic one. However, as Tsibliyev gained his first glimpse of the cargo ship from one of the base block's windows, what he saw filled him with dread and foreboding: Progress M-34 had impacted one of Spektr's solar arrays, tearing a jagged gash in it, buckling a radiator and puncturing the module's hull. Lazutkin quickly moved through to the multiple docking adaptor and, with Foale, began disconnecting umbilicals to prepare Soyuz TM-25 for evacuation, then entered Spektr to hear a loud hiss of escaping oxygen. From Foale's perspective, the impact was quite unlike a similar impact on Earth. "I felt the station *move*, two or three inches, relative to me," he told a Smithsonian interviewer. "Then I felt the shock of the collision through my fingers. That's all. It's not like being in a car crash, where you're all belted in and you feel the impact. Here I wasn't attached to anything, but I felt it in my hands and I knew what had happened. That was the *only* time I had any really serious fear of dying." Immediately, Lazutkin set to work removing cables and the large ventilation tube from the hatch of the module, in which all of Foale's equipment and living quarters were housed.

In the minutes which followed, Mir's sensors registered a steady decline from the normal internal pressure of 780 millimetres of mercury (about 103.9 kPal) to 690, then 685, then 680, then 675, then 670, as it continued to drop precipitously. If it fell lower than 550 millimetres (73.3 kPal), Russian flight procedures demanded an evacuation of the crew from the space station and an emergency return to Earth. The impact, it would be subsequently determined, had sheared away half of one of Spektr's four solar arrays, possibly interfering with the electrical capability of the other three, and had damaged the hull and thermal radiators of the module itself. (Under normal circumstances, Spektr's four solar arrays generated 11 kilowatts of electrical power, providing almost half of the 26-kilowatt output of Mir as a whole.) Tsibliyev's immediate concern, though, was 'feeding the leak' and he quickly set to work locating and opening up portable oxygen tanks from the Kvant-2 module to replenish the oxygen lost from the puncture, perhaps pushing the pressure above 730 millimetres (97.3 kPal) to give them more time. Steadily, the pressure increased. At 12:21 pm Moscow Time, Lazutkin and Foale succeeded in closing Spektr's hatch, sealing off the crippled module from the rest of Mir.

By this stage, Progress M-34 was flying freely, "flying around us," according to Tsibliyev, but the immediate concern was securing Mir itself. Over the next agonising minutes, thanks to the oxygen tanks, the pressure inside the station gradually increased, but the situation was dire, with a power outage causing the gyrodynes to go off-line and the main computer to shut down. (That the gyrodynes failed, perversely, worked in the crew's favour, for they no longer consumed power and decreased the burden on the ailing electricity supply.) Yet the problems sent Mir into a spin of about one degree per second and left the station eerily dark and silent. Due to the orientation of their orbit, they were more often in Earth's shadow than in direct sunlight. "You want to close your ears, so you can't actually hear the sound of silence," Lazutkin remembered later. "It's painful. You experience flight in a completely different way." For Tsibliyev, watching the stars and the polar lights from the darkened station was particularly memorable. At this stage, it was Foale who suggested using Soyuz TM-25's thrusters in an ultimately successful effort to stabilise the station and reorient its solar arrays towards the Sun.

"To stop the spin and face the arrays toward the Sun, the crew needed to know the spin rate of Mir," it was noted in NASA's official history of the programme, *Shuttle-Mir: The United States and Russia Share History's Highest Stage.* "However, the computer and other instruments were out of operation. In the dark and in the silence, Foale went to the windows in the airlock and held his thumb up to the field of stars. Combining a sailor's technique with a scientist's knowledge of physics, Foale estimated the spin rate of the space station. Then, he and Lazutkin radioed the estimates down to the Moscow control centre. The ground controllers fired Mir's engines and that stopped the spin – certainly not perfectly, and in no way permanently, but it showed that it could be done." Using a 45 cm scale model of Mir, with flashlights taped to the sides to mimic Spektr and Priroda, and shining a third flashlight to simulate the Sun, Foale demonstrated to Tsibliyev how he should apply pulses from Soyuz TM-25's thrusters. As the American kept his 'star-watch' at the Kvant-2 window, he would periodically yell directions to Tsibliyev, in the commander's seat aboard the Soyuz.

After a restless few hours of sleep, the three men set to work transferring batteries from the darkened modules into the base block to recharge them. All of Foale's experiments and personal effects were now sealed out of his reach inside the crippled Spektr; Tsibliyev dug behind a panel and found him a new toothbrush and the American worked with ground controllers to arrange for a care package to be launched aboard the Progress M-35 cargo ship in early July. In the immediate aftermath of the collision, Foale set up his temporary quarters in the multiple docking adaptor, but since this would be needed for subsequent EVAs to repair the damaged Spektr, he moved on a more permanent basis into Kvant-2.

In the meantime, after its collision, Progress M-34 was gradually manoeuvred away from Mir and destroyed in the atmosphere on 2 July. Its replacement had already been delayed since earlier in June and duly thundered into orbit from Tyuratam on the 5th. Two days later, it successfully docked at the Kvant-1 aft port, bringing much-needed supplies, including tools which Tsibliyev and Lazutkin would use in an EVA attempt to repair Spektr. Two options were under consideration: firstly, that the cosmonauts would perform an EVA to re-route cables from Spektr's three undamaged solar arrays to the base block to restore electrical power, or, secondly, that they would enter the depressurised module (in a so-called 'intravehicular activity' or 'IVA') to reconnect its power supply to Mir. However, if Tsibliyev and Lazutkin were to perform an IVA inside Spektr, they had to find a way of passing cables through or around the sealed-off hatch. In support of an IVA, the tools carried aboard Progress M-35 included a modified hatch cover, known as a 'hermaplate', which had been hurriedly constructed to allow Tsibliyev and Lazutkin to reattach Spektr's power and other cables. The hermaplate carried 23 holes through which the cables would be threaded. Tentatively scheduled to take place in mid-July, the IVA would involve the two men fully equipped in their Orlan-M space suits, in the depressurised Spektr, and would carry many of the same risks as an 'external' EVA. The condition of Mir remained precarious, with dark, silent, moisture-filled modules, devoid of electrical power, as the three men unloaded Progress M-35 and began to inventory and assemble 30 cables to install a new umbilical adaptor in Spektr's hatch. Foale later recalled that around half of his time was spent mopping up condensation with anything he had at his disposal, including old clothes.

Then, on 13 July, as Tsibliyev exercised on Mir's bicycle ergometer, he recorded an irregular heartbeat. The commander discussed the issue with a physician and plans for him to suit-up and enter Kvant-2 to execute a full-up simulation of his Spektr repair IVA tasks were postponed. "I understood immediately that the reasons were the period after the collision," Alexander Lazutkin was quoted by Bryan Burrough in *Dragonfly*. "I remember distinctly that Vasili didn't have any sleep during this time and if he did he had maybe one or two hours a day. All that time, he worried about the collision and he kept all these feelings inside." Tsibliyev asked for a week to recover from the heart problem and Russian flight controllers asked him to pass the microphone to Foale, whereupon they asked the American to take over the task. Foale stated that he had no objections to performing the IVA and was confident that with Tsibliyev's support he could accomplish it. Either way, the critical work was pushed back by at least ten days, until at least the end of the month. In the intervening days, other problems impacted the mission. On 17 July, during training for the IVA, Lazutkin was tasked with rehearsing all of the required disconnections in order to close the hatches, ahead of depressurisation. However, he accidentally disconnected a power cable connector between a rate sensor in Kristall

4-06: Alexander Lazutkin (left) and Mike Foale grin for Vasili Tsibliyev's camera. For all three men, the summer of 1997 brought precious little to smile about.

and the attitude control computer, which triggered a series of systems failures, including transmitters in Mir's base block and telemetry transceivers. There followed another power outage and another failure of the station's central computer.

A few days later, the seemingly jinxed crew was notified that the Spektr IVA would be deferred to their replacements, Anatoli Solovyov and Pavel Vinogradov, who were due to arrive in early August aboard Soyuz TM-26. It was reported at the time that Tsibliyev, Lazutkin and Foale were acutely disappointed by the news, but it was clear that fatigue had played a key role in the incident and managers decided to allow the two Russians to complete the remainder of their mission with little additional stress. Original plans called for Frenchman Léopold Eyharts to join Solovyov and Vinogradov on Soyuz TM-26. He would have spent about 21 days in orbit, conducting France's $20 million 'Pégase' mission, before returning to Earth with Tsibliyev and Lazutkin. However, with Mir lacking power and unable to support effective science operations, it was decided to defer Eyharts' mission until early 1998. Instead of a three-week handover between the two crews, there would be just a one-week handover, with Tsibliyev and Lazutkin now scheduled to return to Earth on 14 August. In the final days before the arrival of Solovyov and Vinogradov, the crew dried out the Priroda module with warm air in support of the hardware and software which Eyharts would eventually use and Foale resumed work on those of his experiments which did not require Spektr, including the beetle investigations, the greenhouse and its broccoli plants, the crystal growth investigations and the Earth observations with a Hasselblad camera. "The beetle experiment was the only experiment I had really continually in the flight and that was in Priroda," Foale reflected later. "It was without power for 30 hours, but sometime in there I moved the beetles from the Priroda module to the area near the greenhouse in Kristall and that was the configuration for both of those experiments for the rest of my increment. They were being powered off the base block, even though the modules themselves were unpowered."

At 6:35 pm Moscow Time on 5 August, Solovyov and Vinogradov rocketed into orbit aboard Soyuz TM-26. In order to make room for the new arrival, Progress M-35 undocked from the Kvant-1 aft port for two weeks of independent flight. At 8:02 pm Moscow Time on 7 August, Solovyov manually docked his ship to Kvant-1 and Mir's population was temporarily increased to five men. It must have been an exciting moment for Alexander Lazutkin and Pavel Vinogradov to meet in orbit for the first time, because the two men had been selected into the cosmonaut corps together in March 1992. In time, Pavel Vladimirovich Vinogradov would go on to secure the record in 2013, as the oldest Russian cosmonaut and the world's oldest spacewalker, aged 59-60, a title he retains at the time of writing. "I am one of the people of the generation who saw the first flights to space," he told a NASA interviewer, years later, "Gagarin's flight, Glenn's flight. When I was a boy, I dreamed about that, but it seemed kind of a pipe dream at that time, because cosmonauts ... were unique people, as I thought." Born in Magaden, in Russia's Far East, on 31 August 1953, Vinogradov completed high school in Anadyr, deep within the Chukotka Peninsula, and worked as a metal-turner, before entering the Moscow Aviation Institute. The idea of becoming a cosmonaut, or, at least, working with rockets, had already crystallised in his mind, having listened to the news of Neil Armstrong walking on the Moon. To the young Vinogradov, it was like "something from the world of the science fiction". He graduated in 1977 as a mechanical engineer and for the next six years specialised in software development for automated interactive systems of recoverable vehicles, computer graphics and aerodynamic design models. During this period, he also qualified as a computer systems analyst. In 1983, he joined the Energia spacecraft design bureau, working on Soyuz-TM and the Russian Buran shuttle. Following his selection into the cosmonaut corps, Vinogradov underwent two years of basic training and should have flown on Soyuz TM-24, launched in August 1996, but was dropped only days before launch, when his commander, Gennadi Manakov, was grounded by a heart problem. By his own admission, Vinogradov fell victim to space sickness during his early days in orbit, but quickly acclimatised to the new environment.

For Tsibliyev and Lazutkin, the end of their traumatic mission was rapidly approaching. At 11:55 am Moscow Time on 14 August, they undocked from Mir in their Soyuz TM-25 spacecraft and landed safely in Kazakhstan, about 160 km southeast of Jezkazgan, at 3:17 pm. As if cruel fortune desired one final card to play on the unlucky cosmonauts, the descent module's soft-landing rockets failed to fire at the instant of touchdown, giving Tsibliyev and Lazutkin one of the roughest thumps onto *terra firma* ever experienced by a returning crew from Mir. In fact, it left Soyuz TM-25 badly dented and the two cosmonauts felt as if they had been kicked in the back and stomach. Their 184 days in space had been a catalogue of calamities and the two men had somehow managed to keep the aging station operational, and themselves *alive*, against terrifying odds. Yet the ultimate betrayal came in the aftermath of the mission, when the traditional Russian blame game got underway. A commission had been formed, under the chairmanship of Yuri Koptev, Director-General of the Russian Space Agency, into the cause of the Progress M-34 collision and concluded that Tsibliyev was to blame for the incident. Even Russian President Boris Yeltsin, in two quite different addresses, initially blamed the collision on "human error", then replaced it with his thanks for the cosmonauts' "persistence, courage and heroism". In the meantime, Tsibliyev's flight bonus was placed under threat, but he was eventually absolved of responsibility for the collision and in April 1998 he received the Order of Merit for the Fatherland in recognition of his "courage and heroism displayed during prolonged space flight". Although he never flew into space again, Tsibliyev later rose meteorically through the ranks of the Gagarin Cosmonaut Training Centre and eventually served as its commander.

In the aftermath of the departure of Tsibliyev and Lazutkin, the new crew of Solovyov, Vinogradov and Foale boarded Soyuz TM-26 on the morning of 15 August and undocked from Kvant-1. Less than an hour later, they redocked at the forward longitudinal port of the multiple docking adaptor, freeing the aft port for the return of Progress M-35, which, after a delay caused by false instructions, executed a successful docking at Kvant-1 on the 18th. Shortly before the docking, Mir's computer crashed and the station drifted out of attitude, causing the Progress' automatic Kurs rendezvous system to shut down at a distance of 60 m and requiring Solovyov to perform the task manually, via the TORU. On this occasion, the remote-controlled TORU performed well, despite a short dropout in its video image. Mir's computer was repaired and was back up and running about 24 hours later, preparatory to the long-awaited IVA inside Spektr by Solovyov and Vinogradov, two days later than planned, on 22 August. This would hopefully allow them to reconnect three of the module's intact solar arrays to Mir's power supply. Since the station had lost full-pointing capability to the Sun, in the short term the measures would bring the total electrical output up from 50 percent to 70 percent, but would be enough to restore power to the attitude-control system and allow the reactivation of the darkened Priroda and Kristall modules. The IVA would be followed, about two weeks later, by an 'external' EVA by Solovyov and Foale to identify and perhaps repair the puncture on Spektr's hull.

Having Foale perform an EVA at this critical juncture in Mir's history was significant and amply illustrated the confidence of the Russians in his abilities and the extent of his training in the Orlan-M suit. However, it also spelled particularly bad news for Wendy Lawrence, who had been training since the summer of 1996 to replace him as the sixth long-duration NASA crew member aboard Mir. Lawrence had originally been assigned as John Blaha's backup in September 1995, but, as explained in Chapter 2, she had been withdrawn from training only a few weeks later when it became clear that she was too short to fit safely into the descent module of the Soyuz-TM. With her knowledge of the Russian language and systems, she was appointed in March 1996 as NASA's director of operations in Star City. Within weeks of arriving in Moscow, it became clear that two more astronauts would likely fly long-duration missions to Mir, beginning in September 1997 and January 1998, and Lawrence was successfully fitted-out for a Sokol launch and entry suit. Her first hurdle to flying a long-duration Mir mission, it seemed, had been overcome.

Unfortunately for Lawrence, when she came to try the Orlan-M EVA suit for functional testing, it became clear that her hands did not reach to the tips of the gloves. Nevertheless, NASA and the Russians agreed to permit her to fly a long-duration expedition, on condition that she would not be considered for an EVA. In August 1996, Lawrence was consequently assigned to fly the sixth long-duration Mir mission, which was scheduled to launch aboard STS-86 in September 1997 and return to Earth aboard STS-89 in January 1998. She would then be replaced by fellow astronaut Dave Wolf, who would end more than two years of continuous US long-duration experience, returning to Earth aboard the final Shuttle-Mir mission, STS-91, in June 1998.

The Progress M-34 collision in June 1997 proved a game-changer for Lawrence. "For a while there, it looked like *none* of us were going to get a chance to go up to Mir," she explained, as US politicians, media and the public debated the morals and merits of sending Americans to an aging space station which was perceived to be unsafe and unfit for purpose. "It looked like it would probably be the end of the programme, but the Russians persevered. They became focused on repairing Spektr with spacewalks. They were adamant about that. They were very determined that *that* was the course of action they wanted to take and, based on that, both sides felt that all three crew members on board should be able to do an EVA in a Russian suit and it would be an opportunity for US astronauts to get some more spacewalk

4-07: The crippling damage to Spektr's solar arrays from the Progress M-34 collision is evident in this shocking image from the STS-89 Shuttle mission.

experience, particularly in the Russian suit." For Lawrence, this meant that at the end of July 1997, seven weeks before launch, she lost her long-duration expedition. However, to soften the blow was assigned as a mission specialist on not only the seventh Shuttle-Mir flight, STS-86, but also aboard the ninth and final flight, STS-91, a few months later.

During Solovyov and Vinogradov's IVA, Mike Foale was directed to remain in the Soyuz TM-26 spacecraft at the forward port of the multiple docking adaptor, ready for an evacuation, should one become necessary. Meanwhile, Solovyov and Vinogradov depressurised the multiple docking adaptor at 12:59 pm Moscow Time, but the start of their IVA was delayed slightly when a leak was found in one of Vinogradov's space suit gloves and it refused to seal properly. The adaptor was hurriedly repressurised, the glove swapped for a replacement, and the atmosphere was again reduced to vacuum. At 2:14 pm, the hatch into Spektr was opened and, feet first, Vinogradov entered the darkened module. Eleven power and other cables were successfully connected to the hatch hermaplate by 4:00 pm and, as Solovyov passed him tools, Vinogradov began to search for the location of the puncture in the hull. It had been expected that the size of the actual hole was only about 3 cm^2, but the cosmonauts were unable to find it. Fears of the presence of toxic chemicals in Spektr's atmosphere, or even dangerous debris, such as glass splinters, proved groundless and the two men described the interior as being in generally good shape, with a few "white crystals" floating around, possibly from soap or shampoo, together with a thin layer of frosting on several surfaces, which had been exposed to vacuum for eight weeks. Ventilation systems and pumps all appeared to be in satisfactory condition. Solovyov and Vinogradov reconnected power to the two forward solar arrays, and partially reconnected the third, as well as retrieving one of Foale's laptop computers and some photographs. The multiple docking adaptor was repressurised at 6:03 pm Moscow Time.

The success of the IVA enabled power from Spektr's three undamaged solar arrays to be routed to the rest of Mir and brought Priroda and Kristall out of darkness and onto the road to recovery. Over the following days, the three men crisply tackled ongoing problems with Mir, repairing glitches with the primary and backup oxygen generators on 25 August, although the issue of keeping the solar arrays aligned with the Sun remained fraught with difficulty. Temperatures in the base block had risen to more than 30 degrees Celsius, but were as low as 7.2 degrees in Priroda and Kristall. Under Solovyov's command, the crew ducted warmer air from the base block into the modules, which initially caused an increase in condensation as the moist air hit the cold metal, but over time the process of drying began. By mid-September, a week before Atlantis and the STS-86 crew were due to arrive, Foale was overjoyed as Priroda was sufficiently dried-out to switch on its power.

On 6 September, Solovyov and Foale ventured outside Mir on their EVA to identify the location of the puncture in Spektr's hull. Seven possible location points had been identified by ground controllers, but, in *Dragonfly*, Bryan Burrough described Solovyov and Foale's task as akin to finding a lost coin in a junk heap. Their spacewalk required Foale to crank the controls of the Strela boom to manoeuvre Solovyov over to the damaged module to conduct his inspections. "I then basically sat at the base of the crane for most of the EVA, which lasted for six hours," Foale recalled later, "moving him left, right, up a bit, down a bit, using the two handles on the crane. Once or twice, I would translate to Anatoli actually at the work site and hold his feet while he would try and dig in underneath the insulation." Solovyov used what Foale described as "a raisin knife" to cut away a section of insulation to inspect the area around the damaged radiator and solar array mounting point, but although the two men noticed severe buckling of exterior panels and struts, they could find no evidence of the physical puncture. Solovyov manually rotated Spektr's three undamaged solar arrays into a better orientation to capture sunlight. The installation of a cap for the outlet valve for a Vozdukh carbon dioxide scrubber was postponed until a subsequent EVA and Foale successfully dismantled a US radiation dosimeter for retrieval.

Work to bring Mir back to near-full operations continued. The hatch to Progress M-35 was finally reopened, more than three weeks after it had redocked at Kvant-1, and it was intended that the crew would begin loading waste material aboard the cargo craft for destruction in the atmosphere. However, on 10 September, Mir's main computer failed. Although the crew managed to bring it back to life, they were unable to find the cause of the failure and were forced to switch it off to begin repairs. At the same time, Solovyov and Vinogradov worked to gradually bring the station's gyrodynes back online. On the 15th they received instructions to replace the computer with a spare, only to have this hiccup a few days later. A replacement computer had been scheduled to arrive aboard Progress M-36 in October, but was loaded aboard Atlantis to be delivered by STS-86.

The decision to press ahead with the next Shuttle-Mir mission was made by NASA Administrator Dan Goldin himself at 10:00 am EDT on 25 September, only hours before Atlantis was scheduled to lift off, following a final safety review. In spite of the apparent wind of change in Mir's fortunes, there remained much scepticism in the United States about whether a continued presence by long-duration NASA astronauts was appropriate. Immediately after the Progress M-34 collision on 25 June, there had been calls to bring Foale home and cancel the follow-on missions by Wendy Lawrence and Dave Wolf. Shortly after the collision Congressman James Sensenbrenner of Wisconsin, in his position as chair of the US House Science Committee, had paid an urgent visit to Goldin, seeking answers. JSC Director George Abbey established a Mir-Progress Mishap Investigation Team, led by STS-81 astronaut Mike Baker, whilst Shuttle-Mir Program Manager Frank Culbertson oversaw a comprehensive Flight Readiness Review, ahead of STS-86 and Dave Wolf's impending mission. For Culbertson, a veteran Shuttle astronaut, with two missions under his belt, it was a frustrating time. "It was very discouraging to me when there seemed to be so many people in the media and in Congress who just wanted to walk away when it got tough," he reflected. "That's certainly an easy way out and you can guarantee that nobody's going to die in a plane crash if they don't fly, but you don't accomplish anything much by staying on the ground if you want to accomplish things in space." Other teams, led by former astronauts Fred Gregory and Tom Stafford, shed further light on the safety issues, focusing not only on Mir and Progress, but also upon the viability of Soyuz itself as an emergency escape vehicle for the crew. It was highlighted that holding pay bonuses over cosmonauts' heads was a key factor, as it encouraged them to perform the experimental Progress TORU dockings, which carried unnecessary high risks.

On 18 September, with Atlantis on the pad and just one week away from her scheduled liftoff on STS-86, Culbertson testified before the House Science Committee. "When you are exploring new territory or preparing yourself to take a major step into the unknown," he told the committee," who can say when you have learned enough?" Tom Stafford's group found Mir's life-support platform to be "satisfactory ... at this time", with the station generating sufficient power (around 23 kilowatts) to support useful and worthwhile science, whilst other investigators concluded that NASA's response to the problems and its corrective actions were an "acceptable risk". For his part, Dan Goldin admitted that questions of the safety of the NASA astronauts aboard Mir had left him unable to sleep, but it was Dave Wolf who argued a successful middle ground. "It put our leadership ... in a tough spot," he said, "because they had to say something which is inherently never fully safe is *safe enough*. I applaud their leadership in this. I also applaud the good questions that people like Mr Sensenbrenner and other critics brought forward, because if we couldn't accurately address those questions ... then we really didn't have any business flying." Philosophically, Wolf and Culbertson met eye-to-eye:

4-08: Originally assigned to support two long-duration missions, fate and an issue of height would snatch the opportunities from Wendy Lawrence (left) and Scott Parazynski. However, both would fly on STS-86.

"It's easy to be good partners with the Russians when things are going easy," the next long-duration Mir crewman said, "but it's when things are difficult that we really can show what good partners we will be."

ARRIVAL OF 'THE WOLFMAN'

David Alexander Wolf, the man who would take over from Mike Foale aboard Mir, was born in Indianapolis, Indiana, on 23 August 1956. The son of a doctor, in his early childhood he aspired to be a garbageman. "Mr Peacock, our garbageman, let me run the garbage truck," he told the NASA oral historian. "It was an old-time thing, with big levers." His parents encouraged him, as long as he worked hard to be the best garbageman, but in time Wolf's interests changed and focused upon the space programme. After high school he entered Purdue University to study electrical engineering. He received his degree in 1978 and was accepted into medical school at Indiana University, from which he graduated in 1982. Years later, he paid tribute to his family – "my father liked electronics, my mother liked sports, my uncle flew aerobatic airplanes" – for having fostered his interest in the sciences and engineering. Yet it was watching Ed White making America's first spacewalk in June 1965 which really captured Wolf's attention and changed his career aspiration from garbageman to spaceman. "I was nine years old," he later told a NASA interviewer. "It inspired me into technological fields, eventually medicine and engineering, and I always wanted to work at NASA from that point." Shortly after receiving his medical degree, in 1983 Wolf did just that. After completing his medical internship at Methodist Hospital in Indianapolis, he trained as a flight surgeon at Brooks Air Force Base in San Antonio, Texas, and remained on reserve duty with the Air National Guard until 2004, rising to the rank of lieutenant-colonel. He joined NASA as a member of the Medical Sciences Division at JSC, working on the engineering development of the American Flight Echocardiograph and later the bioreactor. Even before entering the halls of the space agency, his experience was considerable: whilst working on his medical degree, Wolf had established himself as a pioneer in ultrasonic image-processing practices. He was accepted into NASA's astronaut corps in January 1990 and first flew aboard the Shuttle in late 1993.

By the time of STS-86, Wolf had about 150 hours of EVA training time and it made sense for him to replace Wendy Lawrence. A humorous anecdote arose in July 1997, when Wolf's own mission (or so he thought) was still six months away. He approached Frank Culbertson one day to ask if he could be considered for further EVA training. "I hadn't come up with a good way to break it to him that I'd like to do that," he said, when all of a sudden Culbertson invited him into his office and offered him the chance to take Lawrence's seat on the September mission. "Dave, we've got to know by tomorrow at noon," Culbertson told him, "whether you can fly on the next increment and fly in maybe five or six weeks from now and, on top of that, put the whole EVA training course in your programme before that." For Wolf, the answer was simple and of a single syllable: *Yes*.

Yet there was still much to do before he could fly the mission. "We squeezed a whole EVA programme that might take six months or more into a *month*," Wolf told the NASA oral historian, "which took basically morning through evening, including weekends, in the water tank. Almost every day, in laboratories and pressurised suits and vacuum chambers, or in the classroom." This proved easier said than done, for blocks of four days had normally to be scheduled between each underwater or vacuum chamber test, to eliminate the risk of Wolf suffering the bends. However, in order to meet the schedule, Wolf was performing the fully-suited tests on alternating days, pushing his own capabilities to their limits.

When the STS-86 crew was announced in December 1996, it originally included six members, most of whom had flown or trained to fly to Mir in some capacity. In command was STS-63 veteran Jim Wetherbee and the cadre of three mission specialists included Russian cosmonaut Vladimir Titov, Frenchman Jean-Loup Chrétien and 'Too Tall' Scott Parazynski, together with Lawrence for NASA's sixth long-duration expedition. "This is your crew of the future," Wetherbee said at one point, making reference to the fact that living and working aboard the ISS would be similarly international and multi-cultural in its nature and in its scope. With the decision in July 1997 to replace Wendy Lawrence, it was judged that her expertise required her to remain on the crew, albeit travelling to and from Mir aboard the

Shuttle. Rounding out the crew was pilot Michael John Bloomfield, the sole member of the STS-86 team who had not flown into space before.

Bloomfield was born in Flint, Michigan, on 16 March 1959 and received his early schooling in his home state. He earned a degree in engineering mechanics from the Air Force Academy in 1981, serving a year as the captain of the academy's Falcons football team. "I got into the Air Force kind of on a fluke," he admitted. "I really wanted to play major college football." His size and weight meant he was not approached by Michigan or Michigan State Universities, but the Air Force Academy fixtures included games with Notre Dame, Georgia Institute of Technology and Boston College, which in Bloomfield's mind were "all the major universities". Upon receipt of his degree, on the advice of the Air Force general father of one of his footballing buddies, Bloomfield underwent flight instruction and graduated at the top of his class at Vance Air Force Base in Oklahoma in 1983. He was selected to fly the F-15 Eagle and until 1991 he served as a combat-ready pilot and instructor pilot. Bloomfield later completed the F-15 United States Fighter Weapons Instructor Course in 1987 and was also selected for test pilot school at Edwards Air Force Base in California. Bloomfield already knew that all Shuttle pilots had progressed through test pilot school and NASA was his goal, although a friend had gone on to fly the F-22 Raptor. "At that point, I decided ... to go to test pilot school and if I don't become an astronaut, then at least I can fly the F-22. So it's a win-win situation." He emerged from test pilot school as a Distinguished Graduate in 1992 and remained at Edwards, conducting test work on the F-16 Fighting Falcon. Bloomfield gained a master's degree in engineering management from Old Dominion University in 1993 and was selected by NASA as an astronaut in December 1994. Being in the hallowed halls of the space agency was an enlightening, exciting experience for Bloomfield. "They have a *dream* out there," he told a NASA interviewer, years later. "They have this vision of going out and exploring space and the neat thing is they're not just experts in one little thing; they're experts in a *lot* of different things."

Jean-Loup Jacques Marie Chrétien, a brigadier-general of the *Armée de l'Air*, the French Air Force, became the first Frenchman to have flown both aboard US and Russian spacecraft. In June 1982, aboard Soyuz T-6, he became France's first man in space. Born on 20 August 1938 in La Rochelle, Chrétien's middle names honour those of his parents: his father, Jacques, was a sailor in the French Navy, whilst his mother, Marie-Blanche, was a housewife. As a child, he lived in Brittany and grew up in a country still occupied by the forces of Nazi Germany: an airfield lay just a few kilometres from his home and he vividly remembers watching fighter aircraft taking off and landing. "I was living under a part of the theatre of fighter airplanes of World War II," he told a NASA oral historian. "I had a *permanent* 3D movie under my eyes." These sights, sounds and experiences laid the foundations of a desire to become a pilot and, after assembling model aircraft in his youth, finally learned to fly in 1953 and gained his private licence a year later. Chrétien received his education at L'École Communale à Ploujean, the Collège Saint-Charles à Saint-Brieuc and the Lycée de Morlaix, before entering France's prestigious Air Force Academy – the *École de l'Air* – at Salon-de-Provence in 1959. Two years later, he received a master's degree in aeronautical engineering. He earned his military pilot's wings in 1962 and

4-09: Jim Wetherbee (right) and Anatoli Solovyov shake hands after the arrival of Atlantis on 27 September 1997.

served in a fighter squadron in Orange, in the south of France, for the next seven years. During this period, Chrétien flew the Dassault Super-Mystère fighter-bomber and the Mirage III interceptor and, in 1970, entered the French test pilot school, the École du Personnel Navigant d'Essais et de Réception, and served at the Istres Flight Test Centre, working on the Mirage F-1. Chrétien was appointed as deputy commander of the South Air Defence Division in Aix en Provence in 1977 and was selected as an Intercosmos candidate by France's national space agency, the *Centre National d'Études Spatiales* (the French National Space Centre, CNES), alongside fellow pilot Patrick Baudry, in June 1980. Three years later, he became France's first man in space on Soyuz T-6 and spent a week aboard the Salyut 7 station. Chrétien later flew to Mir aboard the Soyuz TM-7 mission in November-December 1988, spending 25 days in orbit (the longest flight by a non-Russian and non-American spacefarer at the time) and becoming France's first spacewalker. In March 1995, Chrétien was selected to join NASA for mission specialist training.

STS-86 would involve the second EVA to be conducted by Shuttle crew members outside Mir, but would be notable in that Russian cosmonaut Vladimir Titov would serve as EV2, working with Scott Parazynski. In doing so, this would make him the first non-US crew member to perform an EVA in a US space suit, coming less than two months before Japanese astronaut Takao Doi was scheduled to make his own spacewalk on STS-87. Titov, making his fourth space flight and with more than 376 days of experience from his earlier missions, had already performed three EVAs aboard Mir in 1988, whilst for Parazynski it was to be his first spacewalk. Their primary tasks during the five-hour excursion were to retrieve the four suitcase-sized Mir Environmental Effects Payload (MEEP) experiments, which were attached to the Docking Module by Linda Godwin and Rich Clifford on STS-76, more than 18 months earlier. "This is a very delicate task," Parazynski explained before launch. "As Vladimir and I approach these experiments, we have to be very, very careful not to touch the surfaces of these MEEP pallets. If we do, it we brush our glove against them or our tethers inadvertently come in contact with them, we've compromised 18 months of science." After folding up the MEEP boxes and stowing them in side-wall carriers in Atlantis' payload bay, Parazynski and Titov would tether a large Solar Array Cap to be installed onto the damaged Spektr module during a subsequent EVA by Solovyov and Vinogradov. The cap, which weighed 55 kg, was intended to provide a hermetic seal to isolate the base of the damaged array if and when it was jettisoned at some point in the future.

In addition to this work, Parazynski and Titov were to evaluate ISS tools and evaluate a critical piece of equipment first trialled by STS-64 in September 1994. Unlike the Manned Manoeuvring Unit (MMU) that was used in 1984, the $7 million Simplified Aid for EVA Rescue (SAFER) was not intended for satellite repairs, but as a design solution to the Shuttle programme's requirement to offer a means of 'self-rescue' in the event that an EVA crew member became untethered during the course of a spacewalk. True to the sentiment behind its acronym, SAFER's design centred on the fact that in the event of a tether failure during an EVA, it would be hugely difficult to undock the Shuttle from the space station, rendezvous and retrieve the lost crew member and then redock in a safe manner. SAFER quite literally provided a *safer* means of going about EVA and, on STS-64, supported the United States' first untethered spacewalk in almost ten years. Developed by the Automation and Robotics Division at NASA's Johnson Space Center, it weighed 37.7 kg (less than one-third of the MMU), carried 24 fixed-position compressed nitrogen thrusters and was attached to six 'hard-points' at the base of the life-sustaining backpack. Powered by a 28-volt battery pack, SAFER's attitude-control system included an automatic attitude-hold and six degrees of freedom. Unlike the MMU, it had no bulky arms for hand controllers, instead being equipped with thruster 'towers' extending up the sides of the backpack, and on STS-64 Lee and Meade operated it from controls hard-secured to their suit torsos. (In the current ISS configuration, the SAFER's controller is embedded within one of the thruster towers and is swung out by the simple pulling of a lanyard.) During STS-86, it was intended that Parazynski and Titov would evaluate the first flight production model of SAFER and would test its hand controller, its automatic attitude-hold feature, its valves and cold-gas thrusters. They would conduct these tasks whilst secured in foot restraints; unlike STS-64, no free-flight testing of the SAFER would be undertaken on STS-86. Lastly, the men were to work with a Universal Foot Restraint for ISS, designed to accommodate the boots of both US and Russian space suits, as well as a common tool carrier.

In spite of the complexity of the mission, it still remained unclear until the final hours before launch whether STS-86 would actually deliver Dave Wolf as the next long-duration NASA crew member or just collect Mike Foale and bring him back to Earth. That decision finally came from Administrator Goldin himself, early on 25 September. "It is only after carefully reviewing the facts, thoroughly assessing the input of independent advisors ... that I approve the decision to continue the next phase of the Shuttle-Mir mission," he said in a statement. "Tonight, the Shuttle Atlantis will launch, sending David Wolf to replace Mike Foale and to continue the American presence on-board Mir. We have heard the calls of some who say it's time to abandon Mir ... However, a decision should not be based on emotion or politics. It should not be based on fear. A decision should be based, and *is* based, on a scientific and technical assessment of the mission safety and the agency's ability to gain additional experience and knowledge that cannot be gained elsewhere."

Later that evening, at 10:34:19 pm EDT, Atlantis rocketed away from KSC, turning night into day across the marshy Florida landscape. With the completion of the Orbital Manoeuvring System (OMS) burn to establish the crew in low-Earth orbit, Capcom Scott 'Doc' Horowitz radioed with good tidings. "Atlantis, we see a good OMS-2, good residuals, no trim required," he told them. "Welcome to orbit."

"Thank you," replied Wetherbee, then added: "Scott ... Bloomer keeps saying things like *Incredible!*"

"Tell Bloomer it's real now," replied Horowitz.

Over the course of the next two days, the seven crew members activated the Spacehab double module in the Shuttle's payload bay and executed the now-standard series of phasing burns to rendezvous with Mir. Wetherbee and Lawrence installed the centreline camera in the ODS, whilst Bloomfield thoroughly checked the laser ranging equipment and other rendezvous tools and Parazynski and Chrétien extended and tested the docking ring. With an EVA planned for later in the flight, Parazynski and Titov also checked out their space suits and equipment. On the morning of the 27th, having set up the VHF radio link, fluent Russian-speaker Chrétien took charge of the initial ship-to-ship communications with Solovyov, Vinogradov and Foale aboard Mir.

At 3:58 pm EDT, working from the aft flight deck and assisted by Bloomfield in the commander's seat and Parazynski in the pilot's seat, Jim Wetherbee flawlessly executed the seventh docking between the Shuttle and Mir. At the time of contact and capture, the two space vehicles were high above the northern reaches of the Caspian Sea, to the east of the Russia-Kazakhstan border. Within two hours, following customary leakage and pressure checks, the hatches were opened and the crew were engulfed in bear hugs and laughter as ten astronauts, cosmonauts and a *spationaut* cavorted, tumbled into the narrow passages of the space station. As he shook Solovyov's hand, Wetherbee carried something else in his other hand ... the replacement computer for Mir. Before the flight, it was supposed to be transferred on the second day of docked operations, but Solovyov had asked Wetherbee to bring it over as soon as the two crews shared their welcome ceremony. "So I decided to go him one better," recalled Wetherbee, "to give it to him *during* the handshake ceremony." Vladimir Titov had helpfully drawn a smiling face on the computer box, which Wetherbee proudly presented to a beaming Solovyov.

"When you go visit somebody, you like to bring a gift to let them know you're happy to have the hospitality," joked Frank Culbertson. "I think that's the best gift you could think of for the crew, to bring them a new computer, because they had so many problems with the one that was up there!"

After a safety briefing and a joint meal, the two crews plunged into their five-day period of operations to deliver more than 3,600 kg of equipment and supplies, including 770 kg of water. The STS-86 crew delivered a new gyrodyne, batteries, three air-pressurisation devices and Mir's new attitude-control computer. They also loaded a large quantity of broken and unneeded gear, including the failed Elektron oxygen generator, and US science experiments, aboard the Spacehab. Dave Wolf moved his custom-moulded seat liner over to Mir, officially joining Solovyov's crew, whilst Mike Foale transferred over to become a Shuttle crewman for his final few days in orbit. During this period, Wendy Lawrence played a key role in enabling the long-duration mission which, barely a few months earlier, she had confidently expected to be flying herself. With her original mission cancelled, Lawrence's responsibilities had altered when she became a full member of the STS-86 crew. "Because I got added to that crew so late," she told a NASA oral historian, "there weren't a lot of responsibilities that I could be given and have time to train for, so my focus on that flight was to get the US laboratory set up for Dave Wolf." That *laboratory*, and that living space, was in the Kvant-2 module, for Spektr had been closed off and sealed for more than three months at this stage. For Wolf, the inclusion of Lawrence on STS-86 was critical. "I could not have succeeded at this without her," he said. "She knew more about the planning of the transfer and placing gear, and what all the gear was, than I did. She went up and just made things work fast and got me organised and off on a great start." Wolf also worked extensively with Foale for the customary handover of operations. At one stage,

Foale and Vinogradov harvested crops from Mir's greenhouse and transferred samples over to the Spacehab for return to Earth. With Spektr out of action, Wolf's quarters were in the airlock at the end of the Kvant-2 module. "Talk about a room with a *view*," he told Mission Control. "There's windows on four quadrants with shades or curtains you can close." At one stage, he and Foale spent 15 minutes there, watching in silence as the Shuttle-Mir combination passed over a darkened United States.

Elsewhere, Wetherbee and Bloomfield supported the ongoing Mir Structural Dynamics Experiment (MiSDE) by pulsing Atlantis' RCS thrusters to measure the disturbances to the station's modules and solar arrays, in readiness for ISS operations. Preparations for the EVA by Parazynski and Titov on 1 October soon took precedence and at 9:30 am EDT the hatches between Atlantis and Mir were closed to support the spacewalk. The two men spent about 2.5 hours pre-breathing pure oxygen, in order to rid their bodies of nitrogen and eliminate the risk of the bends, and the EVA officially began when Parazynski and Titov transferred their space suits' life-support utilities to internal battery power at 1:29 pm. The fact that this was Parazynski's first spacewalk, and Titov had not performed an EVA in almost nine years, was hardly noticeable, as the pair speedily set to work on their first task to retrieve MEEP and install its four boxes to two locations on the ODS in Atlantis' payload bay. They turned next to the Solar Array Cap for Spektr, which they fitted to the DM, and worked on their planned SAFER evaluations, before concluding the spacewalk at 6:30 pm, after five hours and one minute outside the Shuttle-Mir combination.

Parazynski's test of the SAFER backpack produced an excited exchange between himself and Mike Bloomfield, who was choreographing the spacewalk from Atlantis' cabin.

"Can you hear the thrusters firing?" asked Bloomfield at one stage.

"Yes, I can," replied Parazynski, who was testing SAFER whilst secured in foot restraints in Atlantis' payload bay, "and I think we'll end the test there. We have 98 percent gaseous nitrogen, 93 percent power."

"Understand you could hear the thrusters firing," pressed Bloomfield. "Could you feel them pushing at all?"

"Not at all."

"You can't feel any of the force from them?"

"Not at all."

Aside from the heavy workload, Parazynski and Titov spared a few moments for some levity. At one stage, they were left awestruck as the Shuttle-Mir combination passed over the South Pacific. Later, Titov was asked to batten down some loose insulation on the Docking Module, which prompted Capcom Marc Garneau to tell Bloomfield to ask the spacewalkers to hold off on returning to the airlock until the task was concluded. "I think you could talk us into that, Marc," Parazynski joked. The task completed, Titov was startled to see Anatoli Solovyov's face, and camera, at one of Mir's windows. He called to Parazynski to take a photograph from their side of the glass, too, a truly unique perspective of two spacefarers, taken by two spacefarers, in two very different locations.

As Parazynski and Titov spacewalked outside Mir, work of equal, if not greater, importance was ongoing inside the aging station, as Solovyov and Vinogradov set to work installing the new attitude-control computer. In readiness for the procedure, the gyrodynes were spun down and the Shuttle provided attitude control for the entire 'stack' to maintain electricity production through Mir's solar arrays. Solovyov and Vinogradov fitted the new computer in the base block and checked it out, allowing flight controllers to upload fresh software over the next few hours, in blocks during four ground station passes. When this was completed, attitude controllability could be restored, a new navigation base established and the gyrodynes spun back up to full capability. The next event was for the two crews to go their separate ways. At 6:45 pm EDT on 2 October, Wetherbee and Solovyov shook hands for the last time in orbit and closed the hatches between their respective spacecraft. The STS-86 crew then bedded down to rest, prior to a picture-perfect undocking at 1:28 pm EDT on the 3rd. The undocking was executed about 90 minutes later than planned, in order to allow Solovyov and Vinogradov additional time to finish installing and testing a new data-relay unit in the Kvant-1 module.

At the instant of separation, the Atlantis-Mir combination was high above south-eastern Russia, close to the northern border of Mongolia, and after Wetherbee had withdrawn to a safe distance he passed control over to Bloomfield for the flyaround inspection. The rookie pilot flew the Shuttle to a point about 180 m 'beneath' and 'ahead' of the station, in order to gather further data from ESA's Proximity Operations Sensor in readiness for ATV missions to the ISS. He then moved in to about 75 m to begin the 46-minute flyaround. It was at this stage that an important attempt to identify the location of the puncture in Spektr's hull was made. At 4:41 pm EDT, Solovyov opened a pressure equalisation valve inside the multiple docking adaptor to admit a very small amount of pressure of air into the damaged module. Although it was clear that there would not be a great deal of air issuing through the leak site, it was hoped that it might displace some thermal blankets or release a flurry of ice crystals, thereby betraying its exact location.

Aboard the Shuttle, Vladimir Titov trained his eyes and cameras on Spektr, whilst Pavel Vinogradov watched from Mir, and both were able to see particles seeping from the base of the damaged solar array, moving away into space at a 45-degree angle. A second pulse of air, about 20 minutes later, at 5:03 pm EDT, proved less conclusive, with only a single particle drifting from an unspecified area of Spektr. Overall, their combined efforts were not enough to pinpoint the exact location and extent of the damage. Frank Culbertson subsequently told journalists that, although it provided more information than had been available to Solovyov and Foale during their 6 September EVA, he cautioned that "particles drifting away from a spacecraft don't change my view of the ability to man-rate a repair. We're a long way from that."

From the perspective of the STS-86 crew, the flyaround was dramatic and awe-inspiring. "The flyaround ... will always be very memorable," recalled Wendy Lawrence in a NASA oral history. "It's one of those situations where you just step back and take it all in. [Mir's] modules are white and, with the Sun shining on them, they absolutely glisten, and then *that*, set against the blackness of space, is a strikingly beautiful sight. I think that's a memory I'll have for all of my life; just the beauty of that moment."

Finally, Bloomfield fired Atlantis' thrusters for the final time at 4:16 pm EDT to depart Mir's vicinity for good. With the final two Shuttle-Mir missions, STS-89 and STS-91, scheduled to be carried out by Endeavour and Discovery, respectively, this was the last occasion for Atlantis to approach the Russian space station. The landing of STS-86 was originally planned for 7:59 pm EDT on 5 October, after a ten-day mission, but the presence of broken clouds at KSC forced NASA managers to wave-off both of the opportunities for that day. Edwards Air Force Base in California was not activated to provide backup support. In anticipation of a 24-hour postponement, Edwards was brought online, with two landing opportunities in Florida and two in California on 6 October.

At 7:30 pm EDT on the 5th, shortly after Atlantis' first landing opportunity had been called off, the crew was notified that STS-86 had officially set a new record, by giving the Shuttle programme a cumulative two years of time in space since the first flight in April 1981. "We thought you'd like to know that we're about to come upon a new milestone for the Shuttle fleet," radioed Capcom Scott 'Doc' Horowitz. "In three seconds, we will have flown two years in space. Ready ... *mark* ... "

"Well, that's quite an accomplishment, Doc," replied Mike Bloomfield, "and we're all proud to be part of it."

Although the KSC weather remained close to acceptable limits, with low cloud cover and significant crosswinds, Wetherbee was given a "Go" for the de-orbit burn and brought Atlantis to a touchdown at 5:55 pm EDT, completing a mission of almost 11 days. As Wetherbee and his crew prepared for their return to Earth, their hard work was seemingly rewarded by a unique perspective of Progress M-36 roaring into orbit from Tyuratam on 5 October. *"Tallo Ho, Houston!"* exulted Wetherbee, as he glimpsed the last two minutes of the ascent, whilst Atlantis orbited in darkness about 320 km 'behind' the Progress and high above the Sinai Peninsula. In readiness for the arrival of the new visitor, aboard Mir, Solovyov, Vinogradov and Wolf prepared to cut loose Progress M-35, although their first attempt was unsuccessful, due to the crew having forgotten to remove one of 16 clamps from the docking mechanism. The old craft was separated shortly afterwards and, three days later, on 8 October, Progress M-36 docked uneventfully at the vacated port. In addition to standard equipment, food, water, fuel and supplies, it carried a spare computer for Mir and a special glue, known as 'germetik', to be used during the upcoming attempt to repair Spektr.

Original plans called for Solovyov and Vinogradov to perform an EVA as early as 16 October, but this was cancelled in favour of an internal IVA on the 20th to replace the hatch between Spektr and the multiple docking adaptor and reconnect the solar array cable with Mir's power buses to increase solar power capability by 15-30 percent. A few days before the IVA, Wolf sealed himself inside the Soyuz TM-26 spacecraft, as a fully-suited Solovyov and Vinogradov ran through a full rehearsal of their every move. It had been hoped that they would swap the steering interface for the solar arrays from Kristall to Spektr, but this was considered too difficult in bulky space suits. In the alternate plan, Vinogradov would deploy three cables from the

4-10: Anatoli Solovyov and Pavel Vinogradov work on the effort to repair the damaged Spektr module.

connection for Spektr's solar array servomotors to the Kristall interface. At 12:40 pm Moscow Time on 20 October, about 45 minutes behind schedule, due to minor suit problems, the two men set to work. Vinogradov entered the depressurised Spektr, then Solovyov, to find a great deal of clutter floating around, including several experiment bags, a bicycle trainer, loose panels and a muddle of cables. Working his way through the mess, Vinogradov set to work connecting the trio of cables between the solar panels' servomotors and contacts on the vacuum plate between Spektr and the multiple docking adaptor. It was tough work and the cosmonaut had to fight against stiff, unyielding cables with a spanner. He fastened two cables, but struggled with the third for an entire orbit of about 90 minutes. On this occasion, spanners and long screwdrivers proved useless and the cumbersome nature of Vinogradov's suit exacerbated the difficulty. (Later, Solovyov would also report that the coating on the inside of the spanner's head was too smooth to gain a firm grip on the bolt.) At 6:25 pm Moscow Time, the cosmonauts were notified that they were approaching the endurance limit of their suits' oxygen and consumables and, although they pressed on and cut into their emergency supplies in a bid to finish the work, they were forced to leave the third cable 'hanging' and the partially successful IVA ended at 7:18 pm after six hours and 38 minutes.

Over the course of the next few days, the crew's efforts were almost exclusively dedicated two 'external' EVAs by Solovyov and Vinogradov on 3 and 6 November. The cosmonauts received video training instructions from the ground, together with copious quantities of printed documentation, and were able to report that two of Spektr's three functional solar arrays were capable of being turned to the most effective angle towards the Sun. One of the tasks of the first EVA was to deploy 'Mini-Sputnik', a small replica of the Sputnik-1 satellite, the 40th anniversary of whose historic launch had been marked on 4 October 1997. Mini-Sputnik had been part of the Progress M-36 cargo and it was to be released from Solovyov's gloved hands. In the final days before the EVA, the cosmonauts spoke to Sergei Samburov, the grandson of the great Russian rocketry pioneer, Konstantin Tsiolkovsky, who reminded them to activate the miniature satellite's radio transmitter prior to deployment. Wolf made sure that he was in position to videotape the Mini-Sputnik deployment. It also gave him the opportunity to reflect upon how quickly humanity had established itself in space exploration.

The EVA got underway at 6:32 am Moscow Time on 3 November, a little later than planned, due to the failure of a telemetry unit in Solovyov's suit. The Mini-Sputnik was deployed and the cosmonauts set to work on their other tasks. They dismantled and removed an old solar panel from Kvant-1 and temporarily 'parked' it on the exterior of Mir's base block, whereupon Dave Wolf used a keypad to electrically open the clamps between its array segments. Next, Solovyov and Vinogradov installed an exterior cap to enable them to subsequently hook up an additional Vozdukh carbon dioxide removal system for the station. The cosmonauts returned inside Mir at 12:36 pm Moscow Time, after six hours and four minutes, but the outer hatch of Kvant-2's airlock failed to seal properly. Solovyov and Vinogradov were forced to use an inner compartment of Kvant-2 as an impromptu airlock. Three days later, at 3:12 am Moscow Time on 6 November, they went out again to install a replacement array, delivered two years earlier by STS-74, onto Kvant-1. Returning inside Mir after six hours and 17 minutes, Vinogradov took a look at the rubber gasket on the outer ring of the airlock hatch and discovered some minor damage, noting that the mechanism did not react smoothly, even though the *Hatch Closed* indicator was positive. Again the cosmonauts retreated to the inner compartment, tightening clamps and latches around the circumference of its hatch to hold pressure.

This required Dave Wolf to move his sleeping quarters from Kvant-2 into the Kristall module. Living, working and sleeping, whilst surrounded by bags of food and equipment, did not depreciate the reality that Wolf's four months aboard Mir were far calmer than those of his immediate predecessors, Jerry Linenger and Mike Foale. "I would characterise it as long-term, very hard work, and by that I mean nine in the morning to midnight, essentially every day of the whole mission," he recalled. "There were a few breaks, afternoon on Christmas, one or two Sundays I took the afternoon off ... but Anatoli and Pavel were determined to bring this spacecraft up the snuff, top working order, and I wasn't going to let them work any harder than me and *that* meant taking another job off the job list at ten o'clock at night ... and it was an *infinite* jobs list! There was also a full science programme, eight or nine hours a day, and so between that and the job list it was an extremely interleaved day of scientific work and systems, both maintenance and handling." Those jobs included his acceptance of one of Vinogradov's tasks to mop up condensate from one of the Elektron heat exchangers; it was a singular task which would occupy him for between four and eight hours, every day, for most of his mission, cleaning up globs "of gooey, slimy, ice-cold fluid", whose sizes varied from that of a bowling ball to a beach ball. Contributing in such a fashion assured Wolf that he was taken seriously by Solovyov and Vinogradov as a full member of the crew. "I think that went a way to their putting me on the team," he said. "There's no small or unimportant job on the space station. *All* of it has to get done and that was the best thing I could come up with to free up their time." In late October, a month into his mission, Wolf had also cast the first US election ballot from orbit, using an email system developed by NASA and the County Clerk's Office in Harris County, Texas. It made him feel valued and attached to the ground, "like I didn't feel before".

Although the situation aboard Mir was far calmer in the final weeks of 1997, the problems remained. On 13 November, the station suffered a temporary power loss during a test of the new Kvant-1 solar array, which caused the attitude-control computer to shut itself down and interrupted several of Wolf's scientific experiments. Solovyov and Vinogradov moved fully-charged batteries from Kristall into the base block and quickly restored power to five of Mir's gyrodynes, but this proved insufficient and the three men alternated shifts to monitor the myriad systems. Whenever Mir entered an attitude that was favourable for collecting solar energy, whoever was on duty would temporarily power up the battery chargers. The slow air leak from Kvant-2's improperly sealed outer hatch persisted, although it was not considered a threat to the crew.

In addition to his full scientific and maintenance workload, the pull of the Earth in the window was irresistible at times for Wolf. "Ghostly outlines of continents, just illuminated by the half-Moon," he wrote in one of his letters home. "At an unfelt five miles per second, we blow out of Earth's shadow and into the harsh, unattenuated sunlight. Solar arrays alertly take notice and rotate precisely into position to capture a bit of this fortuitous energy. We blaze over that moving line on Earth that separates night from day. The dominant features on the planet below are two tectonic plates, one holding the Tibetan Plateau and the other India. The plates are clearly smashing together, incidentally elevating the great Himalayan mountain range. Eyes now adjusting, looking real close, *there*: snow-covered Mount Everest and Kathmandu." He saw clear skies over England, France and Italy, haze and smog over China and southern Siberia and the gemstone-like azure-blue glint of Lake Baikal, the largest body of freshwater in the world.

These moments were punctuated, of course, by the realities of living and working in space. Wolf had a full workload, including a bioreactor which grew tissue cultures, and which he had helped to design in his pre-astronaut days, together with investigations into muscular and bone cellular loss. At 12:30 am Moscow Time on the morning of 22 November, the new computer brought to Mir by STS-86 malfunctioned and the station was again plunged into power loss and near-total darkness. For the next 48 hours, with no systems, pumps or fans running, Wolf experienced "the incredible quiet of space", whose "surreal" nature was intensified tenfold. The replacement computer carried aboard Progress M-36 was installed in its place and normality was restored by 24 November, albeit with no further redundancy. Solovyov and Vinogradov then worked to fit the new Vozdukh carbon dioxide removal system and, as December dawned, set about troubleshooting further coolant leaks. At this stage, the crew learned that the launch of the next Shuttle-Mir mission, STS-89, had been postponed by five days until no earlier than 20 January 1998, in order to provide sufficient time for three EVAs aboard Mir in late December and the first week of the New Year. This was subsequently moved back an additional two days, to the 22nd, due to ongoing work to prevent flaking insulation from the External Tank (ET) and to provide the KSC workforce with time off during the Christmas holidays.

A new cargo craft, Progress M-37, arrived shortly before Christmas, bringing the crew's presents, letters and cards from family and friends and other supplies, including a small festive tree and traditional sweets from the Red October confectionary factory. The new Progress also carried a replacement seal for the Kvant-2 outer hatch. Its arrival followed the departure of Progress M-36, which, once clear, deployed a subsatellite, called 'Inspektor', built by Daimler-Benz Aerospace for a joint Russian-German project. This was supposed to start by automatically circling Progress M-36 in order to verify its manoeuvring system and navigational capabilities, and then Solovyov would 'fly' it, via the TORU controls, in an elliptical orbit around Mir itself, at distances of up to 100 m. In doing so, it would become a satellite of another satellite. Unfortunately for the 70 kg Inspektor, its star-sensor overheated and it quickly drifted away from the space station. Watching through binoculars, the crew noted that the satellite was pointing in the wrong direction and the mission was cancelled. Inspektor subsequently burned up in the atmosphere. This proved particularly disappointing for

Solovyov, but aboard Mir there was never an idle moment and within days his attention was fully occupied by working with Vinogradov on a leaking air-conditioner. The efforts of the Russians were not lost on Dave Wolf. "Tolya and Pasha are master craftsmen as they handle this ship," he said, using Solovyov and Vinogradov's nicknames. "Occasionally, Tolya, flashlight in teeth, will disappear behind a wall panel, tools and parts in tow. Hours later, as the sounds of drilling and wrenching subside, he emerges." To Wolf, *this* was what it meant to be living and working aboard a real space station.

By the end of December, it was becoming increasingly likely that two, rather than three, EVAs would be performed in the early part of January, the first by Solovyov and Vinogradov and the second by Solovyov and Wolf. Plans called for the first spacewalk on the 9th to inspect the Kvant-2 airlock hatch, with the second spacewalk on the 15th to evaluate tools and inspect the exterior of Mir. At 2:08 am Moscow Time on 9 January, Solovyov and Vinogradov used the inner compartment of Kvant-2 as an airlock and were soon able to identify the cause of the problem on the outer hatch of its actual airlock: one of its bolt was loose and had introduced a small, 10 mm gap. They secured the bolt, but did not succeed in totally eliminating the slow leak and after retrieving NASA's Optical Properties Monitor the two men returned inside after three hours. Several days later, at 12:12 am Moscow Time on 15 January, Solovyov and Wolf began the final EVA of the mission, emerging from the outer Kvant-2 airlock after a delay caused by the need to unfasten the defective bolt. Embarking on his first spacewalk, Wolf found that he was like a fish *in* water, with no orientation or other problems, alongside Solovyov, who was making a record-breaking 16th career EVA. In order to allow Solovyov the time to work on the hatch, plans to inspect the condition of the exterior of Mir's base block with a US photo-reflectometer were instead restricted to the exterior of Kvant-2, although it proved difficult to install at certain locations and its handling was also awkward. The device also suffered a display failure, which required intensive co-ordination with Pavel Vinogradov inside Mir. After returning to the airlock, Solovyov performed a thorough check of the hatch integrity and closed it at 4:04 am Moscow Time. However, it soon became evident that there was not a totally airtight seal. The atmospheric pressure was too high for the cooling systems in the two men's suits to operate, but far too low to permit them to doff their bulky ensembles. They were forced to retreat into the impromptu airlock, whose small size required them to connect each other's backup cooling umbilicals. With the EVA complete, Solovyov and Wolf had spent three hours and 52 minutes in vacuum.

THE FINAL VOYAGES

Repairing the Kvant-2 outer hatch would have to await Solovyov and Vinogradov's replacement crew, who were due to launch from Tyuratam at the end of January, just after the departure of Mir's Shuttle visitors. STS-89 marked the eighth of nine Shuttle-Mir docking missions and was added to the manifest on 30 January 1996, following a summit in Washington, DC, between US Vice President Al Gore and

Russian Prime Minister Viktor Chernomyrdin. "The eighth and ninth Mir flights will use the Space Shuttle to reduce a significant logistics shortfall on Mir, conduct vital engineering research and expand our knowledge and experience of the effects of long-duration weightlessness," NASA's Office of Space Flight announced after of the meeting of the Gore-Chernomyrdin Commission. "In addition, these extended Mir operations will assist Russia in its objective to extend the Mir on-orbit lifetime through Fiscal Year 1999."

Twelve months later, in January 1997, veteran astronaut Andy Thomas arrived in Star City to begin training as Dave Wolf's backup for the final long-duration mission. Like Jim Voss, he was assigned as what NASA described as a non-flying backup crewman. "The plan was I'd stay there a year and Dave Wolf would go fly the last increment on Mir and I would come back and do something else," Thomas reflected in his NASA oral history. "I actually undertook that mostly because I was curious about the Russian environment, not expecting that I would get a flight out of it." When he told his family that he was going to Star City as a non-flying backup crew member, Thomas was greeted with some surprise, but by his own admission when he updated them, later in 1997, that he would actually be *flying* a long-duration mission, not long after a traumatic few months of fires and collisions and depressurisations and computer failures, "the conversation got decidedly quiet". Yet his family trusted his judgement; they knew that Thomas was not the kind of person to immerse himself into something foolhardy or impetuous.

In March 1997, the core NASA crew of STS-89 was announced. Its two veteran members, Terry Wilcutt in command and Bonnie Dunbar as the payload commander, had both flown previous Shuttle-Mir missions, whereas the other three astronauts, Joe Edwards, Mike Anderson and Jim Reilly, were first-time spacefarers. The pilot, Joe Frank Edwards Jr, came from Richmond, Virginia, where he was born on 3 February 1958. He received his secondary schooling in Alabama and entered the Naval Academy, earning a degree in aerospace engineering in 1980. Two years later, after initial flight instruction, Edwards was designated a naval aviator and was assigned to Fighter Squadron 143. He flew the F-14 Tomcat on fighter-escort and reconnaissance missions over Lebanon at the height of its bloody civil war in 1983. Edwards completed the Navy's 'Top Gun' Fighter Weapons School the following year and graduated from Naval Test Pilot School in 1986, then worked as a project fight test officer and pilot for the F-14A and F-14D aircraft. During this period, he flew the Navy's first mission in the F-14D and a high angle-of-attack/departure-from-controlled-flight test programme for the F-14 airframe and F-110 engine integration. Subsequently assigned as an operations and maintenance officer, Edwards was serving with Fighter Squadron 142 in October 1991 when the radome separated from his aircraft, impacted with and destroyed his canopy. With a blinded eye, collapsed lung, broken arm and no communications or functional flight instruments, Edwards recovered his "convertible" Tomcat with his radar intercept officer, Scott Grundmeier, hunkered down in the aft cockpit and landed safely aboard the USS *Dwight D. Eisenhower*. At the time of the impact, Edwards and Grundmeier were at an altitude of 8,800 m and flying at Mach 0.9. For his "superior display of airmanship under grave circumstances", Edwards received the Distinguished Flying Cross. He

4-11: Anatoli Solovyov peers around the hatch to welcome the STS-89 crew in January 1998.

was honoured five times as Fighter Pilot of the Year and three times as Carrier Airwing Pilot of the Year between 1984 and 1992. In his subsequent career, Edwards was detailed to the Joint Chiefs of Staff in Washington, DC, from 1992 until his selection by NASA in December 1994. During this period, he earned a master's degree in aviation systems from the University of Tennessee in Knoxville.

Serving as STS-89's flight engineer, Michael Philip Anderson tragically stands out today in the hallowed ranks of spacefarers as a member of Columbia's final crew. Born on Christmas Day in 1959 in Plattsburgh, New York, the son of an Air Force pilot – "an Air Force brat", he once told a NASA interviewer – Anderson grew up on bases around the United States. He earned a degree in physics and astronomy from the University of Washington in Seattle in 1981 and, following in his father's footsteps, was commissioned into the Air Force. "I had a wide interest," he told a NASA interviewer. "I was interested in everything and I tried to take advantage of everything from the sciences to music, to writing, to literature." However, it was science which really captured his imagination. "I picked physics because, out of all the different scientific fields, I think physics is probably the broadest. It covers basically everything." After technical training at Keesler Air Force Base in Mississippi, Anderson served as chief of communication maintenance for the 2015th Communication Squadron at Randolph Air Force Base in Texas and later as director of information system maintenance for the 1920th Information System Group. "But my real interest was flying airplanes," he admitted, "so I put in my application for flight school and got selected." Selected for pilot training in 1986, he flew Boeing's EC-135, a command and control variant of the Stratolifter, in support of the Strategic Air Command's Operation Looking Glass to provide 24-hour flying command platforms during global military events. He gained a master's degree in physics from Creighton University in Nebraska in 1990 and subsequently supported roles as aircraft commander, instructor pilot and tactics officer, prior to selection by NASA as an astronaut candidate in December 1994.

Rounding out the STS-89 crew was James Francis Reilly II, nicknamed 'J.R.' in the astronaut corps. He was born at Mountain Home Air Force Base in Idaho on 18 March 1954, but completed high school in Texas. His fascination with the space programme was inspired by the flight of John Glenn in February 1962. The seven-year-old Reilly was sitting in the dentist's chair at the time. "The dentist was a big space fan, listening to John's ground passes every time he went by," Reilly told a NASA interviewer. "After one that lasted about 15 minutes and he had stopped work and I was just laying there with all this stuff in my mouth, he asked me if I'd like to be an astronaut." Sat in the dentist's chair, the young Reilly's reaction was unsurprising (*I'd like to be anything but here*), but in his own words, "*That* was the trigger." For the rest of his childhood, he described himself as "one of the kids that wrote to NASA and had a pile of stuff my mother kept trying to throw away and I kept recovering". Through his college years, he hoped to become a fighter pilot, as a way into NASA, but with the end of the Vietnam War, that no longer seemed a viable option. He gained a degree in geosciences from the University of Texas at Dallas in 1977. Reilly entered graduate school and participated for three months as a research scientist in stable isotope geochronology ("the age-dating of rocks") as part of the 1977-1978 expedition to Marie Byrd Land in West Antarctica, then was employed by Santa Fe Minerals, Inc., in Dallas as an exploration geologist. He later worked as an oil and gas exploration geologist for Enserch Exploration, Inc., also in Dallas, and rose to become chief geologist of the Offshore Region. In addition to these duties, Reilly was actively involved in new imaging technologies for industrial applications in deep-water engineering projects and biological research and spent 22 days in deep-submergence vehicles operated by Harbor Branch Oceanographic Institution and the US Navy. In 1987, he earned a master's degree in geosciences from the University of Texas at Dallas. He was in the process of completing his PhD in geosciences there at the time of his selection into the astronaut corps and was awarded his doctorate in 1995.

Originally baselined as a six-member crew, including rotating NASA long-duration astronauts, STS-89 expanded to seven members with the October 1997 assignment of Russian cosmonaut Salizhan Shakirovich Sharipov. Although part of Russia's cosmonaut corps, Sharipov was actually born in the ancient town of Uzgen, within today's Osh Province of south-western Kyrgyzstan, on 24 August 1964. At the time of his birth, Kyrgyzstan was a socialist satellite republic of the Soviet Union and although Sharipov officially became the 'first' Kyrgyz national to fly into space, he was actually an ethnic Uzbek ... and a Russian citizen. As a child, he was captivated by aviation and space exploration. "I remember when I was four years old and I saw an aircraft in the sky that was flying very low," he said. "At that time, I realised how it could be to be in the air. I knew that it would take a lot from me, both emotionally and physically." It was a challenge that the young Sharipov was keen to embrace. He graduated from the Soviet Air Force's Kharkov Higher Military Pilot School in 1987 and received a diploma as an engineer-pilot. He then worked as a pilot-instructor of military cadets, flying the MiG-21 and L-39 aircraft. Selected as a cosmonaut in May 1990, Sharipov later earned a degree in cartography from Moscow State University and trained as a Soyuz-TM crew commander, prior to his assignment to STS-89. Upon arrival at JSC in Houston, Texas, Sharipov spoke very little English, but Bonnie Dunbar took him under her wing and encouraged him to talk about his ancestral homeland. In time, Sharipov became a fully-fledged member of the STS-89 crew and would fulfil a critical role in enabling the transfer of equipment and supplies between the Shuttle and Mir.

At the Flight Readiness Review for the mission on 8 January, NASA managers settled on the 22nd as the first opportunity to launch Endeavour, which was marking a return to space after more than a year of modifications in Palmdale, California. She had departed KSC on 30 July 1996 and spent eight months on the West Coast, which included the implementation of 63 modifications, including the removal and transfer of her airlock from inside the middeck to outside in the payload bay, as well as the installation of the ODS to make her capable of docking with both Mir and the future ISS. "The airlock installation included significant structural, avionics, electrical and [environmental-control and life-support] modifications," wrote Dennis Jenkins in *Space Shuttle: The History of the National Space Transportation System*, "and is configured so that it may be installed in Bay 2 or Bay 3 [of the forward payload bay] as needed." At Palmdale, the airlock was installed in Bay 3, since this was the configuration required for Endeavour's next three missions, STS-89 to Mir in January 1998 and the STS-88 and STS-96 flights to the ISS, planned for July and December of the same year. Provisions were also implemented to allow a ground coolant mixture of freon and water into the payload bay to cool the Multi-Purpose Logistics Module (MPLM), an Italian cargo facility to be carried on several ISS missions. Endeavour's Extended Duration Orbiter (EDO) capability was removed, leaving her sister Columbia as the only vehicle in the Shuttle fleet capable of long-duration missions, and her thermal-protection materials received significant enhancements. Endeavour returned to KSC on 27 March 1997 to begin processing for STS-89, early the following year.

The mission was significant as the first to include three upgraded Block IIA main engines, part of an interim effort as difficulties were worked in the certification of Pratt & Whitney's High-Pressure Fuel Turbopump (HPFT) before the introduction of the next-generation $1 billion Block II configuration. However, tests of a new Large-Throat Main Combustion Chamber (LTMCC) successfully demonstrated reduced operating pressures and a redesigned hot-gas manifold. It prompted NASA to develop and fly the interim Block IIA, with all the original enhancements planned for Block II, save the new HPFT. By the time STS-89 flew, less than a year had elapsed since the decision was taken to implement Block IIA. The engine was certified to function at 104.5 percent rated performance (although it was still referred to as '104 percent'), which offset the slightly higher weight of the Block IIA and enabled the carriage of about 225 kg of additional payload into orbit. It was noted that the new engine was capable of 109-percent power for contingency abort scenarios and that, eventually, it could support up to ten Shuttle launches without the need to replace any components or perform detailed inspections. With at least two future ISS construction flights requiring high performance to orbit, it was hoped to employ the Block IIA and Block II engines at up to 106 percent rated thrust and perhaps as high as 109 percent. "We're out there on the edge," admitted engineer Chris Singer of NASA's Marshall Space Flight Center in Huntsville, Alabama, "so we run our test programme ... to demonstrate we've got margin against that edge, but we don't like to walk along it." It was pointed out that a 109-percent capability would substantially aid the ascending orbiter in terms of reaching transoceanic abort landing sites, in addition to supporting the payload requirements of downstream ISS missions.

With only a 40-percent chance of acceptable weather conditions on the morning of 22 January 1998, the seven-strong crew nevertheless stepped smartly through their final checks and donned their orange pressure suits. Led by Terry Wilcutt, they departed the Operations & Checkout Building a few minutes before 6:00 pm EST and were all strapped aboard Endeavour by 7:25 pm. A brief problem with the ground-based data-processing system forced a slight hold in the countdown, requiring controllers to shorten a planned 46-minute hold at T-9 minutes to just 25 minutes and 15 seconds, in order to meet the tight, ten-minute window that opened at 9:48:15 pm. As circumstances transpired, everything came together in the final minutes and at T-5 minutes Joe Edwards was instructed to active the Shuttle's three APUs. "We could hear the units spinning up to speed, deep below us," remembered Andy Thomas, who was seated in Endeavour's darkened middeck. "Then came the call to close and lock our visors and to initiate our oxygen flow, a protection in the event of a depressurisation during the climb-out ... The three of us on the middeck shook hands together and wished ourselves well for the flight. Then the cabin became quiet. At six seconds before launch, a deep rumble started, shuddering the orbiter as its three engines were ignited and run up to full speed ..." Amidst all the din, Thomas heard over the intercom as Edwards reported *"Three at 104!"* to verify that all three engines were running at full power. At T-0, the SRBs ignited and STS-89 roared spectacularly into the darkened Florida sky, precisely on time. Even though he had no windows, Thomas said later, "you did not *need* a window to know what was happening!"

Within nine minutes, Endeavour and her seven souls had attained low-Earth orbit and over the course of the next two days, Wilcutt and Edwards oversaw a series of thruster firings to gradually narrow the distance between themselves and Mir. By the morning of 24 January, the two spacecraft were about 310 km apart and closing at a rate of about 100 km with each 90-minute circuit of Earth. A few hours later, they caught sight of Mir, glinting in the distance. "After sunset, it became as a star," Sharipov recalled of his first glimpse of the space station, emerging from the orbital darkness. "It's unforgettable. It's so beautiful, I'll never forget this view." Under Wilcutt's deft control, the eighth Shuttle mission successfully docked with Mir at 3:14 pm EST, beginning five days of joint operations to transfer Andy Thomas to the station and bring Dave Wolf back to Earth. A significant modification to the rendezvous process on STS-89 called for Wilcutt to guide Endeavour toward Mir in a nose-forward orientation, as part of preparations for future ISS dockings.

Wolf was clearly overjoyed to see the Shuttle and waved heartily from Mir's windows during the final approach. Bonnie Dunbar joked that Andy Thomas had forgotten his suitcases and they would have to take him back. For Wolf, his four months in space had passed neither too quickly, nor too slowly; it *felt* like four months, he said. Long before he launched on STS-86, he had already mentally moved into space. "That was my tool for lasting the time," he said, "and *someday* I'll move back to Earth. I didn't feel like I was moving back to Earth until the Shuttle launched to come get me." Now, on 24 January 1998, he could finally begin the mental process of preparing to move back home.

"You guys look great," Wolf exulted, as he was greeted by Thomas and Bonnie Dunbar. "This is a *lot* of fun!" It was exciting, too, for Dunbar, who met her former backup Soyuz crewmate and STS-71 buddy Anatoli Solovyov again. (Later in the mission, the pair would joke about Dunbar staying behind with Solovyov and Vinogradov and even kidded Mission Control to the effect that she was undecided whether to return aboard STS-89 or not.) Hugs and laughter were shared and Endeavour's crew offered gifts of fresh oranges, chocolate Space Shuttles, new notebooks, ink pens and a set of Swiss Army knives, emblazoned with Shuttle emblems. By the end of the first day, no fewer than five large bags of drinking water had been transferred from the Spacehab double module over to Mir. One early problem followed the transfer of the Soyuz seat liners, when it became clear during a pressure check that Thomas' Sokol launch and entry suit did not fit. Its torso section was too short, even thought it had been checked before launch.

"The problem is I cannot pull it up over my shoulders," he reported. "It's either not sized correctly or there's not been adequate allowance for the growth of my height in zero-gravity. And I suspect it's probably both." He tried Wolf's suit for size and successfully passed a pressure check, but the problem was that when it was under pressure the arms extended about 15 cm from the end of Thomas' hands. "It's basically unusable," he said, "in that configuration, at least." Since Thomas would need the Sokol in the event of an emergency evacuation aboard Soyuz TM-26, it was imperative that the issue was resolved *before* he could become a member of Mir's crew. For that reason, Wolf spent an extra night (the 25th) aboard the space station and the Americans' seat liners were temporarily swapped back. Next morning, aided by Solovyov, the two Americans set to work modifying Wolf's old suit. "The straps on the legs and the underarms are ... cinched up, very tight, and stitched that way," Thomas explained later, "so we can't just release them. We actually have to cut the non-critical stitching." They detached internal straps at the shoulders and groin of Wolf's suit, which resized it for a comfortable fit under both pressurised and non-pressurised conditions and both US and Russian officials granted their approval for the crew exchange to officially go ahead.

There was some talk in the media that Russian Flight Director Viktor Blagov was incensed at the suggestion that his space suit tailors had screwed up in some fashion. "There were no objective problems with his space suit," Blagov is said to have exploded to journalists, as quoted by CBS News journalist Bill Harwood. "The astronaut [Thomas] simply turned out to be somewhat capricious. For us, it's a symptom that the astronaut may remain capricious all through the flight." However, Jim van Laak, NASA's Shuttle-Mir deputy manager, flatly denied that Blagov had made such a remark and asserted that the Russians had "great confidence in Andy".

In addition to the logistics transfer, a wide range of research payloads were housed aboard Endeavour's middeck and inside the Spacehab. These included the Closed Equilibrated Biological Aquatic System (CEBAS), a German-provided experiment, which included a habitat for various organisms, including pregnant swordtail guppy fish, snails and plants, to explore the influence of microgravity upon their development and behaviour. The Microgravity Plant Nutrient Experiment evaluated a delivery mechanism to supply water, food and lighting to a package of wheat seedlings and the Mechanics of Granular Materials (MGM) sought to understand the behaviour of sand and salt at very low confining pressures. Previously flown aboard STS-79, the MGM promised to offer significant insights into earthquake engineering, coastal and offshore engineering, mining, the transportation of granular materials, soil erosion, off-road vehicles and the geology of Earth and other rocky celestial bodies. Elsewhere, the Astroculture plant growth apparatus, a frequent Shuttle flier, was delivered to Mir on its first long-duration mission to support the development of wheat seedlings. It would be returned to Earth, together with the mature wheat plants, aboard the final Shuttle-Mir mission, STS-91, in mid-1998. Also destined for an extended stay on Mir was the Diffusion-Controlled Crystallisation Apparatus for Microgravity (DCAM), which was used to transfer over 160 protein samples between the Shuttle and the station, and the Gaseous Nitrogen Dewar, with a payload of 19 frozen protein samples. The latter was a vacuum-jacketed container with an absorbent inner liner, saturated with liquid nitrogen, which gave it the capacity to remain frozen for two weeks; ample time for it to be transferred to Mir for the start of the protein crystallisation process. And four Getaway Special (GAS) payloads were carried in Endeavour's payload bay, containing experiments into liquid-drop formation and glass melting, a German investigation into gravity-induced convection processes and a Chinese crystal growth study.

Endeavour herself performed admirably, with the exception of an erroneous reading of an RCS thruster leak late on 25 January, which was resolved following an uplinked software 'patch' from Mission Control in Houston, Texas. The problem forced Shuttle managers to temporarily pass attitude control capability over to Mir and, when the station's thruster propellant ran low, control was returned to Endeavour. In total, more than 3,600 kg of equipment, logistics and 16 large bags of water were transferred between the orbiter and Mir and, after five days, the hatches were swung shut at 5:34 pm EST on 28 January, with undocking scheduled for the following morning. However, unlike previous missions, the DM was kept open, in order that the station's oxygen supply could be replenished and the STS-89 crew generated and left a final, 16th bag of water for Solovyov, Vinogradov and Thomas' use. After a final night's sleep, Endeavour undocked from the station at 11:57 am EST on the 29th. In fine fashion, and just half an hour before local sunset, Wilcutt guided his ship onto Runway 15 at KSC at 5:35 pm EST on 31 January, completing a mission of slightly less than nine days.

About 20 minutes after touchdown, ground personnel began cranking open the hatch. "There's a knock at the door and the hatch handle's turning," one of the crew remarked over the audio circuit.

"I'm pretty excited about this," came Wolf's reply. "And the hatch is *open*! Oh, the *smell* ... and the *air* from the Earth!"

He was home. With the landing of STS-89, Dave Wolf had accrued 128 days in orbit. A well-known 'party animal' of an astronaut, he had joked about yearning for a hot pepperoni pizza, "a few beverages" and attending a post-landing beach party at the Wakulla Motel in Cocoa Beach. "If he's in good physical shape," said Sam Poole, JSC's chief flight surgeon, "I don't think we would put restrictions on him." However, after agreeing to be stretchered off Endeavour to assist medical researchers with their data collection about his readaptation to gravity, Wolf decided that discretion was the better part of valour and opted against an immediate post-landing party.

Less than 30 minutes before Endeavour undocked from Mir, at 7:33 pm Moscow Time on 29 January 1998, the Soyuz TM-27 mission roared into orbit from Tyuratam. The liftoff was 12 seconds later than planned. On-board were Mir's next resident crew, consisting of Talgat Amangeldyevich Musabayev in command and Nikolai Budarin as the flight engineer, together with Frenchman Léopold Eyharts, finally setting off on his long-delayed, 21-day 'Pégase' scientific research mission. Musabayev had become only the second ethnic Kazakh from an independent, post-Soviet Kazakhstan to venture into space, when he made his first flight to Mir in 1994. He came from the north-western district of Kargaly, born on 7 January 1951. As a citizen of the Soviet Union, Musabayev graduated from the Engineering Institute of Civil Aviation in Riga, within the territory of today's Latvia, in 1974, and later completed Higher Military Aviation School in Akhtubinsk in 1983, from where he obtained an engineering diploma. He was selected for cosmonaut training in May 1990, then promoted to the rank of major and transferred to the Soviet Air Force cosmonaut group in March of the following year. Musabayev served as backup to his fellow Kazakh countryman, Toktar Aubakirov, on Soyuz TM-13, after which he entered training for the long-duration Soyuz TM-19 mission to Mir, which flew in July-November 1994.

Nikolai Budarin was also making his second space mission, having launched aboard STS-71 and served with Anatoli Solovyov for two months in the summer of 1995. Rounding out the crew was French Air Force colonel Eyharts, who came from Biarritz, on the Bay of Biscay in south-western France, where he was born on 28 April 1957. "I decided that I would like to become an astronaut when I was about 12 years old," he recalled, "and I saw the first US astronauts stepping on the Moon and walking on the Moon. That kind of triggered my motivation for my professional

4-12: Dave Wolf (left) works with Andy Thomas during their five-day handover of operations aboard Mir.

career and everything later on that I did at school or in my professional life as a test pilot was driven by the idea of becoming an astronaut one day." He joined the French Air Force Academy at Salon-de-Provence in 1977 and graduated as an engineer two years later, then became a fighter pilot, flying in an operational Jaguar low-altitude ground-attack squadron from Istres Air Base, near Marseilles. In 1985, Eyharts was assigned as a wing commander at Saint-Dizier Air Force Base in north-eastern France and shortly thereafter completed training as a test pilot. He entered the Bretigny flight test centre, near Paris, and flew numerous military and civilian aircraft, with a focus on radar and equipment testing. Selected by CNES in February 1990, alongside fellow French Air Force candidates Jean-Marc Gasparini and Philippe Perrin and French Navy pilot Benoit Silve, Eyharts initially supported ESA's ill-fated Hermes spaceplane in Toulouse. He served as backup to Claudie André-Deshays on the 'Cassiope' mission in August-September 1996 and was then assigned as the prime crewman for the 'Pégase' flight, originally targeted for August 1997, but postponed until early 1998 following the Progress M-34 collision.

It was originally hoped that Soyuz TM-27 could be docked with Mir *before* the departure of STS-89, thus creating a record of 13 individuals aboard a single spacecraft for the first time in history. In fact, had Endeavour's launch been delayed much beyond 22 January, it would have been postponed until after Musabayev, Budarin and Eyharts had docked. However, the French vetoed this idea, on the grounds that it would adversely affect Eyharts' ambitious programme of scientific research. "The agreement is if we *had* to arrive during that time period, they would stand down from the French programme for four days, while we're docked and we would conduct the transfer operations and get Andy set up and Dave transferred back," explained Frank Culbertson. "Then, after the Shuttle had left, they would add back the number of days they gave up on the research programme and bring the Soyuz down three or four days later than originally planned." Fortunately, STS-89 flew on time and no interference between the two missions was necessary. In readiness for the arrival of Musabayev, Budarin and Eyharts, the Progress M-37 spacecraft was undocked from Kvant-1. Following a standard two-day rendezvous profile, Musabayev successfully docked at the Kvant-1 aft port at 8:54 pm Moscow Time on 31 January, less than five hours before STS-89 made landfall at KSC in Florida.

For Andy Thomas, who had arrived at Mir with freshly crew-cutted hair, the departure of Endeavour and the arrival of Soyuz TM-27 brought mixed emotions. "On the one hand, I was sorry to see my colleagues leave," he said, "but on the other it meant that I was now able to get on with the mission." He quickly set up his quarters in the Priroda module, activated his laptop computer and unpacked his bags with personal books, music stationery, art supplies and toiletries. On 31 January, he watched as the spacecraft bearing Musabayev, Budarin and Eyharts appeared over the horizon, "first as a small point of light that slowly grew to its identifiable shape with its attached habitation module and protruding solar panels". The *thump* of docking heralded a three-week period in which Mir would play host to six men from three discrete sovereign nations. Thomas would later remark that this was a particularly tough time, as it left Mir "very crowded", although spirits remained high and a Dutch observer noted that "there is a lot of joy and they do not complain about their modest housing". In addition to his scientific workload, Eyharts seized the opportunity to view his homeland from orbit. "I saw it from relatively far away, but that was a big emotion," Eyharts later told a NASA interviewer. "There was, I still remember, very good weather one day in February of 1998 and I could see the whole coast of France and Spain and I saw a little bit of my hometown. *That* was a *big* emotion!"

Eyharts' time aboard Mir was relatively calm, with the only incident of note occurring on 4 February, when the attitude-control system failed and the station entered a condition of free drift. The crew transferred to the regime of 'reduced power consumption' and were obliged to switch off the Elektron oxygen generators, but the station's main computer remained fully functional, the gyrodynes continuing to spin and normality was soon restored. Early on the 19th, Solovyov, Vinogradov and Eyharts boarded Soyuz TM-26 and undocked from the forward longitudinal port of the multiple docking adaptor at 8:52 am Moscow Time. A little more than three hours later, at 12:10 pm, the descent module executed a perfect, parachute-assisted touchdown, near the Kazakh city of Arkalyk. The weather was particularly bad in the landing zone, with harsh blizzards, although conditions improved slightly in the final hour before touchdown. Only one Mi-8 rescue helicopter was despatched, in order to avoid a possible collision with the descending spacecraft. The helicopter followed their descent from 4.5 km to touchdown and, after landing close by, kept its engine running, lest the snow should freeze on the spacecraft. It was the end of Solovyov's fifth space mission and marked the end of his cosmonaut career; with more than 651 cumulative days in orbit, even in 2014 he retains his place as one of the top ten most experienced spacefarers in history and, with 16 EVAs, is still by far the world's most experienced spacewalker.

Aboard Mir, Musabayev, Budarin and Thomas boarded their Soyuz TM-27 spacecraft and undocked from Kvant-1 at 11:47 am Moscow Time on 20 February. Their intention was to redock at the forward longitudinal port of the multiple adaptor, but instead of performing a flyaround, they held position as Russian ground controllers rotated the entire Mir complex. In spite of a communications malfunction, they redocked at the forward port about 45 minutes later, thereby clearing Kvant-1 for the return of Progress M-37. The cargo craft executed a perfect, Kurs-guided docking on 23 February. Unfortunately, after opening up the Progress, a foul odour issued into the station, caused by a burst rubbish container, and the cosmonauts decided to leave the hatch ajar.

In late February, almost exactly a year since the Kvant-1 fire, the crew were rudely surprised when thick smoke began to emerge from a device which was intended to remove contaminants from the module's atmosphere. It subsequently became clear that misconfigured switches had caused the unit to overheat and started a small fire, which blew fumes into the Kvant-1 cabin. Musabayev quickly turned it off and it was fortuitous that the device itself contained the fire, which ultimately burned itself out. Worryingly, no fire alarms sounded and carbon monoxide levels reached 20 times above their recommended safety levels. For several days afterwards, the crew regularly took contamination readings, as Mir's atmospheric scrubbers gradually cleansed the air.

Several days later, on 4 March, Musabayev and Budarin were scheduled to venture outside Mir on the first of five EVAs which they would perform during the course of their six-month mission. Key tasks for this first spacewalk were the installation of handrails and an attempt to brace the Spektr module's damaged solar array for future repair work. However, upon entering the outer airlock of Kvant-2, they found themselves out of communications range with ground controllers and had ten bolts to unlock from around the circumference of the hatch; they managed to release nine of them, but the tenth proved particularly tight and unyielding. Budarin used (and broke) the heads of all three available spanners, trying to remove the stubborn lock, prompting Musabayev to remark that it was "not a festive day". The two men returned inside Mir and were advised during a subsequent communications session that they were to wait until new spanners were delivered by Progress M-38 in mid-March before making another attempt at the EVA.

The new cargo ship thundered aloft from Tyuratam on the 14th and successfully docked at Kvant-1 two days later, shortly after the departure of its predecessor, Progress M-37. Among its cargo was a 700 kg thruster package, known as the *Vynosnaya Dvigatelnaya Ustanovka* (VDU, or 'External Engine Unit'). It was to be mounted atop Kvant-1's 14-metre-long Sofora girder in order to provide better roll control of the station and would replace a previous VDU, installed in 1992, which had almost run out of propellant. Also aboard was a new lock for the Kvant-2 hatch, together with spanners and the usual food, water, letters, parcels, experiments and other supplies. Thomas received a computerised family album of photographs, a new CD player and a trio of two-volume sets of Beatles music.

It had been a challenging few weeks for Thomas, since his initial sighting of Mir as a distant star on 24 January to the enormous, spider-like assemblage which he called home for more than four months of his life. "My first views of the station were a little daunting," he admitted, "and it was very confining as we floated down the Kristall module to the base block. There was a lot of equipment stowed on all the panels and in every available location, but it did open out at the base block, which is more spacious by comparison … it was a bit of a shock just how crowded it is and how much stuff was in there, but you get used to it. At no time did I feel claustrophobic up there." Nevertheless, the lack of stowage space was a concern and it was not at all difficult to lose sight of items, including toothbrushes and combs, as they floated away and became lost somewhere in the clutter. Yet the fundamental difficulty for Thomas was the profound isolation from Earth. "The view is always there, and it's an amazing view," he admitted, "but each day tends to roll into the next and there comes a certain monotony. You have to use your own resources to make life interesting, to keep your motivation going." Much of Thomas' time was focused upon 27 experiments in advanced technology, Earth sciences, human life sciences, microgravity research and risk mitigation in readiness for ISS operations. These included an X-ray cosmic background radiation detector and the activation and monitoring of Astroculture and its wheat seedlings, which would have fully matured by the time STS-91 arrived. Thomas worked extensively with the Biotechnology System Co-culture Experiment, making sure that it was rotating correctly and that proper levels of media and nutrients reached its reactor chamber. At one stage, he experienced difficulties as air bubbles kept forming in the reactor chamber and NASA ground controllers advised him to reduce the rate of media and nutrient delivery, which appeared to solve the problem. Interspersed with these tasks, Thomas acquired urine and blood samples and worked with Musabayev and Budarin to repair and replace failed equipment aboard Mir, including two water reclamation systems, an air conditioning unit, thermal loops and the Elektron oxygen generator.

Socialising with his Kazakh and Russian crewmates was also important and allowed the three men to bond as a crew. "We spend a lot of time together in a confined space," Thomas said, "not just working as professionals, but around the dinner table." He shared jokes with Musabayev and, as well as telling "a lot of war stories and talking a lot about music", the two men cued one another on their respective skills in Russian and English. "Talgat himself has got a very effusive personality, to say the least," said Thomas, "and I'm expecting it to be entirely entertaining." Before the flight, Thomas had expressed concern about his lack of full fluency in Russian, which he feared might preclude the kind of personal relationship with his crewmates that he wanted. "I would have liked to have had more extensive language training before I went to Russia," he said, "but within the schedules we had back then, it simply wasn't possible." Although the three men frequently worked apart on their respective tasks, they all made a point of congregating at the dinner table in the base block for afternoon tea and evening meals. In addition to eating in the base block, they frequently watched videos together and talked to the ground together. For his own private time, in addition to email, Thomas took a selection of his own books into orbit, including Mark Twain's *Huckleberry Finn*, which he had never read, and art materials to sketch his views from space.

A month later than planned, on 1 April, Musabayev and Budarin performed the first EVA of their mission. It started 15 minutes later than planned, at 4:35 pm Moscow Time, due to Budarin's struggle with the Kvant-2 hatch locks. They successfully installed handrails and foot restraints, but the work took longer than anticipated and they had to defer the assembly of a new workstation and the bracing of Spektr's damaged solar array until a later spacewalk. A heart-stopping moment occurred at 10:12 pm, near the end of the EVA, when Musabayev accidentally turned off his space suit's power supply, which temporarily cut his ventilation and cooling and communications for around 20 seconds. The two men returned inside Mir after six hours and 26 minutes. Five days later, at 2:27 pm on the 6th, they were back outside and managed to stabilise Spektr's damaged array, but during the EVA a problem developed with Mir's solar attitude and they were called back inside after four hours and 23 minutes to rectify the situation with Priroda's thrusters. Another five days elapsed before Musabayev and Budarin ventured outside the station for the third time at 12:55 pm on 11 April. They spent six hours and 25 minutes in the vacuum of space, removing the old VDU from atop Kvant-1's Sofora girder and, at 3:25 pm, physically pushing it away from Mir. Next, on 17 April, despite a troublesome communications transceiver malfunction in Budarin's suit, they were outside for six hours and 33 minutes to begin the delicate process of preparing the mounting apparatus for the VDU. Finally, on the 22nd, they completed their mammoth series of five spacewalks by successfully installing the new VDU onto the Sofora girder. By mid-May, the new thruster had been fully implemented into Mir's attitude-control system. With the duration of the final spacewalk having totalled six hours and 21 minutes, Musabayev and Budarin had spent a cumulative 30 hours and eight minutes working in the harshest environment known to humanity. It was an achievement that filled Andy Thomas, watching and photographing from Mir's windows, with admiration and upon which he reflected in his NASA oral history. At that point in his career, he had yet to perform an EVA. "I would like to have tried it," he admitted, "but I have to say that 30 hours of EVA work is hard, hard work."

A GLIMPSE OF THE FUTURE

By the end of April 1998, it was becoming clear that the ninth and last Shuttle-Mir mission, STS-91, would be delayed by five days from 28 May until early June, due to problems with the final processing of the orbiter Discovery at KSC. Thomas took the delay in his stride and pressed on with his research, processing samples for the Canadian-provided Queens University Experiment in Liquid Diffusion (QUELD) and completing several sessions of a study into the risk of renal stones in microgravity. He also photographed startling natural phenomena on Earth, including huge Saharan dust storms and vast fires in Honduras and Mexico's Yucatan Peninsula, whose smoke was so dense that it darkened the sky as far away as Houston, Texas, and even made coastlines difficult to discern. As April wore into May, Thomas set to work packing away his personal belongings and conducted a full inventory of all US-furnished equipment aboard Mir.

Progress M-38, laden with trash and unneeded items, undocked from Kvant-1 on 16 May and burned up in the atmosphere shortly afterwards. Its place was taken by Progress M-39, which had launched from Tyuratam two days earlier and which arrived only hours after the departure of its predecessor. One of the items aboard the new craft, a new guitar for Talgat Musabayev, brought a beaming smile to the commander's face. Then, on the 30th, just three days before STS-91 was scheduled for launch, Mir's main computer crashed and triggered a power outage. Within 24 hours, however, Musabayev and Budarin had replaced the computer with a spare and begun to test it. In the meantime, they were required to use Soyuz TM-27's thrusters to position the station's solar arrays appropriately in order to recharge the batteries and restore power. Ironically, for Andy Thomas, the computer failure occurred shortly before he was supposed to give a space-to-ground press conference. "I thought it was *unbelievably* bad timing," he said later. "After 20 weeks this happens, and just before a press conference, because one of the things I was going to say in the press conference was how problem-free the flight had been."

The crew composition for STS-91, the final Shuttle-Mir mission, was an unusual one. The crew assignments were made in October 1997, just seven months before launch, rather than the standard 12 months or more. A few weeks before the announcement, on 17 September, *Flight International* told its readers that veteran spacefarers Frank Culbertson and Valeri Ryumin, directors of the US and Russian sides of the Shuttle-Mir programme, would fly aboard the mission. Culbertson, an experienced Shuttle commander with two flights to his credit, was tipped to lead STS-91, with Ryumin, a three-time cosmonaut, with a cumulative 360 days of space experience, serving as a mission specialist and observer to determine the current state of Mir. In *Dragonfly*, Bryan Burrough also noted that Culbertson had stepped down from his Shuttle-Mir managerial post in the hope of assignment to the final docking mission. As circumstances transpired, when the STS-91 crew was assembled between October 1997 and January 1998, it *did* include Ryumin, but rather than Culbertson NASA opted to assign Shuttle-Mir veteran Charlie Precourt as commander. He had previously piloted STS-71 and commanded STS-84 and would become the only US astronaut to fly as many as three times to the Russian space station.

As explained earlier in this chapter, Wendy Lawrence had already been advised by Chief Astronaut Bob Cabana in July 1997 that she would fly *both* STS-86 and STS-91, having been dropped from her long-duration mission, only weeks before launch. "He felt pretty adamant about making sure that my participation in the programme

4-13: Valeri Ryumin, the blustery cosmonaut and manager, who was assigned to join the STS-91 crew to determine the condition aboard Mir.

would be rewarded," Lawrence recalled in her NASA oral history. "What most people didn't know at that time, what *I* knew, was that I would fly on 86 and I would fly on 91, so that's why a lot of reporters just couldn't understand why I wasn't devastated! I knew that I was getting two flights out of it." The difference on STS-91 was that she would be a 'full' member of the Shuttle crew, right from the outset. "On 91, I served as the flight engineer during ascent and entry," said Lawrence, "and then, while we were docked, I was in charge of the transfers." In fact, within five hours of landing from STS-86 on 6 October 1997, she had received a telephone call from Precourt to invite her to the first STS-91 crew meeting. Lawrence asked for a couple of weeks to complete her post-flight debriefing report from STS-86, after which she committed herself to plunge straight back into training for STS-91.

Two of her future crewmates on the mission, Dom Gorie and Janet Kavandi, had served as family escorts for STS-86. Dominic Lee Pudwill Gorie, who would serve as pilot of STS-91, was born in Lake Charles, Louisiana, on 2 May 1957, the son of a US Air Force B-47 Stratojet pilot. His father, Captain Paul Pudwill, died in an aircraft crash when Gorie was only six years old. "Sitting on the hood of a Buick at the airport," he remembered, "watching big bombers fly over our heads ... was the most exciting thing that I could imagine." Even though his natural father had been an Air Force pilot, and Gorie was divided for a time over which service to follow, he ultimately decided on a Navy career. "The Naval Academy won out, mainly because they had the most exciting airplanes to fly and getting to fly off aircraft carriers," he said. "I don't think you can top that as far as an aviation choice." After receiving his schooling in Florida, Gorie therefore entered the Naval Academy and earned a degree in ocean engineering in 1979, then trained as a pilot and was designated a naval aviator in 1981. He flew the A-7E Corsair aboard the USS *America* and the F/A-18 Hornet aboard the USS *Coral Sea* and was selected for test pilot school in 1987. Upon graduation, Gorie worked as a test pilot at the Naval Air Test Center until 1990. In that same year, he received his master's degree in aviation systems from the University of Tennessee Space Institute. Subsequent assignments included flying the F/A-18 aboard the USS *Theodore Roosevelt* and supporting 38 combat missions over Iraq as part of Operation Desert Storm. He applied unsuccessfully for admission into the 1992 astronaut class and spent the next two years working with the US Space Command in Colorado Springs, before undertaking F/A-18 refresher training and was eventually selected by NASA in December 1994. "As a test pilot, you experiment with different airplanes and different systems," he told an interviewer, "and NASA likes to see that background in their pilots, so they pretty much have that as a requirement. Once you have that in your resumé, you can apply to come to NASA."

Selected alongside Gorie in December 1994, and flying her first mission with him, was Janet Lynn Kavandi. She came from Springfield, Missouri, where she was born on 17 July 1959, with a father who always told her that she could, and would, succeed in life; it was not a question of *if* she could do something, but a question of *when*. "I think I can attribute 90 percent of my success to his memory," she reflected, years later. And for Kavandi, the desire to explore space began early. "Space has been a subject that's intrigued me since I was a child," she told a NASA interviewer.

"I've always been interested in space and astronomy." As a young girl, she watched the Apollo lunar landings with fascination, but until the first women astronauts were selected at the dawn of the Shuttle era she simply did not believe it was a career option for her. Kavandi attended high school in Carthage, in her home state, and graduated as *valedictorian*, with the highest ranking among her peers, in 1977. Kavandi went on to earn a degree in chemistry from Missouri Southern State College in 1980, followed by a master's from Missouri University of Science and Technology in 1982. She then joined Eagle-Picher Industries as an engineer in battery development for defence applications and in 1984 was employed by Boeing as a power systems technology engineer, working on the secondary power supply for the Short Range Attack Missile-II and as a technical representative on the design and development of thermal batteries for Sea Lance and the Lightweight Exo-Atmospheric Projectile. Whilst at Boeing, Kavandi entered the University of Washington at Seattle to pursue her PhD in analytical chemistry, which she received in 1990. With her eye on a career in NASA, she carefully selected her doctoral focus. "I wanted to make it an interdisciplinary-type thesis," she said, "so I found a professor that was willing to work with me and we found a topic that NASA was sponsoring, where I would use aerodynamics combined with chemistry to do a research process. I think that helped, maybe, get me an edge in with NASA and working with NASA personnel."

Rounding out the six-person STS-91 crew was Costa Rica-born Franklin Chang-Díaz, the third person to record a sixth space mission, and Russia's Valeri Viktorovich Ryumin. The latter was born on 16 August 1939 and grew up in the large industrial city of Komsomolsk-on-Amur, several hundred kilometres north-east of Khabarovsk in the Soviet Far East. His education took him into engineering and he graduated from a technical college in Kaliningrad in 1958, specialising in the cold working of metals. At the Department of Electronics and Computing Technology of Moscow's Forestry Engineering Institute, he later worked on spacecraft controls. He also served three years in the army as a tank commander, during which time he heard over the radio that Yuri Gagarin had become the first man in space. Ryumin was electrified by the exciting news. From 1966, however, his career brought him firmly within the sphere of the space programme: he served as a ground electrical test engineer, deputy lead designer for the Salyut orbital stations and ultimately deputy general designer for testing. By the end of 1969, Ryumin was working exclusively on space station design. With the possible exception of Konstantin Feoktistov, he was perhaps more intimately involved in the design and definition of each Soviet space station than any other cosmonaut. He was chosen as a civilian cosmonaut in March 1973, and flew three times: aboard Soyuz 25 in October 1977, which failed to dock with the Salyut 6 station, and aboard Soyuz 32 and 35, launched in February 1979 and April 1980, both of which lasted for six months. At the end of his third mission, Ryumin had accrued a total of 360 days in orbit, which made him the world endurance record-holder for almost six years, until he was surpassed by another cosmonaut, Leonid Kizim. From 1981 until 1989, Ryumin served as flight director for the Salyut 7 space station and from 1992 directed the Russian side of Shuttle-Mir. Within the pages of Bryan Burrough's

Dragonfly, Ryumin as a manager was painted as a "blustery" individual whom some cosmonauts came to dislike. Precourt's perspective was quite the opposite. "He's got a crazy sense of humour," Precourt recalled of Ryumin, whom he was glad to have aboard as an additional Russian-speaker. "It's really a shame he couldn't be as funny in English! He just has a real natural sense of humour that is good at cutting through tension and making people feel good about what they're doing." Yet the calamitous events of 1997 had convinced many within the Shuttle-Mir senior leadership to consider flying a managerial observer aboard one of the missions and in January 1998 Ryumin was assigned as a crew member on STS-91.

"After my three flights" to Salyut 6 in 1977-1980, Ryumin later told a NASA interviewer, "I was thinking it would be nice to fly for the fourth time. People just told me to work for a while and *we'd see*. Then I was so busy, life was so hectic, that I stopped thinking about flights." With Shuttle-Mir, Ryumin was faced with a stark question: *should* he fly again and his response was *yes*, that he had designed spacecraft and space stations for most of his professional life and he was one of the world's most experienced cosmonauts. Yet he had to pass intensive medical examinations. "For 17 years, I haven't been trained and prepared for any flights," he said, "and I didn't know if I would be able to pass through all this again, but I decided to try. First of all, I had to lose 25 kg, pass through all the medical boards, and the requirements on the boards in Russia are more serious and more rigid." He had to receive approval from his direct superiors and from the Russian Space Agency itself.

The External Tank (ET) for STS-91, though outwardly similar to its predecessors, was quite different in that it marked NASA's first Super-Lightweight Tank (SLWT). Weighing about 26,000 kg, more than 3,400 kg lighter than earlier ETs, its development began in 1991, when NASA's Marshall Space Flight Center (MSFC) contracted with Martin Marietta to incorporate new weight savings into the existing design. The space agency was acutely aware that the Shuttle would need to deliver large payloads into high-inclination orbits for ISS construction and, with the cancellation of the Advanced Solid Rocket Motor, it was keen to seek alternative options for increasing the reusable orbiter's cargo capacity. "Each pound we can take from the External Tank is one more pound we can take to orbit," explained ET Project Manager Parker Counts. "This becomes especially important when launching the International Space Station into its proper orbit." Under a $172.5 million development effort, Martin Marietta developed and patented an aluminium-lithium alloy for the new tank, which was 40 percent stronger and ten percent less dense than the aluminium alloy used in earlier ETs. The walls were machined in an orthogonal, waffle-like pattern, which provided the additional strength and stability. Each new tank was reportedly produced at a cost of about $43 million. "It would be akin to taking your car as it sits today, removing all four doors *and* the engine and still have something that drives down the street," explained Charlie Precourt. "They've done a great job pulling off a large percentage of that weight."

However, the SLWT endured a troubled development process. Prime manufacturer Lockheed Martin encountered problems with cracking where circumferential welds intersected vertical ones. The cracks appeared when the welds were reheated during the process of joining sections. "We've learned a lot about lithium-aluminium," said Counts. "It's definitely the material of the future, but it's definitely *not* as easy to work with as the old stuff." Nevertheless, a successful series of certification tests were concluded in July 1996, followed by further tests in September, and the first SLWT, destined for STS-91, was rolled out of NASA's Michoud Assembly Facility in New Orleans, Louisiana, in January 1998. The tank arrived by barge at KSC in early February and was transported to the Vehicle Assembly Building (VAB) for processing. On 19 May, mated to Discovery and the SRBs at Pad 39A, it underwent a full tanking test to assess its responses and tensional loads during the transfer of cryogenic propellants.

Early on 2 June, after receiving confirmation that both Discovery and Mir were ready for their respective roles in the mission, NASA managers gave the go-ahead to begin fuelling the SLWT with liquid oxygen and hydrogen propellants. By midday, the tank was fully loaded and in 'stable replenish' mode. Later that afternoon, the crew of Precourt, Gorie, Chang-Díaz, Lawrence, Kavandi and Ryumin departed the Operations & Checkout Building for Pad 39A and shortly thereafter were strapped into their seats aboard the orbiter. Liftoff was scheduled to occur within a ten-minute window starting at 6:04:59 pm EDT, but based upon updated tracking of Mir's position in space was timed to take place at 6:06:24 pm. The ascent was a surprise for Janet Kavandi. "Until I felt the SRBs go off, I didn't really fully anticipate what was going to happen," she said later. "It brought tears to my eyes; I was actually *crying* on the way up, not because I was scared, but because it was just a flood of emotion as I felt all the power and everything thrusting us into space."

Arriving in orbit, Precourt was amazed at how quickly Ryumin readapted to the microgravity environment, after an absence from space of almost 18 years. "He went right to work," Precourt said, "like he'd just come home yesterday from the last one." Over the course of the next two days, the carefully choreographed symphony of rendezvous and phasing manoeuvres ran crisply, with the only trouble of note being a problem with Discovery's Ku-band antenna, which was unable to transmit television images or experiment data to the ground. The system was able to receive uplinked transmissions, however, and remained in action as a rendezvous radar.

Aboard the payload bay, STS-91 carried a Spacehab single module, rather than the double variant. It housed a double rack for Russian logistics, including food and water containers, clothing and sleeping materials and personal hygiene items, as well as three storage batteries for Mir. The mission would also evaluate the Spacehab User Communication System (SHUCS) to send and receive telephone voice calls and faxes, as well as video images of the crew, as part of tests of improved payload uplink and downlink hardware. With a round-trip time of 0.7-1.2 seconds, SHUCS permitted file transfers, commanding and uplink and downlink of voice and fax communications by means of three ground stations and the International Maritime Satellite Organisation (INMARSAT).

At the aft end of the bay was the $33 million Alpha Magnetic Spectrometer (AMS), a prototype for an astrophysics instrument, destined for the ISS, to investigate the presence of 'dark matter' in the Universe. It was conceived by high-energy particle physicist Samuel Ting, who jointly won the Nobel Prize for Physics in

1976 for his co-discovery of the 'J-particle', and developed and constructed by the Department of Energy, which provided over 90 percent of federal support for high-energy physics in the United States. Co-operating in this expansive research effort were physicists from 37 institutions in China, Finland, France, Germany, Italy, Portugal, Romania, Russia, Spain, Switzerland, Taiwan, the United Kingdom and the United States. AMS was described by NASA as "the first of a new generation of space-based experiments which use particles, instead of light, to study the Universe". In addition to exploring the birth and evolution of the Universe, AMS was also designed to answer a pair of critical astrophysical questions. Firstly, if equal amounts of matter and antimatter were produced at the beginning of the Universe, as described by the Big Bang hypothesis, and the galaxies that we see today are composed only of 'ordinary' matter, where has the antimatter gone? Secondly, and of equal importance, was the puzzle of why the mass of a given galaxy seemed to be greater than the visible sum of the mass of all of its stars, gas and dust, which implied the presence of an unknown 'dark matter' which had eluded detection. AMS carried sufficient sensitivity to detect minute amounts of antimatter in cosmic rays originating from beyond the Milky Way Galaxy and could measure one anti-helium particle for every 100 million 'ordinary' helium nuclei. In anticipation of a permanent AMS instrument aboard the ISS, to be continuously operated for at least three years, the prototype aboard STS-91 was to perform a complete systems checkout to demonstrate its functionality and carry out a basic search for anti-helium and anti-carbon nuclei and measure the spectrum of anti-protons.

Weighing 3,170 kg, the AMS comprised five major elements: a permanent magnet, time-of-flight scintillators, a silicon microstrip tracker, anti-coincidence counters and an aerogel threshold counter, together with supporting electronics and interfaces to Shuttle payload computers. The permanent magnet was cylindrical in shape and was surrounded by blocks of neodymium ferrous boron to create a magnetic field. When charged particles or anti-particles entered AMS, the time-of-flight counters would provide a primary 'trigger' and, together with the 1,921 silicon sensors of the six-layered microstrip tracker, would initiate measurements of their velocities and trajectories. It was explained by NASA that the Ku-band malfunction would not impair AMS's data collection, which was stored aboard the instrument and returned to Earth at the end of the mission. Late on 3 June, the crew performed an in-flight maintenance procedure by setting up a bypass system to allow AMS data to be downlinked through the Shuttle's secondary S-band communications network.

At 12:58 pm EDT on 4 June, some 42 hours into the STS-91 mission, Charlie Precourt guided Discovery to a textbook docking with Mir, as the two spacecraft flew high above the border between Russia and Kazakhstan, to the north-west of the Caspian Sea. Ninety minutes later, following standard pressure and leak checks, the hatches were opened and the newcomers were engulfed in hugs and laughter from Musabayev, Budarin and Thomas. For Precourt, his third visit to Mir was quite different to his experiences on STS-71 and STS-84. For starters, Spektr was totally sealed off, but the station as a whole seemed in better condition than on his previous missions. "The air was cleaner, it was better controlled temperature, it was dried," he reflected. "The walls of the surfaces of the structures everywhere were dry. There had been lots of humidity on my previous two visits that was evident everywhere." With four days of joint operations anticipated, the two crews set to work transferring logistics and supplies between the orbiter and the space station. The Space Acceleration Measurement System (SAMS), a US instrument aboard Mir, was dismantled and loaded into the Spacehab for its return to Earth, and scientific research samples were also transferred.

Elsewhere, Precourt tested the Ku-band antenna to verify that its signal processor was receiving electrical current in order to allow it to send a transmit-enable signal. Mission Control hoped to find no current in the system, which would have allowed the Shuttle crew to establish a bypass in order to restore high-data-rate transmission and video capabilities, but Precourt reported that the meter showed good current flow to the signal processor, implying that the malfunction existed in a component which was inaccessible to the astronauts.

In order to support AMS operations, several remote tracking stations were called up to assist with scientific data collection, although it was clear that anything not received in this manner would be stored aboard the Shuttle for retrieval after landing. However, the plan for STS-91 was that AMS would run through five or six hours of optimisation and calibration tests, then gather a minimum of 100 hours of scientific data. By 6 June, it had acquired about 50 hours' worth of data, but the Ku-band malfunction meant that barely *one hour* of that data had actually made it to ground stations. "Clearly, one hour is *much* less than five hours," said Italian AMS researcher Roberto Batiston, "so if by the end of the mission the time we have to optimise and calibrate the instrument is much less than five hours, the possibility of a second flight is a serious possibility." With almost 100,000 channels, AMS was one of the most complex scientific instruments ever placed above the atmosphere and its performance had to be understood at very fine levels of detail to ensure that it was properly optimised to identify what Batiston described as "very rare events". By the time the dedicated AMS was aboard the ISS, nothing could be done to modify or repair it, which made it essential that the STS-91 tests ironed out all potential problems.

As AMS operated in the background, the activity of the final Shuttle-Mir mission continued at a hectic pace. With Valeri Ryumin aboard Discovery, the opportunity was presented for him to see with his own eyes the situation about which so many astronauts and cosmonauts had complained over the past few years. His inspections focused on the condition of Mir's hull, its electrical cabling and the jumbled thoroughfare between the modules; "in short," Ryumin said, "those items of the station that cannot be replaced. Approximately 90 percent of the equipment we bring up into space can be replaced. However, the remaining ten percent cannot and it is these ten percent that ultimately determine the lifetime of the station." He was particularly upset about the clutter, which he felt made work far more difficult for Mir's crews, especially first-time fliers. On one occasion, Ryumin found rubbish behind a wall panel and, upon asking Russian flight controllers for permission to remove it, his request was flatly denied on the grounds that the area included critical cables which could not be moved. Every so often, during space-to-ground conversations, Musabayev would actually defer to Ryumin, including an instance regarding the installation of cameras to document Discovery's departure from Mir. The brackets to hold the cameras did not fit correctly and Ryumin, in no uncertain terms, told the ground that this problem had to be noted in the mission report in *red capital letters*.

"Charlie, this place is in *bad* shape," he told Precourt at one stage.

"What do you mean?"

"I don't know *how* they live up here. This is *awful*. This is worse than I imagined. This is unbelievable; this is *unsafe*."

"Well, this is the way it's been for the last four years," said Precourt. "I've been *telling* you this for the last two missions I debriefed."

"Yeah, but when you don't get to see it for yourself, you hear these stories, you don't imagine how bad it really is."

From Ryumin's managerial perspective, the Russians had lost control on stowage and inventory aboard Mir by about 1989, after about three years of operations. He stressed that the only reason cosmonauts had been able to close the hatch to Spektr and seal it off following the June 1997 collision was because it was a relatively new module and had not yet become cluttered with equipment. Unlike the Shuttle, in which equipment was stowed in lockers and racks, Ryumin found cameras, tools, notebooks and other materials hanging from Mir's walls and attached to every available surface. To an outsider, it might have seemed illogical to suppose that stowage aboard a space station would be so difficult to manage, but to Precourt it was akin to loading and reloading items into one's garage, year after year, and only after a decade or so having the opportunity to dig through the detritus and find things that had been long forgotten. Aside from his irritation, Ryumin was clearly enjoying his return to space. Early in the flight, on 5 June, he surpassed 365 cumulative days in orbit and the Shuttle and station crews celebrated his achievement. Musabayev pulled out his guitar and strummed a few Beatles numbers, together with traditional Russian and Kazakh folk-songs and even the Soviet-era singer-songwriter Vladimir Vysotsky.

With STS-88, the first ISS construction mission, only months away, Wendy Lawrence and Janet Kavandi evaluated the movement of the RMS mechanical arm around Mir's structure to test new electronics and software for upcoming flights, as well as showcasing its dexterity around large objects. On 6 June, Musabayev and Budarin released about 6.3 kg of a coloured tracer gas of acetone and biacetyl into the Spektr module through an air pressurisation unit on the hatch, as part of efforts to locate the still-elusive leak site. The astronauts and cosmonauts monitored the process through various Shuttle and Mir windows, but were unable to see anything escaping from any area of Spektr. This did not prove surprising, as solar viewing angles were hardly optimum for such an observation, but it was anticipated that the STS-91 crew would observe the crippled module closely after their undocking. "We have tried it in vacuum chambers and we see a green glow," said STS-91 Lead Flight Director Paul Dye, before the mission. "There are a lot of unknowns about whether we'll really see it. If you take the black background of space, the white blankets on the surface of the Mir and you're backlighting this whole thing from behind with sunlight, the question is whether the fluorescence of the gas will be bright enough. There's some thought that this gas might actually glow a little bit at night, so we're going to see what we can see."

By the time Discovery prepared to undock from Mir on 8 June, the astronauts and cosmonauts had transferred 500 kg of water and almost 2,100 kg of cargo, experiments and supplies between the two spacecraft. Hatches were closed at 9:07 am EDT and Precourt executed a smooth separation at 12:01 pm, after which Dom Gorie performed the final Shuttle-Mir flyaround inspection. At a distance of about 75 m, he positioned Discovery in a nose-forward orientation and initiated a period of station-keeping, whereupon Musabayev released the tracer of acetone and biacetyl into Spektr. The release occurred about 20 minutes after undocking, and three minutes ahead of orbital sunrise, which it was hoped would render its dull-green, luminous presence visible to the STS-91 crew. Unfortunately, they were unable to see any evidence of the gas.

"Once in a while we think we might be seeing something," Precourt radioed to Capcom Marc Garneau in Mission Control. "Franklin's got the intensified image on the camcorder, but with the small viewfinder it's very hard to tell if anything discernible is there. We'll have to review that film."

"Thanks, Charlie," replied Garneau.

"It sure is a pretty view of the station, though, from here, with the Sun coming," continued Precourt. "We're just sitting here in position and they're tracking the Sun and it's just *glowing*. It's really something!"

"Enjoy the view," offered Garneau, "because you're probably the last crew to have a good look at Mir."

At 1:27 pm EDT, Dom Gorie manoeuvred away from Mir for the final time to begin the journey back to Earth. Departing his five-month home for the last time was a bittersweet experience for Andy Thomas. "Perhaps one of the most moving moments ... was as we drew further and further away," he recalled. "We went into the night side of the planet and I could see stars and the running lights of the station were on. You couldn't see the station, all you could see was lights flashing and they were just going off into the distance, these flashing points of light, fading out, slowly. That was kind of an emotional moment, because I knew that would be the *last* time I would see it ... *ever!*"

With their Mir commitments behind them, the crew settled down to their final days in orbit. Chang-Díaz and Kavandi worked extensively with AMS. The instrument was powered down at one stage on 9 June, following indications that its electronics were getting too warm. However, Precourt and Gorie made a slight adjustment to Discovery's orbital attitude and thermal conditions improved and AMS operations resumed. Samuel Ting explained that the loss of the Ku-band antenna had impaired their chances of getting the required five or six hours of optimisation and calibration data, but pointed out that the team understood the response from the instrument and with the use of ground-based accelerators it would be possible to extrapolate the results from space. "We have many, many data points," he said, "so we're pretty confident we can do that. It will take a little longer, that's all." In fact, the sensitivity and stability of AMS had been amply demonstrated during Discovery's ground-shaking liftoff on 2 June ... for the instrument's silicon components had not moved any more than a single *micron* ...

Meanwhile, Gorie and Lawrence worked with the Solid Surface Combustion Experiment (SSCE) in Discovery's middeck and the Space Experiment Module (SEM) housed several payloads from high school students. These included investigations into genetic changes within fungi and crystal growth to the impact of magnetic fields on computer chips and included as specimens soil, water, seeds, grass, peas, popcorn, yeast, flowers and pieces of orange peel. Others explored the development of the eggs of various non-shelled freshwater organisms, whilst the Boy Scouts and Girl Scouts provided their own experiments. Commercial protein crystal growth investigations were also carried out in a locker in Discovery's middeck, as part of efforts to combat the parasite-spread Chagas disease, which was particularly prevalent in Latin America, affecting 16-18 million people and killing as many as 20,000 per year.

Although the STS-91 mission had run smoothly, the gremlins were not far away. One of Discovery's General Purpose Computers (GPCs) suffered a software glitch, which indicated a growing discrepancy between where the global-positioning system reported the Shuttle to be situated and where it was calculated to be. Mission managers decided to utilise a different GPC for guidance and navigation and the crew deactivated the GPS component of the software program. Lawrence and Kavandi also used the RMS to identify possible ice build-up near a valve from which fuel cell water was being vented overboard. Precourt reported that there was no obvious indication of ice.

Early on 10 June, the crew was awakened to the theme of the movie *Superman*, in honour of Franklin Chang-Díaz, who had eclipsed the achievement of fellow astronaut Jeff Hoffman to become the new record-holder for time in orbit aboard the Shuttle. Hoffman's record of 1,211 hours had been attained at the end of STS-75 in March 1996 and by the end of STS-91 Chang-Díaz would have added another 48 hours to that total.

Ten days after leaving Florida, at 2:00 pm EDT on 12 June 1998, Precourt guided his ship to a smooth touchdown on KSC's Runway 15, bringing Andy Thomas back to Earth. In total, Thomas had spent 141 days in orbit. Discovery's landing brought to an end 977 total days spent in space by seven NASA astronauts since Norm Thagard, of which 907 days had been spent aboard Mir itself. Thomas' landing also marked the end of a continuous 812-day occupation of the station, since the arrival of Shannon Lucid aboard STS-76 in March 1996. There existed some hope that at least one more Shuttle-Mir mission might be attempted, with Valeri Ryumin having expressed his desire in January 1998 that another flight should go ahead. "To tell you the truth, I feel very sorry that ... we will end our programme," he said. "I don't want to leave it at the ninth, so maybe we need to think about how we can extend this phase to include another, tenth, flight." Such hopes were inevitably inspired by Russia's unwillingness to rid itself of Mir, which would become a virtual inevitability without the lifeline provided by the Shuttle, and which many Russians saw as a shining beacon of national pride. NASA expected Mir to be deorbited by the summer of 1999, although as circumstances transpired the old station would remain circling Earth for far longer. However, there would be no tenth Shuttle-Mir mission, as stated categorically by NASA Administrator Dan Goldin, who was keen for both the United States and Russia to set about the construction of the ISS, without the distraction of keeping the expensive and problem-plagued Mir operational for longer than was necessary. For his part, Frank Culbertson, agreed with Goldin. "There really is no place in the schedule right now to put an additional flight to the Mir for logistics support," he explained, adding that the only possibility would be to assist with deorbiting the station. "With *that* type of approach, we might be able to have some discussions and come up with some kind of a joint plan," he said, "but that's really the only thing I think would be considered by anybody."

The return of STS-91 brought down the final curtain on the remarkable years of the Shuttle-Mir programme. Although the last scientific results would not return to Earth until August 1998, aboard the Soyuz TM-27 descent module, with Musabayev and Budarin, the transition to the ISS era had been greatly eased thanks in no small part to one of the most audacious and difficult exercises in international co-operation ever undertaken. Many astronauts, cosmonauts, engineers and managers were unanimous in their agreement that the challenges of Shuttle-Mir, including the delays, the fires, the collision, the breakdowns and the malfunctions, were a critical preparation tool for two old foes to build a joint human space programme. Even though Mir was eventually deorbited in March 2001, its ultimate legacy, the ISS, remains operational, a glistening star, fashioned by human hands, which nightly graces Earth's skies.

Six months after Charlie Precourt brought Discovery back to Earth and closed the final chapter on Shuttle-Mir, the construction of the ISS began. It would require more than a decade to assemble its multiple pressurised modules, its vast solar arrays and its immense power, communications and life-support networks, and would be toiled upon by several hundred astronauts and cosmonauts from the United States, Russia, France, Germany, Italy, Canada, Japan, the United Kingdom, Spain, Sweden, Brazil, Belgium and the Netherlands. It would be visited by men and women of multiple backgrounds and faiths, including the first Malaysian astronaut, the first South Korean astronaut, the first Iranian-born astronaut, the first South African-born astronaut and even the first 'official' space tourist. It would play host to dozens of EVAs for its construction and maintenance, including the first national spacewalks ever performed by representatives of Canada, Sweden and Italy. It would prove itself nothing less than a masterpiece of human ingenuity. It would represent arguably our species' greatest scientific accomplishment and the loftiest endeavour ever undertaken in engineering and technology.

Perhaps more importantly, less than six decades since the vicious conflict and slaughter of the Second World War, it would bring a multitude of nations together in the largest and most far-reaching example of peace, scientific co-operation and technological advancement in human history. As Wendy Lawrence remarked in the hours after STS-91 landed, both she and Dom Gorie were active-duty commanders in the US Navy. "We trained to *fight* against Russia," she said. "I know when I was at the Naval Academy, some 20 years ago, I never thought in my lifetime that I would have an opportunity to *go* to Russia, much less an opportunity to go to

Russia and work side by side with Russian cosmonauts and realise they were just like me." To Lawrence, the Russians were no longer enemies, or even comrades, or even working partners. Rather, they were friends.

Yet there was one final, tangible link in the chain between Mir and the ISS. On 8 June 1998, as Talgat Musabayev and Charlie Precourt traded farewells before closing the hatches, the Kazakh grinned and offered a leaving gift. It was a 60 cm wrench.

"Charlie, take this wrench," Musabayev grinned. "It's sort of a relay stick from the *Old Lady*, Mir, to the International Space Station."

"We're going to need this," Precourt replied, "for all the work that we have ahead of us."

5

New millennium

BRAVE NEW WORLD

It is one of the ironies of history that at the dawn of the year 2000, more precise radar-generated maps existed of Venus (courtesy of NASA's Magellan imaging spacecraft) than of our Home Planet. Although some areas of Earth, including the United States, much of Europe, Australia and New Zealand, had digital maps at the 30 m resolution level, the vast majority of our world lacked such reliable data, primarily due to cloud cover in equatorial regions, which precluded imaging by optical satellites or aircraft. "Humans always want to know about their environment," said German astronaut Gerhard Thiele, "and for the first time we will generate a coherent data set; a *picture* of the Earth. We don't *have* that yet, which is pretty surprising!" On the very cusp of a new millennium, Shuttle mission STS-99, with Thiele as one of its crew members, sought to redress the balance. Building upon a wealth of experience gained from two highly successful Space Radar Laboratory (SRL) missions in April and September 1994 and two earlier Shuttle Imaging Radar (SIR) missions in November 1981 and October 1984, STS-99 would go one step further on the ambitious $220 million Shuttle Radar Topography Mission (SRTM). For the first time on a human space flight, they would utilise an innovative radar-mapping technique, called 'single-pass interferometry', to assemble the most comprehensive, three-dimensional maps of our world ever acquired, with surface resolutions down to about the 30 m level. Although it has since been enhanced by more recent technologies, by early June 2011 there were 750,000 users of the SRTM topography dataset, worldwide, and its resources had been accessed by people in no fewer than 221 countries.

SRTM was a truly international affair, spearheaded by NASA and the Department of Defense's National Imagery and Mapping Agency (NIMA) and supported by the German and Italian space agencies. Operating from a high-inclination orbit of 57 degrees to the equator, the STS-99 crew of commander Kevin Kregel, pilot Dom Gorie and mission specialists Gerhard Thiele; Janet Kavandi, Janice Voss and Mamoru Mohri would be required to map everything 'beneath' them, from about 60 degrees North to 56 degrees South latitude, which constituted roughly 80 percent of Earth's surface and represented 'home' to about 95 percent of their fellow human beings. To achieve their objective, the crew would perform more than a thousand radar 'data-takes' during their 11-day flight, operating around the clock in two 12-hour shifts, using 13,600 kg of instruments mounted both 'inside' and 'outside' the Shuttle's payload bay. "We're not aiming for specific targets," admitted Thiele. "Whenever we fly over land, the radar will be on to map the Earth's terrain. We take some calibration data-takes over ocean, so the radar actually is turned on 15 seconds before we hit a land mass. If, for instance, we run across the North Sea or the Baltic or some smaller ocean areas, we just keep the radar running. There *will* be some ocean data-takes, but not to map the ocean; they serve the purpose of calibrating the radar."

The two main SRTM instruments, affixed to a Spacelab pallet, were NASA's third-generation Shuttle Imaging Radar (SIR)-C and the joint German/Italian X-band Synthetic Aperture Radar (X-SAR), both of which flew on the two SRL missions. These instruments would transmit microwaves at C- and X-band wavelengths towards the surface and gather the reflected echoes on two separate receivers. SIR-C offered multi-frequency, multi-polarisation imagery and represented the first spaceborne radar with the ability to transmit and receive horizontally and vertically polarised waves at both the L- and C-band wavelengths. Measuring 12 m long and 4 m wide and weighing 10,500 kg, SIR-C was the most massive piece of flight hardware ever built at the Jet Propulsion Laboratory (JPL) in Pasadena, California. It was a 'synthetic-aperture' radar and, like a huge dining table, consumed almost half of Endeavour's payload bay for STS-99. Originally assembled from spares left over from NASA's 1978 Seasat mission, SIR-C was affixed to its own truss support structure, which, in turn, was mounted onto a Spacelab pallet, providing a 'side-looking' viewing angle, some 47 degrees from the nadir. As a result, during data-gathering operations, the Shuttle was required to reorient itself to precisely direct SIR-C at its ground targets. Its radar beam was formed from hundreds of transmitters, embedded within the surface of the antenna. By properly adjusting the energy from these transmitters, the beam could be electronically 'steered' and, when combined with Shuttle manoeuvres, offered the scope to acquire images from various directions. However, SIR-C was not the only radar in the payload bay for SRTM. Running like a strip along the uppermost edge of SIR-C was X-SAR, a 12 m × 0.4 m synthetic-aperture device, built by a partnership of organisations, including the German Dornier and Italian Alenia Spazio companies, together with the German and Italian space agencies, which offered a single-polarisation radar in the X-band. It was designed to be mechanically aligned with the L-band and C-band beams of SIR-C.

With the SIR-C/X-SAR combo serving as one of SRTM's two receivers (dubbed the 'Main Antenna'), the second receiver (known as the 'Outboard Antenna') was mounted at the end of a 60 m mast, which extended 'sideways' into space over the Shuttle's payload bay wall. Radar 'echoes' from the instruments were captured, amplified and their data recorded on digital tapes for post-flight analysis. It was also expected to be possible to downlink data from one of the radar frequencies in real time to ground-based researchers, through NASA's Tracking and Data Relay Satellite (TDRS) constellation.

Interferometry as a radar-mapping technique was not entirely new ground for the Shuttle, having first been demonstrated at the end of the SRL-2 mission. However, since that mission only carried the SIR-C/X-SAR payload and was not equipped with a mast-mounted outboard antenna, the SRL-2 crew had to fly over the same patch of ground track and acquire two radar images over separate orbital passes. In effect, it was only possible to achieve 'multiple-pass interferometry', rather than acquiring the data in a single orbital pass. Moreover, the Shuttle's altitude and the relative instability of its attitude meant that it could only collect useful interferometric data with the L-band capability of SIR-C. Nevertheless, the seed of an idea was planted for a fixed-baseline, single-pass interferometer and in July 1996 the concept of SRTM had crystallised into a future Shuttle mission proposal.

The payload carried the elements of the first fully-equipped single-pass interferometer ever carried aboard the Shuttle, capable of taking two simultaneous images of Earth from slightly different locations and at different frequencies, to be combined into high-resolution, three-dimensional topographical maps for a wide variety of civilian, military and scientific purposes. The presence of the outboard antenna on SRTM allowed this interferometric data to be gathered with greater efficiency at C- and X-band frequencies. Since it would be possible to acquire both images in the same orbital pass, the potential data return was expected to be significantly higher. In the words of Kevin Kregel, operating *two* antennas in this manner, was akin to acquiring *two* radar images, 60 m apart, "and *that* gives us the capability to get the 3D image". The outboard antenna and its 60 m mast were to be deployed from a 2.9 m canister, which was also mounted onto the Spacelab pallet, next to the main antenna, and would represent the real novelty of SRTM. Its design was based upon the same technology used for the truss members destined for the International Space Station (ISS) and it was fabricated by AEC-ABLE Engineering of Goleta, California, which also built a mast to support the Tethered Satellite System (TSS) on two earlier Shuttle missions. It consisted of carbon-fibre-reinforced plastic, stainless steel, alpha titanium and an iron-nickel alloy that is known as 'Invar'. Officially designated the 'ABLE Deployable Articulated Mast' (ADAM), when fully extended the 290 kg device represented the longest rigid structure ever deployed from the Shuttle and, in fact, was *longer* than Russia's Mir space station. During its launch and re-entry phases, the mast was folded up in its canister in a manner not dissimilar to a giant accordion. Once described by Gerhard Thiele as nothing short of a "miracle", the mast was deployed in space by means of a rotating nut, which grasped the edges and corners and pushed them outwards into a succession of 87 almost-perfect 'cubes'. It was designed to deploy successfully without twisting, which allowed it to house 220 kg of electrical wiring, fibre-optic cables and gas thruster lines along its entire length to connect the outboard antenna with the instruments in the payload bay. AEC-ABLE also guaranteed that the mast would remain stable to within 2 cm of total tip displacement, which was considered essential if the radar-mapping standards demanded of SRTM were to be met. At the mast's tip was the 360 kg outboard antenna itself, which had been assembled from spares left over from the two SRL missions, and included 8 m C-band and 6 m X-band passive receiver panels to collect radar echoes from the SIR-C/X-SAR transmissions.

In spite of the exciting nature of its scientific potential, STS-99 was a long time coming. When its six-strong crew were announced by NASA in October 1998, they confidently anticipated launching in September of the following year, on an 11-day flight. By the middle of July 1999, the SRTM hardware had been loaded into Endeavour's payload bay in the Orbiter Processing Facility (OPF) at the Kennedy Space Center (KSC) in Florida, ready for the rollover to the Vehicle Assembly Building (VAB) in August for stacking onto her External Tank (ET) and Solid Rocket Boosters (SRBs). Unfortunately, as described in Chapter 3 of this volume, theirs and several other Shuttle missions met with significant delays in the summer and winter of 1999, following extensive wiring defects which impacted the entire fleet of orbiters Columbia, Discovery, Atlantis and Endeavour. In order to inspect Endeavour's wiring, the SRTM hardware had to be removed from the payload bay and the STS-99 launch was postponed until at least early October. A bent freon line on SRTM, later repaired with a brace, added to the problems, but by early December the wiring inspections were completed and on the 13th the Shuttle stack was rolled out to Pad 39A, tracking an opening launch attempt in mid-January 2000.

This date soon became untenable, following lengthy delays to STS-103, which eventually flew over Christmas 1999. A new launch target of 31 January was established and countdown operations ran smoothly, the crew suited up and were strapped into their seats. Unfortunately, they were thwarted by doubtful weather conditions, including low clouds and rain at KSC, and by a potential problem with an on-board Master Events Controller (MEC). The latter was housed in avionics boxes within the Shuttle's main engine compartment and the countdown clock was held at the T-20-minute mark, then scrubbed for 24 hours when engineers could not replicate the exact cause of the problem. At length, the Mission Management Team opted to play it safe and replace the MEC. Although removal and replacement was fairly straightforward, the function of the MEC was entwined with a number of pyrotechnic operations, which firstly required the safing of pyrotechnics, then the removal and replacement procedure itself and *then* the reinstallation and reactivation of the pyrotechnics. This pushed back the STS-99 launch until no earlier than 9 February. This was also moved 48 hours to the right, due to the countdown demonstration test on the 10th of a Titan IV booster at nearby Cape Canaveral Air Force Station, which required support from the Eastern Range. Eventually, on the 11th, despite minor technical issues associated with cabin leak checks, a hydraulic recirculation pump and a liquid hydrogen manifold tank heater, Endeavour passed smoothly through the final stages of the countdown and rocketed into orbit at 12:43 pm EST.

Within hours of entering space, and establishing themselves in a 230 km orbit, the crew divided into its two teams, with the 'blue shift' of Gorie, Voss and Mohri setting up a network of laptop computers, then heading to bed for an abbreviated, six-hour sleep period. Meanwhile, the 'red shift' of Kregel, Kavandi and Thiele began activating the activation of the SRTM payload, its instruments and deploying its outboard antenna. Since both Kavandi (EV1) and Thiele (EV2) had been specifically trained to perform a contingency spacewalk to manually unfurl or retrieve the mast in the event of problems, it made sense for Kregel to assign them to the same shift, as well as to the shift which began the deployment process. (Had problems arisen with the mast during this critical phase, Kavandi and Thiele would have embarked on an EVA to decouple its motors and manually drive it open or closed with a power wrench.) The trio initiated the extension of the mast at 6:27 pm EST and all 87 cube-shaped bays were successfully deployed from their canister within 17 minutes, reaching a maximum length of 60.95 metres from Endeavour's port side. The astronauts manoeuvred Endeavour into the correct mapping attitude, with the orbiter's tail facing into the direction of travel, and under Kregel's direction executed a series of thruster firings to test the ability of dampers to absorb the force of manoeuvres and keep both the main and outboard antennas correctly aligned with their targets. By the time Kregel, Kavandi and Thiele retired to sleep at about 10:45 pm, about ten hours after launch, they had triumphantly executed the STS-99 mission's most complex task.

It came as an enormous relief for German physicist Gerhard Paul Julius Thiele, who served as 'Payload Lead' for the red team and was responsible for SRTM science operations during his 12 hours on shift. Thiele came from Heidenheim an der Brenz in the southern state of Baden-Württemberg, where he was born on

5-01: The mast of the Shuttle Radar Topography Mission (SRTM) extends away from Endeavour during STS-99.

2 September 1953. "I wanted to become an astronaut since I was a boy, aged ten," he later told a NASA interviewer, having watched the early space missions of Project Mercury and Gemini with admiration and awe. "Space always fascinated me; not space only, but the Universe in general." After his early schooling, Thiele joined the German Navy and served as an operations and weapons officer aboard fast patrol boats, then attended the University of Munich and the University of Heidelberg from 1976 until 1982 to study astronomy and physics. He received his doctorate from the University of Heidelberg in 1985, with a focus upon environmental physics, and undertook postdoctoral research in global oceanic circulation and its implications for climatic change at Princeton University. In 1988, Thiele was selected into then-West Germany's astronaut team and commenced training in support of the Spacelab-D2 mission aboard the Shuttle, for which he was selected as a backup payload specialist. During the course of the mission in April-May 1993, Thiele served in the Payload Operations Control Center (POCC) of the *Deutsche Agentur für Raumfahrtangelegenheiten* (DARA), the German Aerospace Agency from 1989-1997, predecessor of today's *Deutsches Zentrum für Luft- und Raumfahrt* (DLR). In 1996, he was selected as a German astronaut for Shuttle mission specialist training and, two years later, in August 1998, he officially joined the European Space Agency (ESA) astronaut corps.

For Thiele, the deployment of the outboard antenna and its mast was critical for the success of SRTM. "Everything has to go like clockwork and the red shift will deploy the mast, calibrate the radar, verify that both antennas ... look at the same point on Earth," he explained before the flight. "We will calibrate it with the help of the ground, of course, and then at the end of our shift, the mission will be rolling." If the mast-based antenna did not deploy correctly, the mission would be significantly impaired; "not so much that we could not live with a shorter [radar] baseline," admitted Thiele, "but the key thing here is safety." The mast needed to be firmly locked in place in order to compensate for Shuttle motions and it could *not* be firmly locked into place if it was improperly deployed. The criticality of the work to be done by Thiele and his red shift crewmates in the first few hours after Endeavour reached space was, therefore, acute.

However, it soon became evident that the mast's damping system, which acted as a shock absorber, was not behaving as expected. Mission Control opted to leave the dampers in their 'locked' position, since calculations indicated that the mast was at no risk in this configuration and it would not impair the science capability. By the morning of 12 February, less than 18 hours into the STS-99 mission, the SRTM radars had begun their first 'swath' over southern Asia and by the end of their first full day in orbit had mapped a total of 4.5 million km^2, which was roughly equivalent to an area half the size of the contiguous United States. X-band imagery of White Sands in New Mexico was met with delight from investigators at its level of detail and quality. The breadth of their radar mapping covered all of Earth's southern areas, save Antarctica, and all northern areas to the south of the southern tip of Greenland, southern Alaska and central Russia. During his second shift, Kregel performed a series of thruster pulses to measure the movement of the mast and noted that the tip moved, as predicted, by a matter of less than 2 cm. Later, on

13 February, the crew also performed their first so-called 'flycast manoeuvre', which was designed to reduce strain on the mast and maintain the most optimum conditions for radar mapping. It was named in honour of the casting technique adopted by a fly fisherman and sought to limit the loads on the mast during frequent Shuttle attitude-control manoeuvres.

"Once a day, because we're at a fairly low altitude, we have to adjust our orbital altitude," explained Dom Gorie before launch. "We do that with a couple of reaction control jets in the aft end of the Space Shuttle. If we were to do that without thinking of the ramifications on the mast, we would pulse on the back end of the orbiter with the reaction control jets and the mast would deflect backwards, because of its mass and inertia. Once we stop the burn, if we let it go, the mast would swing back and forth. What we have designed is a manoeuvre that will minimise the deflections on the mast and minimise the loads at the tip. If we did this *without* the flycast manoeuvre, we would have a [90 cm] deflection at the mast tip and we would approach the mast limit loads at the base. So we pulse for just a short time, one-sixth of the natural frequency of the mast, and the mast will deflect 'back' slightly. Right when it starts springing backwards, we will pulse again and, in essence, 'catch' it in its deflected state. We are then burning the reaction control jets for the duration of the burn that it takes to adjust our altitude. When we stop the burn, the mast wants to start swinging back, and, if left alone, it would start flycasting again; it would start twanging back and forth, just like that fly fisherman's rod. As it reaches back to its neutral position, we do one more pulse at one-sixth the frequency of the mast for that duration and catch it in its neutral position. That way we take those [90 cm] of deflection that we would expect without a flycast manoeuvre and we halve it to [45 cm] and the mast loads at the tip are halved as well. What we've done is designed a manoeuvre that achieves our end parameters, but greatly reduces the risk of damage to the mast." During each of the flycast manoeuvres performed during the mission, Endeavour was reoriented into a nose-first attitude, with the mast extending away into space, after which the Reaction Control System (RCS) thrusters were pulsed to deflect the mast backwards and then rebound it forwards. When it reached 'vertical', a stronger thruster was applied to arrest its motion and increase the orbiter's velocity, thereby keeping Endeavour at the most optimum altitude for radar mapping.

Overall, SRTM operations ran with exceptional smoothness, with flight controllers' attention drawn only to a minor propellant consumption issue associated with the nitrogen gas thruster at the end of the mast. The thruster was designed to keep the mast from 'righting' itself in response to Earth's gravity and remove the need for additional thruster firings to keep the antenna in its data-collection position. Although propellant *was* flowing to the thruster, it appeared that little or no thrust was being produced, in spite of the crew periodically cycling its valve open and closed. In the meantime, Kregel and Gorie were required to use orbiter thruster firings to maintain the fine position, which increased propellant expenditure by about 0.07 to 0.15 percent per hour. However, efforts to conserve propellant were explored over the following days, including a relaxation of the requirement to maintain the mast's attitude, which was more stable than predicted.

Other measures included changing the manner in which waste water was dumped overboard and limiting the use of non-critical equipment. On the 17th, Gorie cycled the nitrogen line and Voss reported seeing the dislocation of a small, white object, which was presumably a small piece of ice and flight controllers subsequently noticed that the thruster began to generate some thrust. Later that day, propellant conservation measures paid off and the crew was advised that they would be granted a ninth full day of science, as originally intended in the mission plan. "That's great news," exulted Gorie. "They're getting some fantastic data on this mission!"

Fantastic, indeed, for SRTM was uncovering the Home Planet as never seen before. By the evening of 13 February, more than 45.8 million km^2, about equal to 38 percent of Earth's total land masses, had been mapped in unprecedented detail. "We're starting to see the first 'quick-look' results from the X-band and C-band antennas and the details are fantastic," said SRTM Project Scientist Michael Kobrick of JPL. "Even in this lower resolution ... we can see many topographic features that were completely invisible in the best maps we have today." Kobrick later noted that Endeavour's crew were mapping approximately 100,000 km^2 *every minute*, more than tripling the world's supply of digital terrain elevation data. By the fifth day of the flight, 15 February, more than half of the targeted land surface had been mapped and around 20 percent had been mapped twice, covering regions the size of Africa, North America and Australia combined. Total mapping coverage of the world had reached 75 million km^2 by this stage and continued to grow: 90.6 million km^2 by the 16th, then 94.3 million km^2 by the 17th, then 108.8 million km^2 by the 18th. Dozens of sites were mapped two and three times and images of Brazil, South Africa and New Zealand's South Island were so spectacular in their topographical detail that enthusiastic scientists were confident the data would be used for decades to come. Although 'quick-look' data was available to researchers, the bulk of the SRTM results were stored on about 270 high-density tapes, which contained approximately the same amount of information as 13,500 CDs.

At one stage, STS-99 crewman Mamoru Mark Mohri spoke to Japanese Prime Minister Keizo Obuchi and Minister of State for Science and Technology Hirofumi Nazkasone. This was Mohri's second space flight, coming almost eight years after he became the first 'professional' Japanese astronaut aboard the STS-47 Spacelab-J mission in September 1992. Mohri was born in the town of Yoichi, famed from its fruits, its wines and its whiskies, on Japan's second-largest island, Hokkaido, on 29 January 1948. He attended school in the local area and later studied chemistry at Hokkaido University, from where he received his bachelor's degree in 1970 and his master's credential in 1972. Mohri earned his doctorate in chemistry from Flinders University in Adelaide, South Australia, in 1976, and spent the following decade as a faculty member at Hokkaido University's Department of Nuclear Engineering. By the early 1980s he had risen to the position of associate professor, with extensive interests in surface physics and chemistry, high-energy physics, ceramic and semiconducting thin films, environmental pollution and biomaterials spectroscopy. At the time of the naming of the principal investigators for Spacelab-J in July 1984, a call was issued for astronaut candidates. Mohri applied and in July of the following year he and aerospace engineer Takao Doi and physician and physiologist Chiaki

Naito (later Mukai) were selected from 533 qualified candidates reviewed by NASDA. Four months later, NASA's final pre-Challenger manifest listed Spacelab-J on Mission 81G in February 1988. The loss of Challenger led many flights to be suspended indefinitely and in 1987 Mohri was appointed as an adjunct professor of physics in the Center for Microgravity and Materials Research at the University of Alabama at Huntsville. During his two years in Huntsville, he was involved in microgravity experiments in alloy solidification and liquid behaviour at the Marshall Space Flight Center's drop tower. In April 1990 he was named as the prime Japanese payload specialist for Spacelab-J. Following his first flight on STS-47 in September 1992, Mohri was selected for NASA mission specialist training in August 1996.

As STS-99 wore on, additional flycast manoeuvres continued to fine-tune Endeavour's orientation. The sixth such manoeuvre of the mission was executed early on 18 February and lasted slightly longer than previous burns, in order to conserve propellant by eliminating another flycast planned for the 20th. By this stage, one week into the 11-day flight, over 83 percent of the pre-launch scientific targets had been mapped on at least one occasion, with over 50 percent of that figure surveyed two or more times. NASA explained that the rate of mapping coverage allowed SRTM to cover the equivalent of the state of Alaska in just 15 minutes and Rhode Island in less than two *seconds*. Images of the San Andreas Fault and the Rose Bowl area of southern California, together with the Kamchatka Peninsula in Russia's Far East and the Hawaiian island of Oahu were judged as so significant that they could lead to new insights into wildfires, lava flows, tsunamis and localised flooding. An additional nine hours of mapping time, and, later, on the 20th, an extra ten *minutes*, was granted to the crew for additional radar observations. Although a seemingly small addition to the timeline, those extra ten minutes allowed SRTM to radar-map Flinders Island on the north-east corner of Tasmania, which completed the mission's coverage of Australia and its neighbouring islands in their entirety. It also brought up the total mapping coverage to 99.96 percent of the pre-launch targeted sites. Elsewhere, the EarthKAM digital camera acquired more than 2,700 images of the Home Planet on behalf of student investigators from 75 middle schools around the world.

As they worked, the sight of Earth in all its glory in their windows provided an ever-present source of inspiration for the crew. "I can't see how anyone who's spent *any* time in space would say it would be boring," said Kregel before the flight. "I can spend two weeks easily, just looking out the window. We're going to be kept pretty busy. We've got to do manoeuvres every 45 minutes to maintain the accuracy that's required. We have to change out the tapes. We're there to back up when ground commanding is not available, to go ahead and start taking the data." For Janice Voss, who served as payload commander and 'Payload Lead' for the blue shift, it was "beyond my comprehension" that space flight could ever become mundane. "If you ever have a little bit of extra time and ... need to change your perspective," she said, "you would just look out the window and freshen your mind up for a few seconds and come back to what you're doing."

With the additional hours of radar time, Endeavour's mapping period ended at 6:54 am EST on 21 February, about 24 hours ahead of the scheduled return to Earth.

The SRTM instruments and the STS-99 crew had imaged almost all of their targeted areas, save small portions of the United States, which were already well characterised in terms of topography. The scientists who oversaw the mission were united in their praise. "A magnificent accomplishment" was Program Scientist Earnest Paylor's judgement, with Tom Hennig of NIMA adding that its success was "absolutely wonderful". After 222 hours and 23 minutes of uninterrupted radar-data collection, the planned 18-minute process of retracting the 60 m mast began at 8:17 am EST on 21 February. The procedure ran crisply, until the mast had been retracted to just 2.5 cm shy of the canister … then stopped. One of the three latches on the canister's lid had failed to engage properly, although Kregel radioed that from his perspective the protruding segment of mast posed no obstruction and he did not think there would be any difficulty in safely closing Endeavour's payload bay doors at the end of the mission. Nor did he consider it necessary to awaken Kavandi and Thiele for a contingency EVA which did not seem necessary. Mission Control agreed. "Appreciate your concerns for your crew members," replied Capcom Steve Robinson from Houston. "Our current plans do not include them heading out the door, so I think you can leave them as is."

Suspicion on the cause of the problem centred on the possibility that the mast had endured cold temperatures whilst fully deployed and that this had reduced its flexibility. At first, the astronauts waited for higher temperatures to warm up any stiffened components inside the canister, then drove open the latches and turned on the retraction motors in the hope that the mast would retract and the latches would re-engage. They did not. "We saw perhaps just a tiny movement at initial power-up, but then nothing else," reported Kregel, glumly, "and, of course, you saw the currents were pretty high." Even switching the retraction motors to their maximum torque only served to move the mast by an additional 0.6 cm, then stalled. Finally, on the fourth attempt, the mast was successfully folded into its canister.

That almost ten full days, or 223 hours, of mapping time was acquired was remarkable. Original plans called for ten days and ten hours (a total of 250 hours) of mapping operations, with the shutdown of the radar instruments and the retraction of the mast on the day before landing. However, in late January 2000 NASA managers had opted to halt operations about 24 hours earlier, giving the SRTM teams only nine days for science. The reason was linked to nervousness about how well the mast would retract at the end of the mission and to preserve the option for a contingency spacewalk. If Janet Kavandi and Gerhard Thiele were required to perform an EVA to manually crank the mast closed, they would need the additional time to prepare their space suits, pre-breathe pure oxygen and perform the spacewalk itself. Leaving the retraction until the final day before landing offered no time in the flight plan to handle unforeseen problems. Also, since the radar consumed so much energy during the mission (about 900 kilowatt-hours, or enough to power an average home on Earth for two months), an EVA would not have been possible after a full 250 hours of mapping. "This is a device we've never deployed before," explained STS-99 Lead Flight Director Paul Dye. "It's the longest space structure anybody's ever put out and we would like to be able to get it back. If we had a problem on the stow day, we'd have had no opportunity to try to fix that problem with an EVA or

5-02: Dom Gorie demonstrates the SRTM radar imaging concept with a model aboard Endeavour.

other means. We had stretched the mission as long as we possibly could." Dye compared taking 24 hours off the mapping time as the difference between "a glass that's 10 percent empty or 90 percent full". With the extension of mapping activities by the additional nine hours, then an additional ten minutes on top of that, most of SRTM's originally baselined radar time was met.

When judging the success of SRTM, the numbers spoke for themselves: eight *terabytes* of data collected, 332 high-density data tapes (equivalent to more than 20,600 CDs) filled, 99.98 percent of targeted sites mapped on at least one occasion and 94.6 percent of it mapped on at least two occasions. In fact, several sites, with a total land area larger than the United States, were mapped as many as *four* times during the flight. In years to come, it was expected that the SRTM data would support a variety of civilian and military applications, including the enhancement of ground collision-avoidance systems for aircraft, civil engineering projects, land-use planning, disaster recovery, line-of-sight determination for cellular communications, the development of better flight simulators, logistical planning, air traffic manage-ment, missile and weapons guidance systems and battlefield management tactics. "There are a lot of benefits for our lives," explained Mamoru Mohri, drawing particular attention to a huge earthquake which hit the harbour city of Kobe, on the southern side of the Japanese island of Honshu in January 1995. "More than 5,000 people were killed. This tragedy happened because of the fact that some houses were built on a fault ... and this fault could not be predicted to cause such serious damage. If we had known the land more precisely, thousands of lives could have been saved. There are many unstable ground areas in the world and in the future we may be able to alert people who live in such areas about imminent dangers."

Landing day for STS-99, on 22 February 2000, brought news of questionable weather conditions. High winds and potential heavy cloud cover seriously threatened the orbiter's return to KSC. On the West Coast of the United States, Edwards Air Force Base in California had been activated and placed on standby to support the landing if necessary. The first KSC opportunity, which would have produced a landing on Runway 33 at 4:50 pm EST, was waved off, due to ongoing concerns about the weather, but the Mission Management Team eventually opted for the second opportunity. Kregel and Gorie guided Endeavour to a smooth touchdown on Runway 33 at 6:22 pm, closing out a spectacularly successful 11-day mission. They had uncovered the Home Planet as never before and had established themselves as the first humans to journey into space in the new millennium. With the Mir space station having been vacated by its most recent long-duration crew in August 1999, Year 2000 had begun with not a single member of mankind in orbit around Earth. The STS-99 crew gained the honour of being first purely by happenstance, but there surely could have been no better mission to begin a new millennium than one which turned our gaze back to the cradle from which we came and enhanced our understanding of its beauty, its majesty and its mystery.

THE END OF MIR

With the departure of Andy Thomas from Mir in June 1998, the aging space station reverted to a crew of two cosmonauts for the first time in more than two years. Yet it remained a busy time for Soyuz TM-27 crewmen Talgat Musabayev and Nikolai Budarin, who participated in a unique 'tele-bridge' communications link with Kazakhstan's new capital, Akmola, and spoke to the new president, Nursultan

Nazarbayev. (Musabayev was the second ethnic Kazakh cosmonaut from an independent Kazakhstan and the first in history to command a space mission.) The tele-bridge was organised to coincide with the day on which Akmola was inaugurated as the new nation's capital. Speaking in Kazakh and Russian, Musabayev congratulated Nazarbayev, who in years to come, would be harshly criticised for his increasingly authoritarian regime, and expressed his admiration for the rapid pace at which the Central Asian country's post-Soviet economy and infrastructure were being rebuilt.

During the course of the conversation, Musabayev noted that he had returned the guitar back to Earth with STS-91, because there would be no other future method for this to occur. The Soyuz-TM spacecraft had insufficient volume and Discovery's departure had long been recognised to be the final Shuttle-Mir docking mission, although hopes remained alive for a time that the highly successful programme might be extended. Upon its return to Earth, Musabayev's guitar was taken to his house to await his return to Earth.

The reality of life aboard Mir continued and much of Musabayev and Budarin's time was spent working on problems with the Elektron oxygen generators, a malfunctioning gyrodyne and replacing part of the Antares transmitter for space-to-ground communications through the Altair-2 satellite. A faulty smoke alarm also required replacement. On the other hand, Mir's life-support system experienced fewer difficulties, perhaps due to the reduced crew size of two men. Several scientific experiments were, however, conducted, with particular emphasis upon a Kazakh-provided alloy processing investigation with the Gallar furnace and even reactivating the Mariya spectrometer aboard the Kvant-1 module to study electromagnetic processes in the upper atmosphere. However, the future of the station continued to hang very much in the balance, with insufficient funding from the Russian government offering up a serious threat that the Soyuz TM-28 replacement crew of cosmonauts Gennadi Padalka and Sergei Avdeyev in August 1998 might not be launched. NASA was pushing strongly for Russia to deorbit Mir by mid-1999 in order to focus exclusively upon the construction of the International Space Station (ISS), but was greeted with a reluctance to dispose of what many Russians saw as a symbol of their national pride.

Deorbiting Mir in a safe manner was always going to be a difficult and intricate procedure, complicated by the fact that the station was an asymmetrical shape, described by one journalist as "porcupine-like". It was expected that it would be disposed of in the Pacific Ocean, probably either to the east of New Zealand in the south or the east of Siberia in the north. In theory, several of the more recent modules could be undocked and deorbited separately, but the lack of manoeuvring capabilities in most components and the damaged and depressurised nature of Spektr made it impractical to bring Mir back to Earth in pieces. It was recognised that the engines of several Progress craft would be needed in order to lower the perigee (low point) of the station's orbit to about 150 km, at which point a final 'burn' could be executed to bring it down into a relatively uninhabited stretch of the Pacific.

On 19 June, Musabayev spoke briefly with Yuri Mikhailovich Baturin, who had been selected in surprisingly short order to join Padalka and Avdeyev as a crew member aboard Soyuz TM-28 for a mission to Mir. Baturin's assignment was an intriguing one, for he was a lawyer and senior Russian politician. Born on 12 June 1949 in Moscow, he graduated from the Moscow Institute of Physics and Technology in flight and dynamics control in 1973. He subsequently attended Moscow State University's Law Institute in 1980 and earned a doctorate of laws degree. In the aftermath of the dissolution of the Soviet Union, he served as Russia's head of national security and as a member (and later secretary) of the Defence Council from 1991 until February 1998. He was also a noted expert in constitutional law. Following the Progress M-34 collision in June 1997, Baturin was quoted by *Flight International* as describing "rumours of the death of the Russian space programme" as being "not only exaggerated, but wrong", prompting suggestions that his assignment to a space mission was designed to assess Mir's viability for continued use. He was selected for cosmonaut training, "for political reasons", in August 1997 and completed a very brief period of training in February of the following year.

In early July 1998, the Russian government announced that it had assigned the necessary financial resources to continue operations aboard Mir until June 1999 with as many as three Soyuz-TM spacecraft. It was revealed that these missions would be partly financed by Russian oligarch Mikhail Khodorkovsky's Bank Menatep. This would include Soyuz TM-28, with Padalka, Avdeyev and Baturin, followed by the commercial Soyuz TM-29 flight involving Slovakia's first cosmonaut and a French *spationaute*. However, it was noted that if the Slovak and French crew members were launched aboard the same mission, only *two* spacecraft, rather than three, would need to be funded. With the French and Slovak crewmen joined by commander Viktor Afanasyev, Soyuz TM-29 would be the first Russian mission to include two 'guest' cosmonauts ... and *that* meant that Sergei Avdeyev would be required to remain aboard Mir for almost a full year, until mid-1999. (In fact, *Flight International* reported in August 1998 that Afanasyev's crew would "remain until June, just five days before [Mir] plunges to destruction" in the upper atmosphere.) Afanasyev's assignment to command Soyuz TM-29 was by no means coincidental; already a veteran of two long-duration missions, he was named in the summer of 1998 as commanding officer of Russia's detachment of cosmonauts. It was in recognition of this appointment that Musabayev relayed his congratulations from aboard Mir.

Originally scheduled for launch on 3 August 1998, Soyuz TM-28 finally left Earth at 12:43 pm Moscow Time on the 13th, carrying Padalka, Avdeyev and Baturin. In anticipation of the arrival of the new crew, the Progress M-39 cargo craft had earlier been undocked from the aft port of Kvant-1. The docking of Soyuz TM-28 was expected to be automatic, but at a distance of 20 m, Mission Control expressed uncertainty about the functionality of the Kurs system and directed the crew to perform a manual docking. Consequently, with human hands at its controls, the spacecraft docked successfully at 1:56 pm Moscow Time on 15 August, a little more than two days after leaving Tyuratam. In command for the docking was a man who was embarking on his first space mission, but at the time of writing in early 2014 would have established himself within the top five list of most experienced spacefarers in the world. Gennadi Ivanovich Padalka was born on 21 June 1958 in Krasnodar, on the Kuban River, about 150 km north-east of Russia's Black Sea port of Novorossiysk. He entered Eisk Military Aviation College in 1975 and graduated four years later, then served for a decade as a pilot and senior pilot in the Soviet (and later Russian) Air Force, reaching the rank of colonel and performing more than 300 parachute jumps. Padalka was selected as a cosmonaut candidate in January 1989 and completed two years of basic training. Until 1994, he worked as an engineer-ecologist at UNESCO's International Center of Instruction Systems and later participated as an investigator on the US-funded Advanced Diagnostic Ultrasound in Microgravity (ADUM) experiment to apply diagnostic telemedicine techniques to spacefarers. Unlike Padalka and Baturin, Sergei Avdeyev was embarking on his third space mission. By the time he returned to Earth in August 1999, Avdeyev would have secured a record for himself as the world's most experienced spacefarer. It was a record that he would hold for more than six years.

Under normal conditions, the first contact between the new crews would have been a handshake between the two commanders, but the presence of Yuri Baturin meant that it was he, the politician, who was accorded the honour of being the first Soyuz TM-28 crewman to float aboard Mir at 3:30 pm Moscow Time. With Baturin scheduled to return to Earth after ten days, alongside Musabayev and Budarin, his custom-moulded seat liner was transferred into Soyuz TM-27.

Ten days after the arrival of the new crew, Musabayev, Budarin and Baturin boarded Soyuz TM-27 and undocked from Mir at 5:05 am Moscow Time on 25 August. Aboard the spacecraft were the final experiment samples from the Shuttle-Mir programme, thus bringing down the final curtain on the highly successful period of Phase 1 US-Russian co-operation. A little more than three hours later, at 8:22 am, they touched down in Kazakhstan, concluding 207 days in orbit for Musabayev and Budarin and 12 days for Baturin. Unsurprisingly, the newly returned crew lauded Mir's performance. Talgat Musabayev described the old station as being in an "excellent state" and that its scientific potential was "only just beginning to be discovered". He added that, by mid-1998, it was in much better condition than it had been during his first flight in July-November 1994. For his part, responding to suggestions that deorbiting Mir would result in the loss of as many as 100,000 jobs, Yuri Baturin explained in October 1998 that it required at least two more years in orbit and needed only preventative maintenance to keep it in service. "One cannot just throw it away," he was quoted by *Flight International*. "At the very least, it should be transferred to the International Space Station ... at least part of it should be."

In the meantime, aboard Mir, Padalka and Avdeyev relocated their own Soyuz TM-28 to the forward longitudinal port of Mir's multiple docking adaptor on 27 August and welcomed the automatic, Kurs-guided return of Progress M-39 to the Kvant-1 aft port on 1 September. With two manual Soyuz dockings accomplished, and two salary bonuses, within two weeks of beginning his first space mission, Gennadi Padalka had already established a reputation for himself as a top-notch cosmonaut commander. The two men then settled down to a range of geophysical, astronomical and medical experiments, including making checks of their own cardiovascular systems.

At 11:00 pm Moscow Time on 15 September, the two men embarked on an 'internal spacewalk', properly termed an 'Intravehicular Activity' (IVA), inside the damaged Spektr module and began reconnecting cables and checking and repairing sockets for the solar array steering mechanism. One of the three undamaged Spektr arrays could not be turned to the correct Sun-angle and delivered a tiny fraction of its potential power and, with energy at a premium, its restoration into Mir's power grid was imperative. By this stage, it had long been concluded that no further attempts would be made to repair Spektr's punctured hull and it would remain depressurised and sealed off until the end of the station's life. Estimated to last three hours, the IVA work was completed in just 30 minutes, after which Padalka and Avdeyev closed the Spektr hatch and repressurised the multiple docking adaptor. Their effort to re-establish commanding capability over the arrays was successful and power was restored to Mir. As many as four EVAs had been planned for their expedition, but as the economic situation in Russia worsened, only one took place.

Two months after their IVA, on the night of 10-11 November, Padalka and Avdeyev performed an EVA of five hours and 54 minutes to install a French-built detector in anticipation of the upcoming Leonids, the meteoroidal stream of dust particles left in the wake of Comet Tempel-Tuttle. They also hand-deployed the small Sputnik-41 amateur radio satellite, whose bleeps, transmitted every five seconds, served to demonstrate it was operating satisfactorily. Some of the particles from the Leonids event, which peaks each November, have been known to measure upwards of 10 mm in diameter and impact at velocities of around 72 km/sec. As a precaution against potentially hazardous impacts, the cosmonauts retreated into Soyuz TM-28 for a few hours at the peak of the storm on 17 November. With their spacecraft situated at the multiple docking adaptor, Padalka and Avdeyev were thus protected by the entire axial length of Mir, including the base block and Kvant-1 at the aft port, together with the radial 'arms' of the Spektr, Priroda, Kristall and Kvant-2 modules.

By this stage of the mission, Sergei Avdeyev was already aware that there existed a strong likelihood that he would remain aboard Mir until the summer of 1999. The next visiting mission, Soyuz TM-29, had Slovak and French crew members, and only a single Russian, which required an experienced Russian flight engineer to remain aboard Mir for the two, back-to-back, six-month expeditions. The troublesome Elektron generators consumed much of the cosmonauts' time and, despite telemetry indications of high carbon dioxide levels, they reported that nothing seemed amiss with Mir's atmosphere. On 25 October, Progress M-39 undocked from Mir and its replacement, Progress M-40, lifted off from Tyuratam and docked two days later, carrying equipment and supplies, including experiments for the upcoming Slovak and French visitors. One of its key payloads was the 'Znamya' ('Banner') reflector, which would be deployed from the 'nose' of the Progress after undocking from Mir at the end of its mission. It was to expand to a maximum diameter of 25 m, and was expected to reflect sunlight over a 6 km area onto selected cities, with a particular focus on Western Europe, including Liege in Belgium and Frankfurt in Germany.

The experiment was part of ongoing research into methods of developing 'space mirrors' to provide lighting for Siberian towns in the long, dark Arctic winter. It had already drawn much criticism from environmentalists and astronomers, on the grounds of light pollution, and as circumstances transpired it snagged on a rendezvous antenna shortly after deployment on 4 February 1999. The experiment was abandoned and Progress M-40 was directed to burn up during re-entry a few hours later.

With Progress M-40 in residence from late October 1998 until early February 1999, Padalka and Avdeyev unpacked its contents and worked on a programme of maintenance activities and scientific research. They refuelled the Kvant-2 hydraulic thermoregulation system, repaired several power units, electrical interfaces and cables, replaced a gas analyser, fixed the water purification system and continued to carry out plant growth experiments in the Svet greenhouse, photometric studies of Earth's atmosphere, magnetic observations of high-energy cosmic particles and microgravity materials processing investigations in the Krater, Gallar and Optizon furnaces. On 8 February 1999, as their six-month mission drew towards its conclusion, Padalka and Avdeyev boarded Soyuz TM-28 and undocked from the forward longitudinal port of Mir's multiple docking adaptor at 2:23 pm Moscow Time, then redocked at Kvant-1 at 2:39 pm in order to free up the forward port for the arrival of Soyuz TM-29.

Two weeks later, at 7:18 am Moscow Time on 20 February, the new crew rocketed into orbit from Tyuratam on what was expected to be the final long-duration mission to Mir. In command was 50-year-old Viktor Mikhailovich Afanasyev, the recently appointed head of Russia's cosmonaut detachment, who was making his third space flight. Afanasyev was born on 31 December 1948 in Bryansk, some 400 km south-west of Moscow, and entered the Soviet Air Force shortly after graduating from Kachynskoye Military Pilot School. He spent six years as a pilot, senior pilot and aircraft flight commander and in 1976 attended the Test Pilot Training Centre, subsequently serving as a test pilot and senior test pilot and receiving a Class 1 military test pilot certification. In 1980, he graduated from Moscow's Ordzhonikidze Aviation Institute. Several years later, Afanasyev began basic cosmonaut training at Star City, proceeded to advanced training in 1988 and in 1989 became backup commander for Soyuz TM-10. In his book *Dragonfly*, Bryan Burrough originated an informal cosmonaut tradition with Afanasyev: the checking of all household appliances, before launch. "The joke," wrote Burrough, "and there is some truth to it, is the moment a cosmonaut blasts into space, his family's apartment begins falling into disrepair." Before each of his launches, Afanasyev checked the refrigerator, the stove, the stereo, the lamp bulbs, the radiators and the television. It was a tradition followed by several subsequent Mir crews. Afanasyev flew two long-duration missions to Mir in the early 1990s.

Speaking in December 1998, two months before launching on Soyuz TM-29, Afanasyev was adamant that his flight would not be the last long-duration expedition to Mir. He stressed that his crew would likely remain from February until June, but that a further two missions would be required in 2000 to prepare the station for deorbiting. At about the same time, *Flight International* told its readers that attempts by the Russian Space Agency to seek private funding to continue Mir operations "appear to have floundered", raising the likelihood of a deorbiting scenario later in 1999. By March, steadily dwindling budgets would have cancelled two of the four EVAs planned during Afanasyev's mission.

Serving as the flight engineer aboard Soyuz TM-29 was Frenchman Jean-Pierre Haigneré, who was making his second space flight, having spent three weeks aboard Mir as a member of the Soyuz TM-17 crew in July 1993. Interestingly, his backup on that mission, Claudie André-Deshays, later became his wife. Both had been selected into the CNES astronaut corps in September 1985 and, years later, both would be honoured by having Main Belt Asteroid 135268 named *Haigneré* for them. Born in Paris on 19 February 1948, Haigneré graduated in engineering from the French Air Force Academy at Salon de Provence and later qualified as a fighter pilot at Tours. He rose to become Squadron Leader on the Mirage V and Mirage IIIE aircraft. In 1981, he completed the Empire Test Pilots School at Boscombe Down in England, winning the Hawker Hunter and Patuxent Shield awards, and subsequently served as project test pilot (later chief test pilot) for the Mirage 2000N. As a pilot, Haigneré's accomplishments and qualifications were impressive: a colonel in the French Air Force, he flew 105 different types of aircraft during his career, with test-piloting and air-transport professional licences, Airbus A300 and A320 credentials, a helicopter private licence and mountain and seaplane ratings. Upon admission into the CNES astronaut corps, he headed the Manned Flight Division of the Hermes and Manned Flight Directorate, developing a parabolic flight programme, and in December 1990 was assigned as Michel Tognini's backup on the Franco-Soviet Soyuz TM-15 mission. In the wake of this flight, Haigneré was assigned to the TM-17 flight in July 1993, after which he served as backup for André-Deshays' mission in August 1996 and was then named as prime crewman for Soyuz TM-29, which was named in honour of the constellation 'Perseus'. Aboard the mission, Haigneré became the first Frenchman to serve in the capacity of a fully-fledged Soyuz-TM flight engineer.

Rounding out the crew was Major Ivan Bella of the Slovak Air Force, who came from the ancient town of Brezno, on the banks of the River Hron, within the central region of today's Slovakia. He was born on 25 May 1964. At the time of his birth, Brezno was part of eastern Czechoslovakia and the young Bella grew up in troubled times, as Soviet tanks and troops rumbled into his homeland in August 1968 to crush the liberal reforms of Alexander Dubček and forcibly establish the country as a satellite republic. Two decades later, in 1989, the peaceful Velvet Revolution brought Czechoslovakia's era of Communist rule to an end and in December 1992 the country was dissolved into two successor sovereign states: the Czech Republic and Slovakia. However, the two nations remained close partners and in 2004 Slovakia went on to join NATO and the European Union. Ivan Bella grew up in this surprisingly peaceful maelstrom of change for his country, graduating from the military school at Banska-Bystritsa in 1983 and from the military academy in Koice in 1987. He served as a pilot and navigator at the 33rd Air Force Base in Malacky in western Slovakia and was selected in March 1998, alongside fellow Slovak Air Force pilot Michal Fulier, to train for Soyuz TM-29. Although Vladimir Remek had flown a mission with the Soviets in March 1978, representing Czechoslovakia, the flight of

Bella in February 1999 marked the first occasion on which an ethnic Slovak had ventured into space.

Under Afanasyev's deft command, Soyuz TM-29 docked smoothly at the forward port of Mir's multiple docking adaptor at 8:36 am Moscow Time on 22 February 1999. Aboard the station for the next week, Bella supported a number of scientific experiments, including one on the long-term survival of Japanese quails in space. He transported a number of eggs to Mir and, despite accidentally crushing four of them, succeeded in transferring most into incubator modules aboard the station. Several of the quails which grew inside these incubators began accepting food and water, whilst others in a slowly spinning centrifuge succumbed to severe stress, caused by the darkness and low temperatures. After Bella's return to Earth, the quails were tended by Afanasyev, Avdeyev and Haigneré. Slovakia's first cosmonaut also took blood samples in support of a hormone production and endocrine physiology experiment and examined the radiation interaction with various materials, including biological tissues and integrated circuits. He was also able to carry out research into space motion sickness, having suffered nausea-like headaches during his first few days in orbit, and engaged in conditioning his cardiovascular system to identify changes in his physiology before, during and after the flight. Bella's seven-day mission, named Štefánik', paid tribute to Milan Rastislav Štefánik (1880–1919), a Czech-born Slovak politician, fighter pilot, diplomat and astronomer who served as a French Army general in the First World War and as the Czechoslovak Minister of War and was subsequently a leading voice in the call for Czechoslovak sovereignty. Early on 28 February, Soyuz TM-28 undocked from Mir, carrying Padalka and Bella, and touched down about 100 km to the south of the Kazakh city of Kijma at 5:14 am Moscow Time. This completed Padalka's first space mission, after 198 days in orbit.

With Haigneré aboard Mir for what was expected to be a mission of about three months, ending in the early summer of 1999, the station hosted a variety of research payloads which had been sponsored by France. Part of his work continued the activities of previous French missions to Mir, including Jean-Loup Chrétien's 'Aragatz' mission in November-December 1988, Michel Tognini's 'Antares' mission in July 1992 and Haigneré's own 'Altair' mission in July 1993. On his first flight, almost six years earlier, Haigneré had shown himself to be an enthusiastic amateur radio ham and it was a pastime in which he would also indulge during his second mission. In early April, Progress M-41 was launched, on what was expected to be the final flight of the cargo spacecraft family to Mir. Together with its payload of propellant, food, water and crew provisions, it carried 18 live lizards for the Russian-French Genesis experiment.

The mission would also prove significant in that it featured only the second joint Franco-Russian EVA in history, coming more than a decade after Jean-Loup Chrétien's spacewalk with Soviet cosmonaut Alexander Volkov, also aboard Mir, in December 1988. Scheduled to last a little more than five hours, the EVA was expected to involve Afanasyev and Haigneré installing and retrieving experiments outside the station and hand-deploying the small Sputnik-99 satellite. The latter had been delivered by Progress M-41. On 13 April, the two men spent much of their working day preparing their equipment and Afanasyev discovered a problem with his suit's telemetry system, which prevented the transmission of biomedical data. He suggested ignoring the problem, but Russian physicians on the ground refused to authorise the EVA until it was resolved. Their rationale was that if a medical issue arose *during* the spacewalk, it could be endured, but it was too risky to *begin* the EVA with faulty telemetry of such a critical data point. Consequently, Afanasyev was ordered to use another suit and this required further arduous preparations and equipment checks. Afanasyev and Haigneré ventured outside Mir at 7:37 am Moscow Time on 16 April 1999 and completed their work in six hours and 19 minutes.

It was clear by the early summer that when Soyuz TM-29 returned to Earth, a curtain would fall on almost ten years of continuous occupation of Mir. In June, *Flight International* noted that due to financial difficulties, Sergei Zalyotin and Alexander Kaleri would not be launched aboard Soyuz TM-30 in August to replace Afanasyev, Avdeyev and Haigneré. Russian President Boris Yeltsin had expressed his support to continue Mir operations, but only if they were privately financed and did not shift attention or funds away from the ISS. Welsh-born British businessman Peter Llewelyn had expected to fly into orbit for a week as a paying 'tourist' with Zalyotin and Kaleri, at a reported fee of $12 million, but his inability to guarantee the finance led to the mission being shelved in May 1999.

Despite the gloom cast by the impending closure of the station, the early summer of 1999 proved far calmer for Mir than in previous years, although overnight on 8-9 July the electrical power supply in the Priroda module malfunctioned and all lights, ventilators, experiments and the amateur radio equipment were temporarily shut down. Only Haigneré's battery-powered computer remained functional. However, power was soon restored. Although it was clear that the station would be left unoccupied after the departure of Afanasyev, Avdeyev and Haigneré in late August, it was also obvious that the Russians had no intention of disposing of Mir until they were left with absolutely no other alternative. Progress M-41 would *not* be the final cargo flight and a new craft, Progress M-42, was launched from Tyuratam on 16 July, carrying among its payload a new navigation computer to autonomously control the station during its forthcoming period of unmanned activity. Whilst the new vehicle was in transit to Mir, the old one was undocked from Kvant-1's aft port on 17 July, opening up a spot for its successor to dock on the 18th. In the days which followed, the crew installed and tested the new computer.

Two EVAs were performed by Afanasyev and Avdeyev in late July. On the 23rd, the cosmonauts departed the airlock at 2:06 pm Moscow Time, nine minutes ahead of schedule. Their tasks for the planned 5.5-hour EVA were substantial and not all of them could be fully accomplished. They toiled to install an experimental parabolic antenna, measuring between 5.2 and 6.4 m in length, onto the Sofora girder. This was described as an engineering test to evaluate a remote-controlled antenna deployment mechanism for use on future navigation satellites. Unfortunately, Afanasyev was not able to unfurl it fully. The joint Russian-Georgian antenna reached about 80 to 90 percent 'open' and would move no further. Flight controllers decided to wait until the next EVA on 28 July to open it fully; if *that* failed, they would abandon the attempt and jettison the antenna. Elsewhere, the two men retrieved the Exobiology and Dvikon experiments from Mir's exterior, which sought to expose organic samples to the harsh environment of low-Earth orbit and determine how various materials were affected by Mir's thruster plumes. An overheating filter in Afanasyev's suit forced the cosmonauts to return to the airlock after six hours and seven minutes.

Five days later, at 12:37 pm Moscow Time on 28 July, the cosmonauts ventured outside for the final EVA of their mission. The absence of Haigneré on either of the two spacewalks was notable, for many observers had expected him to perform at least two excursions during the mission. Instead, he remained aboard Mir and relayed instructions and commands to his comrades. Afanasyev and Avdeyev's first port of call was the experimental antenna, which they successfully deployed to its full elliptical shape. When the test had been completed, they disconnected it and pushed it away into space. The cosmonauts then pressed on with their other EVA work, which included the installation and retrieval of several experiments and the changing of cassettes in the Migmas ion spectrometer. Afanasyev and Avdeyev returned inside Mir for what seemed likely to be the final spacewalk of the station's life at 5:59 pm Moscow Time, after five hours and 22 minutes.

It was anticipated that the three men would return to Earth on either 23 or 28 August, with French officials expressing their preference for the latter date. The final weeks of the mission proceeded smoothly, with Haigneré highlighting a minor attitude control issue and computer failure, apparently caused by improper instructions from the ground. In the meantime, his colleagues installed the new navigation computer (officially termed the 'Unit for Control, Docking and Orientation') in Mir's propulsion control system and tested it extensively on 2 August. Operating alongside the station's main computer, the new unit would ensure safe and successful flight in autonomous mode, commanding the VDU roll thrusters to restore attitude control and solar array positioning and, in the event of an outage, switching off all energy-consuming systems to conserve power until such time as a new team of cosmonauts arrived.

And that possibility seemed very much in flux. By August 1999, it was hoped that a new long-duration crew might return to Mir aboard Soyuz TM-30 as early as February 2000, although the financing of their mission was expected to come from commercial means, rather than from the Russian government. Certainly, Avdeyev and Haigneré expressed uncertainty as to whether or not such a mission could occur, with even flight controllers describing the chance of launching a new crew as "unlikely", so desperate was the financial situation. It was anticipated that a final cargo flight, Progress M-43, carrying four full tanks of propellant, would arrive early in the year to support the controlled destruction of Mir in the upper atmosphere.

Throughout August, Afanasyev, Avdeyev and Haigneré prepared their bodies for a return to terrestrial gravity, completed experiments and loaded Progress M-42 with waste materials to clear as much clutter from the station as possible. They attempted to watch a solar eclipse on the 11th. It was noted that after landing all three men would go their separate ways for post-flight rehabilitation, with Haigneré returning to France, Afanasyev expected to visit the Black Sea coast and Avdeyev vacationing to the North Caucasus spa city of Kislovodsk. Finally, on the 23rd, the hatch to the

Kristall module was closed for the final time, followed by that of Priroda two days later, effectively ending scientific operations aboard both facilities. With a small, but persistent, air leak from Kvant-2, the crew regularly checked its pressure level and were obviously under a great amount of stress themselves to secure the station for autonomous operations. On the late evening of 27 August, the three men piled into Soyuz TM-29 and undocked shortly after midnight, at 12:17 am Moscow Time. They touched down in Kazakhstan, 80 km east of Arkalyk, about three hours later at 3:35 am on the 28th. This ended a mission of 379 days for Avdeyev (who had now racked up a cumulative 747 days in space, across three flights, eclipsing the 678-day cumulative record established in March 1995 by Valeri Polyakov) and almost 189 days apiece for Afanasyev and Haigneré. This also marked France's longest single human space mission, an achievement which, at the time of writing in early 2014, remains unbroken.

For Mir, the departure of Soyuz TM-29 on 28 August 1999 ("with grief in our souls", according to Viktor Afanasyev) brought the curtain down on almost a full decade of continuous occupation. Since the arrival of Soyuz TM-8 cosmonauts Alexander Viktorenko and Alexander Serebrov on 7 September 1989, dozens of representatives from many nations had called Mir their home for periods as short as just one week to durations of up to 14 months. Among their number was Valeri Polyakov, who established the longest single space flight ever conducted (437 days) between January 1994 and March 1995, which remains an unbroken world record to this day. They also included Sergei Avdeyev and Sergei Krikalev, who both flew *two* back-to-back expeditions and ended up spending about a full year in space on a single mission. They included Germany's Thomas Reiter and France's Jean-Pierre Haigneré, who both secured six-month flights and established themselves high on the table for the most experienced nations in space. They included a Japanese journalist, a British woman, an Austrian, four Germans, three Frenchmen, an ethnic Kazakh, a Slovak and the first ethnic Kazakh ever to command a space mission. They included the crews of no fewer than nine Shuttle missions. By the time Viktor Afanasyev, Sergei Avdeyev and Jean-Pierre Haigneré returned to Earth, the construction of the ISS had already commenced; its first Russian and US components, linked together, were already circling Earth. Yet for more than 13 years, the *first* International Space Station had already been in business.

And Russia certainly hoped that its 'business' with Mir was not yet over. A group of entrepreneurs, led by CEO Jeffrey Manber, telecommunications and space investor Walt Anderson and space advocate Rick Tumlinson, created 'MirCorp' in 1999 as part of efforts to utilise the station as a commercial platform for science, technology and 'space tourism'. In February 2000, MirCorp reached a $30 million agreement with Russia's RSC Energia company, which carried the rights to Mir, on the first commercial lease of an orbiting space station. Under the terms of the agreement, Mir would remain operational until at least the summer of 2000 and would include the first privately funded manned mission to the station (Soyuz TM-30) in April-June, the first privately funded EVA (by cosmonauts Sergei Zalyotin and Alexander Kaleri) and the first privately funded Progress cargo craft. Under the terms of the agreement, RSC Energia owned 60 percent of MirCorp, with the remainder in the hands of its investors, including venture capital firm Gold & Appel and Internet and telecommunications entrepreneur Chirinjeev Kathuria. Under this partnership, the Russians would run the 'space' portion of the venture, whilst the private investors would handle the business, in an arrangement which Anderson described as "based on trust, pure and simple". MirCorp was to be headquartered in Amsterdam, because, according to Anderson, the country offered a "more ethical" environment in which it could thrive.

Unsurprisingly, the formation of MirCorp and its support from RSC Energia was greeted by NASA with surprise, dismay and a certain amount of ridicule, particularly from Administrator Dan Goldin. The agency announced that it was "not pleased with the performance and attitude" of RSC Energia. Having already agreed with the Russians that Mir *would* be deorbited in the summer of 2000, NASA strongly opposed any extension of its operational lifetime, on the grounds that it would divert already scarce Russian funds and attention away from the construction of the ISS. Extending Mir's lifetime received endorsement, though, from the new Russian President, Vladimir Putin, who described the commercial efforts as "convincing" and noted that "space exploration is not simply a matter of prestige and a show of the country's might", but also "a fundamental area of economics and science". However, Putin also added that Russia would "abide by its commitments to the International Space Station". Yuri Semenov, head of RSC Energia, attempted to allay NASA's fears by asserting that Russia would meet its obligations "in full and exactly on schedule", but it was clear that this was a hollow gesture. The critical 'Zvezda' ('Star') service module for the ISS, originally scheduled for launch in 1998, had been extensively postponed, firstly to the spring of 1999 and eventually to the summer of 2000, due to funding problems. This had effectively stalled the US/international assembly sequence of Shuttle construction missions and delayed the arrival of the first long-duration crew from early 1998 until October 2000.

Yet plans for Soyuz TM-30 went ahead, with launch targeted from Tyuratam in early April 2000. Prior to the flight, Mir's environment and pressure integrity were checked remotely from the ground. Original plans called for Zalyotin and Kaleri to spend about 45 days in space, perhaps joined by Russian actor Vladimir Steklov on a commercial contract to film scenes for a movie about a stranded cosmonaut. Although Steklov completed his pre-flight medical requirements, payments for his flight were not forthcoming and he was quietly removed from consideration. In late March 2000, as Zalyotin and Kaleri arrived at Tyuratam for launch, *Flight International* revealed that Soyuz TM-30 would be extended to somewhere between 70-90 days, bringing it back to Earth in the June-July period.

Sergei Viktorovich Zalyotin was born in the industrial city of Tula, about 200 km south of Moscow, on 21 April 1962. He attended the Borisoglebsk Higher Military School, graduated with a diploma of pilot-engineer in 1983 and entered the Soviet (and later Russian) Air Force, where he trained and flew as a fighter pilot. He flew the L-29, L-39, MiG-21, MiG-23 and Su-17 aircraft. Zalyotin was based primarily in the Moscow military region and in May 1990 was selected to join the cosmonaut corps. After completing his training in March 1992, he served with Kaleri and Oleg Kotov on the backup crew for the Soyuz TM-28 mission in August 1998-February

1999. During his time as a cosmonaut, Zalyotin also qualified as an engineer-ecologist from the International Center of Training Systems and earned a master's degree in ecological management. In March 1999, he and Kaleri completed their training for Soyuz TM-30, the next Mir expedition to replace Afanasyev's Soyuz TM-29 crew, but were stood down in June when it became impossible to secure funding for their mission. The two cosmonauts continued training and in March 2000 were declared ready for flight. Although this was Zalyotin's first space mission, he was in command.

Soyuz TM-30 thundered into orbit at 8:01 am Moscow Time on 4 April. Two days later, Zaloytin guided his ship to a smooth docking with the forward longitudinal port of Mir's multiple docking adaptor, taking over manual control at a distance of 9 m when he noticed a slight deviation along one of the approach axes. The crew entered the station without face masks, indicative of Russia's appreciation that the atmosphere was acceptable, and got straight to work activating the main computer and the life-support and cooling systems. They replenished the oxygen supplies, repaired Mir's heating loop and conducted pressure and leakage checks of the various modules. This required them to disconnect the multitude of snake-like cables in the multiple docking adaptor to hermetically seal the Kristall, Kvant-2 and Priroda modules.

Speaking in advance of the anniversary of Yuri Gagarin's flight on 12 April, Zalyotin explained that it had been "very difficult" for him to accustom himself to the microgravity environment. However, there was precious little time to acclimatise, for the work began in earnest to bring Mir's systems back to full functionality and within days the temperature inside the base block had been brought up to about 25 degrees Celsius and a little cooler in the other modules. Betwixt the repair work, the cosmonauts made visual and photographic observations in mid-April of flooded areas in Hungary, Romania and Poland. In response to questions from Russian students, they explained that Mir had been brought back to life and that their only disappointment was that there was no dinner ready for them upon their arrival at the station.

They spent time unpacking the Progress M1-1 cargo craft, which had docked automatically at the Kvant-1 aft port on 3 February. Originally designed for ISS operations, the new-specification Progress was optimised for the carriage of propellant over pressurised cargo and after being unloaded by Zalyotin and Kaleri it was undocked and deorbited on 26 April. Twenty-four hours earlier, Progress M1-2 had been launched on the first MirCorp-funded cargo flight and it docked successfully at Kvant-1, under Kurs-control, at 12:28 am Moscow Time on 28 April. According to Kaleri, its cargo compartment was only half-full on this mission. As events transpired, Progress M1-2 would be the last such cargo mission to dock at Mir whilst a human crew was in residence. It was used to slightly nudge the station's orbit to an altitude of 360×378 km.

In addition to the first privately financed Progress mission, Zalyotin and Kaleri also supported the first privately financed EVA on 12 May, when they spent four hours and 52 minutes outside Mir. On what turned out to be the final spacewalk ever performed from the old station, the two men departed the Kvant-2 airlock at 1:44

pm Moscow Time. They used the specialised 'Germatizator' glue to seal a tiny 'slit' in the base block's hull and performed a panoramic inspection of the exterior of the station, inspected a malfunctioning solar array mechanism on Kvant-1 (one of whose steering cables had suffered a short-circuit and burned out) and executed a panoramic inspection of the general status of Mir.

Following the EVA, the remaining weeks of Zalyotin and Kaleri's mission encompassed repairs of a seemingly endless number of filters, ventilators, cables, pumps, valves and other pieces of equipment, together with commercial scientific and technological experiments. They performed routine pressure and humidity checks throughout the station. However, their precise landing date remained undecided. By the end of May, it seemed likely that Soyuz TM-30 would return to Earth at some point between 9-14 June, but it was actually extended by several more days. At length, Zalyotin and Kaleri boarded Soyuz TM-30 and undocked from Mir at 12:24 am Moscow Time on 16 June for what would be the final time a human crew would ever occupy Mir. A little more than three hours later, at 3:44 am, their descent module touched down safely about 45 km south-east of Arkalyk, concluding a flight of 73 days in space.

Even as Zalyotin and Kaleri's mission came to an end, MirCorp explained that it required in excess of $100 million in investment to support future flights, but stressed that it expected the station to be in "full-scale commercial operations" by 2001, kicking off an ambitious three-phase, ten-year plan to firstly stabilise and maintain its orbit, then install new solar arrays and communications equipment and finally replace its base block. It was envisioned that from 2001 Mir would be occupied continuously by rotating teams of cosmonauts. In the aftermath of Soyuz TM-30, MirCorp claimed a backlog of customer orders for 'space tourism' flights, whose value reached about $70 million. On 19 June 2000, three days after the return of Zalyotin and Kaleri, it announced a $20 million contract with US businessman Dennis Tito to embark on a privately funded, ten-day flight to Mir in the spring of 2001, making him the world's first paying 'Citizen Explorer'. Ironically for NASA, Tito was a former engineer at the Jet Propulsion Laboratory (JPL) in Pasadena, California, who later made millions as founder of the Wilshire Associates investment firm. "Tito's foray into orbit as history's first Citizen Explorer will certainly be a milestone, but whether it will open the floodgates to space is questionable," *Flight International* told its readers in September 2000. "Few people have $20 million to pay for the privilege. In addition, the Explorer must become a cosmonaut and meet strict medical requirements and risk being disqualified from flying as late as the morning of the launch. Explorers would have to undergo months of training and tolerate invasive medicals."

NASA met the announcement of Tito's flight with disdain (and some US politicians even went so far as to label him "unpatriotic"). Finally, in November 2000, the Russians finally caved in to American pressure and decided that Tito's mission would be transferred to the ISS – a switch that further irritated NASA. Tito's presence was met with hostility at the Johnson Space Center (JSC) in Houston, Texas, when he arrived for additional training. He was even reportedly sent home. As circumstances transpired, Tito *did* fly to the ISS for eight days in April 2001.

5-03: Although Dennis Tito (left) eventually flew to the International Space Station in April 2001, alongside cosmonauts Talgat Musabayev (centre) and Yuri Baturin, had circumstances been different he may have participated in a mission to Mir.

Others were less fortunate. In September 2000, MirCorp signed up the NBC television network and Mark Burnett, the producer behind the 'Survivor' reality show. The hope was that a new reality show, entitled 'Destination Mir', would be created, involving 12 contestants undergoing training at Star City. They would be steadily eliminated, week by week, during the course of the show, with the final, two-hour episode to the filmed at the Tyuratam launch site, after which the competition winner, watched by live television cameras, would board the Soyuz spacecraft and launch into orbit with his or her two Russian crewmates.

However, in reality, MirCorp struggled to raise the required funds without state support from the cash-strapped Russians. In spite of the bold proposals, the writing was clearly on the wall for Mir, which was well beyond its intended operational lifetime. Even Yuri Koptev, director-general of the Russian Space Agency, was aware that any of its vital systems could fail at any time. However, late in April 2000, *Flight International* noted that MirCorp had received "private commitments" to attempt a second commercial mission (Soyuz TM-31) by cosmonauts Salizhan Sharipov and Pavel Vinogradov in September 2000. It was anticipated that hardware would be delivered to Mir during their mission to establish the "first-ever Internet portal in space", funded by telecommunications entrepreneur Chirinjeev Kathuria and Gold & Appel. This was expected to carry data and live images of Earth from space, although space policy analyst Dwayne Day poured scorn on this possibility, because Mir's communications network did not place it in continuous contact with the ground. In October 2000, MirCorp financed the Progress M-43 mission, which was launched to an unmanned Mir on the 16th to raise the station's orbital altitude for future operations. However, a lack of funding forced the slippage of plans to launch Sharipov and Vinogradov to November 2000 and then to January 2001, before it was cancelled. So too was a subsequent mission (Soyuz TM-32) by Talgat Musabayev and Yuri Baturin. At the same time, the Russian government expressed publicly its desire to deorbit Mir early the following year. During a trip to India in October 2000, Russian Deputy Prime Minister Ilya Klebanov told journalists that the station "must end its operation".

It was by no means a unanimous conclusion; indeed, it has been estimated that an online opinion poll, solicited by the website Space.com, pointed out that 67 percent of voters around the world wanted Mir to remain in orbit. The Liberal Democratic Party of Russia passed a resolution in the lower house of the *Duma* (parliament) in November 2000 to prevent Mir's destruction, Vladimir Putin was petitioned and implored to save it and Gennadi Zyuganov, first secretary of the Communist Party of the Russian Federation, condemned the act of a "helpless, weak-willed, inefficient and not very responsible" government. Cosmonauts were mixed in their own opinions, with some favouring a policy of moving on with the ISS, others wishing that Mir could be revisited and still others questioning NASA's motives. Was the US space agency planning to increasingly marginalise Russia, asked Soyuz TM-12 cosmonaut and Mir veteran Anatoli Artsebarski, now that the Russians were 'partners', rather than 'owners', of the new ISS? At one point, Iran offered to *buy* Mir and fund it for three years, in exchange for Russian training of its own crew members, but the offer came too late for meaningful dialogue.

MirCorp continued its efforts to save the station, almost until the very end, proposing at one stage that it could be raised to a higher orbit by means of an 8 km dielectric conducting 'tether' in order to prolong its life until new funding could be secured. Early plans envisaged carrying a $14.2 million demonstrator to Mir aboard a Progress craft and installing it onto the exterior of Kvant-1 during an EVA and 'counterweighting' it with the *Sredstvo Peredvizheniy Kosmonavtov* (SPK, or 'Cosmonaut Manoeuvring Equipment'), the equivalent of NASA's MMU, which had been stowed outside Kvant-2. It was envisaged that such a tether might generate upwards of two kilowatts of electricity. Such a plan was stalled when the United States forbade the export of its space tether technology ... until *after* Mir had been removed from orbit.

A short-lived proposal to hand Mir over to the United Nations was also suggested, but went nowhere. Continued funding difficulties and incessant US pressure, together with the unsafe nature of the station itself, obliged the Russian government to authorise the deorbiting process to occur in late February 2001 at a cost of $27 million. Debris from the largest spacecraft ever brought back into Earth's atmosphere was expected to fall into a desolate stretch of the Pacific Ocean, within an impact 'footprint' extending between 1,500 km and 2,000 km from Australia. It was recognised that components as heavy as 700 kg might survive re-entry, producing impact craters up to 2 m in diameter. *Flight International* explained that two Progress cargo craft would be needed to effect the propulsive deorbit manoeuvres. For a time it looked as if Sharipov, Vinogradov and Tito would fly in January 2001 on a short mission to assist with preparing Mir for its demise. However, in December 2000, and with grudging agreement from NASA, MirCorp and the Russians announced that Tito's flight would visit the ISS in April 2001.

No more human beings occupied Mir after the departure of Sergei Zalyotin and Alexander Kaleri on 16 June 2000. Their mission had met stiff opposition from US politicians. "How can the Russians maintain a serious involvement in both Mir and the International Space Station?" asked Republican Dave Weldon, vice-chair of the House Science Committee's Space and Aeronautics Subcommittee. "They are so cash-strapped, I don't see how they can do it." He argued that RSC Energia was "not serious about the ISS".

That seriousness was finally demonstrated in the final weeks of 2000, when the plans to dispose of Mir were set in motion. Progress M-43 was undocked from Mir on 25 January 2001 and deorbited, after which Progress M1-5 (nicknamed 'Hearse') arrived and docked at the Kvant-1 port to begin the preparations to end Mir's life. It carried 2,678 kg of propellant to effect the deorbit 'burn'. Several weeks later, in mid-February, RSC Energia responded to criticism of its decision to deorbit Mir by explaining in an open letter that "the actual condition of the on-board systems ... [does] not make possible the safe and reliable operation" of the station. Moreover, it was clear that efforts to prolong Mir's life "may lead to the loss of control" and could result in "catastrophic consequences". It was obvious that, aside from the politics and the promise of bearing commercial fruit, Mir's time had come. The station was established in a spin-stabilised orientation and it was decided to allow its orbit to succumb to natural gravitational drag, until it reached an altitude of about 265 km in early March. On the 7th, the Russians delayed the deorbit burn until Mir reached 220 km, as this would conserve propellant and offer a broader range of options in case the re-entry did not go to plan. Either way, orbital mechanics and mathematical calculations determined that Mir would naturally enter the atmosphere no later than the 28th.

The first of three 'burns' of Progress M1-5's engines got underway at 3:32 am Moscow Time on 23 March, using its docking and attitude control thrusters. It lasted 21.5 minutes and established Mir in an orbit with a perigee of 188 km and an apogee (high point) of 219 km. This set the conditions for the second burn, lasting 24 minutes, which got underway at 5:24 am and lowered that orbit still further to 158×216 km. Finally, the last burn began at 8:07 am. Originally scheduled to last 20 minutes, it was decided to run Progress M1-5's thrusters to exhaustion in order to ensure re-entry. The final signals from the dying Mir were received at 8:30 am, after a total of 15 years, one month and three days of orbital operations. The complex disintegrated over the South Pacific at about 8:52 am, presumably with the solar arrays, antennas and other appendages being destroyed in the first instance, followed by the main modules. Official Russian sources declared that Mir "ceased to exist" at 8:59:24 am Moscow Time, with debris trails monitored by the US Army site on Kwajalein Atoll in the Pacific. No damage to property or individuals on Earth was caused by the demise of Mir and, with the first long-duration US-Russian crew already aboard the ISS, its Viking-like funeral served to usher out the Old Era and welcome the New.

A NEW DAWN

On 20 November 1998, the new era began when a Russian Proton rocket thundered into a 403×374 km, 51.6-degree-inclination orbit from Tyuratam, carrying 'Zarya' ('Dawn'), the inaugural component of the International Space Station (ISS). Measuring 12.5 m long and 4.1 m wide and weighing 19,300 kg, this large pressurised module was also known as the 'Functional Cargo Block' (in Cyrillic characters, the 'FGB') and was designed to provide electrical power, storage, propulsion and guidance for the nascent space station during the early months of an ongoing mission which, at the time of writing, has spanned 16 years. Zarya was equipped with a pair of solar arrays, each measuring 10.7×3.4 m, and six nickel-cadmium batteries which provided about three kilowatts of electrical power. As the first module of the ISS, Zarya was designed to provide the initial junction point between the station's Russian Orbital Segment (ROS) and the US Orbital Segment (USOS) in what would evolve into the grandest, most audacious engineering accomplishment in human history. Two weeks after Zarya's launch, Space Shuttle Endeavour would deliver Node-1, also known as 'Unity', to connect the first two pieces of Russian and US hardware in orbit and so begin the gargantuan effort to build the ISS, high above the Home Planet. In fact, on no other occasion in history have so many nations – all told, the United States, Russia, Canada, Japan, Austria, Belgium, the Czech Republic, Denmark, Finland, France, Germany, Greece, the Republic of Ireland, Italy, Luxembourg, the Netherlands, Norway, Poland, Portugal, Romania, Spain, Sweden, Switzerland, the United Kingdom and, at one point, Brazil – co-ordinated their efforts in such an astonishing and far-reaching example of peacetime engineering in the interests of science and human advancement.

The road to the ISS, admittedly, was fraught with political and economic difficulty and the bright, permanently inhabited star which can now be seen nightly from many areas of the world came agonisingly close to cancellation on more than one occasion. On 25 January 1984, President Ronald Reagan formally announced that the United States would build a permanent space station, "and do it within a decade", having already discussed such a project informally with NASA Administrator Jim Beggs. "We can follow our dreams to distant stars," Reagan said in his State of the Union Address, "living and working in space for peaceful economic and scientific gain." Within weeks of Reagan's announcement, in April 1984, NASA named former Apollo flight director Neil Hutchinson as the first Space Station Program Manager at the Johnson Space Center (JSC) in Houston, Texas. In making the announcement, Beggs described JSC as the "lead centre" for the new programme. Under Hutchinson's leadership, the Space Station Program Office produced the first design for what the new station might someday look like. The early designs for Freedom were impressive in their size, scope and capability. With a total pressurised volume of 878 m^3, it would be built, piece by piece, some 400 km above Earth, by a succession of Shuttle crews and promised to be the largest spacecraft ever assembled in orbit. The so-called 'Power Tower' design, unveiled in April 1984, contained a long, central keel, with most of its mass situated at one end to exploit the gravity gradient for stability, thereby reducing the need for thruster firings in support attitude control. Most illustrations featured a cluster of pressurised modules at one end of the keel and a set of articulated solar arrays and a satellite servicing bay at the opposite end. By April 1985, the first contracts had been issued for definition studies and preliminary designs. Hutchinson remained in his post for two years, before stepping down in February 1986, to be replaced in an acting capacity by his deputy, John Aaron. Hutchinson retired from NASA the following August. At the time, construction of the Space Station was expected to be well underway by 1992. In the weeks following the loss of Challenger, in March 1986, the Space Station design changed somewhat to feature a pair of keels. The resultant 'dual-keel' moved the pressurised modules to the central truss, thereby placing them at the centre of gravity to offer a more quiescent environment for the sensitive microgravity experiments. Concurrently, the number of pressurised US laboratory modules was reduced from two to one, as Europe and Japan joined the project as international partners with their own home-built modules.

A major review in September 1986 produced an estimated development cost for a dual-keel Space Station at $14.5 billion, with the Challenger disaster having by this time pushed the First Element Launch date from January 1993 to January 1994. This triggered uproar in Congress, where many politicians were already beginning to doubt the programme's viability. At around the same time, NASA explored new Space Station configurations in order to reduce these development costs, with proposals ranging from a Skylab-type 'monolithic' station to a more gradual, phased construction of a dual-keel complex. The latter approach, known as 'Phase 1', required the division of the Shuttle assembly process into two halves and promised to reduce the overall programme cost to about $12.2 billion. A pair of 37.5-kilowatt solar arrays would have provided electricity, although Congress quickly pressed NASA to incorporate a further two arrays in order to generate sufficient power for scientific users.

In August 1988, in the true spirit of Reagan-era politics, the Space Station was renamed 'Freedom'. Shortly afterwards, in September, NASA signed ten-year contracts to begin the development of actual hardware and a Space Station Intergovernmental Agreement was signed to allocate resources between the United States and its European, Japanese and Canadian partners. Under the language of the agreement, 97 percent of the resources in the US laboratory module would be allocated to NASA, with 3 percent assigned to the Canadian Space Agency in return for its contribution of robotic assets to the station. Europe and Japan would each retain 51 percent of their own laboratory modules, with the United States and Canada receiving 46 percent and 3 percent, respectively. Although a crew of only four was planned in the early stages, it was anticipated that this would expand to eight: six Americans and two international astronauts from Europe, Japan or Canada. It was originally planned for station expeditions to last around 90 days, although this was subsequently increased to 120 days (and perhaps as long as 180 days), in order to reduce the number of dedicated Shuttle flights from eight per annum to five or six per year.

Yet in spite of the grandiose imagery of what Freedom would look like, there was little 'freedom' from the harsh budgetary realities of the late 1980s and early 1990s. Congress firstly reduced NASA's Fiscal Year 1988 budgetary allocation for the station from $767 million to $525 million and made only $300 million of this funding available until June 1988, with the remainder provided only after the agency delivered a less costly, 'rescoped' project. This was followed by only $900 million for Freedom in 1989, which was about half as much as NASA had requested in its original planning. The station's design met with a slight redesign in late 1989, when NASA's Fiscal Year 1990 budget for the programme was also reduced from $2.05 billion to $1.75 billion. More troubling was that Freedom remained 23 percent overweight, seriously over-budget, too complicated to build and offered only 66 percent of the electrical power for its research needs. In addition, an External Maintenance Task Team (EMTT), led by astronaut Bill Fisher and JSC engineer Charles Price, found in September 1990 that the assembly and maintenance of the station would require between 2,282 and 3,276 hours of EVA *per year* ... a far cry from NASA's officially stated goal of around 500 hours per year. Another redesign was demanded by Congress in October 1990 and NASA's Fiscal Year 1991 budget included a scathing $550 million shortfall, to just $1.9 billion for Freedom. Furthermore, NASA Administrator Dick Truly was informed that the agency could expect no more than ten-percent annual growth between 1992 and 1996, with peak spending on Freedom not to exceed $2.6 billion. These budgetary ground rules, including the $550 million cut in the Fiscal Year 1991 budget, represented a dramatic $5.7 billion shortfall in what NASA had intended to spend on Freedom over that five-year period. In response to the funding issue, NASA initiated a major restructuring of the station in November 1990.

Four months later, it produced a plan for an extensively redesigned Freedom. The result was a cheaper and smaller space station, with shorter laboratory and habitation modules, that would be less complex to assemble by a fewer number of Shuttle missions. The radiators and attitude-control systems were also simplified. "We've cut costs, simplified the design and reduced the complexity of the project," explained Bill Lenoir, a former astronaut and NASA's Associate Administrator for Space Flight. "At the same time, Freedom will be a quality facility, providing a research laboratory unsurpassed in the world for life sciences and microgravity research and a stepping stone into the future." Under the March 1991 plan, the start of construction would occur in January-March 1996 and would achieve 'man-tended' capability by April-June 1997. "In the man-tended phase, astronauts brought up to Freedom by the Space Shuttle will be able to work inside the US laboratory for periods of two weeks," NASA announced. "They will return to Earth with the Shuttle. At this stage, one set of Freedom's solar arrays will generate about 22 kilowatts of power, with a minimum of 11 kilowatts available to users. Six Shuttle flights will be required to achieve the man-tended configuration." It was anticipated that a four-person permanently manned capability would be achieved by the turn of the millennium, with US, European and Japanese laboratory modules and a habitation module in place, together with Canadian-provided robotic assets and three sets of solar arrays furnishing 65 kilowatts of electrical power, almost half of which would be dedicated to scientific operations. Provisions to expand Freedom still further with an additional solar array, pushing its electrical output to 75 kilowatts, together with a second laboratory module and connecting nodes, were also under consideration, as was the possibility of eventually expanding the station's permanent crew to eight. The US laboratory and habitation modules would be 8.2 m in length (about 40 percent shorter than the original specification) and 4.4 m wide, which was expected to allow them to be fully outfitted and tested on the ground, before being launched into orbit by the Shuttle. The enormous truss-like backbone of the station, which was destined to support the solar arrays and radiators, would also be prefabricated on the ground, which differed substantially from the original plan to assemble them during Shuttle-based EVAs. This was expected to cut the number of spacewalks to build Freedom by about 50 percent, whilst the length of the truss was reduced substantially from 150.2 m to 107.6 m.

Not surprisingly, the international partners were unimpressed by the additional delays, the reduced power capability and the limited science return from the new Space Station Freedom, whose myriad features were gradually being stripped back to the bones. In June 1992, the frustration of the new NASA Administrator, Dan Goldin, was evident in his remarks to the National Space Club. "We can light up the sky with the inspirational work of Space Station Freedom," he said, "or we can stand by and watch the greatest technological bonfire of the century if it's cancelled." Goldin added that the United States had "waited long enough" for a space station, that it "must have a permanent presence in space" and, delivering the final punchline: "We *need* Space Station Freedom ... and we need it *now*!" Research into bone loss, muscle wastage, red and white blood cell reductions and sensory problems could all benefit from the experiments to be performed aboard Freedom, he said, pointing to the potential for developing countermeasures for osteoporosis and other diseases. When challenged by the cost of the station, Goldin retorted: "Sounds like a lot, until compared with the $6.3 billion Americans spend on pet food each year or the $4.3 billion we spend on potato chips or the $1.4 billion for popcorn." In his mind, Space Station Freedom carried the potential to give the United States "reason to hope our future will be forever brighter than our past".

Preparations for the station continued with static firings of the development test article for Freedom's propulsion module in late December 1992 at the White Sands Test Facility in New Mexico. The tests served to validate the concept of the 13-thruster module, which was to be used for attitude control, orientation and space debris avoidance, with Dick Kohrs, head of the Space Station Freedom Program at NASA Headquarters in Washington, DC, describing the tests as evidence that "Freedom is no longer a *paper* station". Two propulsion modules were scheduled for installation during the second Shuttle assembly mission, with two more expected at the point of man-tended capability and an additional pair just prior to the permanently manned stage. It was true, indeed, that Freedom was not a paper station, but its troubles continued into 1993 and its ignominious death drew nearer. The new President, Bill Clinton, who entered office in January, planned to cut

5-04: Rising tower-like from Endeavour's payload bay, the joined Zarya and Unity modules symbolised not only the beginning of International Space Station construction, but the union of two former foes.

NASA's budget by 15 percent over five years in order to reduce the federal deficit. However, Clinton *wanted* the Space Station, albeit in a configuration which would significantly reduce development, operational and utilisation costs, as well as honouring the United States' commitments to its international partners and advancing the nation's science and technology capabilities. In February 1993, NASA was directed to redesign Freedom ... for what turned out to be the final time.

Goldin appointed Joe Shea, a long-time NASA veteran and Acting Chair of the NASA Advisory Council, to oversee the redesign process, and announced new measures to conserve resources and restrict new spending. In early April, John Gibbons, director of the Office of Science and Technology Policy, outlined a trio of five-year (1994–1998) budgetary options, priced at $5 billion, $7 billion and $9 billion, for guidance in the deliberations on Freedom's redesign. By this time, as described earlier in this volume, the Russians were being increasingly courted as potential future partners in the Space Station programme. In his address, Gibbons announced that the United States, Europe, Japan and Canada had decided to give "full consideration" to the use of Russian assets in a consultancy role during the redesign process. Russia had already delayed the launch of its next-generation Mir-2 station until at least 1997, due to funding difficulties in the aftermath of the collapse of the Soviet Union, and it was clear that both sides needed each other in order that their respective space programmes could survive.

Early in June 1993, the redesign team submitted its final report on Gibbons' three budgetary options to the White House Advisory Committee on the Redesign of the Space Station at the Stouffer Hotel in Crystal City, Virginia. Lauding the team's work, Dan Goldin pointed to "three technically viable space stations", each of which reflected "complete and accurate costs". Option A used a combination of Space Station Freedom hardware and flight-qualified systems from other sources, including a self-contained Department of Defense spacecraft, known as 'Bus-1', for propulsion, guidance, navigation and control. It would be assembled in four key phases, delivering photovoltaic array components, the US laboratory, the international modules and additional solar arrays. Option B was derived from mature Freedom designs and made maximum use of current systems and hardware to provide an "incrementally increasing capability", which enhanced payload accommodations for scientific users and adhered to NASA's commitments to its international partners. Like Option A, it would be assembled over four phases. Finally, Option C was a single-launch space station, consisting of a core module some 28 m in length and 7 m in diameter. Its interior would be divided into seven 'decks', connected by a central passageway, and would offer about 736 m^3 of pressurised volume. In all three options, the Russian Soyuz-TM spacecraft was utilised as an Assured Crew Return Vehicle (ACRV), permanently in place to cater for emergencies. Also in June 1993, an amendment by the Indiana Democrat Tim Roemer to remove Space Station funding from NASA's appropriations bill (which would have effectively killed the project) failed by just one vote in the House of Representatives. The station in the minds of politicians, it seemed, had hit rock-bottom. The only way ahead now was *up*. On 17 August, responding to the Station Redesign Team's recommendation that all separate hardware contracts should be consolidated under one prime contractor, Dan Goldin announced the selection of Boeing to fulfil this critical role. "As the prime contractor, Boeing will be responsible for the design, development, physical and analytical integration and test and delivery of the Space Station vehicle," NASA explained. Fifteen months later, after much negotiation, NASA and Boeing signed a $5.63 billion contract to design and develop the new space station.

By November 1993, as discussed earlier in this volume, high-level negotiations brought Russia formally into the new project, which was renamed the 'International Space Station' (ISS). A two-day Systems Design Review in March 1994 evaluated the design status of the new station, serving to "lock in the key technical elements," according to NASA Space Station Program Manager Randy Brinkley, "as well as the schedule and cost". The new station utilised about 75 percent of the hardware originally planned for Freedom, thereby maintaining an existing investment and redesigning it to be both less expensive and more capable. The hardware would include an Integrated Truss Structure (ITS) to support the solar arrays and radiators, a US laboratory and habitation module, the Russian Science Power Platform (SPP), Service Module (SM) and Functional Cargo Block (FGB), along with the international partners' contributions: Europe's Columbus laboratory, the Japanese Experiment Module (JEM) and Canada's Space Station Remote Manipulator System (SSRMS) based on the mechanical arm of the Shuttle. Its operational altitude would be 440 km, inclined 51.6 degrees to the equator, making it reachable by Soyuz-TM spacecraft, launched from Tyuratam. Its permanent crew size would begin with three members and later expand to six, including three Russians, two Americans and one international partner.

The Systems Design Review settled on a provisional manifest for the missions which would create the ISS. Beginning in November 1997, the FGB module, later named 'Zarya', would be delivered into orbit by a Russian Proton rocket from Tyuratam to provide initial command, control and propulsion capabilities. Two weeks later, a US-built connecting node (Node-1) would be delivered by the first Shuttle assembly crew and attached to the FGB during a series of spacewalks. Next would come the Russian-built SM in January 1998, containing crew quarters and life-support systems. Further downstream would come other major components: the SPP, the US laboratory module, the Canadian SSRMS, a joint airlock to support both US and Russian space suits, and the Japanese and European laboratories. Significantly, the arrival of the US laboratory module, then planned for May 1998, would achieve 'man-tended capability' and would be followed by the launch of the first long-duration crew in June. With construction scheduled to require 55 months, the final mission would occur in June 2002, after 13 Russian assembly flights and 16 Shuttle launches. From its inception in early 1994 until its completion in mid-2002, the ISS was anticipated to cost about $17.4 billion.

The future was beginning to brighten for the project and on 29 June 1994, the House of Representatives defeated an amendment to terminate the ISS. In praising the House's decision, Dan Goldin described it as "a vote for America and for the American people and a vote for our future". This was reinforced on 3 August, when the Senate voted on an amendment from the Democratic Senator Dale Bumpers of Arizona to terminate the ISS. The amendment failed by 64-36 votes, prompting Dan Goldin to remark that the Senate's action "cast a vote of confidence in America's space programme, a vote for investment in our nation's future and a vote for continued US leadership in technology and exploration for decades to come".

By this stage, the ISS assembly sequence had crystallised still further. Congress had expressed concern that the position of the US laboratory module, which provided a command and control centre for the station, as well as a scientific facility, at an early stage in the sequence meant that it would be deprived of sufficient electrical power and redundancy for its critical systems. In response to this concern, it was decided to move one of the four sets of US-built solar arrays 'up' in the assembly sequence to be installed on the station *before* the arrival of the laboratory.

The array would be mounted on a small truss, atop Node-1, and would be relocated to its permanent position on the ITS at a later stage in the assembly sequence.

Every journey, though, begins with an initial step; and every construction project with the laying of a foundation stone. For the ISS, that was the connection of Russia's FGB module with America's Node-1. Built by Russia, but paid for and owned by the United States, the FGB was originally intended as a component for Mir, but was unflown and rededicated as the cornerstone of the early ISS. Its assembly began in December 1994 at the Khrunichev State Research and Production Space Centre in Moscow. Several weeks later, in February 1995, NASA Space Station Program Manager Randy Brinkley and the Russian Space Agency's Deputy of Piloted Space Flight Boris Ostroumov signed a government-level protocol for the US purchase of the module. The terms of the protocol called for its design, development, manufacturing, testing and delivery at a cost of $190 million. Six months later, in August 1995, Boeing reached final agreement with Khrunichev to complete and deliver the FGB into orbit. In December 1996, the FGB was completed on schedule and within mandated budgets, with the expectation that it would be transported to the Tyuratam launch site in May 1997 for final preparations, ahead of its Proton launch the following November.

In the meantime, by the summer of 1995, the machining of Node-1, the first US-built element of the ISS, was also completed. Built by Boeing, the cylindrical node, later named 'Unity', consisted of two cylindrical segments, six 'berthing ports' (four radial and two axial) and three large ring frames and, as a central 'hub' for the station, it carried several kilometres of electrical wiring, 216 lines for fluids and gases and 50,000 mechanical items. It measured 5.4 m in length and 4.2 m in diameter and weighed about 11,340 kg. Reflecting upon the successful completion of the module, Wil Trafton, NASA's Acting Associate Administrator for Space Flight, noted wryly that "*this* is what two years of stable funding and hard work will get you". Node-1 was painted in April 1996 and underwent extensive pressure tests at Boeing's plant in Huntsville, Alabama, between August and November, ahead of its final assembly and checkout and delivery to KSC in mid-1997. NASA astronaut Jim Voss, who spent five months aboard the ISS as a member of its second long-duration crew, liked Node-1 from the moment he floated aboard, although he was somewhat sceptical about its colour scheme at first. "I found that being in the node ... was an extremely pleasant experience," he told a Smithsonian interviewer. "Even though it's smaller than the other station modules, it's still much larger than any other volume I'd been inside in space and it has a very pleasant soft lighting. Before this flight, I had always laughed when they talked about using a pinkish, coral colour for the node interior ... but whoever it was did a wonderful job. That was where I went whenever I wasn't able to sleep or when I wanted to be alone to listen to music or write in my notebook."

However, the construction of another critical Russian element, the Service Module, had fallen eight months behind schedule, due to inadequate government and contractor funding. This component was originally the centrepiece of Russia's ill-fated Mir-2 station and, in its ISS incarnation, would provide crew quarters, life-support and propulsion systems. Initially scheduled for launch in March 1998, it was postponed until no earlier than December 1998. "We knew from the outset that building an International Space Station was going to be tremendously challenging," NASA Administrator Dan Goldin reflected in April 1997. "Space exploration is not easy or predictable. We will work through this schedule issue and we undoubtedly will face additional problems in the future, but we are well on our way to the realisation of this world-class facility." In order to mitigate the impact of this delay upon the remainder of the assembly sequence, several steps were taken by NASA management. These included the construction of a US-built Interim Control Module (ICM) in conjunction with the Naval Research Laboratory to provide reboost capability for the infant ISS in the event of further significant delays. Other options were to modify the FGB for on-orbit refuelling and to upgrade its avionics in order to augment the early station's capabilities or to install life-support equipment inside the US-built laboratory module to permit an early human presence. In May 1997, the International Space Station Control Board approved a new assembly plan, which rescheduled the FGB launch for June 1998 and the delivery of Node-1 by STS-88 in early July. "The recent completion of a major Russian general designers review for the Service Module, in which I participated, and full Russian funding of the work, gives us high confidence that the Service Module can meet a revised launch date of December 1998," said Randy Brinkley. "The Russian Space Agency has been extremely forthcoming in its dealings with NASA on this subject and they and their contractors have gone out of their way to demonstrate their resolve to meet their commitment." Only after the SM was in place could the station's first long-duration crew take up residence.

In the shadow of this delay, preparations for Node-1 continued smoothly. On 26 June 1997, the module was delivered from the Marshall Space Flight Center's Space Station Manufacturing Facility in Huntsville, Alabama, to KSC in Florida, aboard an Air Force C-5 Galaxy transport aircraft for processing. Randy Brinkley described the event as an indication that the ISS had "begun moving from the factory floor to the launch pad". A few weeks later, in July, Node-1 was joined in KSC's Space Station Processing Facility (SSPF) by the first of two Pressurised Mating Adaptors (PMAs). The second PMA arrived in October 1997. Built by McDonnell Douglas, these asymmetrical, open-cone-shaped elements were to be affixed to the forward and aft ports of Node-1 to provide connection interfaces with Zarya at one end and (for the first few assembly and logistics missions) the docking system of the Shuttle at the other. They measured about 2.1 m long and between 1.5 m and 2.7 m at their narrowest and widest points and would serve as pressurised passageways. Also accommodated within the PMAs were external multiplexer-demultiplexers and electrical equipment, which STS-88 Payload Manager Glenn Snyder described as "the intelligence for the Node".

By October 1997, when representatives of the nations that were participating in building the ISS met in Houston, Texas, to finalise the assembly sequence, the situation appeared to have stabilised, with all partners confirming that they were on schedule to meet their individual commitments. The Russian Space Agency "reassured" the partners, in the words of Randy Brinkley, that it would launch the Service Module in December 1998, with finances having started to reach manufacturers from five banks as part of an 800-billion-rouble ($140 million) guarantee from the Russian government. The long-delayed module, which marked the first fully-Russian ISS component, had passed a critical milestone on 12 September 1997, when it emerged from the general designers' review in Moscow with flying colours. Also in September, the assembly and testing of the FGB was finished and on 17 January 1998 it was delivered to Tyuratam for launch on 30 June. Although Brinkley excitedly declared 1998 to be the "Year of the Space Station", the celebrations turned out to be premature. It was already becoming clear that the Service Module would still not be ready in time, even for a December launch. In May, at meetings of the Space Station Control Board and Heads of Agency at KSC, the ISS partners were forced to again revise the assembly sequence and reschedule the FGB launch for November and STS-88 for early December. This threw subsequent missions into flux. Under the new plan, the Service Module would fly from Tyuratam in April 1999, followed by the station's first long-duration crew later that summer. Although NASA officially tried to play down the seriousness of the continued Service Module delays, it stressed that "the international partners expressed their concern" and "brought to the attention of [Russia] that it is critical to all participating nations that the station programme schedule is met".

With STS-88 now planned for the end of the year, in July 1998 NASA unusually assigned Krikalev to join the crew. Although his assignment came only months before launch, it was noted by Dave Leestma, the head of Flight Crew Operations, that Krikalev's "experience with both the US and Russian programmes and his familiarity with the Shuttle make him a valuable addition to this crew". By this point, the other five members of the STS-88 crew had been training for almost two years and, without casting doubt upon Krikalev's credentials, it seemed that the decision to include a Russian cosmonaut on the first piloted ISS mission at such a late stage was an exclusively political one. "We could've accomplished the mission without him," admitted STS-88 commander Bob Cabana, "but it's an extremely challenging mission from a time-scheduling point of view and Sergei's going to provide expertise on the Zarya. With him, it just makes it a little easier. He's also going to be backing us up on a lot of the other tasks that we have on the flight ... so Sergei brings a lot of experience, but specifically we've made him 'prime' for the FGB and all its systems during the ingress." As a result, Cabana and Krikalev would be the first two humans to enter the ISS in space.

The STS-88 stack was transferred from the VAB to Pad 39A on 21 October, with a view to launching at 3:58 am EST on 3 December, about two weeks after the liftoff of Zarya. In the meantime, the Unity/PMA combo was delivered to the pad on 26 October and installed into Endeavour's payload bay in mid-November. "There has been a tremendous amount of excellent work done by everyone involved with Unity ... to get to this point," explained Steve Francois, head of Space Station and Shuttle payload processing at KSC. "Unity represents the first new human spacecraft to go to a Kennedy launch pad since the first Space Shuttle launch."

Finally, after so many frustrating delays, the FGB/Zarya achieved orbit on 20 November 1998, clearing the way for the launch of STS-88 and the Unity/PMAs combo on 3 December. Aboard Space Shuttle Endeavour for STS-88, in addition to

Sergei Krikalev, were five US astronauts. In command was Robert Donald Cabana, making his fourth Shuttle flight, who had served between September 1994 and October 1997 as chief of NASA's astronaut corps. Born on 23 January 1949 in Minneapolis, Minnesota, he attended high school in his home town and entered the US Naval Academy, receiving a degree in mathematics in 1971, then joining the Marine Corps. Cabana completed Basic School at Marine Corps Base Quantico in Virginia, then entered naval flight officer training at Naval Air Station Pensacola in Florida. Cabana flew as a bombardier and navigator in the A-6 Intruder attack aircraft, based at Marine Corps Air Station Cherry Point in North Carolina, and at that service's base at Iwakuni in Japan. Returning to Pensacola in 1975, he started pilot training and was designated a naval aviator in September of the following year. As the top Marine to complete pilot training, Cabana was awarded the Daughters of the American Revolution Award. Graduation from the Naval Test Pilot School at Patuxent River, Maryland, followed in 1981 and Cabana served as programme manager for the A-6 and project officer for the X-29 advanced technology demonstrator. He was serving as assistant operations officer of Marine Aircraft Group 12 in Iwakuni when he was selected as a Shuttle pilot candidate by NASA in June 1985. He flew three Shuttle missions in the early 1990s, prior to his stint as chief astronaut, and in August 1996 was named to command STS-88. The mission was originally scheduled to fly in December 1997, but was extensively delayed.

Seated to Cabana's right side on Endeavour's flight was pilot Frederick Wilford Sturckow, the only first-time spacefarer on STS-88. Before launch, he explained that this his role as the 'rookie' was to listen to the stories of his more experienced crewmates. "I counted up my crew and they've got stories from *twelve* different Space Shuttle flights," he said. "When Sergei gets on there, that'll be *thirteen* different Shuttle flights. I've got *zero* to talk about ... so I've done a *lot* of listening!" In time, however, between 1998 and 2009, Sturckow would become the first person to fly four missions to the ISS. Born in La Mesa, California, on 11 August 1961, Sturckow grew up on a ranch just outside the rural town of Lakeside in eastern San Diego County. "We raised turkeys and we had cattle," he recalled to a NASA interviewer, "and we later started a tree-farming business. We had a variety of different trees we sold." Sturckow completed high school in 1978 and worked for two years as a truck mechanic ("a lube boy," as he described himself) at Earley International Harvester (IH) Truck Center, serving with the Goodyear Baja pit crew at the weekends. He then entered California Polytechnic State University to study mechanical engineering and, whilst there, he led the Society of Automotive Engineers (SAE) off-road truck project and built his own stock car. (By his own admission, the SAE racing programme was one of the incentives which drew Sturckow to Cal Poly in the first place.) It had other positive influences on him, too. During this period, one of his professors (a Vietnam War veteran) suggested that Sturckow should join the Marine Corps and become a pilot. Following receipt of his degree in 1984, he was commissioned into the Marines and graduated from Basic School with honours, then gained his aviator's wings. "After you graduate Officer Candidate School, every Marine officer goes to the Basic School," he recalled. "It's six months of infantry officer training at Quantico, Virginia. The theory is that if you're going to support

Marines on the ground from the air, you should know something about what they do on the ground." Sturckow trained to fly the F/A-18 Hornet and was assigned to Marine Corps Air Station Beaufort in South Carolina, during which time he undertook overseas deployments to Japan, South Korea and the Philippines. He was selected to attend the Navy Fighter Weapons School (known as 'Top Gun') in March 1990 and, the following August, headed to Sheik Isa Air Base in Bahrain for eight months in support of Operation Desert Storm, during which time he flew 41 combat missions over Iraq. During his early Marine Corps career, Sturckow earned the nickname 'CJ' (for 'Caustic Junior'), apparently because he resembled a squadron commander who was appropriately known as 'Caustic'. Following his return to the United States, he attended Air Force Test Pilot School at Edwards Air Force Base in California and later served as the project pilot for the E/F models of the F/A-18, flying a variety of classified and other aircraft programmes at Naval Air Station Patuxent River in Maryland. Selected as an astronaut candidate by NASA in December 1994, Sturckow was one of the first members of his class to draw a flight assignment. In fact, announced as a crewman on the *first* ISS assembly mission was *the* Shuttle flight assignment of the late 1990s. "That was a great feeling," he recalled. "Colonel Cabana, who was chief of the office, called me up and told me that I was going to be flying with him the first space station assembly mission, so that was really great news. It was very exciting!"

At the time of the initial STS-88 crew assignment in August 1996, the team was rounded out by three experienced mission specialists: Jerry Ross, who became one of only a handful of human beings to chalk up six space flights, would lead a series of complex EVAs with STS-69 veteran Jim Newman to install the Unity/PMAs and connect electrical, fluid and other provisions, whilst STS-70 veteran Nancy Currie would be in charge of operating Endeavour's Remote Manipulator System (RMS) mechanical arm. Originally planned to last just seven days, the early stages of the mission bore some parallels to the STS-74 flight, which had delivered the Docking Module to Mir in November 1995. In a similar manner, Currie would employ the RMS to lift the Unity/PMAs from the payload bay and attach them to the Orbiter Docking System (ODS), whereupon Cabana would fly Endeavour to within 10 m of Zarya. "On STS-74, they had cameras mounted such that when they docked with the Mir space station they were looking directly through the other end of the docking adaptor and were able to fly the orbiter precisely onto the mating surface of Mir," explained Cabana. "What we're doing, however, is we're going to rendezvous with Zarya up in space and actually fly it *down* into the payload bay, *behind* the Node, where we can't even *see* it out the windows." Zarya would be essentially invisible to their eyes, blocked by the mass of the ODS-Unity stack in the forward payload bay. Using external television cameras and the Canadian-built Space Vision System (SVS), Currie would extend the RMS, grapple Zarya and mate it with Unity. In doing so, STS-88's capture of Zarya would mark the heaviest object ever handled by the RMS, exceeding the previous record-holder (the 17,000 kg Compton Gamma Ray Observatory, launched in April 1991) by about 2,300 kg. Following the mating of Zarya and the Unity/PMAs, Ross and Newman would perform a series of EVAs (originally two, but later increased to three) to establish electrical, fluid and other connections, after which Cabana would undock from the 21 m stack and return to Earth.

Despite a 60 percent likelihood of unacceptable weather conditions in the early hours of 3 December, the six-strong crew boarded Endeavour shortly after midnight and proceeded smoothly through their final tasks and checklists. Launch was timed for 3:58:19 am EST, right in the middle of the ten-minute 'window'. With nine minutes on the clock, all systems were "Go", with the exception of weather, which looked iffy should Cabana and Sturckow be forced to perform a Return to Launch Site (RTLS) abort. Aware that Endeavour had only a narrow period in which to launch, Flight Director John Shannon opted to resume the clock and continue counting down to T-5 minutes, at which point another 'hold' would occur in order to reassess the weather. By this stage, with the weather bordering on acceptable, the countdown resumed. Rick Sturckow activated the three Auxiliary Power Units (APUs) to give Endeavour hydraulic muscle for her mission ... but shortly afterwards the sound of the master alarm pierced the silence of the cockpit. The alarm occurred at T-4 minutes and 24 seconds and was caused by a momentary low-pressure indication from one of the APUs. The countdown continued to T-31 seconds, the critical point at which command was scheduled to be handed over from the Ground Launch Sequencer (GLS) to the 'autosequencer' and the vehicle's on-board computers.

However, by this stage, only 60 seconds remained in the five-minute window. Consensus was reached that all three APUs were in good health and the call "GLS is Go for Autosequence Start" crackled over the airwaves, initiating the final 31 seconds of pre-launch activities.

"Everything looks nominal," reported NASA Test Director (NTD) Doug Lyons.

"NTD, Houston Flight, we think it was System 1," replied John Shannon. "We are Go for Launch." After further confirmations that all seemed normal, Shannon pressed Lyons to resume the countdown.

"Gotta pick it up, NTD," said Shannon.

"That's affirmative," came the response from Lyons. He then requested controller Janine Pape, seated at the GLS console, to pick up the countdown, but three seconds passed with no response. "GLS, pick up the count on your mark," Lyons repeated. By this point, those few missed seconds had caused the available time in the performance window to expire.

"NTD, we're no longer Go for Launch," called Shannon.

"Copy that. And GLS, NTD," said Lyons.

"Yes, sir, we've picked up the count, we're at 24 seconds," replied Pape. "Request cutoff?"

"Please cut off."

"Yes, sir. GLS safing is in progress."

The official call to scrub the launch came at T-19 seconds.

Although the lost seconds undoubtedly contributed to keeping Endeavour on the ground, it was reflected in the hours which followed that the events of 3 December 1998 were one of the most tension-filled countdowns in the Shuttle programme's history. "We wanted to make sure we had the right technical decision," Launch

Director Ralph Roe explained later. "By the time we got to the right technical decision, we missed the performance window ... by between one and two *seconds*." Although an almost imperceptibly small period of time, to have gone ahead with the launch two seconds after the closure of the window would have compromised the optimum conditions for reaching Zarya with the most economical orbital manoeuvres and propellant expenditure. STS-88 was postponed for 24 hours, with the next window requiring the Shuttle to launch between 3:33:07 and 3:40:34 am. Next morning, on 4 December, there were no problems. Endeavour entered and emerged from her built-in holds at T-20 minutes, T-9 minutes and T-5 minutes and no difficulties arose after the activation of the APUs. Six seconds before liftoff, the orbiter's three main engines roared to life and at 3:35:34 am EST the first Shuttle mission of the new era pierced the black Florida sky and thundered perfectly into orbit.

Immediately upon reaching space, the crew set to work preparing for their rendezvous with Zarya on the third day of the mission. By this point, two weeks since its own launch, the Russian-built module was making its 222nd circuit of Earth and was experiencing a relatively problem-free flight, with the minor exception of a glitch with one of its six battery charging systems. Cabana and Sturckow performed the first of six thruster burns to close the distance between themselves and their quarry; by mid-morning on 4 December Endeavour was about 2,090 km 'behind' Zarya and closing at a rate of about 1,090 km with each 90-minute orbit. In the meantime, Ross and Newman worked on their space suits and Simplified Aid for EVA Rescue (SAFER) backpacks and set up and activated the Canadian-built Space Vision System (SVS) which would assist Currie in the grappling and aligning Unity and Zarya. Elsewhere, Currie busied herself with the activation of the RMS mechanical arm in readiness to unstow the Unity/PMAs combo on 5 December. The cabin pressure was also lowered from its normal 101.3 kPal to 70.3 kPal to prepare the spacewalkers for the 29.6 kPal pure oxygen of their suits, as well as clearing nitrogen from their bloodstreams to avoid an attack of the bends.

Late on the afternoon of 5 December, Currie grappled the Unity/PMAs combo and lifted it out of Endeavour's payload bay, carefully positioning it in an orientation perpendicular to the Shuttle's main axis. After PMA-2, on Unity's forward port, had been latched onto the ODS, Cabana fired the RCS thrusters to bring the mechanisms together to form a rigid connection at 6:45 pm EST, as the orbiter flew over eastern China.

"Houston, Endeavour, we've got Unity firmly attached to Endeavour and we're off to a great start on building the International Space Station," Cabana radioed triumphantly. "Thanks to everybody down on the ground for such excellent training."

"Endeavour, Houston, congratulations to all the members of the crew," responded Capcom Chris Hadfield from Mission Control. "It's a beautiful sight."

The following hours were spent pressurising the vestibule between the ODS and PMA-2 and the hatches were opened, whereupon Cabana and Ross entered the adaptor and placed caps over vent valves in anticipation of the long-awaited first ingress into Unity itself. Next day, 6 December, at a distance of about 88 km 'behind'

5-05: Jerry Ross (left) and Jim Newman during their EVAs to install and activate the Unity-Zarya 'stack'.

Zarya, Cabana and Sturckow executed a thruster firing to fine-tune Endeavour's final approach, slowing their rate of closure. Two hours later, at just 14.5 km, the Shuttle crew performed the Terminal Initiation (TI) 'burn' to establish themselves on the 'R-Bar' rendezvous path and arrive in a position about 180 m 'above' Zarya, just one orbit later.

Cabana took manual control of his ship at 5:45 pm EST, at a distance of about 0.8 km. He circled in front and directly 'above' his quarry, in just 45 minutes ("twice the orbital rate"), before flying 'down' toward the Russian-built module. "The reason for the flyaround," Cabana explained, "and not just joining up with it from below is so that we're not blocking Russian command to it from the ground." Since Zarya entered the vicinity of the payload bay 'behind' the ODS-Unity stack, in order for Currie to grapple it with the RMS, the crew could not physically see it. Instead, they had to rely entirely upon the views from the arm's cameras and a prime and backup camera in the keel of the payload bay, for a nail-biting 17 minutes. "We were waiting until we got over a Russian ground communications station, so we could grab it," Cabana told a Smithsonian interviewer, years later. "I had gotten Zarya's grapple fixture about three feet from the end of the arm, so Nancy Currie could move in and grab it. It was perfectly stable, but then the orbiter did an automatic firing of its jets to stay within a set attitude, which moved us *towards* it. All of a sudden, the Zarya ... is heading *down* into the payload bay and toward the arm." Cabana was obliged to back away from the module, "fine-tune everything" and execute another approach. "It got our attention!" he recalled, in a typically understated manner. "Normally, Jim Newman loved to give advice during these kinds of operations: *What do you think about this?* or *Why don't you do that?* or *I think you need to go down some.*" Yet even Newman was rendered speechless in perhaps the most stressful part of the mission so far. "It was just dead quiet in the cockpit," recalled Cabana. "Nobody said *anything.*"

When she was within the 10 m range, Currie extended the RMS and at 6:47 pm grappled Zarya.

"Houston, Endeavour, we have Zarya firmly attached to the orbiter," called Cabana. "We're halfway home for the day."

"A lot of people exhaling down here, that's great," replied Chris Hadfield, "and we had a terrific view from Zarya watching Nancy's great work."

Following careful alignments, the two components were brought into a close, mechanised embrace a little over an hour later. "With the Orbiter Docking System, because it's aft of the [flight deck] windows, it's very difficult to tell things like pitch and yaw and roll" of Zarya, explained Currie before the mission, "so we have a centreline camera that looks straight up through the Orbiter Docking System. Then, on the end of the Pressurised Mating Adaptor is a target ... so accurate that if you're, say, 0.2 degrees off in pitch, you can tell by looking at this target. We feel pretty confident that this is a very good way to get within a very tight tolerance. About two inches and two degrees is the tolerance of how close we have to be, between the two elements, in order to have a successful mating. We actually won't take [Zarya] all the way down the Orbiter Docking System; we'll take it four inches apart, so the petals will overlap, but the rings themselves will not be touching."

At this point, Bob Cabana took over. At 8:07 pm, he fired Endeavour's thrusters to push Zarya and the Unity/PMAs together and the first two modules of the ISS were secured with hooks and latches at 9:48 pm. As Endeavour circled Earth with a 21 m infant space station sprouting from her payload bay, the STS-88 crew celebrated a major first accomplishment of their mission.

"Houston, Endeavour, we have captured Zarya!" called Cabana.

"We copy," replied Hadfield. "Congratulations to the crew of the good ship Endeavour. That's terrific!"

However, there remained much to do and the real work on the interior and exterior of the space station remained ahead. Currie used the RMS cameras to perform a detailed inspection of Zarya, focusing specifically upon a pair of TORU antennas, which had failed to deploy automatically in the aftermath of the module's launch on 20 November. Flight controllers concluded that the pyrotechnic pins which held the antennas in place *had* fired, but that the deployment process had not occurred. It was possible that Ross and Newman would take a closer look during their first EVA. And there was a suspect component in one of Zarya's six batteries, which was not discharging properly in its automatic mode. Perhaps Sergei Krikalev could replace it later in the mission.

At 5:10 pm EST on the 7th, the first spacewalk of the ISS era got underway, when Ross and Newman departed Endeavour's airlock hatch and floated into the blackness of space. Since their assignment to STS-88 in August 1996, the two men had spent 540 hours of training for the EVAs, including more than 240 hours underwater in the Neutral Buoyancy Laboratory (NBL) at JSC in Houston, Texas. Assisted by Rick Sturckow, who served as the Intravehicular (IV) crewman on the Shuttle's aft flight deck, the two spacewalkers worked quickly and ahead of the timeline. With Ross riding at the end of the RMS, his initial task was to install mating plugs and jumper cables to route electrical power through Unity, whilst Newman released the launch locks which had secured wiring along the exterior of PMA-2. In total, the two men mated 40 data and power cables and connectors, running the entire length of Zarya-Unity, and well as EVA handrails and other tools to aid future spacewalkers. At one stage, as the Endeavour-ISS combo flew over Russian ground stations, commands were issue to activate a pair of US-Russian voltage convertors to feed electrical power between Zarya and Unity for the first time, bringing Node-1's systems officially to life at 10:49 pm. This "perfect electrical continuity between the two modules", as NASA described it, enabled the activation of Unity's internal systems, including its avionics and computers. Shortly before returning to Endeavour, and after Unity's own heaters had taken over the role of temperature control, Ross removed thermal covers from a pair of Multiplexer-Demultiplexers (MDMs). Newman was also raised by Currie on the RMS to inspect the stuck TORU antennas. Twenty-seven minutes before the end of the EVA, Ross (who was making his fifth career spacewalk) officially eclipsed the achievement of fellow astronaut Tom Akers as the most experienced US spacewalker. By the time he and Newman returned inside Endeavour at 12:31 am EST on 8 December, after seven hours and 21 minutes, Ross had accumulated a total of 30 hours and eight minutes working in a pressurised suit in the vacuum of space.

Late the following evening, Cabana and Sturckow used Endeavour's thrusters in a 'staccato-fashion' for about 22 minutes to slightly boost the orbit of Zarya-Unity by about 8 km. Meanwhile, the rest of the crew prepared for Ross and Newman's second EVA. This got underway at 3:33 pm EST on 9 December and the spacewalkers moved to install a pair of 'early' S-band antennas on the exterior of Unity in order for ground controllers to monitor its systems and provide basic videoconferencing capabilities for the first long-duration ISS crew. Next, the two men pressed on with removing launch restraint pins on Unity's four 'radial' hatches and fitted a sunshade over the two MDMs. Newman was also lifted atop the RMS to attend to one of the two jammed TORU antennas. On slightly nudging the antenna with an extendable, 3.3 m grappling hook, it deployed. The spacewalkers returned to Endeavour's airlock at 10:35 pm EST, after a little more than seven hours.

Following the completion of the second EVA, the astronauts and Krikalev opened the hatches to enter Unity itself for the first time at 2:54 pm EST on 10 December. To guard against the presence of floating particulates inside the module, they wore goggles and face masks. In total, no fewer than *six* hatches required to be opened to gain full access to the nascent ISS: firstly the hatch of Endeavour's ODS, then the hatch to PMA-2, followed by the hatches into Unity, into PMA-1, into Zarya's spherical pressurised adaptor and, lastly, into the FGB's main section. First aboard the new station, fittingly and symbolically, were Cabana and Krikalev, representing the United States and Russia. They immediately set to work fitting portable fans and lights and completing the installation of the early S-band communications (or 'Early Comm') system. The videoconferencing capability of this system was thoroughly tested. Sturckow removed access panels to unstow hardware for future crews, but at some point a mid-bay pivot fitting, used to tilt an equipment rack, floated out of sight. Ninety minutes after entering Unity, at 4:12 pm, they opened the hatch into Zarya, where Krikalev and Currie set about replacing the failed battery charging unit. For Krikalev, it was a chore with which he was familiar from his long-duration missions aboard Mir. Launch restraints were also removed from several of Zarya's internal panels.

In readiness for STS-88's third and final EVA on the afternoon of 12 December, the crew departed the station, retreating back through the modules and closing each hatch in sequence, before returning to Endeavour. By 7:26 pm EST on the 11th, the final hatch was secured, ending 28 hours and 32 minutes aboard the Zarya-Unity stack. Next day, at 3:33 pm, Ross and Newman floated into the vacuum of space for what turned out to be an EVA of six hours and 59 minutes, primarily focused upon preparing the ground for Tammy Jernigan and Dan Barry, the spacewalkers of the second Shuttle assembly crew, who were to carry out further work during STS-96 in May 1999. The first task for Ross and Newman was the releasing of ties to relieve the tension in a quartet of cables that had been installed earlier in the mission. Mission Control had noticed from television views that leaving the cables 'as is' would offer insufficient 'play' in them to accommodate cyclical heating and cooling during the orbital 'day' and 'night' periods. After checking an insulation cover on PMA-2 they attached a bag of tools, including wrenches, power grippers, ratchets and foot restraints, to the side of PMA-1 ready for Jernigan and Barry. They disconnected cables and Ross used the 3.3 m grappling hook in an effort to free the second jammed TORU antenna. Although the antenna proved a little stubborn, his tapping and nudging prompted it to roll out from its spool into the fully deployed configuration. Finally, the spacewalkers performed a photographic survey of the state of the vehicle in aid of future missions. By the time they returned inside Endeavour, they completed a total of 21 hours and 22 minutes on the first three EVAs of the ISS era, which established Ross as the most experienced US spacewalker, with 44 hours and nine minutes, and Newman in third place with 28 hours and 27 minutes.

The work of STS-88 was drawing inexorably towards its conclusion and shortly after the spacewalkers returned inside the Shuttle, Rick Sturckow depressurised the vestibule between the ODS and PMA-2 to establish the proper conditions for undocking from the ISS. At 11:36 am EST on 13 December, the new International Space Station Flight Control Room at JSC performed its first wake-up call for an orbiting crew when Capcom Mike Fincke played the song 'Goodnight, Sweetheart, Goodnight' in honour of the impending farewell between STS-88 and the ISS. At 3:25 pm, with Sturckow at the controls, Endeavour undocked from the space station and manoeuvred to a position about 140 m 'above' the outpost. He initiated a nose-first flyaround, as his crewmates acquired stunning still and television photographic imagery of the two connected modules, set against the glorious backdrop of Earth. Finally, at 4:49 pm, Sturckow pulsed Endeavour's thrusters for the final time to draw away from the station.

The final days of STS-88 were spent conducting ancillary tasks. Sturckow deployed a pair of small satellites from Getaway Special (GAS) canisters in Endeavour's payload bay. The first was SAC-A, provided by the Argentinian National Committee for Space Activities. Cube-shaped and weighing 267 kg, this small satellite carried five technology experiments, including a differential global positioning system, a magnetometer, a set of silicon solar cells, a charge-coupled device for Earth observations and an experiment to track whales in the South Atlantic. It was deployed at 11:31 pm EST on 13 December, kicking off a planned nine-month mission. Next day, another small satellite, 'MightySat', was also

5-06: Jim Newman (left) and Bob Cabana at work in the Unity node during STS-88.

deployed into space. Slightly heavier than SAC-A, at about 320 kg, MightySat was deployed at 9:09 pm EST. Provided by the Phillips Laboratory at Kirtland Air Force Base in New Mexico, it was designed to demonstrate advanced technologies, including a composite structure, improved solar cells, a microparticle impact detector, upgraded electronics and a shock device.

Scattered to broken clouds and an absence of rain showers offered great promise for an on-time landing on the evening of 15 December and Mission Control selected Wagner's chest-pounding 'Ride of the Valkyries' to awaken the six-member crew from their slumbers on the morning of Landing Day. With a 70-percent likelihood of acceptable weather conditions in Florida (which climbed to 80 percent as the day wore on), Bob Cabana was given the go-ahead to perform the deorbit manoeuvre at 9:47 pm EST to bring his crew back to Earth. A little over an hour later, having swept across the darkened Americas, crossing the Guatemala-Mexico border and entering Florida airspace, Endeavour alighted on Runway 33 at the Shuttle Landing Facility (SLF) at 10:54 pm, concluding a mission of just five hours shy of 12 full days.

EPILOGUE: THE NEW MILLENNIUM

On 12 April 1961, the first human ventured into space. Yuri Alexeyevich Gagarin was a 27-year-old former farmboy, whose dreams and achievements turned him into a hero for all the ages. Yet although his image is now in pride of place aboard the Service Module of the ISS, the arrival of that Service Module at the station proved a long time coming. In spite of earlier Russian pledges to the contrary, it had become clear since the late summer of 1998 that the third major component of the ISS would meet with further significant delay from April 1999 until no earlier than July. Shortly after Endeavour's departure, the station was commanded into a new orientation, with its Unity 'end' facing Earth and its Zarya 'end' pointing out, with commands issued to place it into a slow spin of about one revolution every 30 minutes for optimal thermal conditions. Zarya's motion control system was to be activated each week to ensure its continued functionality.

Nor was that Zarya module, already in orbit, immune from problems. Before STS-88 even launched, it came to light that it had failed in a key area of its Flight Readiness Review, due to excessive internal noise levels. When the second Shuttle assembly mission, STS-96, took to the skies in May 1999, its seven-member team of astronauts and cosmonauts – including a Russian crew member and Julie Payette, the first Canadian to board the ISS – were obliged to wear industrial-specification ear protectors owing to noise levels as high as 72 decibels. They also suffered headaches, nausea, eye irritation and vomiting, reportedly caused by excessive carbon dioxide concentrations. Nevertheless, as Dan Goldin once remarked, the process of building the largest and most audacious piece of human engineering in history was always going to be fraught with difficulty. By the time STS-96 reached the station, the Service Module – now renamed 'Zvezda' ('Star') – had been delayed until no earlier than November 1999 and eventually well into 2000. Another Shuttle

5-07: The Zarya (bottom) and Unity (top) modules, in unison, after the departure of STS-88 in December 1998.

crew, that of STS-101, flew to the station in May 2000 to deliver equipment and logistics aboard a Spacehab double module. Its seven-member crew included Russian cosmonaut Yuri Usachev and NASA astronauts Jim Voss and Susan Helms, who were then training for the second ISS long-duration expedition.

At length, Zvezda rocketed into orbit on 12 July 2000 and two weeks later docked at the aft longitudinal port of the Zarya module. This was followed by two more Shuttle missions. Firstly, STS-106 arrived in September and set about outfitting Zvezda, as well as installing exterior equipment during an EVA by NASA astronaut Ed Lu and Russian cosmonaut Yuri Malenchenko, and secondly, STS-92 flew in October to jumpstart the actual construction by mounting the large Z-1 truss segment atop Unity to provide control moment gyroscopes and other functions for the young station. The STS-92 crew, which included Koichi Wakata as the first Japanese occupant of the ISS, also placed PMA-3 on the truss. Less than ten days after the departure of STS-92, on 31 October 2000, Soyuz TM-31 was launched from Tyuratam, carrying the Expedition 1 crew of Commander Bill Shepherd and Russian cosmonauts Yuri Gidzenko and Sergei Krikalev. Docking at the station two days later, the three men would spend almost five months in orbit, initiating an uninterrupted human presence in space which, at the time of writing in early 2014, looks set to continue throughout the second decade of the 21st century.

During their expedition, Shepherd, Gidzenko and Krikalev welcomed no fewer than three visiting Shuttle crews. In December 2000, STS-97 arrived to install the first set of US-built electricity-generating solar arrays, radiators and batteries. When unfurled, these enormous arrays were the largest and heaviest of their kind ever to be deployed in space, and provided early power for the station and, in particular, the US laboratory module (named 'Destiny'), which itself arrived aboard STS-98 in February 2001. A month later, in March, the STS-102 crew delivered a reusable Italian-built logistics carrier, known as the Multi-Purpose Logistics Module (MPLM), and exchanged Shepherd's crew for the Expedition 2 crew of Usachev, Voss and Helms. Two more Shuttle missions in the summer of 2001 completed 'Phase 2' of ISS construction, with STS-100 in April delivering Canada's SSRMS mechanical arm – and including Chris Hadfield in its crew, who became the first Canadian in history to perform a spacewalk – and STS-104 in July installing the Joint Airlock (known as 'Quest') to enable station-based EVAs.

With Zvezda in place, the rapidity and apparent smoothness of the ISS construction sequence was impressive to witness. As expedition crews cycled in and out of the station, rotated aboard Shuttle missions, and with six-monthly Soyuz 'taxi' crews delivering fresh emergency return vehicles, the ISS steadily grew in size and capability. By the end of 2002, the first three major components of its vast, backbone-like truss structure were in place and the following years, 2003 and 2004, were expected to see the integration of three more sets of electricity-generating solar arrays, batteries and radiators, together with Node-2 (later named 'Harmony') to provide the docking hub for the arrival of the Columbus and JEM modules.

The assembly sequence was stopped abruptly on 1 February 2003, when, with horrifying suddenness, Space Shuttle Columbia was destroyed during re-entry, as she returned from an independent 16-day scientific research mission. All further flights were placed on indefinite hold, as a panel of investigators set to work exploring the cause of the second Shuttle disaster in 17 years. At the time of Columbia's loss, the Expedition 6 crew of commander Ken Bowersox and flight engineers Nikolai Budarin and Don Pettit were ten weeks into their planned four-month mission, anticipating a return to Earth in mid-March. Their expedition was extended by six weeks, as plans were laid to establish a series of two-member 'caretaker' crews, launched via Soyuz, to keep the ISS operational until the Shuttle returned to service and the construction sequence could resume. Expedition 7 crewmen Yuri Malenchenko and Ed Lu arrived in late April 2003, allowing Bowersox, Budarin and Pettit to return home aboard their own Soyuz spacecraft, and so began a sequence of two-man crews for more than three years.

Although the Shuttle returned to flight with STS-114 in July 2005, it demonstrated that not all of the problems which doomed Columbia had been amply addressed and the programme was stood down again until STS-121 flew a second 'Return to Flight' mission in July 2006. At this stage, the ISS – then occupied by Expedition 13 crewmen Pavel Vinogradov and Jeff Williams – was expanded back up to its full, three-member strength for the first time since the departure of Bowersox, Budarin and Pettit. German astronaut Thomas Reiter launched with the STS-121 crew and would remain aboard the ISS until December 2006. Two months after Reiter's arrival, STS-115 officially returned to actual construction work, by delivering the second set of US-built solar arrays. In December 2006, STS-116 transported Sweden's first astronaut (and spacewalker), Christer Fuglesang, into orbit and exchanged Reiter for US astronaut Suni Williams, who joined a new expedition crew of Mike Lopez-Alegria and Mikhail Tyurin. Little by little, the ISS was living up to its billing as a truly 'international' space station. More solar arrays and equipment arrived in 2007, including Node-2, and in February 2008 Europe's Columbus module was installed. At around the same, time, the first of three Shuttle missions to install the components of Japan's JEM – which included the main pressurised laboratory module, a logistics module and an external experiments platform – got underway as the assembly of the ISS entered its final stages. A doubling of the Soyuz flights in 2009 proved significant for the station, because it allowed the permanent crew to expand from three to six. As the Shuttle era entered its twilight, with an official requirement for the fleet to be retired by mid-2011, it transported the final handful of ISS components into orbit, including a multi-windowed observation cupola, the Alpha Magnetic Spectrometer (AMS)-2 and a Permanent Multi-Purpose Module (PMM) for storage and logistics.

And by the 50th anniversary of Yuri Gagarin's flight, in April 2011, the ISS easily rivalled the accomplishments of its predecessor, Mir, having hosted astronauts and cosmonauts from the United States, Russia, Canada, Japan, Italy, France, South Africa, Belgium, Spain, the Netherlands, Brazil, Germany, Sweden, Malaysia and South Korea. Of those nations, the United States, Russia, Canada, Japan, Italy, France, Belgium and Germany had staged long-duration flights, in excess of one month, at the station. Spacewalks to build and maintain the ISS were performed by astronauts and cosmonauts who wore on the arms of their space suits the Stars and Stripes of the United States, the maple leaf of Canada, the blue-and-gold

Scandinavian cross of Sweden, the blood-red rising sun of Japan and the respective tricolours of Russia, France and Germany.

Building on unrealised dreams from the Mir era, the ISS also played host to the first paying 'space tourists', beginning with Dennis Tito in April 2001 and followed by South African software entrepreneur Mark Shuttleworth in April 2002, US engineering entrepreneur Greg Olsen in October 2005, Iranian-born US business-woman Anousheh Ansari in September 2006, US video game developer Richard Garriott (the son of Skylab astronaut Owen Garriott) in October 2008 and Hungarian-born US software executive Charles Simonyi, who flew twice to the ISS in April 2007 and March 2009. The most recent paying space tourist was Cirque du Soleil founder Guy Laliberté, who flew to the station in September-October 2009. All of these missions were organised privately with the Russians, but following Laliberté's mission, the increase in ISS crew size from three to six meant that all available Soyuz spacecraft seats would henceforth be required for professional crew members. This may change in late 2015, with the English soprano Sarah Brightman joining a Soyuz crew for a short-duration mission.

The first five volumes in this six-volume history series focused almost exclusively upon competition, rather than co-operation, between the two major spacefaring powers of the United States and Russia. That competition fostered the capabilities which enabled Yuri Gagarin and Al Shepard, Valentina Tereshkova and John Glenn, Alexei Leonov and Neil Armstrong, and many dozens of others, to accomplish goals which had been only dreams to most of humanity for thousands of years. That competition allowed for the development of rocketry, spacecraft, instrumentation, computers and other technologies, and uniquely in the history of the space programme, it was a competition that was largely endorsed and supported by the two nations' political leaders. By the end of the 20th century, the situation had changed. The collapse of Communism in the former Soviet Union in the late 1980s and early 1990s was the beginning of the end for the two main superpowers and, for Russia, an economic, political, military and ideological struggle ensued throughout the final decade of the second millennium.

It is ironic that politics was allowed for so long to dictate many of the outcomes of human space exploration. Yet the harsh economic downturn of the 1990s precluded Russia from developing its Mir-2 space station, just as dwindling budgets and an increasingly apathetic White House and Congress gradually drove Space Station Freedom from what could have been a shining star of technological achievement into a mere shadow of its original design. Nor could the international partners, Europe, Japan and Canada, support their own human space flight aspirations in a figurative and literal vacuum; their space station components were wholly dependent upon the United States. The decision to merge Mir-2, the remnants of Freedom, together with Europe's Columbus and Japan's JEM modules, as well as Canada's SSRMS mechanical arm, into the new ISS project was the most logical (and indeed only) means for all of the human spacefaring nations to achieve their collective goals.

The journey of the first half-century of human exploration of the cosmos has been filled with triumph and tragedy. The triumphs of the astronaut and cosmonaut heroes of the past, and the achievements of sending our emissaries into Earth orbit

5-08: Photographed in March 2011, just shy of the 50th anniversary of Yuri Gagarin's pioneering mission, the International Space Station is complete and ready to support our species' next phase of exploration into the Universe around us.

and to the dusty surface of the Moon, have been starkly juxtaposed against the tragedies of Apollo 1, Soyuz 1, Soyuz 11, Challenger and Columbia. Nor should we fail to remember the brave souls who cheated the Reaper, whilst placing their lives on the line in pursuit of a greater goal: Friendship 7's John Glenn and Soyuz 5's Boris Volynov, enduring harrowing re-entries which could quite easily have ended in catastrophe, or the crew of Apollo 13, against whom the odds of a successful return from the Moon in a crippled spacecraft were heavily loaded, or Soyuz 18A's Vasili Lazarev and Oleg Makarov and Soyuz T-10A's Vladimir Titov and Gennadi Strekalov escaping death by the skin of their teeth during near-disastrous failed launches, or Gemini VIII's Neil Armstrong and Dave Scott, whose quick thinking in an extremely dire emergency saved not only themselves, but a substantial part of the plan to land a man on the Moon. Nor, indeed, should we forget the others who paid the ultimate price, without ever leaving the launch pad, not least among them Gemini IX's Elliot See and Charlie Bassett, killed in an aircraft crash, weeks before their first space flight. Theirs and many other names are engraved on the Astronauts' Memorial at the Kennedy Space Center in Florida and theirs and many other names, Russian and American alike, were engraved on a commemorative plaque left on the Moon in August 1971 by the Apollo 15 crew as a memorial to heroes who lost their lives, striving for a greater goal.

Today we live in different times. Political differences remain, as highlighted with particular acuteness in 2014, following Russia's annexation of Crimea and the sabre-rattling with the West over the future of Ukraine. Yet although the likelihood of a space race between the United States and Russia appears to have diminished, what is abundantly clear is that few of our species' abundant goals for the next five decades – returning to the Moon, visiting an asteroid, reaching Mars and venturing beyond – are realistically within the financial and technological resources of single nations. The arrival of China on the scene, with its first independent human space flight in October 2003, followed by its first independent EVA in September 2008 and its first woman spacefarer and first experimental space station in June 2011, establishes a new player on the field for the next half-century of operations in space. The first five decades of human space exploration have showcased our capabilities and what our ingenuity can accomplish when the will is present; the next five decades must deliver nothing less and undoubtedly will produce substantially more. "Space exploration's not for any one country anymore," said Bob Cabana. "It's incumbent upon all of us to participate. Space exploration *is* our future. We can choose to participate and be a leader or we can fall behind. And I think we've all chosen to participate. This is *important*. We're going to go beyond the confines of Earth's gravity. We're going to go back to the Moon and on to Mars and beyond and, when we do that, we're going to do it as a united team."

Bibliography

'Space Station Office manager named.' NASA Headquarters, Washington, DC, 9 April 1984

'NASA awards space station testbed contract to Rockwell.' NASA Headquarters, Washington, DC, 5 February 1986

'Hutchinson leaves space station post.' NASA Headquarters, Washington, DC, 28 February 1986

'Neil Hutchinson leaves NASA.' NASA Headquarters, Washington, DC, 1 August 1986

'Space Station shower prototype developed at JSC.' NASA Headquarters, Washington, DC, 28 October 1987

'Space Station Freedom restructuring plan completed.' NASA Headquarters, Washington, DC, 21 March 1991

'Launch of final Mir modules delayed.' *Flight International*, 15 January 1992

'Mir-core replacement delayed until 1997.' *Flight International*, 20 May 1992

'Goldin says America needs Space Station Freedom now.' NASA Headquarters, Washington, DC, 24 June 1992

'Space Station Freedom propulsion firing tests underway.' NASA Headquarters, Washington, DC, 7 January 1993

'NASA budget boosts technology, promises improved Space Station Program.' NASA Headquarters, Washington, DC, 18 February 1993

'Goldin announces key Space Station posts, spending measures.' NASA Headquarters, Washington, DC, 25 February 1993

'Gibbons outlines Space Station redesign guidance.' NASA Headquarters, Washington, DC, 6 April 1993

'Discovery may fly-by the Mir-1.' *Flight International*, 14 April 1993

'Polyakov aims for new flight record.' *Flight International*, 28 April 1993

'Station Redesign Team to submit final report.' NASA Headquarters, Washington, DC, 4 June 1993

'Space Station host center and prime contractor announced.' NASA Headquarters, Washington, DC, 17 August 1993

'Russians at sea over Atlantis.' *Flight International*, 18 August 1993

'USA/Russia agree space deal.' *Flight International*, 5 January 1994

'Cameron to manage NASA activities at Star City, Russia.' NASA Headquarters, Washington, DC, 23 February 1994

'Super lightweight External Tank to be used by Shuttle.' NASA Headquarters, Washington, DC, 28 February 1994

'Space station system design review completed.' NASA Headquarters, Washington, DC, 24 March 1994

'Cosmonauts tackle Shuttle training.' *Flight International*, 1 June 1994

'Crew named for first Space Shuttle, Mir docking mission.' NASA Headquarters, Washington, DC, 3 June 1994

'Astronaut chief to lead first Mir flight.' *Flight International*, 15 June 1994

'NASA selects payload specialists for Spacelab mission.' NASA Headquarters, Washington, DC, 20 June 1994

'NASA selects scientists for Neurolab Shuttle mission.' NASA Headquarters, Washington, DC, 22 June 1994

'NASA and Russian Space Agency sign space station interim agreement and $400 million contract.' NASA Headquarters, Washington, DC, 23 June 1994

'NASA Administrator releases statement on GAO report.' NASA Headquarters, Washington, DC, 24 June 1994

'NASA chief hails House vote preserving space station.' NASA Headquarters, Washington, DC, 29 June 1994

'NASA claims Mir-1 input will cut costs.' *Flight International*, 6 July 1994

'Astronaut Readdy to replace Cameron as NASA manager in Russia.' NASA Headquarters, Washington, DC, 12 July 1994

'Station Control Board ratifies improved assembly sequence.' NASA Headquarters, Washington, DC, 15 July 1994

'Goldin hails solid Senate vote on space station.' NASA Headquarters, Washington, DC, 3 August 1994

'NASA and Boeing reach agreement on space station contract.' NASA Headquarters, Washington, DC, 1 September 1994

'Cabana appointed chief of Astronaut Office.' NASA Headquarters, Washington, DC, 6 September 1994

'Second Endeavour date set.' *Flight International*, 28 September 1994

'Spektr module delayed.' *Flight International*, 5 October 1994

'Astronaut Sega to replace Readdy as NASA manager in Russia.' NASA Headquarters, Washington, DC, 1 November 1994

'Astronauts Blaha and Lucid to train for flight on Mir.' NASA Headquarters, Washington, DC, 3 November 1994

'Astronauts Chilton, Readdy to command Shuttle/Mir missions.' NASA Headquarters, Washington, DC, 8 November 1994

'Shuttle scheduled for Mir rendezvous.' *Flight International*, 14 December 1994

'NASA receives first new Shuttle engine.' *Flight International*, 4 January 1995

'NASA, Boeing sign agreement for International Space Station.' NASA Headquarters, Washington, DC, 13 January 1995

'Harris becomes first African-American to walk in space.' NASA Headquarters, Washington, DC, 2 February 1995

'NASA/Russian Space Agency reach agreement on key station element.' NASA Headquarters, Washington, DC, 8 February 1995

'NASA awards $481.6 million contract to Hughes.' NASA Headquarters, Washington, DC, 23 February 1995

'Discovery paves way for Mir space station docking.' *Flight International*, 22 February 1995

'Shaky partnership.' *Flight International*, 1 March 1995

'Nagel moves from Astronaut Office to SR&QA.' NASA Headquarters, Washington, DC, 3 March 1995

'Astronaut Baker to replace Sega as NASA manager in Russia.' NASA Headquarters, Washington, DC, 10 March 1995

'New Space Shuttle Main Engine ready for flight.' NASA Headquarters, Washington, DC, 21 March 1995

'First American flies to board the Mir.' *Flight International*, 22 March 1995

'Training facility will honour astronaut's memory.' NASA Headquarters, Washington, DC, 29 March 1995

'Lucid prime for second Mir stay; Linenger selected for third.' NASA Headquarters, Washington, DC, 30 March 1995

'Exterior of space station module completed; first IDR held.' NASA Headquarters, Washington, DC, 6 April 1995

'Crews selected for third, fourth Shuttle/Mir docking missions.' NASA Headquarters, Washington, DC, 14 April 1995

'More US astronauts head for Mir date.' *Flight International*, 26 April 1995

'Shuttle schedule may switch as Mir docking is delayed.' *Flight International*, 3 May 1995

'Thagard breaks US space record.' *Flight International*, 17 May 1995

'Spektr launched towards June Mir rendezvous.' *Flight International*, 31 May 1995

'100th manned mission nears.' *Flight International*, 7 June 1995

'Discovery pushed down launch pecking order.' *Flight International*, 14 June 1995

'Improvements are needed for Shuttle/Mir missions.' *Flight International*, 19 July 1995

'Boeing, Khrunichev sign contract for space station element.' NASA Headquarters, Washington, DC, 15 August 1995

'Successful docking starts second EuroMir mission.' *Flight International*, 13 September 1995

'Astronauts Precourt, Lawrence head to Russia.' NASA Headquarters, Washington, DC, 20 September 1995

'Space station funding passes major milestone.' *Flight International*, 11 October 1995

'Parazynski discontinues Mir training.' NASA Headquarters, Washington, DC, 14 October 1995

'Astronaut Lawrence to remain in United States.' NASA Headquarters, Washington, DC, 24 October 1995

'Spacewalk challenge.' *Flight International*, 25 October 1995

'Double or quits.' *Flight International*, 13 December 1995

'Astronauts Foale, Voss join colleagues in Star City, Russia.' NASA Headquarters, Washington, DC, 16 January 1996

'Exteior of US space station modules completed; flight hardware on track for launch in 1997.' NASA Headquarters, Washington, DC, 30 January 1996

'NASA and RSA agree to extend Shuttle-Mir activities.' NASA Headquarters, Washington, DC, 30 January 1996

'Crew named for fifth Shuttle/Mir docking; commander for the sixth.' NASA Headquarters, Washington, DC, 2 February 1996

'Lawrence to replace Precourt as NASA manager in Russia.' NASA Headquarters, Washington, DC, 13 March 1996

'Third success is achieved for Shuttle-Mir programme.' *Flight International*, 10 April 1996

'Mir Co-operative Solar Array is deployed/ISS power hardware being built and tested.' NASA Headquarters, Washington, DC, 25 May 1996

'NASA delays launch of Space Shuttle.' NASA Headquarters, Washington, DC, 12 July 1996

'Crew named to sixth Shuttle-Mir docking mission.' NASA Headquarters, Washington, DC, 15 July 1996

'Shuttle super lightweight fuel tank completes test series.' NASA Headquarters, Washington, DC, 18 July 1996

'Booster snags mar Mir plans. *Flight International*, 24 July 1996

'Crew named to first space station assembly flight.' NASA Headquarters, Washington, DC, 16 August 1996

'Lawrence and Wolf to train for Mir missions.' NASA Headquarters, Washington, DC, 16 August 1996

'Cosmonaut Kondakova named to STS-84 crew.' NASA Headquarters, Washington, DC, 22 August 1996

'Russia sends backup crew to the Mir space station.' *Flight International*, 28 August 1996

'Space Station Node-1 and laboratory modules successfully complete proof pressure tests.' NASA Headquarters, Washington, DC, 28 August 1996

'Atlantis moved to VAB; STS-79 launch delayed.' NASA Headquarters, Washington, DC, 4 September 1996

'Atlantis rolled onto Kennedy pad for next Mir mission.' *Flight International*, 4 September 1996

'Military mapper.' *Flight International*, 4 September 1996

'Atlantis moved back out to launch pad; September 16 set as new launch date.' NASA Headquarters, Washington, DC, 5 September 1996

'Lopez-Alegria to replace Lawrence as NASA manager in Russia.' NASA Headquarters, Washington, DC, 5 September 1996

'Shuttle super lightweight fuel tank completes tests.' NASA Headquarters, Washington, DC, 11 September 1996

'Transcript of excerpted remarks by NASA Administrator Daniel S. Goldin following the landing of Space Shuttle Atlantis.' NASA Headquarters, Washington, DC, 26 September 1996

'NASA delays another Shuttle launch.' *Flight International*, 13 November 1996

'First US space station module successfully completes final pressure test; launch just one year away.' NASA Headquarters, Washington, DC, 20 November 1996

'Wetherbee to lead international crew on seventh Shuttle/Mir mission.' NASA Headquarters, Washington, DC, 6 December 1996

'Station's first module assembled; ready for testing.' NASA Headquarters, Washington, DC, 9 December 1996

'NASA harvest of Mir space wheat marks US-Russian first.' NASA Headquarters, Washington, DC, 12 December 1996

'International Space Station faces Service Module crisis.' *Flight International*, 18 December 1996

'Russian programme in crisis.' *Flight International*, 8 January 1997

'Space Shuttle Atlantis modification work to be performed at Palmdale facility.' NASA Headquarters, Washington, DC, 16 January 1997

'Astronaut Cady Coleman begins training as backup for STS-83.' NASA Headquarters, Washington, DC, 18 February 1997

'Small fire extinguished on Mir.' NASA Headquarters, Washington, DC, 24 February 1997

'Kurs docking system fails again.' *Flight International*, 26 February 1997

'NASA pays Russia $20 million.' *Flight International*, 26 February 1997

'Wilcutt to lead crew on eighth Shuttle/Mir docking mission.' NASA Headquarters, Washington, DC, 4 March 1997

'ISS is placed under new pressure.' *Flight International*, 12 March 1997

'Russian Soyuz TM-24 spacecraft returns.' *Flight International*, 12 March 1997

'Rominger to replace Ashby as STS-85 pilot.' NASA Headquarters, Washington, DC, 18 March 1997

'Advanced X-ray telescope mirrors provide sharpest focus ever.' NASA Headquarters, Washington, DC, 20 March 1997

'Shuttle's new lighter, stronger External Tank completes major pressure tests.' NASA Headquarters, Washington, DC, 28 March 1997

'NASA revised International Space Station schedule.' NASA Headquarters, Washington, DC, 9 April 1997

'Commander, pilot, flight engineer round out STS-90 crew.' NASA Headquarters, Washington, DC, 18 April 1997

'USA gets criticised over exaggerated Mir comments.' *Flight International*, 23 April 1997

'Russian cash fails to halt space station delay.' *Flight International*, 23 April 1997

'Microgravity Science Laboratory mission set for July; remaining 1997 Shuttle manifest adjusted slightly.' NASA Headquarters, Washington, DC, 25 April 1997

'US astronaut ready for milestone spacewalk.' NASA Headquarters, Washington, DC, 25 April 1997

'Payload specialists selected for future Shuttle mission.' NASA Headquarters, Washington, DC, 28 April 1997

'First US-Russian spacewalk is completed in ISS rehearsal.' *Flight International*, 7 May 1997

'Space Station Control Board approves new assembly sequence.' NASA Head-quarters, Washington, DC, 15 May 1997

'First US space station component begins launch preparations.' NASA Head-quarters, Washington, DC, 26 June 1997

'Ukrainian payload specialists selected for Shuttle mission.' NASA Headquarters, Washington, DC, 16 May 1997

'First US space station components begins launch preparations.' NASA Head-quarters, Washington, DC, 26 June 1997

'Progress will go to aid Mir crew.' *Flight International*, 2 July 1997

'First space station modules prepared for 1998 launch.' *Flight International*, 9 July 1997

'Second US space station component begins launch preparations.' NASA Head-quarters, Washington, DC, 25 July 1997

'NASA announces revised plan for Mir staffing.' NASA Headquarters, Washington, DC, 30 July 1997

'Spacewalk in August will be used to restore Mir power.' *Flight International*, 30 July 1997

'A risky business.' *Flight International*, 6 August 1997

'Soyuz docks for new Mir repair mission.' *Flight International*, 13 August 1997

'Power is restored to Mir Kristall module.' *Flight International*, 3 September 1997

'Chiefs plan to fly on final SMM.' *Flight International*, 17 September 1997

'Panels give astronaut a 'Go' for launch to Mir.' NASA Headquarters, Washington, DC, 25 September 1997

'Statement of The Honorable Daniel S. Goldin, Administrator, National Aero-nautics and Space Administration.' NASA Headquarters, Washington, DC, 25 September 1997

'Control board reports International Space Station on target, finalised assembly sequence.' NASA Headquarters, Washington, DC, 1 October 1997

'Thomas will continue American presence on Mir.' NASA Headquarters, Washington, DC, 10 October 1997

'Cosmonaut rounds out STS-89 crew.' NASA Headquarters, Washington, DC, 15 October 1997

'Mir restoration work continues after Atlantis returns to Earth.' *Flight International*, 15 October 1997

'Precourt to lead final Shuttle/Mir docking mission.' NASA Headquarters, Washington, DC, 23 October 1997

'Tognini named to STS-93 crew.' NASA Headquarters, Washington, DC, 13 November 1997

'US/Ukrainian students collaborate on Shuttle experiment.' NASA Headquarters, Washington, DC, 17 November 1997

'Advanced X-ray Astrophysics Facility delivery delayed.' NASA Headquarters, Washington, DC, 5 December 1997

'New Space Shuttle External Tank ready to launch Space Station era.' NASA Headquarters, Washington, DC, 15 January 1998

'First station element to be shipped to Russian launch site.' NASA Headquarters, Washington, DC, 16 January 1998

'Sen. Glenn gets a 'Go' for Space Shuttle mission.' NASA Headquarters, Washington, DC, 16 January 1998

'Veteran cosmonaut nominated to fly on final Shuttle/Mir mission.' NASA Headquarters, Washington, DC, 21 January 1998

'Space station project moves on as FGB goes to Baikonur.' *Flight International*, 28 January 1998

'Brown to command STS-95 mission.' NASA Headquarters, Washington, DC, 13 February 1998

'Collins named first female Shuttle commander.' NASA Headquarters, Washington, DC, 5 March 1998

'Assembly of NASA's X-ray telescope completed.' NASA Headquarters, Washington, DC, 12 March 1998

'NASA announces contest to name X-ray observatory.' NASA Headquarters, Washington, DC, 16 April 1998

'International Space Station partners adjust target dates for first launches, revise other station assembly launches.' NASA Headquarters, Washington, 1 June 1998

'Future space station resident joins assembly crew.' NASA Headquarters, Washington, DC, 30 July 1998

'Training begins for crew of next Hubble Space Telescope servicing mission.' NASA Headquarters, Washington, DC, 30 July 1998

'Last men to Mir.' *Flight International*, 19 August 1998

'Russian activity continues despite crisis.' *Flight International*, 9 September 1998

'NASA delays shipment of X-ray telescope to Kennedy Space Center to allow additional testing.' NASA Headquarters, Washington, DC, 13 October 1998

'NASA sets November ISS date.' *Flight International*, 14 October 1998

'First International Space Station module moves to launch pad.' NASA Headquarters, Washington, DC, 26 October 1998

'Russian Parliament makes plea for Mir reprieve.' *Flight International*, 28 October 1998

'Crew members named for Earth-mapping mission.' NASA Headquarters, Washington, DC, 26 October 1998

'Troubled Endeavour mission falls under the spotlight.' *Flight International*, 18 November 1998

'NASA renames telescope and sets new launch date.' NASA Headquarters, Washington, DC, 21 December 1998

'More Mir missions planned after June.' *Flight International*, 23 December 1998

'NASA announces delay in shipment of Chandra observatory.' NASA Headquarters, Washington, DC, 20 January 1999

'New crew gets ready for trip to Mir station.' *Flight International*, 27 January 1999

'Last of the Mir crews are put into orbit.' *Flight International*, 3 March 1999

'NASA plans an early servicing mission to Hubble Space Telescope.' NASA Headquarters, Washington, DC, 10 March 1999

'Shuttle veterans complete Hubble servicing crew.' NASA Headquarters, Washington, DC, 12 March 1999

'Mir budget cuts see spacewalks cancelled.' *Flight International*, 24 March 1999

'Neurolab team to discuss results, potential health benefits.' NASA Headquarters, Washington, DC, 13 April 1999

'Mir August mission likely to be scrapped.' *Flight International*, 2 June 1999

'Farewell Mir mission extended five days.' *Flight International*, 4 August 1999

'Hubble Telescope placed into safe mode as gyroscope fails.' NASA Headquarters, Washington, DC, 15 November 1997

'Mir hopes.' *Flight International*, 22 February 2000

'NASA shocked by commercial Mir plans.' *Flight International*, 29 February 2000

'Soyuz-TM cosmonauts to bring Mir out of mothballs.' *Flight International*, 14 March 2000

'Soyuz go-ahead to Mir.' *Flight International*, 28 March 2000

'Cosmonauts board Mir as USA raises money concerns.' *Flight International*, 11 April 2000

'Second Mir mission set to fly with private funding.' *Flight International*, 18 April 2000

'Russia seeks $10 million to keep Mir on station.' *Flight International*, 9 May 2000

'MirCorp needs more funds for Mir flights.' *Flight International*, 20 June 2000

'US businessman prepares for week-long visit to Mir.' *Flight International*, 27 June 2000

'Board go-ahead will lead to manned Mir.' *Flight International*, 1 August 2000

'Cashing in.' *Flight International*, 12 September 2000

'Mir flight offered as TV show prize.' *Flight International*, 19 September 2000

'MirCorp mission to raise space station's orbit and extend life.' *Flight International*, 3 October 2000

'Russia government seeks to end Mir operation.' *Flight International*, 10 October 2000

'Mir faces February deorbit after Russian government cuts lifeline.' *Flight International*, 21 November 2000

'MirCorp to focus on Space Station.' *Flight International*, 19 December 2000

Burrough, Bryan (1998) *Dragonfly: NASA and the Crisis Aboard Mir*. London: Fourth Estate

Clark, Phillip (1988) *The Soviet Manned Space Programme*. London: Salamander

Cosmo, M.L. and Lorenzini, E.C., *Tethers in Space Handbook* (Third Edition). Smithsonian Astrophysical Observatory, prepared for NASA George C. Marshall Space Flight Center, Huntsville, Alabama, December 1997

Evans, Ben (2005) *Space Shuttle Columbia: Her Missions and Crews*. Chichester: Springer-Praxis

Evans, Ben (2009) *Escaping the Bonds of Earth*. Chichester: Springer-Praxis

Evans, Ben (2010) *Foothold in the Heavens*. Chichester: Springer-Praxis

Evans, Ben (2011) *At Home in Space*. Chichester: Springer-Praxis

Evans, Ben (2012) *Tragedy and Triumph in Orbit*. Chichester: Springer-Praxis

Evans, Ben (2013) *Partnership in Space*. Chichester: Springer-Praxis

Froelich, Walter (1984) *Spacelab: An International Short-Stay Orbiting Laboratory*. NASA Headquarters, Washington, DC

Glenn, John, with Taylor, Nick (1999) *John Glenn: A Memoir*. New York: Bantam Books

Hall, Rex D. and Shayler, David J. (2003) *Soyuz: A Universal Spacecraft*. Chichester: Springer-Praxis

Heppenheimer, T.A. (1999) *The Space Shuttle Decision*. NASA Office of Policy and Plans, NASA Headquarters, Washington, DC

Jenkins, Dennis R. (2001) *Space Shuttle: The History of the National Space Transportation System - The First 100 Missions*. Hinckley: Midland Publishing

Jones, Tom (2006) *Skywalking: An Astronaut's Memoir*. New York: HarperCollins

Linenger, Jerry M. (2000) *Off the Planet: Surviving Five Perilous Months Aboard the Space Station Mir*. New York: McGraw-Hill

Lord, Douglas R. (1987) *Spacelab: An International Success Story*. Scientific and Technical Information Division, NASA Headquarters, Washington, DC

Morgan, Clay (2001) *Shuttle-Mir: The United States and Russia Share History's Highest Stage*. NASA History Series, Lyndon B. Johnson Space Center, Houston, Texas

Portree, David S.F. and Trevino, Robert C. (1997) *Walking to Olympus: An EVA Chronology*. NASA History Office, NASA Heaquarters, Washington, DC

Reichhardt, Tony (ed.) *Space Shuttle: The First 20 Years*. London: DK Publishing, Inc., 2002

Index

Afanasyev, Viktor, 466, 469–474, 476
Allen, Andy, 68, 73–74, 76–77, 79–80, 83–84
Altman, Scott, 318–321, 324, 326, 331
Akers, Tom, 241–243, 373, 497
Anderson, Mike, 426–428
Andre-Deshays, Claudie, 235–237, 393, 434, 470
Apt, Jay, 241–243, 245
Ashby, Jeff, 288–289, 291, 350–352, 359–362
Avdeyev, Sergei, 197, 200–202, 211–215, 465–469, 471–474

Baker, Ellen, 183, 190, 192, 194
Baker, Mike, 248–249, 252–254, 410
Barry, Dan, 62–66, 498
Baturin, Yuri, 465–467, 478–479
Bella, Ivan, 470–471
Blaha, John, 216–221, 234–235, 237–239, 242–248, 252–254, 391, 399
Bloomfield, Mike, 410–420,
Bowersox, Ken, 38–53, 255–274
Brady, Chuck, 97–109
Brown, Curt, 84–96, 288–300, 332–348, 362–377
Buckey, Jay, 318–331
Budarin, Nikolai, 166, 178–179, 183, 189–192, 195–197, 201, 215, 433–439, 445–450, 464–467, 503
Bursch, Dan, 84–96

Cabana, Bob, 487–500
Cameron, Ken, 202–212
Casper, John, 84–96
Chandra X-ray Observatory, 349–361

Chang-Diaz, Franklin, 67–83, 438–451
Chawla, Kalpana, 300–317
Cheli, Maurizio, 67–83
Chiao, Leroy, 62–66
Chilton, Kevin, 215–230
Chretien, Jean-Loup, 410–420
Clervoy, Jean-Francois, 362–377, 392–399
Clifford, Michael 'Rich', 215–230
Cockrell, Ken, 19–37, 110–123
Coleman, Cady, 38–53
Collins, Eileen, 125–149, 392–399
Crouch, Roger, 275–287
Curbeam, Bob, 288–300
Currie, Nancy, 1–18, 487–500

Davis, Jan, 288–300

Dezhurov, Vladimir, 149, 160, 165–183, 187–195
Doi, Takao, 300–317
Duffy, Brian, 62–66
Dunbar, Bonnie, 426–428
Duque, Pedro, 332–348

Edwards, Joe, 426–428
EuroMir, 195–214
Eyharts, Leopold, 235, 405, 433–435

Favier, Jean-Jacques, 97–109
Foale, Mike, 125–149, 362–377, 381–405
Fuglesang, Christer, 195–214

Garneau, Marc, 84–96
Gernhardt, Mike, 19–37, 275–287

Gibson, Robert 'Hoot', 181–194
Gidzenko, Yuri, 96, 197, 200–201, 211–215, 502
Glenn, John, 332–348
Godwin, Linda, 215–230
Gorie, Dom, 438–451, 453–464
Grunsfeld, John, 248–254, 362–377
Guidoni, Umberto, 67–83

Hadfield, Chris, 202–212
Haignere, Jean-Pierre, 351, 393, 470–474
Halsell, Jim, 202–212, 275–287
Harbaugh, Greg, 181–194, 255–274
Harris, Bernard, 125–149
Hawley, Steve, 255–274
Helms, Susan, 97–109
Henricks, Tom, 1–18, 97–109
Hire, Kay, 318–331
Hoffman, Jeff, 67–83
Horowitz, Scott 'Doc', 67–83, 255–274
Hubble Space Telescope, 255–274, 362–377

Inertial Upper Stage, 1–18, 349–361
Inflatable Antenna Experiment, 84–96
Ivins, Marsha, 248–254

Jernigan, Tammy, 110–123
Jett, Brent, 54–66, 248–254
Jones, Tom, 110–123

Kadenyuk, Leonid, 300–317
Kaleri, Alexander, 235–237, 243, 246–247, 252–254, 379–386, 472–480
Kavandi, Janet, 438–451, 453–464
Kelly, Scott, 362–377
Kondakova, Yelena, 142–143, 150–151, 170–172, 234, 392, 395–396
Korzun, Valeri, 235–237, 243, 246–247, 252–254, 379–386
Kregel, Kevin, 1–18, 97–109, 300–317, 453–464
Krikalev, Sergei, 96, 131, 155, 474, 490–502

Lawrence, Wendy, 410–420, 438–451
Lazutkin, Alexander, 381–405
Lee, Mark, 255–274
Leslie, Fred, 38–53

Life and Microgravity Spacelab, 97–109

Lindsey, Steve, 300–317, 332–348
Linenger, Jerry, 248–254, 379–400

Linnehan, Rick, 97–109, 318–331
Linteris, Greg, 275–287
Lopez-Alegria, Mike, 38–53
Lu, Ed, 392–399
Lucid, Shannon, 215–246

McArthur, Bill, 202–212
Microgravity Science Laboratory, 275–287
Mohri, Mamoru, 453–464
Mukai, Chiaki, 332–348
Musabayev, Talgat, 425–451
Musgrave, Story, 110–123

Neurolab, 318–331
Newman, Jim , 19–37, 487–500
Nicollier, Claude, 67–83, 362–377
Noriega, Carlos, 392–399

Parazynski, Scott, 410–420, 332–348
Pawelczyk, Jim, 318–331
Precourt, Charlie, 181–194, 392–399, 438–451

Readdy, Bill, 237–246
Reilly, Jim, 425–433
Robinson, Steve, 288–300, 332–348
Rominger, Kent, 38–53, 110–123, 288–300
Ross, Jerry, 202–212, 487–500
Runco, Mario, 84–96

Sacco, Al, 38–53
Scott, Winston, 54–66, 300–317

Searfoss, Rick, 215–230, 318–331
Sega, Ron, 215–230
Shuttle Pallet Satellite, 110–123, 288–300
Shuttle Radar Topography Mission, 453–464
Smith, Steve, 255–274, 362–377
Space Flyer Unit, 54–66

Space Station Freedom, 481–487
SPARTAN, 19–37, 54–66, 84–96, 125–149, 300–317, 332–348
Still, Susan, 275–287
STS-63, 125–149
STS-69, 19–37

STS-70, 1–18
STS-71, 181–194
STS-72, 54–66
STS-73, 38–53
STS-74, 202–212
STS-75, 67–83
STS-76, 215–230
STS-77, 84–96
STS-78, 97–109
STS-79, 237–246
STS-80, 110–123
STS-81, 248–254
STS-82, 255–274
STS-83, 275–287
STS-84, 392–399,
STS-85, 288–300
STS-86, 410–420,
STS-87, 300–317
STS-88, 487–500
STS-89, 425–433
STS-90, 318–331
STS-91, 438–451
STS-93, 349–361
STS-94, 275–287
STS-95, 332–348
STS-99, 453–464
STS-103, 362–377
Sturckow, Rick, 487–500

Tanner, Joe, 255–274

Tethered Satellite System, 67–83
Thagard, Norm, 149–181
Thiele, Gerhard, 453–464
Thirsk, Bob, 97–109
Thomas, Andy, 84–96, 425–451
Thomas, Don, 1–18, 275–297
Thornton, Kathy, 38–53
Titov, Vladimir, 125–149, 410–420
Tognini, Michel, 349–361
Tracking and Data Relay Satellite, 1–18
Tryggvason, Bjarni, 288–300

United States Microgravity Laboratory,
    38–53
United States Microgravity Payload, 67–83,
    300–317

Voss, Janice, 125–149, 275–297, 453–464
Voss, Jim, 19–37

Wakata, Koichi, 54–66
Wake Shield Facility, 19–37, 110–123
Walker, Dave, 19–37
Walz, Carl, 237–246
Weber, Mary Ellen, 1–18
Wetherbee, Jim, 125–149, 410–420
Wilcutt, Terry, 237–246, 425–433
Williams, Dave, 318–331
Wisoff, Jeff, 248–254
Wolf, Dave, 410–433

CPSIA information can be obtained at www.ICGtesting.com
Printed in the USA
LVOW01s1449191214

419624LV00001B/1/P

9 781493 913060